Progress in Group Field Theory and Related Quantum Gravity Formalisms

Progress in Group Field Theory and Related Quantum Gravity Formalisms

Special Issue Editors

Steffen Gielen
Sylvain Carrozza
Daniele Oriti

MDPI • Basel • Beijing • Wuhan • Barcelona • Belgrade

Special Issue Editors
Steffen Gielen
University of Sheffield
UK

Sylvain Carrozza
Perimeter Institute for
Theoretical Physics
Canada

Daniele Oriti
Ludwig Maximilian University
of Munich
Germany

Editorial Office
MDPI
St. Alban-Anlage 66
4052 Basel, Switzerland

This is a reprint of articles from the Special Issue published online in the open access journal *Universe* (ISSN 2218-1997) from 2018 to 2020 (available at: https://www.mdpi.com/journal/universe/special_issues/GFT).

For citation purposes, cite each article independently as indicated on the article page online and as indicated below:

LastName, A.A.; LastName, B.B.; LastName, C.C. Article Title. *Journal Name* **Year**, *Article Number*, Page Range.

ISBN 978-3-03936-178-6 (Hbk)
ISBN 978-3-03936-179-3 (PDF)

Cover image courtesy of Tomás Saraceno
Algo-R(h)i(y)thms, 2019, Exhibition View Esther Schipper, Berlin
Courtesy the artist and Esther Schipper, Berlin
Photo © Studio Tomás Saraceno 2019

© 2020 by the authors. Articles in this book are Open Access and distributed under the Creative Commons Attribution (CC BY) license, which allows users to download, copy and build upon published articles, as long as the author and publisher are properly credited, which ensures maximum dissemination and a wider impact of our publications.

The book as a whole is distributed by MDPI under the terms and conditions of the Creative Commons license CC BY-NC-ND.

Contents

About the Special Issue Editors vii

Sylvain Carrozza, Steffen Gielen, Daniele Oriti
Editorial for the Special Issue "Progress in Group Field Theory and Related Quantum Gravity Formalisms"
Reprinted from: *Universe* 2020, 6, 19, doi:10.3390/universe6010019 1

Astrid Eichhorn, Tim Koslowski and Antonio D. Pereira
Status of Background-Independent Coarse Graining in Tensor Models for Quantum Gravity
Reprinted from: *Universe* 2019, 5, 53, doi:10.3390/universe5020053 9

Vincent Lahoche and Dine Ousmane Samary
Progress in Solving the Nonperturbative Renormalization Group for Tensorial Group Field Theory
Reprinted from: *Universe* 2019, 5, 86, doi:10.3390/universe5030086 45

Lisa Glaser and Sebastian Steinhaus
Quantum Gravity on the Computer: Impressions of a Workshop
Reprinted from: *Universe* 2019, 5, 35, doi:10.3390/universe5010035 76

Goffredo Chirco
Holographic Entanglement in Group Field Theory
Reprinted from: *Universe* 2019, 5, 211, doi:10.3390/universe5100211 98

Philipp A. Höhn
Switching Internal Times and a New Perspective on the 'Wave Function of the Universe'
Reprinted from: *Universe* 2019, 5, 116, doi:10.3390/universe5050116 123

Isha Kotecha
Thermal Quantum Spacetime
Reprinted from: *Universe* 2019, 5, 187, doi:10.3390/universe5080187 144

Jakub Mielczarek
Spin Foam Vertex Amplitudes on Quantum Computer—Preliminary Results
Reprinted from: *Universe* 2019, 5, 179, doi:10.3390/universe5080179 166

Andreas G. A. Pithis and Mairi Sakellariadou
Group Field Theory Condensate Cosmology: An Appetizer
Reprinted from: *Universe* 2019, 5, 147, doi:10.3390/universe5060147 187

Bekir Baytaş, Martin Bojowald, Sean Crowe
Equivalence of Models in Loop Quantum Cosmology and Group Field Theory
Reprinted from: *Universe* 2019, 5, 41, doi:10.3390/universe5020041 210

Killian Martineau and Aurélien Barrau
Primordial Power Spectra from an Emergent Universe: Basic Results and Clarifications
Reprinted from: *Universe* 2018, 4, 149, doi:10.3390/universe4120149 218

Marco de Cesare
Reconstruction of Mimetic Gravity in a Non-Singular Bouncing Universe from Quantum Gravity
Reprinted from: *Universe* 2019, 5, 107, doi:10.3390/universe5050107 234

Suddhasattwa Brahma and Dong-han Yeom
On the Geometry of No-Boundary Instantons in Loop Quantum Cosmology
Reprinted from: *Universe* **2019**, 5, 22, doi:10.3390/universe5010022 254

Iarley P. Lobo and Michele Ronco
Rainbow-Like Black-Hole Metric from Loop Quantum Gravity
Reprinted from: *Universe* **2018**, 4, 139, doi:10.3390/universe4120139 271

Max Joseph Fahn, Kristina Giesel and Michael Kobler
Dynamical Properties of the Mukhanov-Sasaki Hamiltonian in the Context of Adiabatic Vacua
and the Lewis-Riesenfeld Invariant
Reprinted from: *Universe* **2019**, 5, 170, doi:10.3390/universe5070170 289

About the Special Issue Editors

Steffen Gielen studied physics and mathematics in Hannover and Berlin before moving to Cambridge, where he completed Part III of the Mathematical Tripos in 2007 and his Ph.D. in the Relativity and Gravitation Group in 2011. After postdoctoral positions at the Albert Einstein Institute, Perimeter Institute for Theoretical Physics, Imperial College London, and the Canadian Institute for Theoretical Astrophysics, in 2017 he was awarded a University Research Fellowship by the Royal Society. He held the first part of this fellowship (2018–19) at the University of Nottingham, and in October 2019 moved to the University of Sheffield. He received Second Prize in the Buchalter Cosmology Prize competition in 2017. His research currently mostly deals with the application of quantum gravity to early universe cosmology, in particular the possible resolution of the Big Bang singularity through quantum gravity.

Sylvain Carrozza studied physics and mathematics at École Normale Supérieure de Lyon. He received a Ph.D. in theoretical physics from Paris XI University in 2013, for work completed in collaboration with the Albert Einstein Institute in Potsdam, where he was a long-term visiting graduate student. Since then, he has held postdoctoral positions at Aix-Marseille University, the University of Bordeaux, and Perimeter Institute for Theoretical Physics. His research focuses on quantum gravity, quantum field theory and renormalization, and, more broadly, explores the interplay between mathematical physics and combinatorics. He has more specifically contributed to the renormalization program of group field theory, and to recent developments in the theory of random tensors.

Daniele Oriti got his Ph.D. from the University of Cambridge and held research positions at Utrecht University and Perimeter Institute for Theoretical Physics. He directed a research group on quantum gravity at the Albert Einstein Institute for Gravitational Physics, Potsdam, until 2019. He is currently a senior researcher and group leader at the Arnold Sommerfeld Center for Theoretical Physics of the Ludwig-Maximilians-University in Munch, Germany. His research interests span quantum gravity, broadly intended, with specific focus on group field theory formalism, and related subjects, like discrete quantum gravity and loop quantum gravity, as well as fundamental cosmology and quantum black holes. He maintains also an active interest in the foundations of physics and the philosophy of science.

Editorial

Editorial for the Special Issue "Progress in Group Field Theory and Related Quantum Gravity Formalisms"

Sylvain Carrozza [1], Steffen Gielen [2,*] and Daniele Oriti [3]

1. Perimeter Institute for Theoretical Physics, 31 Caroline St. N., Waterloo, ON N2L 2Y5, Canada; scarrozza@perimeterinstitute.ca
2. School of Mathematics and Statistics, University of Sheffield, Hicks Building, Hounsfield Road, Sheffield S3 7RH, UK
3. Arnold-Sommerfeld-Center for Theoretical Physics, Ludwig-Maximilians-Universität München, Theresienstraße 37, 80333 München, Germany; daniele.oriti@physik.lmu.de
* Correspondence: s.c.gielen@sheffield.ac.uk

Received: 16 January 2020; Accepted: 17 January 2020; Published: 20 January 2020

Abstract: This editorial introduces the Special Issue "Progress in Group Field Theory and Related Quantum Gravity Formalisms" which includes a number of research and review articles covering results in the group field theory (GFT) formalism for quantum gravity and in various neighbouring areas of quantum gravity research. We give a brief overview of the basic ideas of the GFT formalism, list some of its connections to other fields, and then summarise all contributions to the Special Issue.

Keywords: quantum gravity; group field theory

1. The Group Field Theory Formalism for Quantum Gravity

Group field theory (GFT) sits at the intersection of various formalisms within the wider field of quantum gravity [1–3]. The basic idea behind GFT is to extend the framework of random matrix and tensor models, where a sum over triangulations is generated as the perturbative expansion of a theory of matrices or tensors, by including additional group-theoretic data to be interpreted as the discrete parallel transports of a connection formulation for gravity. These are the same variables that are fundamental to the definition of loop quantum gravity and spin foam models. GFT are thus a proposal for formulating the dynamics of quantum states built out of the kinematical data of loop quantum gravity, and thus for completing and extending the loop quantisation programme.

A straightforward example of a GFT that illustrates these aspects is the Boulatov model [4] for three-dimensional Riemannian quantum gravity. This model is defined by the action

$$S_{\text{Boul}}[\varphi] = \frac{1}{2} \int d^3 g \, \varphi^2(g_1, g_2, g_3) \\ - \frac{\lambda}{4!} \int d^6 g \, \varphi(g_1, g_2, g_3) \varphi(g_1, g_4, g_5) \varphi(g_2, g_5, g_6) \varphi(g_3, g_6, g_4), \qquad (1)$$

where the GFT field φ is a real-valued function on three copies of $SU(2)$ with an additional permutation symmetry,

$$\varphi : SU(2)^3 \to \mathbb{R}, \quad \varphi(g_1, g_2, g_3) = \varphi(g_2, g_3, g_1) = \varphi(g_3, g_1, g_2), \qquad (2)$$

and "gauge invariance" under the diagonal left action of the group on all the arguments of the field,

$$\varphi(g_1, g_2, g_3) = \varphi(hg_1, hg_2, hg_3) \quad \forall h \in SU(2). \qquad (3)$$

The action consists of a quadratic "kinetic" term, with trivial propagator, and an interaction term with a somewhat unusual ("nonlocal") pairing of arguments. In fact, the group nature of the domain of the dynamical fields and such non-local pairing of arguments in the interactions (shared with matrix and tensor models) can be understood as defining properties of the formalism, the other ingredients (e.g., symmetries, choice of group, kinetic and interaction terms) being a specification of models within the general framework. If one now considers the perturbative expansion of the partition function

$$Z_{\text{Boul}} = \int \mathcal{D}\varphi \, e^{-S_{\text{Boul}}[\varphi]}, \tag{4}$$

in powers of the coupling λ, due to this peculiar structure of the interaction term, the Feynman graphs arising in such an expansion are dual to three-dimensional simplicial complexes, i.e., discrete combinatorial spacetimes. Concretely, one finds

$$Z_{\text{Boul}} = \sum_{\Gamma} \lambda^{V(\Gamma)} \sum_{\{j_f\} \in \text{Irrep}} \prod_{f \in \Gamma} (2j_f + 1) \prod_{v \in \Gamma} \begin{Bmatrix} j_{v_1} & j_{v_2} & j_{v_3} \\ j_{v_4} & j_{v_5} & j_{v_6} \end{Bmatrix}, \tag{5}$$

which is a sum over graphs Γ and, for each Γ, over assignments of irreducible representations j_f of SU(2) to each face of Γ. For each such assignment of j_f one finds an amplitude which is a product over 'face amplitudes' $(2j_f + 1)$ and 'vertex amplitudes' given by a Wigner 6j-symbol for the six faces that meet at a vertex (the number six arising from the six group elements integrated over in the interaction).

Each graph Γ is dual to an oriented 3d simplicial complexes C, where each vertex $v \in \Gamma$ is dual to a tetrahedron $T \in C$, and each face $f \in \Gamma$ dual to a link $l \in C$. The interesting observation is now that the amplitude appearing in the expansion (5) is nothing but the Ponzano–Regge state sum [5] of the triangulated manifold C, multiplied with an overall weight $\lambda^{V(\Gamma)} \equiv \lambda^{N_T(C)}$ depending on the number $N_T(C)$ of tetrahedra in C. The Ponzano–Regge state sum defines a discrete path integral for three-dimensional quantum gravity on a given triangulation C, written on a basis which is the analog of spherical harmonics; see e.g., [6] for details and a discussion of how to rigorously define such a state sum. In these variables, one obtains what is known as a spin foam model in the loop quantum gravity literature, i.e., a covariant definition of the quantum dynamics of spin networks. The same Feynman amplitudes can also be expressed directly in group variables, where they take the form of a lattice gauge theory for 3d BF theory (equivalent to pure 3d gravity with no cosmological constant). An expression in which the same amplitudes coincide with the discrete path integral for 3d quantum gravity in triad and connection variables can also be given.

In summary, the perturbative expansion of the Boulatov model generates a sum over discrete (simplicial) spacetimes with a discrete quantum gravity path integral assigned to each spacetime, augmented by a sum over discrete topologies:

$$Z_{\text{Boul}} = \sum_{C} \lambda^{N_T(C)} Z_{\text{PR}}(C). \tag{6}$$

The Ponzano–Regge state sum defines a topological field theory which is triangulation independent for fixed topology, so that the sum over triangulations of the same 3d manifold merely leads to repeated factors of the same state sum appearing in this expansion. However, in models that are not topological, summing over all simplicial complexes would restore discretisation independence.

This correspondence between perturbative expansions of appropriate GFT models and discrete path integrals for quantum gravity, as well as the connection to spin foam models, extend to other cases, in particular candidate models for quantum gravity in four spacetime dimensions. Spin foam models of immediate interest for loop quantum gravity [7], for which a more detailed understanding in terms of simplicial geometry is available, can be obtained from the expansion of a GFT for a field with four arguments valued in the Lorentz group or SU(2), with additional geometricity conditions imposed on the kinetic or interaction kernels and combinatorially nonlocal interactions of φ^5 type [8].

In what we have presented so far, the GFT approach appears to give simply a reformulation of known expressions for spin foam models and other discrete quantum gravity path integrals, which can be obtained by other means. However, being able to define them in terms of a quantum field theory—albeit an unusual one in that the field φ does not live on spacetime but on an abstract group manifold—provides further avenues to explore spin foam and other models, beyond studying the GFT perturbative expansion.

In particular, one can study perturbative and non-perturbative renormalization of GFTs, and look for theories that can be defined consistently at all scales, and hence become candidates for a fundamental theory. Relying on results and methods developed in the context of tensor models [9] which share the same basic combinatorial structure as GFTs, a tentative but mathematically precise GFT renormalization framework has been developed [10,11]. It has allowed us to demonstrate the perturbative renormalizability—that is, the consistency and predictivity—of simple but non-trivial 'tensorial' GFT actions, and has led to further investigations of the phase structure of GFTs at the non-perturbative level. While the precise physical interpretation of such abstract and background-independent fixed points remains to be elucidated, it is hoped that these technological advances will find suitable extensions to realistic four-dimensional GFT models of quantum gravity.

Regarding the perturbative expansion of GFT itself, tensorial GFT actions are of even broader interest because they admit a $1/N$ expansion. As in the widely studied case of matrix models for two-dimensional quantum gravity, the $1/N$ expansion allows to partially re-sum the perturbative expansion, and thereby provides crucial control over the critical regime of the theory. As a result, the study of tensorial GFT models has seen a number of interesting developments in recent years [12].

The renormalization analysis is also related to the search for a continuum limit in GFTs which can be pursued with quantum field theory methods, addressing the key open question of a continuum limit (or sum over discretizations) in spin foam models and loop quantum gravity. This continuum limit may be given by a non-perturbative phase (often suggested to be of condensate type) in which the GFT field acquires a nonvanishing expectation value, which would be where relevant continuum physics is found (in particular, the number of building blocks diverges). The possible condensate phase of GFTs has been studied with methods coming from condensed matter theory, and has been applied to the description of cosmology and black holes within GFT [13,14]. For example, the emergent cosmological dynamics of the universe, whose microscopic description is given by a GFT model, is extracted from the condensate hydrodynamics rephrased in terms of suitable geometric observables. These effective cosmological dynamics show both the correct classical limit at large volumes and a rather generic bouncing dynamics in place of the classical big bang singularity. Moreover, under further assumptions, they match the effective dynamics found in loop quantum cosmology.

Motivated by these applications to cosmology, formal investigations of the algebraic structure of GFTs have been initiated, aiming at a more refined account of GFT condensate states, and of the condensation mechanism itself. Even more ambitiously, this research direction lays the groundwork for a reformulation and extension of thermal physics to background-independent quantum gravity [15,16].

This Special Issue consists of contributions related to the different avenues of research within the GFT program and to neighboring areas of interest. As we have made clear, loop quantum gravity, spin foam models, and more generally discrete and combinatorial approaches to quantum gravity are closely related to GFT and thus work in these fields has direct implications for GFT. Vice versa, results in the GFT formalism could be of both inspiration and direct application in other quantum gravity formalisms. Looking further afield, submissions from research fields with relevance to more specific aspects of GFT research were also encouraged; these included, for instance, fundamental cosmology, quantum information or condensed matter theory, but also mathematical and formal aspects.

2. Contributions to the Special Issue

The Special Issue consists of 14 published manuscripts; ten research articles and four review articles. The research articles (listed in chronological order of publication) cover the following topics:

- A number of symmetry-reduced models of loop quantum gravity (LQG) have indicated that the fine structure of the LQG quantum state space may naturally lead to deformations of the constraint algebra of general relativity at the semiclassical level. This can, in turn, be interpreted as a quantum deformation of general covariance, required by the existence of a new invariant length scale, the Planck scale. In *Rainbow-Like Black-Hole Metric from Loop Quantum Gravity* [17], Iarley P. Lobo and Michele Ronco investigate spherically-symmetric black hole solutions predicted by effective models of LQG. They show that their quantum-deformed covariance leads to a modified dispersion relation for the total radial momentum, which they then analyze within the paradigm of rainbow gravity.
- *Primordial Power Spectra from an Emergent Universe: Basic Results and Clarifications* [18] by Killian Martineau and Aurélien Barrau discusses a non-standard scenario for the beginning of the universe, known as the emergent universe. In the emergent universe, the Big Bang (or big bounce) is replaced by a transition from a static to an expanding universe. The authors investigate features of the primordial power spectrum of tensor perturbations, or gravitational waves from the early universe. They study the conditions required for a scale-invariant spectrum from an emergent universe scenario and show how features of the spectrum depend on the details of the scale factor evolution near the transition from static to expanding phase.
- One of the most ambitious hopes for quantum gravity is that it can teach us something about the initial state of the universe. *On the Geometry of No-Boundary Instantons in Loop Quantum Cosmology* [19] takes up one of the most prominent ideas of this type, Hawking's no-boundary proposal, and incorporates quantum corrections from loop quantum cosmology into it. Suddhasattwa Brahma and Dong-han Yeom study semiclassical instanton solutions to the LQC path integral. They find that, in contrast to calculations in pure semiclassical general relativity, these instantons have a characteristic infinite tail, and they tend to close off in a regular way as was one of the original ideas behind the no-boundary proposal.
- In *Equivalence of Models in Loop Quantum Cosmology and Group Field Theory* [20], Bekir Baytas, Martin Bojowald, and Sean Crowe observe that the emergent GFT dynamics of homogeneous isotropic universes filled with a massless scalar, which form the basis of the application of GFT to cosmology, can be understood in terms of the algebraic structure of the Lie algebra $\mathfrak{su}(1,1)$. The same algebra structure is known to underlie the most studied models of loop quantum cosmology. The similarities seen between cosmological features of GFT and loop quantum cosmology are then explained in algebraic terms. Furthermore, this underlying algebraic structure suggests possible generalizations of GFT cosmology.
- In *Status of Background-Independent Coarse Graining in Tensor Models for Quantum Gravity* [21], Astrid Eichhorn, Tim Koslowski, and Antonio D. Pereira explore applications of the functional renormalization group to tensor models. They review recent efforts attempting to leverage non-perturbative methods to probe the existence of new large-N limits in tensor models. Once rephrased in the appropriate renormalization group language, in which the size of the tensor plays the role of abstract scale, the existence of such a scaling limit manifests itself by the presence of a non-trivial renormalization group fixed point. The Wetterich equation then provides an elegant and powerful discovery tool, which allows us to scan the theory space of tensor models within larger and larger truncations. From the point of view of quantum gravity, any new large-N limit will translate into a new way of taking the continuum limit. Such investigations are therefore crucial for assessing the viability of tensor and GFT models of quantum gravity in dimension higher than two.
- *Reconstruction of Mimetic Gravity in a Non-Singular Bouncing Universe from Quantum Gravity* [22] by Marco de Cesare deals with bouncing cosmologies such as have been found in the

application of GFT to cosmology. Such bouncing cosmologies have also been seen in models of (limiting curvature) mimetic gravity, in which one modifies gravity by including a scalar field; therefore, the precise relation of mimetic gravity and the cosmological sector of quantum gravity has recently attracted interest. This paper presents a reconstruction procedure by which, starting from a given cosmological effective dynamics from quantum gravity, one can obtain a classical mimetic gravity action (given in terms of a particular function $f(\Box\phi)$) that reproduces this cosmological solution, in the isotropic and homogeneous sector. This might then be seen as a candidate for an effective field theory for quantum gravity approaches such as GFT. The effective field theory is then used to study anisotropies and inhomogeneities.

- Philipp A. Höhn's article *Switching Internal Times and a New Perspective on the 'Wave Function of the Universe'* [23] discusses the fundamental question of how to extend the notion of general covariance from classical to quantum gravity. The central question is how to switch between descriptions given by different observers of what should be the same physics; in other words, between quantum reference frames. Such a relational definition of the quantum dynamics is commonly employed in quantum gravity, and, for example, in GFT cosmology, to define evolution of geometric quantities in a fully diffeomorphism-invariant, thus physical, manner. The paper formulates a general method for relating reduced quantum theories (theories defined after a choice of reference system) to the perspective-neutral framework of the Dirac quantization, akin to the passage from a given coordinate system to generally covariant expressions in classical general relativity. This is then applied to simple models of quantum cosmology where it provides a new angle on the 'wave function of the universe', which becomes a global, perspective-neutral state, encoding all descriptions of the universe relative to different choices of reference system.

- The study of cosmological perturbations is important in the application of quantum gravity models to the early universe, including, for example, in the context of GFT cosmology. *Dynamical Properties of the Mukhanov–Sasaki Hamiltonian in the Context of Adiabatic Vacua and the Lewis–Riesenfeld Invariant* [24] by Max Joseph Fahn, Kristina Giesel and Michael Kobler aims to define suitable initial quantum states for inflation in a near-de Sitter geometry using Hamiltonian methods. The dynamics of cosmological perturbations in an expanding universe can be written in the form of harmonic oscillators with time-dependent frequency. For finite-dimensional systems with such dynamics, an important role is played by the Lewis–Riesenfeld invariant, a constant of motion. One of the main aims of this paper is to extend the application of the Lewis–Riesenfeld invariant to the infinite-dimensional case of field theory. The states thus generated as candidates for an initial state for inflation are then compared to well-known initial states such as the Bunch–Davies vacuum.

- *Spin Foam Vertex Amplitudes on Quantum Computer—Preliminary Results* [25] by Jakub Mielczarek outlines first steps of an ambitious project: the use of quantum algorithms to understand spin foam vertex amplitudes, one of the key ingredients in defining the dynamics of spin foam models (and hence indirectly, of GFT models). In this article, the focus is on a simple spin network (a complete graph of five vertices representing five tetrahedra forming the boundary of a four-simplex) with all spins set equal to $\frac{1}{2}$. The paper discusses how to calculate absolute values of vertex amplitudes for this process, and the approach is tested by comparing the results obtained by existing quantum algorithms with known exact results.

- In *Thermal Quantum Spacetime* [26], Isha Kotecha discusses an extension of equilibrium statistical mechanics and thermodynamics to background-independent systems that is then applied to discrete quantum gravity approaches, such as GFT. A generalised notion of Gibbs equilibrium is characterized in information-theoretic terms, where entropy plays a more fundamental role than energy. This then forms the basis for a framework of a statistical mechanics of discrete quantum gravity in the absence of standard notions of time and energy. Covariant GFT is shown to arise as an effective statistical field theory of generalized Gibbs states. The paper presents also a conceptual review of these and other results in this context and an extensive outlook of further work in this important direction.

The Special Issue also includes four review articles, namely

- In *Quantum Gravity on the Computer: Impressions of a Workshop* [27], Lisa Glaser and Sebastian Steinhaus summarise the outcome of the workshop they organized in March 2018 at NORDITA, in Stockholm. Spanning a rather wide array of distinct approaches (including loop quantum gravity and spin foams, as well as GFT), this article reviews recent and ongoing contributions of computational physics to open problems in discrete quantum gravity, such as those related to the challenging question of the restoration of the diffeomorphism symmetry in the continuum limit. The review concludes with an insightful roadmap, which, among other targets, advocates the creation of open data science infrastructures and online repositories dedicated to numerical investigations of quantum geometry.
- Functional renormalization group (FRG) techniques have recently been successfully applied to GFT models. *Progress in Solving the Nonperturbative Renormalization Group for Tensorial Group Field Theory* [28] by Vincent Lahoche and Dine Ousmane Samary gives an overview over three previous papers by these authors, in which the FRG is applied to Abelian GFT models based on gauge group $U(1)^d$, without a closure/gauge invariance constraint (such a constraint is usually imposed for the geometric interpretation of these models, as it introduces a gauge connection and turns GFT models into a quantization of gauge theories or gauge-theoretic gravitational models). A quartic interaction term of the melonic type is studied in these models. An effective vertex expansion method is introduced in order to solve the FRG and study the resulting renormalization flow, in particular with the aim of identifying non-Gaussian fixed points; these fixed points may be associated to phase transitions that can be interpreted as describing the formation of a GFT condensate (see above). Ward–Takahashi identities provide additional constraints that have to be taken into account when finding approximate solutions to the flow equations.
- In recent years, the GFT formalism has permitted the emergence of a new approach to quantum cosmology, based on the general paradigm of condensation in GFT, as we discussed above. Thanks to the quantum field theory language underlying GFT, the idea that cosmological spacetime structures may be the result of the condensation of a large number of pre-geometric and quantum degrees of freedom has been concretely realized and thoroughly investigated in simple GFT models. In *Group Field Theory Condensate Cosmology: An Appetizer* [29], Andreas G. A. Pithis and Mairi Sakellariadou provide a gentle and pedagogical introduction to this fast-developing area of research. After reviewing how isotropic and homogeneous cosmology can be recovered from a GFT condensate, they summarise recent efforts aiming at including anisotropies and cosmological perturbations, paving the way towards the derivation of observable consequences.
- A number of recent developments in quantum gravity suggest that the Einstein equations might be best understood as a reflection of the entanglement structure of fundamental and yet-to-be-discovered quantum gravity degrees of freedom. In the context of the AdS/CFT correspondence, this idea is beautifully captured by the Ryu-Takayanagi formula, which relates the entanglement entropy of regions in the boundary CFT to the area of extremal surfaces in the bulk. In *Holographic Entanglement in Group Field Theory* [30], Goffredo Chirco reviews the realization of such ideas in the context of GFT, where candidate microscopic degrees of freedom are available. Relying on a general dictionary allowing to view GFT many-body states as tensor networks, a pedagogical introduction to the computation of Rényi entropies by means of the replica method is proposed. This allows the author to derive a GFT analog of the Ryu-Takayanagi equation, which is fully compatible with the geometric interpretation of the GFT fundamental degrees of freedom: the area term entering the formula is consistently given by the expectation value of the corresponding GFT area operator.

Funding: The work of S.G. is supported by the Royal Society under a Royal Society University Research Fellowship (UF160622) and a Research Grant for Research Fellows (RGF\R1\180030). The work of D.O. is supported by the Deutsche Forschung Gemeinschaft (DFG).

Acknowledgments: The guest editors would like to thank all the authors for their contributions and the reviewers for the constructive reports. Their work helped the editors to collect this Special Issue. S.C. acknowledges support from Perimeter Institute. Research at Perimeter Institute is supported in part by the Government of Canada through the Department of Innovation, Science and Economic Development Canada and by the Province of Ontario through the Ministry of Economic Development, Job Creation and Trade.

Conflicts of Interest: The authors declare no conflict of interest. The funders had no role in the writing of the manuscript, or in the decision to publish the results.

References

1. Oriti, D. The group field theory approach to Quantum Gravity. In *Approaches to Quantum Gravity—Toward a New Understanding of Space, Time and Matter*; Oriti, D., Ed.; Cambridge University Press: Cambridge, UK, 2009; pp. 310–331, ISBN 978-0521860451.
2. Freidel, L. Group Field Theory: An Overview. *Int. J. Theor. Phys.* **2005**, *44*, 1769–1783. [CrossRef]
3. Krajewski, T. Group field theories. *PoS QGQGS 2011* **2011**, 005.
4. Boulatov, D.V. A Model of three-dimensional lattice gravity. *Mod. Phys. Lett. A* **1992**, *7*, 1629–1646. [CrossRef]
5. Ponzano, G.; Regge, T. Semiclassical limit of Racah coefficients. In *Spectroscopic and Group Theoretical Methods in Physics*; Bloch, F., Cohen, S.G., de-Shalit, A., Sambursky, S., Talmi, I., Eds.; North-Holland: Amsterdam, The Netherlands, 1968; pp. 1–58, ISBN 978-0720401400.
6. Barrett, J.W.; Naish-Guzman, I. The Ponzano-Regge model. *Class. Quant. Grav.* **2009**, *26*, 155014. [CrossRef]
7. Perez, A. Spin foam models for quantum gravity. *Class. Quant. Grav.* **2003**, *20*, R43. [CrossRef]
8. Reisenberger, M.P.; Rovelli, C. Spacetime as a Feynman diagram: The connection formulation. *Class. Quant. Grav.* **2001**, *18*, 121–140. [CrossRef]
9. Gurau, R. *Random Tensors*; Oxford University Press: Oxford, UK, 2016; ISBN 978-0198787938.
10. Rivasseau, V. Quantum Gravity and Renormalization: The Tensor Track. *AIP Conf. Proc.* **2012**, *1444*, 18–29.
11. Carrozza, S. Flowing in Group Field Theory Space: A Review. *SIGMA* **2016**, *12*, 070. [CrossRef]
12. Rivasseau, V. The Tensor Track, III. *Fortsch. Phys.* **2014**, *62*, 81–107. [CrossRef]
13. Gielen, S.; Sindoni, L. Quantum Cosmology from Group Field Theory Condensates: A Review. *SIGMA* **2016**, *12*, 082. [CrossRef]
14. Oriti, D. The universe as a quantum gravity condensate. *Comptes Rendus Phys.* **2017**, *18*, 235–245. [CrossRef]
15. Kotecha, I.; Oriti, D. Statistical Equilibrium in Quantum Gravity: Gibbs states in Group Field Theory. *New J. Phys.* **2018**, *20*, 073009. [CrossRef]
16. Chirco, G.; Kotecha, I.; Oriti, D. Statistical equilibrium of tetrahedra from maximum entropy principle. *Phys. Rev. D* **2019**, *99*, 086011. [CrossRef]
17. Lobo, I.P.; Ronco, M. Rainbow-like Black Hole metric from Loop Quantum Gravity. *Universe* **2018**, *4*, 139. [CrossRef]
18. Martineau, K.; Barrau, A. Primordial power spectra from an emergent universe: Basic results and clarifications. *Universe* **2018**, *4*, 149. [CrossRef]
19. Brahma, S.; Yeom, D.-H. On the geometry of no-boundary instantons in loop quantum cosmology. *Universe* **2019**, *5*, 22. [CrossRef]
20. Baytas, B.; Bojowald, M.; Crowe, S. Equivalence of models in loop quantum cosmology and group field theory. *Universe* **2019**, *5*, 41. [CrossRef]
21. Eichhorn, A.; Koslowski, T.; Pereira, A.D. Status of background-independent coarse-graining in tensor models for quantum gravity. *Universe* **2019**, *5*, 53. [CrossRef]
22. De Cesare, M. Reconstruction of Mimetic Gravity in a Non-Singular Bouncing Universe from Quantum Gravity. *Universe* **2019**, *5*, 107. [CrossRef]
23. Höhn, P.A. Switching Internal Times and a New Perspective on the 'Wave Function of the Universe'. *Universe* **2019**, *5*, 116. [CrossRef]
24. Fahn, M.J.; Giesel, K.; Kobler, M. Dynamical Properties of the Mukhanov-Sasaki Hamiltonian in the context of adiabatic vacua and the Lewis-Riesenfeld invariant. *Universe* **2019**, *5*, 170. [CrossRef]

25. Mielczarek, J. Spin Foam Vertex Amplitudes on Quantum Computer—Preliminary Results. *Universe* **2019**, *5*, 179. [CrossRef]
26. Kotecha, I. Thermal Quantum Spacetime. *Universe* **2019**, *5*, 187. [CrossRef]
27. Glaser, L.; Steinhaus, S. Quantum Gravity on the computer: Impressions of a workshop. *Universe* **2019**, *5*, 35. [CrossRef]
28. Lahoche, V.; Samary, D.O. Progress in the solving nonperturbative renormalization group for tensorial group field theory. *Universe* **2019**, *5*, 86. [CrossRef]
29. Pithis, A.G.A.; Sakellariadou, M. Group Field Theory Condensate Cosmology: An Appetizer. *Universe* **2019**, *5*, 147. [CrossRef]
30. Chirco, G. Holographic Entanglement in Group Field Theory. *Universe* **2019**, *5*, 211. [CrossRef]

© 2020 by the authors. Licensee MDPI, Basel, Switzerland. This article is an open access article distributed under the terms and conditions of the Creative Commons Attribution (CC BY) license (http://creativecommons.org/licenses/by/4.0/).

Article

Status of Background-Independent Coarse Graining in Tensor Models for Quantum Gravity

Astrid Eichhorn [1,2]**, Tim Koslowski** [3] **and Antonio D. Pereira** [2,*]

[1] CP3-Origins, University of Southern Denmark, Campusvej 55, DK-5230 Odense M, Denmark; eichhorn@sdu.dk
[2] Institut für Theoretische Physik, Universität Heidelberg, Philosophenweg 16, 69120 Heidelberg, Germany
[3] Instituto de Ciencias Nucleares, UNAM, Apartado Postal 70-543, Coyoacán 04510, Ciudad de México, Mexico; koslowski@nucleares.unam.mx
[*] Correspondence: a.pereira@thphys.uni-heidelberg.de

Received: 30 November 2018; Accepted: 30 January 2019; Published: 5 February 2019

Abstract: A background-independent route towards a universal continuum limit in discrete models of quantum gravity proceeds through a background-independent form of coarse graining. This review provides a pedagogical introduction to the conceptual ideas underlying the use of the number of degrees of freedom as a scale for a Renormalization Group flow. We focus on tensor models, for which we explain how the tensor size serves as the scale for a background-independent coarse-graining flow. This flow provides a new probe of a universal continuum limit in tensor models. We review the development and setup of this tool and summarize results in the two- and three-dimensional case. Moreover, we provide a step-by-step guide to the practical implementation of these ideas and tools by deriving the flow of couplings in a rank-4-tensor model. We discuss the phenomenon of dimensional reduction in these models and find tentative first hints for an interacting fixed point with potential relevance for the continuum limit in four-dimensional quantum gravity.

Keywords: quantum gravity; renormalization group; discrete quantum gravity models

---

1. Invitation to Background-Independent Coarse Graining in Tensor Models for Quantum Gravity

The path integral for quantum gravity takes center stage in a diverse range of approaches to quantum spacetime. It is tackled either as a quantum field theory for the metric [1–6], or in a discretized fashion with a built-in regularization [7–19]. The latter approach, relying on unphysical building blocks of space(time), provides access to a physical space(time) only when a universal continuum limit can be taken. Universality [20–22] is key in this setting, as it guarantees independence of the physics from unphysical choices, e.g., in the discretization procedure, i.e., the shape of the building blocks. To discover universality, background-independent coarse-graining techniques are a well-suited tool as universality arises at fixed points of the coarse-graining procedure.

The notion of "background-independent coarse graining" at a first glance appears to be an oxymoron and suggests this review should be extremely short. After all, to coarse grain, one first needs to define what one means by "coarse" and by "fine". Intuitively one would expect these notions to rely on a background. In particular, a definition of ultraviolet and infrared, key to the setup of Renormalization Group (RG) techniques, seems to require a metric, i.e., a geometric background. Yet, with RG techniques now playing an important role in different quantum-gravity approaches, coarse-graining techniques suitable for a setting without distinguished background have successfully been developed [23–38] and applied to various quantum-gravity models. In this review, we will focus on the developments kicked off in [13,39–42], and introduce the key concepts behind a background-independent RG flow and the associated notion of coarse graining. In particular, we will focus on the development and application of these tools to tensor models.

Tensor models are of interest for quantum gravity both as a way of exploring the partition function [13,43–45] directly as well as through a conjectured correspondence of specific tensor models to aspects of a geometric description in the context of the SYK-model [46–48]. In both settings, the large N' limit, where N' is the tensor size, is of key interest, and physical results are extracted in the limit $N' \to \infty$. In their simplest version that is of particular interest to quantum gravity, tensor models are 0-dimensional theories, i.e., there is no notion of spacetime in the definition of the models. Instead, the *dual* interpretation of the interactions in tensor models is that of discrete building blocks of space(time), cf. Figure 1. In this interpretation of tensor models through the graphs dual to the Feynman diagrams, the building blocks are interpreted as pieces of flat space(time). Curvature is accordingly localized at the hinges. The dual representation of tensors is in terms of building blocks of geometry. Closely related to tensor models are tensorial (group) field theories which are characterized by the same non-trivial combinatorial structure of tensor model interactions. In addition, a non-trivial kinetic term is present, and group field theories are defined on a group manifold, see, e.g., [49–56]. This extra group data may be associated with intrinsic geometric data of the associated simplices.

The d indices of a rank-d tensor are associated with the $(d-2)$-subsimplices of a $(d-1)$-simplex. For instance, for rank 3, the indices are associated with the edges ($(d-2)$ subsimplices) of a triangle ($(d-1)$ simplex), cf. Figure 1. Correspondingly, in the rank-4-case, each index is associated with one of the four faces of a tetrahedron, cf. Figure 2. When two tensors are contracted along one index, the corresponding $(d-1)$-simplices share a $(d-2)$ simplex, e.g., for the rank-3 case, two triangles are glued along an edge, cf. Figure 3. In the rank-4-case, two tetrahedra are glued along a face. Allowed interaction terms are positive powers of the tensors that contain no free indices. This means that each $(d-2)$-subsimplex is glued to another $(d-2)$-subsimplex. Therefore they correspond to d-dimensional building blocks of space(time), e.g., tetrahedra for a fourth-order interaction in the rank-3 case, cf. Figure 1. The propagator of the theory identifies all d indices of two tensors, corresponding to a gluing of one $d-1$ simplex to another, e.g., gluing of two triangles along their faces. Accordingly, the terms in the Feynman diagram expansion of tensor models have a dual interpretation as simplicial pseudomanifolds. In other words, the combinatorics of tensor models encode dynamical triangulations. In the simplest case, when no additional rules are imposed on the gluing, Riemannian pseudomanifolds are generated. The inscription of local lightcones inside the building blocks, such that a consistent notion of causality can emerge and the pseudo-manifold is Lorentzian, requires additional rules for the gluing and more than one type of building block [9–11,57].

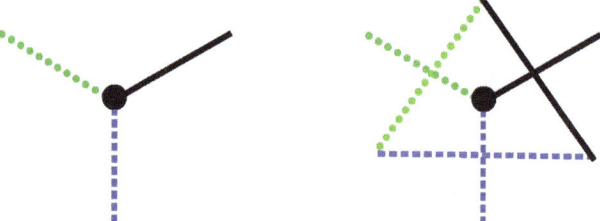

Figure 1. The three indices of a rank-3-tensor are associated with the three lines of a triangle.

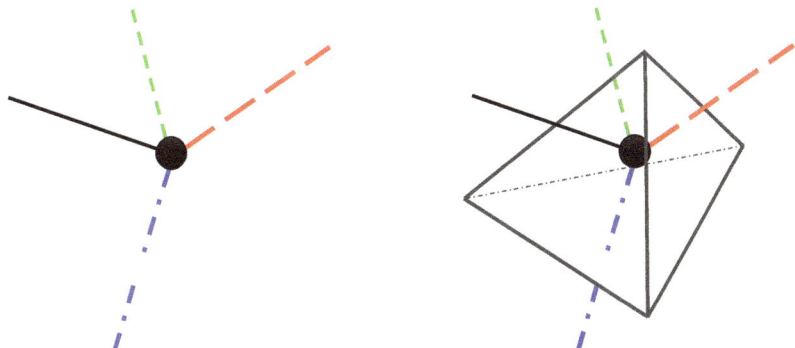

Figure 2. The four indices of a rank-4-tensor are associated with the three triangles of a tetrahedron.

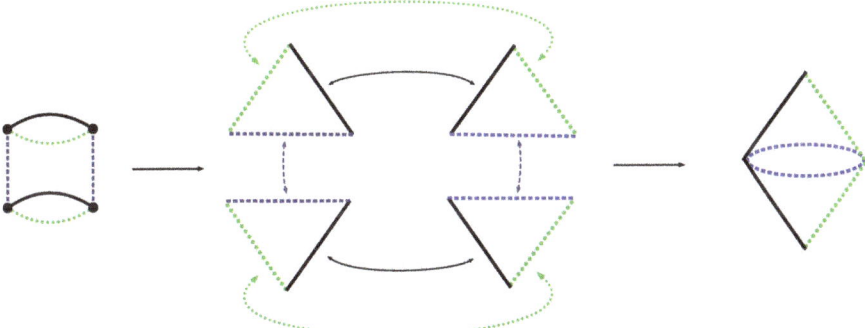

Figure 3. The invariant $T_{ijk}T_{ijl}T_{mnl}T_{mnk}$, depicted on the **left**, is associated with the gluing of four triangles (**center**) into a building block of 3-space (to the **right**). The contraction of common indices is associated with the gluing of triangles along common edges.

There are no experimental hints that indicate that spacetime is a simplicial pseudo-manifold, accordingly it is assumed to be a continuum manifold. In particular, while the presence of physical discreteness close to the Planck scale could be compatible with all observations to date, one would not expect a naive discretization as it arises from tensor models, to actually be physical. Instead, this form of discreteness should be regarded merely as a regularization of the path integral. To take the continuum limit in tensor models, the number of degrees of freedom, encoded in the tensor size N', must be taken to infinity. In [42,43,58–62] it was shown that models of real (complex) tensors with a $O(N') \otimes O(N') \otimes ... \otimes O(N')$ ($U(N') \otimes U(N') \otimes ... \otimes U(N')$) symmetry[1] admit a $1/N'$ expansion, where N' is the size of the tensors. Here, each symmetry group in the above product acts on exactly one of the indices of the tensor. Due to the existence of a $1/N'$ expansion, these are viable candidates to search for a physical continuum limit by taking $N' \to \infty$. Yet, simply taking $N' \to \infty$ is not sufficient to obtain a physical continuum limit: The microscopic properties and structure of the building blocks in the model is not taken to be physical, but only a discretization/regularization. Different microscopic choices can be made that should not leave an imprint on the continuum physics, such as, e.g., the shape of the building blocks. Accordingly, the continuum limit should be *universal*. Universality is achieved at fixed points of the RG flow. Therefore, an RG flow must be set up for these models.

---

[1] Here we use the notation $G \otimes G$ to denote the direct product of group actions as linear transformations of different indices of tensors.

Unlike in quantum field theories defined on a background, no local, i.e., geometric notion of scale is available. In fact, the only notion of scale is the size of the tensors, N'. In fact, using the tensor size N' as a scale agrees with the intuitive notion of coarse graining, also underlying formal developments such as the a-theorem [63]: Coarse graining leads from many degrees of freedom (large N'), to fewer, effective degrees of freedom (small N'). Therefore, a pregeometric RG flow is set up in the tensor size N', where a universal continuum limit can then be discovered as an RG fixed point. This point of view was advocated in [13,39] and formally developed and benchmarked in [40,41,64,65].

In the dual picture, the lattice spacing needs to be taken to zero in such a way that the correlation length on the lattice diverges. Then, microscopic details of the setup become irrelevant. This is possible at a higher-order phase transition, linked to a fixed point in the space of couplings. The intuition behind these ideas can be tested in the two-dimensional case, where the double-scaling limit of matrix models [66–69], which is a universal continuum limit, is obtained by taking $N' \to \infty$ while tuning the coupling to a critical value as a power of N'. This is completely analogous to the case of continuum RG flows, where universal critical behavior with diverging correlation length is tied to RG fixed points, near which couplings scale with particular powers of the scale. Specifically, the double-scaling limit in matrix models with coupling g is achieved by taking $N' \to \infty$ and $g \to g_{\text{crit}}$, while holding

$$(g - g_{\text{crit}})^{\frac{5}{4}} N' = \text{const}, \tag{1}$$

which can be rewritten in the form

$$g(N') = g_{\text{crit}} + \text{const.}^{4/5} N'^{-4/5}. \tag{2}$$

This immediately brings to mind the linearized scaling of couplings close to RG fixed points, which is given by the scale raised to the power $-\theta$, with the critical exponent θ.

Please note that there are arguments suggesting that quantum gravity should be discrete. One might interpret this as implying that there is no need to take the continuum limit in tensor models, and one can instead even work at finite N'. Yet, discreteness is actually a subtle issue in quantum gravity. As discussed in more detail, e.g., in [70], kinematical and dynamical discreteness are not the same thing in quantum gravity, and discreteness can be an *emergent* property of the physical continuum limit. On the other hand, a simple implementation of discreteness in the sense of a cutoff potentially features the same breakdown of predictivity at scales near the cutoff that effective field theories do. Specifically, the interaction terms compatible with the symmetries of a model are infinitely many for tensor models. The continuum limit is a way of imposing predictivity in a model by reducing the number of free parameters characterizing its dynamics to finitely many. In the RG language, this is linked to the fact that fixed points feature only finitely many relevant directions. In the language of critical phenomena, one must tune only finitely many parameters to approach criticality in the sense of a higher-order phase transition. In this spirit, we aim at discovering a universal continuum limit in tensor models for quantum gravity such that both independence of unphysical microscopic details as well as predictivity is guaranteed. We leave open the question whether these models feature *emergent* discreteness once the continuum limit is taken, but merely point out that taking the continuum limit does in fact not preclude the possibility of emergent, physical discreteness.

In summary, to discover a universal continuum limit, at which a physical spacetime could emerge from discrete building blocks of spacetime, we must discover universal critical points. These are linked to RG fixed points. In the absence of a background, the only scale available for coarse graining is the tensor size N'. As we will explain in the next sections, setting up an RG flow in N' is both conceptually meaningful as well as feasible in practice.

This review is structured as follows. In Section 2 we introduce the conceptual basics of background-independent coarse graining. We provide an overview of how to implement these ideas in practice and how to set up a flow equation in Section 3. In Section 4 we discuss in detail how scaling dimensions can be derived in a setting without a background, translating to the absence of physical length scales and corresponding units that would define canonical dimensions. We provide an overview of the benchmark case of two dimensions in Section 5, where quantitatively robust results on the well-known continuum limit can be achieved using our flow equation. In Section 6 we summarize results in the rank-3-case, where several RG fixed points give access to a dimensionally reduced continuum limit. We also highlight a recently discovered candidate for a fixed point which might potentially turn out to be relevant for three-dimensional quantum gravity. To provide a step-by-step instruction in how to set up and evaluate RG flows in tensor models, we present the first study of a rank-4-model with these tools in Section 7. We discover several universality classes featuring dimensional reduction. As a hint of the promise our method could have, we unveil tentative indications for a universality class that might potentially be linked to four-dimensional quantum gravity. In the Outlook and Conclusions 8 we advocate that progress towards a comprehensive understanding of quantum gravity could be accelerated by strengthening the effort to bridge the gap between different approaches to quantum gravity. We discuss in particular how continuum studies of asymptotic safety, Monte Carlo simulations of (causal) dynamical triangulations and FRG studies of tensor models could provide a link to phenomenology and particle physics, while allowing to probe features of emergent geometries and enabling us to link the discrete and continuum side via a universal transition.

2. Conceptual Basics: Background-Independent Renormalization Group Flow in Gravity

RG techniques are playing a role in several different approaches to quantum gravity. This includes the asymptotic safety program [4,5], the continuum limit in spin foams [23–29,32–36] and Hamiltonian RG flows in canonical loop quantum gravity [37], tensorial (group) field theories [52,71] as well as holographic RG flows in the context of the AdS/CFT conjecture [72]. Yet, at a first glance, quantum gravity would appear to be the one of the fundamental interactions to which RG techniques are not easily applicable. The reason lies in the dichotomy of background independence and local coarse graining. While the results obtained with a local coarse-graining formulation can be made background-independent, see, e.g., [30], the RG flow itself necessarily relies on an (auxiliary) background, if the flow has the interpretation of a local coarse graining. A more direct reconciliation of RG techniques with background independence is provided by a non-local form of coarse graining: RG flows, in agreement with the a-theorem [63], connect descriptions with many degrees of freedom with effective descriptions of the same system based on fewer degrees of freedom. This idea can be realized both in a local as well as a non-local form. The latter is directly applicable to tensor models for quantum gravity. These are defined without any notion of spacetime, metric or locality. Yet they come with a measure of the number of degrees of freedom, namely the tensor size N'. Coarse-graining therefore corresponds to integrating out subsequent "layers" of the tensors (rows and columns in the matrix-model case), thereby connecting a description at large N' with an effective description at small N'. In particular, such coarse-graining techniques allow us to search for a well-defined large N'-limit, where the dynamics stays invariant under the step from N' to $N' + 1$, such that the limit $N' \to \infty$ can be taken. In this limit, one can hope for quantum space(time) to emerge from tensor models.

Note also that while local coarse-graining techniques typically rely on Riemannian signature, raising the difficulty of connecting back to the Lorentzian case of interest for physics, a non-local coarse graining does not rely on a momentum cutoff. Accordingly, a more direct search for a universal continuum limit for Lorentzian models could become possible in this setup. This includes applications of the FRG to tensor models dual to causal dynamical triangulations [73] as in [74], as well as the application of coarse-graining techniques to the link matrix in causal sets [75].

We will now explain how to implement these ideas in practice in the form of a flow equation. One can view the flow equation as a reformulation of the path integral in terms of a functional differential equation. The search for a continuum limit in the path integral then becomes the search for a well-defined ultraviolet (in an appropriate sense) solution of the flow equation. At a completely general and formal level, the derivation of the flow equation from the path integral works as follows: One introduces a new term into the exponential in the generating functional that is quadratic in the field and depends on some external parameter which we will call $\mathcal{K}$ here. For now, we leave this parameter completely general, and do not provide any physical interpretation associated with it. It is simply to be thought of as a "sieve" on the space of field configurations, letting through only a subset of configurations. The generating functional depends on $\mathcal{K}$ and is denoted by $Z_\mathcal{K}$, schematically

$$Z_\mathcal{K} = \int \mathcal{D}\varphi\, e^{-S[\varphi] + \text{Tr} J \cdot \varphi - \frac{1}{2} \text{Tr} \varphi \cdot R_\mathcal{K} \cdot \varphi}, \tag{3}$$

where $S[\varphi]$ is a given microscopic action, J is an external source and φ denotes the random fields. The trace is to be interpreted in a suitable way for the model at hand, i.e., it signifies a momentum integral and trace over internal indices in standard QFTs on a background, and an appropriate summation over indices in the discrete case, e.g., for tensor models. We do not write indices for simplicity, but the fields are not necessarily scalars. As a function of the parameter $\mathcal{K}$, a subset of configurations in the generating functional are suppressed, such that in the limit $\mathcal{K} \to \infty$, all configurations are suppressed. Conversely, in the limit $\mathcal{K} \to 0$ the unmodified generating functional is recovered. Since the suppression term is quadratic in the field, $\partial_\mathcal{K} Z_\mathcal{K}$ can be expressed in terms of the two-point function,

$$\partial_\mathcal{K} Z_\mathcal{K} = -\frac{1}{2} \int \mathcal{D}\varphi\, \text{Tr}\, \varphi \cdot (\partial_\mathcal{K} R_\mathcal{K}) \cdot \varphi\, e^{-S[\varphi] + \text{Tr} J \cdot \varphi - \frac{1}{2} \text{Tr} \varphi \cdot R_\mathcal{K} \cdot \varphi}. \tag{4}$$

For the modified Legendre transform

$$\Gamma_\mathcal{K}[\phi] = \sup_J \left(\text{Tr} J \cdot \phi - \ln Z_\mathcal{K} \right) - \frac{1}{2} \text{Tr} \phi \cdot R_\mathcal{K} \cdot \phi, \tag{5}$$

with $\phi = \langle \varphi \rangle$, this implies

$$\partial_\mathcal{K} \Gamma_\mathcal{K}[\phi] = \frac{1}{2} \text{Tr}\left[\left(\frac{\delta^2 \Gamma_\mathcal{K}[\phi]}{\delta \phi^2} + R_\mathcal{K} \right)^{-1} \partial_\mathcal{K} R_\mathcal{K} \right], \tag{6}$$

which is known as the functional renormalization group (FRG) equation. For the case of a continuum QFT on an (auxiliary) background it was derived in [76], see also [77,78], pioneered for gauge theories in [79] and gravity in [3]. Up to here, the derivation of the flow equation from the path integral is just a formal "trick" that can be performed with any (functional) integral: Instead of performing the integral "all at once", one introduces the exponential of a quadratic term that depends on an external parameter. This allows to derive a differential equation that encodes how the result of the integral reacts to changes in the parameter. As long as the suppression term is quadratic in the field, an equation which is structurally of the form Equation (6) follows directly from the definition Equation (3). The question to address in a physics setting is whether any physical meaning can be given to the external parameter and consequently to the ensuing differential equation.

For instance, in local field theories introducing an external parameter that does not lead to a notion of local coarse graining is not expected to be fruitful. In such cases, the modes that remain after integrating out some "shells" of modes do not contain physically relevant degrees of freedom. Thus deriving effective field theories for those degrees of freedom might be an interesting computational exercise, but is presumably not useful for answering physical questions. The notion of UV/IR is therefore key to make the effective field theories obtained by renormalization useful for practical computations. Thus, although both in QFTs with and without a background, different choices for $\mathcal{K}$ are possible, "non-local" choices have not yet been tested for their usefulness in the setting with a background. Accordingly, in the case with a background it turns out to be the most powerful tool to relate $\mathcal{K}$ to a momentum scale. This choice allows to implement a notion of local coarse graining: Decomposing configurations into eigenfunctions of an appropriate Laplacian, $R_\mathcal{K}$ suppresses configurations with eigenvalues of the Laplacian smaller than k^2. In this case, the flow equation has the interpretation of providing the response of the effective dynamics to a local coarse-graining step.

The quest for a well-defined path integral, which exists as all configurations are taken into account becomes the question for a well-defined solution of Equation (6) for $\mathcal{K} \to \infty$. Specifically, in tensor models, it is useful to choose the suppression term as a function of the number of components of the tensor, N, e.g., in the form

$$\Delta S_N = \frac{1}{2} \mathrm{Tr}\, T_{a_1...a_d} R_N(a_1,...,a_d) T_{a_1...a_d}, \qquad (7)$$

such that

$$\partial_t \Gamma_N[T] = N \partial_N \Gamma_N[T] = \frac{1}{2} \mathrm{Tr} \left[\left(\frac{\delta^2 \Gamma_N}{\delta T_{a_1...a_d} \delta T_{b_1...b_d}} + R_N(a_1...a_d) \delta_{a_1 b_1}...\delta_{a_d b_d} \right)^{-1} \partial_t R_N(a_1...a_d) \right]. \qquad (8)$$

In a slight abuse of notation, we use T both for the tensors that are integrated over in the generating functional, as well as for their expectation value on which the effective average action Γ_N depends.

As we search for a phase transition in these models, we will employ the FRG to search for infrared (IR) fixed points. The relevant directions correspond to the number of parameters that require tuning to reach criticality. As we aim at approaching such IR fixed points in the limit of large tensors, we will set up beta functions in the large N limit.

The general structure of the flow equation was first derived and benchmarked in the case of matrix models in [40,41] and applied and further developed for rank-3 tensor models in [64,65]. Similarly, the FRG has been employed in the context of tensorial (group) field theories in [71,80–90]. See also [91,92] for related studies using the Polchinski equation.

In the following, we will review how to use Equation (8) for practical calculations and to search for candidates for a universal continuum limit in quantum gravity.

3. Lightning Review of the Setup: Theory Space, Regulator and How to Calculate in Practice

3.1. Theory Space

The flow equation provides the scale-dependent change of coefficients of the dynamics, spanned by the infinitely many terms compatible with the symmetries. For instance, starting with a quartic interaction term in $\Gamma_N[T]$, Equation (8) takes the schematic form

$$\partial_t \Gamma_N[T] \sim \frac{\#}{\# + T^2}, \qquad (9)$$

which admits a Taylor expansion with non-vanishing coefficients not only of the quartic but generically also of all T^{2n}, $n \in \mathbb{N}$. The is the analogue of the well-known observation that Wilsonian coarse-graining flow generates all quasi-local interactions that are compatible with the symmetries[2]. Even though in the case at hand we are not dealing with a local coarse-graining flow, the analogous observation holds and all interactions with positive powers of tensors that obey the symmetries, are generated. Accordingly, to implement the flow equation in practice requires the following steps

(a) understanding which interactions are part of the (infinite-dimensional) theory space,

---

[2] In the local case, the well-known non-locality of the full effective action $\Gamma_{k \to 0}$ is expected to arise through resummation of quasi-local terms, i.e., terms with arbitrary high but positive powers of derivatives, see, e.g., [93]. If non-local terms, i.e., terms with negative powers of derivatives and/or fields are included in theory space as independent basis elements, predictivity is expected to break down, even at interacting fixed points, as these interaction terms have increasingly positive canonical dimension. Although there is no notion of quasilocality in spacetime in the tensor-model interactions, terms with negative powers of tensors are expected to suffer from the same problem. Moreover, terms with inverse powers of tensors do not directly provide an interpretation in terms of building blocks of geometry in a dual picture. In fact, if a "quasi-local" truncation of theory space is chosen, no interactions which cannot be written in a quasi-local form are generated by the flow at finite scales. Accordingly, this restriction of the theory space is both well-motivated as well as self-consistent.

(b) selecting a criterion according to which truncations of the theory space to a (finite-dimensional) subspace can be chosen,
(c) truncating theory space to a subspace in which Equation (8) can be evaluated in practice,
(d) finding solutions of Equation (8) and checking whether they satisfy the criterion in (b).

Steps (c) and (d) are then iterated and only fixed-point solutions which reach stability under the steps in the iteration procedure are kept.

The tensor models[3] typically of interest for quantum gravity feature an independent symmetry group for each index. position, e.g., a product of d copies of an $O(N')$ symmetry for the real rank-d model. Accordingly, interactions cannot have an explicit index-dependence, and no tensors with open indices can occur. All allowed interactions $\mathcal{O}^{(n)}$ of n tensors can therefore be cast in the form

$$\mathcal{O}^{(n)} = T_{a_1 b_1 \ldots d_1} \cdots T_{a_n b_n \ldots d_n} \mathcal{C}_{a_1 \ldots a_n, b_1 \ldots b_n, \ldots, d_1 \ldots d_n}, \tag{10}$$

where d is the rank. The contraction pattern $\mathcal{C}_{a_1 \ldots a_n, b_1 \ldots b_n, \ldots, d_1 \ldots d_n}$ is a product of Kronecker deltas, in which a's can only be contracted with a's, b's with b's and so forth. All possible permutations of the labels 1 to n must be taken into account independently for each index set a, b, etc. Some of the resulting $\mathcal{O}^{(n)}$ will be combinatorially equivalent, in which case only one representative is taken into account. In Table 1 we list combinatorially distinct structures up to sixth order in the tensors for the real and complex rank-3 models.

Please note that the theory space includes multi-trace interactions. This name derives from the rank-2, i.e., matrix-model case, where interactions take the form $\operatorname{Tr} T_{a_1 b_1} \cdots T_{a_n b_n} \cdot \ldots \cdot \operatorname{Tr} T_{a_1 b_1} \cdots T_{a_m b_m}$. In the case of higher rank, similarly combinatorially disconnected interactions are part of the theory space. These are generated by the flow, even if they are not included in a truncation. There is no symmetry principle (that we are aware of) that allows to set the corresponding couplings to zero.

Table 1. Graphical representation of all invariants allowed by $O(N')^{\otimes 3}$ ($U(N')^{\otimes 3}$) symmetry for a rank-3 real (complex) tensor model up to sixth order in the tensors. For the real model, all tensors are represented by black vertices while in the complex model, black and white vertices are used to distinguish the tensor T and its complex conjugate $\bar{T}$. In this case, the algebraic representation of the invariants must be written as contractions of T-tensors with $\bar{T}$-tensors and takes the analogous form to the real expressions provided explicitly.

Number of Tensors	Invariant	Graphical Representation (Real Model)	Graphical Representation (Complex Model)
2	$T_{abc} T_{abc}$		
4	$T_{a_1 a_2 a_3} T_{b_1 a_2 a_3} T_{b_1 b_2 b_3} T_{a_1 b_2 b_3}$		
4	$T_{a_1 a_2 a_3} T_{a_1 b_2 b_3} T_{b_1 a_2 b_3} T_{b_1 b_2 a_3}$		—
4	$T_{a_1 a_2 a_3} T_{a_1 a_2 a_3} T_{b_1 b_2 b_3} T_{b_1 b_2 b_3}$		
6	$T_{a_1 a_2 a_3} T_{b_1 a_2 a_3} T_{b_1 b_2 b_3} T_{c_1 b_2 b_3} T_{c_1 c_2 c_3} T_{a_1 c_2 c_3}$		
6	$T_{a_1 a_2 a_3} T_{a_1 b_2 a_3} T_{b_1 b_2 b_3} T_{c_1 c_2 b_3} T_{c_1 c_2 c_3} T_{b_1 a_2 c_3}$		

---

[3] In this review, we call tensor models 0-dimensional theories of random tensors. The kinetic term is a product of Kronecker deltas, i.e., there is no non-trivial kinetic operator which breaks the $O(N')^{\otimes d}$ (or $U(N')^{\otimes d}$) symmetry.

Table 1. Cont.

Number of Tensors	Invariant	Graphical Representation (Real Model)	Graphical Representation (Complex Model)
6	$T_{a_1a_2a_3}T_{a_1b_2b_3}T_{b_1b_2a_3}T_{b_1c_2c_3}T_{c_1c_2c_3}T_{c_1a_2b_3}$		–
6	$T_{a_1a_2a_3}T_{b_1a_2b_3}T_{b_1c_2c_3}T_{c_1c_2a_3}T_{c_1b_2b_3}T_{a_1b_2c_3}$		
6	$T_{a_1a_2a_3}T_{b_1a_2b_3}T_{b_1b_2c_3}T_{c_1b_2b_3}T_{c_1c_2a_3}T_{a_1c_2c_3}$		–
6	$T_{a_1a_2a_3}T_{a_1a_2a_3}T_{b_1b_2b_3}T_{b_1b_2b_3}T_{c_1c_2c_3}T_{c_1c_2c_3}$		
6	$T_{a_1a_2a_3}T_{b_1a_2a_3}T_{b_1b_2b_3}T_{a_1b_2b_3}T_{c_1c_2c_3}T_{c_1c_2c_3}$		
6	$T_{a_1a_2a_3}T_{b_1a_2b_3}T_{b_1b_2a_3}T_{a_1b_2b_3}T_{c_1c_2c_3}T_{c_1c_2c_3}$		–

3.2. Regulator & Symmetry-Breaking

The key ingredient to set up the flow equation is the regulator, or "infrared" suppression term. In this context, infrared means low values of indices. Accordingly, the regulator should satisfy the two limits

(1) $R_N(\{a_i\}) \to 0$ for $N/\sum_{i=1}^d a_i \to 0$,
(2) $R_N(\{a_i\}) > 0$ for $\sum_{i=1}^d a_i/N < 1$,
(3) $R_N(\{a_i\}) \to \infty$ for $N \to N' \to \infty$.

The first condition ensures that "UV" modes are unsuppressed. It also ensures that no modes are suppressed once the IR cutoff scale N is lowered to zero. The second condition enforces that "IR" modes are suppressed. The third condition ensures that in the limit of infinite cutoff, the effective action essentially reproduces the classical action. The three conditions can be achieved with different so-called shape functions, i.e., different choices of $R_N(\{a_i\})$. The arguably simplest choice is

$$R_N(\{a_i\}) = \left(\frac{N^r}{\sum_{i=1}^d a_i^p} - 1\right)\theta\left(\frac{N^r}{\sum_{i=1}^d a_i^p} - 1\right), \quad (11)$$

where $r, p > 0$. While there are optimization criteria for a similar shape function at lowest order in the derivative expansion in the continuum [94], it has not yet been investigated what form an optimized cutoff takes for tensor models.

As a generalization, one might consider the argument of the regulator to be $N^r/(a^{r_1} + b^{r_2} + c^{r_3})$, which should result in three combinations of the four parameters r, r_1, r_2, r_3 to appear in the beta functions. Demanding a discrete symmetry of the indices' fixes $r_1 = r_2 = r_3 = p$.

The introduction of the regulator term necessarily breaks the symmetry of the model, as the $O(N') \otimes ... \otimes O(N')$ (or $U(N') \otimes ... \otimes U(N')$) symmetry requires all index positions to be treated on an equal footing. Setting up the RG flow is therefore incompatible with the unbroken symmetry, leading to an enlargement of the theory space. Specifically, the invariants in Equation (10) are generalized and include

$$\mathcal{O}_{SB}^{(n)} = f(a_1, ..., d_n)T_{a_1b_1...d_1}...T_{a_nb_n...d_n}\mathcal{C}_{a_1...a_n,b_1...b_n,...,d_1...d_n}, \quad (12)$$

with functions $f(a_1, ..., d_n)$ encoding the explicit index-dependence. Yet there is an important difference to a setting where the symmetry is broken from the outset, and which features the same theory space. It lies in a modified Ward identity that accounts for the symmetry-breaking introduced by the regulator. It selects a hypersurface in the larger theory space on which the full symmetry is recovered at the IR endpoint of the flow. Although the regulator vanishes in this limit, this is not sufficient to restore the symmetry, since the regulator has introduced symmetry violations in the flow at all finite scales. To compensate these, the initial condition for the flow, set in the UV needs to break the symmetry in a specific way that is dictated by the Ward identity. Therefore, a fixed point of the RG flow simultaneously needs to solve the modified Ward identity to lead to a symmetric IR limit. This requirement cannot necessarily be imposed on truncations: While the exact flow equation and Ward identity are compatible, the Ward identity in general requires other terms to be present in the truncation than the flow equation provides.

In matrix models, a simple solution of the Ward identity was discovered [41]: As symmetry-breaking is not introduced through tadpole diagrams (i.e., the leading-order contributions to the beta functions in an expansion in couplings) in matrix models, the theory space is not enlarged in the tadpole approximation. Beyond rank 2, such a simple solution is no longer possible, as even the tadpole approximation generates symmetry-breaking terms.

3.3. Bootstrap Strategy for Consistent Truncations

To characterize a universality class, at least all non-irrelevant critical exponents must be calculated. Accordingly, the set of all couplings which have a significant overlap with a relevant or marginal direction must be included in a minimal truncation. A priori, this set is not determined at an interacting fixed point. In practice, the following strategy is available: one starts with an assumption about a systematic division of theory space into relevant and irrelevant directions. A reliable truncation should at least include all couplings which are expected to be relevant as well as the leading irrelevant ones. If the beta functions in this truncation feature a fixed point, the critical exponents at the fixed point indicate whether the initial assumption about relevant couplings holds. If this is the case, terms beyond the truncation are expected to most likely only provide subleading corrections to the relevant critical exponents.

A particularly useful assumption is that of near-canonical scaling which allows one to use the canonical dimension as a guiding principle. This assumption works very well for a large class of fixed points and implies essentially that low orders in a vertex expansion are sufficient to obtain quantitative estimates of the critical exponents. The underlying reason is that for these cases, the mechanism that induces the fixed point is a balance between canonical scaling and leading-order quantum corrections. This mechanism is at work as soon as one departs from the critical dimension of a particular interaction, and generates a UV (IR) attractive fixed point if the coupling is asymptotically free (trivial) in its critical dimension. Examples include Yang-Mills in $d = 4 + \epsilon$, the Gross-Neveu model in $d = 2 + \epsilon$ for the former and the Wilson-Fisher fixed point for the latter, see, e.g., [95–100].

For the search of a quantum gravity fixed point in the tensor-model theory space, the canonical dimension could be a useful guiding principle. The motivation for this comes from the hope that the universality class discovered for quantum gravity in the continuum where metric fluctuations are summed over appears to be near-canonical. To match the corresponding spectrum of scaling exponents, one would expect a near-canonical scaling also on the tensor-model side[4]. The continuum asymptotic safety regime has been studied intensively and there is mounting evidence that the non-Gaussian fixed point explored in that approach features near-canonical scaling, the largest anomalous scaling is about 2 while the difference of quantum to canonical scaling goes to zero for the couplings of $\sqrt{g}R^n$ with $n > 3$, see, e.g., [101–104]. It is, therefore, a well-motivated starting point to assume that no operator with, e.g., canonical dimension -4 (or slightly more negative) can have significant overlap with a

---

[4] Please note that if the continuum limit in tensor models can be taken and provides a well-defined continuum space(time), this is equivalent to a version of quantum gravity being asymptotically safe. We stress that asymptotic safety is a general scenario for path integrals. As such it is not tied to one particular choice of configuration space and might even be realized in several distinct configuration spaces. Therefore, the continuum path integral corresponding to tensor models might well be one that includes a summation over (a subset of) topologies and is therefore not the same asymptotically safe gravity model that appears to exist according to continuum studies summing over metric fluctuations only.

relevant direction at the quantum-gravity fixed point in tensor models. This leads to a truncation ansatz in which one includes all operators up to this scaling. One can then search for a fixed point with quantum-gravity characteristics and check explicitly whether the near-canonical scaling assumption is justified.

Having found such a semi-perturbative fixed point in a truncation, one needs to check whether this fixed point is a truncation artifact, i.e., that the RG flow at that point simply is parallel to the projection onto the truncation. One can obtain hints about this by (1) varying the regulator and the way in which one projects onto the truncation ansatz and (2) enlarging the truncation. If a fixed point is stable under variations of the regulator and projection rule and if it appears with the same features in larger truncations, then it is unlikely that the fixed point is a truncation artifact. A larger truncation also allows one to re-check the assumption of near-canonical scaling. Ideally one finds that the deviation from canonical scaling decreases for the new operators which suggest that canonical scaling becomes a better and better assumption for the operators not included in the truncation. A similar strategy was successfully applied to the semi-perturbative UV-attractor of the Grosse-Wulkenhaar model [105], where it was indeed possible to bound the deviation from canonical scaling. Deriving such a bound for tensor models would complete the bootstrap approach.

3.4. In Practice: The $\mathcal{P}\mathcal{F}$ Expansion

The FRG Equation (8) is an equation for the effective action functional, which involves inverting a field-dependent operator and taking the regulated trace over the eigenvalues of the field- and index-dependent two-point function A very useful strategy to perform these two operations is the $\mathcal{P}\mathcal{F}$-expansion, which is a Taylor expansion of the RHS of the flow equation in the tensor T_{abc} around the vanishing field configuration $T_{abc} \equiv 0$. To obtain the expansion, we rewrite the regularized inverse two-point function that enters the flow Equation (8) as

$$\Gamma^{(2)}_{N,abcdef}[T] + R_{N,abcdef} = \underbrace{\Gamma^{(2)}_{N,abcdef}[T=0] + R_{N,abcdef}}_{\mathcal{P}} + \underbrace{\Gamma^{(2)}_{N,abcdef}[T] - \Gamma^{(2)}_{N,abcdef}[T=0]}_{\mathcal{F}}, \qquad (13)$$

where we use a shorthand notation $\Gamma^{(2)}_{N,abcdef} = \delta^2 \Gamma_N / \delta T_{abc} \delta T_{def}$. Thence, the flow Equation (8) is expressed as

$$\partial_t \Gamma_N = \tfrac{1}{2}\text{Tr}\left[(\mathcal{P}+\mathcal{F})^{-1}\partial_t R_N\right] = \tfrac{1}{2}\text{Tr}\left[(\partial_t R_N)\,\mathcal{P}^{-1}\right] + \tfrac{1}{2}\sum_{n=1}^{\infty}\text{Tr}\left[(-1)^n (\partial_t R_N)\,\mathcal{P}^{-1}\left(\mathcal{P}^{-1}\mathcal{F}\right)^n\right], \qquad (14)$$

where we suppressed the tensor indices for simplicity and expanded the inverse two-point function as a geometric series. This way of writing the RHS of the flow equation is very useful when one considers finite polynomial truncations in T_{abc}, because in this case one can truncate the sum at finite order. All further terms of the sum would possess more tensors than the monomials in the truncation.

4. Large N Scaling Dimensions

In settings with a background, where the RG flow corresponds to a local coarse graining, one RG step is literally a scale transformation. Accordingly, the canonical scaling dimensions of couplings are their mass dimensions. These can be determined prior to studying the actual RG flow. In the background-independent setting, there is no notion of locality or spacetime and accordingly all couplings are dimensionless in terms of units of length or mass, and no notion of mass dimension exists. Yet, mass dimension is not the notion of dimensionality that is relevant to a pregeometric RG flow anyway. Instead, it is a consistent scaling with N that is central here. This scaling is not determined a priori. Nevertheless, one can determine it in two steps:

1. Since the purpose of the FRG setup is the investigation of the large N-behavior of the tensor model, we need to scale the coupling constants in such a way that the beta functions admit a $1/N$ expansion. This gives a stack of coupled inequalities, which exclude most scaling prescriptions. Imposing the additional requirement that no interactions should be artificially decoupled from the system uniquely fixes all but one scaling dimensions.
2. A further condition comes from the geometric interpretation of tensor models. Specifically, the interpretation in terms of the Regge action of the triangulation that is associated with each tensor-model Feynman graph is only possible for a particular scaling of the associated coupling constant with N.

We will now present these two steps in more detail and determine the scaling dimension for the tensor models of quantum gravity.

For the first step, let us briefly return to the background dependent continuum setting. There, the flow equation automatically provides a scaling dimension. It arises by demanding that the beta functions form an autonomous system, such that after an appropriate rescaling of the couplings, the explicit dependence on the scale drops out. As a specific example, consider the beta function for the Newton coupling $\bar{G}$, which reads

$$\beta_{\bar{G}} = \#k^{d-2}\bar{G}^2, \tag{15}$$

to leading-order in $\bar{G}$ with $\# < 0$, [1,106–110]. Demanding independence from k provides the scaling dimension and agrees with the mass dimensionality. The dimensionless coupling takes the form

$$G = \bar{G}k^{d-2}. \tag{16}$$

Without knowing anything about mass dimensionality, one can thus alternatively fix the canonical scaling dimensions of couplings by demanding that the beta functions form an autonomous system at large N. This strategy is applicable to the pregeometric setting. For instance, the coupling $\bar{g}_{4,1}^{2,1}$ of the interaction $T_{abc}T_{ade}T_{fde}T_{fbc}$ in a real rank 3 model has the beta function

$$\beta_{\bar{g}_{4,1}^{2,1}} = \#N^{\frac{2r}{p}} \left(\bar{g}_{4,1}^{2,1}\right)^2 + \mathcal{O}(N^0), \tag{17}$$

resulting in

$$g_{4,1}^{2,1} = N^{\frac{2r}{p}} \bar{g}_{4,1}^{2,1}. \tag{18}$$

Please note that fixing the scaling dimensions in this way is possible in the large N limit, but not at finite N. This is a consequence of the fact that at any given order in the couplings, different orders in N appear. As we explicitly use the large N-limit, this only results in an *upper bound* on the scaling dimensions. Choosing scaling dimensions below this upper bound also results in autonomous beta functions in the large N limit[5]. Yet, for this choice the corresponding interactions decouple from the beta functions. The "most interacting" system, where no interactions are suppressed artificially, is achieved when the scaling dimensions are chosen as the upper bounds.

As is evident from Equation (18), the thus-determined scaling dimensions depend on the parameters r and p in Equation (11). Insight into the physics allows to fix the ratio r/p. For instance, for the geometric interpretation of tensor models, the interpretation of the dual picture in terms of dynamical triangulations results in a relation of the couplings of the tensor model and the scale N to the couplings of the Regge action. This relation only works for a specific choice of canonical scaling for the leading coupling (i.e., one of the quartic couplings). In turn, this scaling dimension fixes

---

[5] For some couplings, however, the set of inequalities to be fulfilled for a well-defined large-N limit also provide lower bounds. A consistent assignment of scaling dimension for a coupling therefore generically requires the inspection of several beta functions which depend on this coupling.

r/p. It turns out that this is $r/p = 1$ for a rank d tensor model, if d is the dimension entering the corresponding Regge action in the continuum picture. Yet, for those fixed points in the tensor model that show dimensional reduction to a matrix model[6], $r/p < 1$ is the correct choice.

As a specific example for how the geometric interpretation fixes r/p, consider the possibly simplest quantum-gravity tensor model, the so-called rank 3 colored complex model [43] defined through the action

$$S(T,\bar{T}) = \sum_{i=0}^{3} T^i_{abc} \bar{T}^i_{abc} + N^{-3/2} \left(\lambda\, T^0_{abc} T^1_{ade} T^2_{fbe} T^3_{fdc} + c.c. \right). \tag{19}$$

The Feynman diagram expansion of this model yields the amplitude

$$A(\gamma) = N^{N_1 - 3/2 N_3} (\lambda\bar{\lambda})^{N_3/2}, \tag{20}$$

where N_3 denotes the number of 3-cells[7] in the triangulation $\Delta(\gamma)$ associated with the colored Feynman graph γ and where N_1 denotes the number of 1-cells. Comparing this with the Regge action $S_R[\Delta] = \kappa_3 N_3 - \kappa_1 N_1$ of a triangulation Δ allows us to identify the coupling constants as $\kappa_1 = \ln(N)$ and $\kappa_3 = \frac{3}{2}\ln(N) - \frac{1}{2}\ln(\lambda\bar{\lambda})$. The uncolored models that we investigate with the FRG are obtained by integrating out all but the last color. The scaling $N^{-3/2}$ of the coupling constant λ then implies the scaling N^{-2} for the cyclic-melonic interactions, i.e., $T_{abc}T_{ade}T_{fde}T_{fbc}$ and its color permutations. In other words the geometric compatibility condition fixes $r/p = 1$.

5. Benchmarking the FRG in Matrix Models

In two-dimensional quantum gravity, the relevant critical exponent of the double-scaling limit is known. In this limit, the continuum limit in dynamical triangulations can be taken in such a way that all topologies contribute. For reviews and introductions, see, e.g., [21,111–114]. The matrix model that is dual to dynamical triangulations can be chosen to be Hermitian $N \times N$ matrices φ, with the generating functional given by

$$Z = \int \mathcal{D}\varphi\, e^{N\left(-\frac{1}{2}\mathrm{Tr}\varphi^2 + \frac{g_4}{4}\mathrm{Tr}\varphi^4\right)}. \tag{21}$$

The double-scaling limit requires taking $N \to \infty$, while holding

$$(g_4 - g_{4\,\mathrm{crit}})^{\frac{5}{4}} N = \mathrm{const}, \tag{22}$$

where $g_{4\,\mathrm{crit}}$ is the critical value of the coupling. This can be rewritten in the form

$$g_4(N) = g_{4\,\mathrm{crit}} + c\, N^{-\frac{4}{5}}. \tag{23}$$

This is structurally similar to the leading-order scaling of couplings in the vicinity of a fixed point of the RG flow. Accordingly, one is led to identify

$$\theta = \frac{4}{5}, \tag{24}$$

as a relevant critical exponent. This similarity prompted the authors of [39] to set up a pregeometric RG flow in matrix size N. In that paper as well as the follow-up works [115–120], the coarse graining

---

[6] See Sections 6 and 7.
[7] An n-cell of a triangulation is an n-dimensional simplex that appears as an elementary building block of the triangulation of a d-dimensional pseudo-manifold. For instance, a 3-cell is a tetrahedron, a 2-cell a triangle, which appears in the boundary of a tetrahedron, a 1-cell an edge, which appears in the boundary of a triangle, and so on.

was implemented explicitly by integrating out the outermost rows and columns of the matrices in a Gaussian approximation.

In [40], the flow Equation (8) in the pregeometric setting was first derived. Applying it to truncations of a single-trace form, $\Gamma_N = \sum_{i=2} g_{2i} \text{Tr}\phi^{2i}$, yielded a critical exponent that approaches $\theta = 1$ from above. Extending the truncation to multi-trace operators does not improve the estimate, but instead makes it worse. The critical exponent $\theta = 0.8$ for gravity is first reproduced at the first multicritical point [41], which corresponds to gravity coupled to conformal matter [121].

Instead of reviewing these results in greater detail, here we explore an alternative prescription to calculate the critical exponents that leads to a significant improvement in the estimate. This prescription was already explored in [64] for tensor models, and has been put forward for continuum QFTs in [122]. It consists in keeping the anomalous dimension $\eta = -N \partial_N \ln Z_N$ constant while calculating the stability matrix, i.e.,

$$\tilde{\theta}_I = -\text{eig}\left(\left(\frac{\partial \beta_{g_i}}{\partial g_j}\right)_\eta\right)\bigg|_{\vec{g}=\vec{g}_*}. \tag{25}$$

The notation $()_\eta$ indicates that the derivative is taken at fixed η. The alternative, more standard prescription differs by including derivatives of η and will be denoted by θ_I to clearly differentiate between the two. As a specific example, consider the case where a coupling g_i already corresponds to an eigendirection at a fixed point. No off-diagonal elements of the stability matrix contribute to its critical exponent, such that

$$\tilde{\theta} = -\left(\left(\frac{\partial \beta_{g_i}}{\partial g_j}\right)_\eta\right)\bigg|_{\vec{g}=\vec{g}_*}, \tag{26}$$

$$\theta = -\left(\frac{\partial \beta_{g_i}}{\partial g_i}\right)\bigg|_{\vec{g}=\vec{g}_*} = \tilde{\theta} + \left(\frac{\partial \beta_{g_i}}{\partial \eta}\frac{\partial \eta}{\partial g_i}\right)\bigg|_{\vec{g}=\vec{g}_*}. \tag{27}$$

In [122] it was observed that a scaling relation for critical exponents in the $O(N) \oplus O(M)$ model, which is known to hold for the epsilon-expansion [98,123,124], is only satisfied for the FRG in truncations of the full flow to the local potential approximation plus anomalous dimension for the prescription in Equation (25). The more standard prescription leads to small violations of the scaling relation in those truncations.

Here, we show that the $\tilde{\theta}$-prescription gives improved results for the critical exponent of the double-scaling limit, resulting in only 14% deviation already in a calculationally very straightforward truncation, cf. Figure 4.

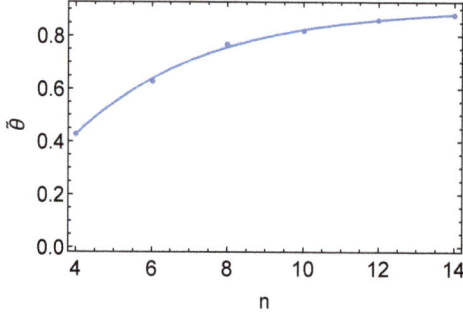

Figure 4. The relevant critical exponent in the matrix model according to the prescription Equation (25) as a function of truncation order in a truncation of the form $\Gamma_N = \sum_{i=1}^n g_{2i}\,\text{Tr}\phi^{2i}$.

The results for the critical exponent as a function of truncation order in Figure 4 appear to be fit well by a function of the form

$$\tilde{\theta}(n) = a - b e^{-cn}, \tag{28}$$

with fit parameters $a = 0.91$, $b = 1.54$ and $c = 0.29$. An extrapolation to $n \to \infty$, which is the complete single-trace subsector of theory space, yields $\tilde{\theta}(n \to \infty) = 0.91$, which is only a 14 % deviation from the exact result $\theta = 0.8$. Whether this is accidental, or whether there is a deeper reason the $\tilde{\theta}$ prescription works better for matrix and potentially also tensor models, remains to be explored in the future.

One source of systematic errors for the critical exponent is the breaking of the $U(N)$ symmetry of the matrix model through the regulator [41]. This can be seen by the fact that the $U(N)$-Ward identity obtains a non-vanishing RHS through the introduction of the regulator:

$$\mathcal{G}_\epsilon \Gamma_N = \epsilon \operatorname{Tr} \left(\frac{[A, R_N]}{\Gamma_N^{(2)} + R_N} \right), \tag{29}$$

where A is the generating matrix of an infinitesimal unitary transformation, which generates the transformation $\mathcal{G}_\epsilon$. By generating we mean that a unitary transformation $U = \exp(i\epsilon A)$ transforms the matrix ϕ as $\phi \mapsto U^\dagger.\phi.U = \phi + i\epsilon[A, \phi] + \mathcal{O}(\epsilon^2)$. This implies that the RG flow generates symmetry-breaking operators even if the initial condition is a $U(N)$ symmetric action. In particular, the relevant directions will acquire contamination by these symmetry-breaking operators. Hence, when investigating the large N-limit with the FRG, one must include these symmetry-breaking operators to find accurate critical exponents.

Including the symmetry-breaking operators into a truncation and distinguishing them from the symmetric operators by a projection on the truncation is a technically rather challenging task. Fortunately, there is a self-consistent work-around in the case of matrix models that gives surprisingly stable results [41]: It is based on the observation that tadpole diagrams of $U(N)$-symmetric operators do not generate symmetry-breaking operators for the rank-2-case. In other words, the tadpole approximation to the broken $U(N)$-Ward identity is solved by a symmetric effective average action. Using the tadpole approximation in a single-trace truncation allows one to find the infinite series of so-called multicritical points. The m-th multicritical point is a fixed point with m non-vanishing couplings at the fixed point whose fixed-point values occur with alternating sings and whose critical exponents are $\theta_n^{(m)} = \frac{n}{m}$. We see that the largest critical exponent (the pure-gravity exponent) is still $\theta_m^{(m)} = 1$ in the single-trace truncation. However, including multi-trace operators in the truncation yields improved results for the largest critical exponents $\theta_m^{(m)} = 0.80...0.82$ at the $m = 2, 3, 4, ...$ multicritical fixed points. For this leading critical exponent, one therefore obtains an estimate that deviates from the exact value by only 3%, which is a rather high precision. The subleading relevant critical exponents at the multicritical points are not reproduced with comparable precision in this truncation. Nevertheless, we interpret the precision of the leading relevant exponent as a signature that the FRG can successfully pass the benchmark test posed by rank 2 models. Only at the double-scaling limit, i.e., the $m = 1$ fixed point, one obtains $\theta_1^{(1)} = 1$. This relatively large discrepancy of the critical exponents from 0.8 can be explained by the fact that the tadpole approximation does only capture effects from the tree-level truncation, which contains only one coupling. Given such a small truncation, it is actually remarkable to obtain the values of the critical exponent with 25% accuracy.

As a consequence of universality, specific fixed points in tensor models can also reproduce the matrix-model results. This is due to the fact that the shape of the building blocks is not relevant for the continuum limit. Therefore even higher-dimensional building blocks can reproduce a lower-dimensional continuum limit, at a point in theory space where the effective dynamics "flattens" these building blocks in an appropriate way. To recover the lower-dimensional scaling, the canonical scaling dimensions of the model must be adjusted by choosing $r/p < 1$ in Equation (11). In that choice, and for the prescription Equation (25), the matrix-model exponent is approximately recovered from the fixed points in tensor models, see Sections 6 and 7.

6. Charting Three Dimensions from a Tensor-Model Point of View

An important motivation for RG studies of the large-N- behavior of tensor models is the search for a continuum limit that can be associated with quantum gravity. The first step in the systematic program that can lead to the confirmation or refutation of the conjecture that there might exists a continuum limit in tensor models which corresponds to quantum gravity, is a systematic investigation of theory spaces. Varying the number of tensor fields, the rank of the tensors and symmetry-structures provides several different theory spaces which one can then investigate with the FRG. The first step in the FRG investigation of a theory space consists of finding tentative candidates for universal fixed points. This provides insight into which interaction structures could be of particular importance for a continuum limit. Below, we discuss the status of this systematic program in more detail for rank-3 tensor models.

In summary, by investigating a complex uncolored model, i.e., a model with $U(N') \otimes U(N') \otimes U(N')$ symmetry, and a real uncolored model, i.e., a symmetry group of the form $O(N') \otimes O(N') \otimes O(N')$, we discover that certain classes of fixed points are shared. In particular, we find fixed points that exhibit a form of dimensional reduction and evidence that these fixed points are not truncation artifacts. Crucially, the real model features a new, tetrahedral interaction, cf. the third entry in Table 1, introduced by Carrozza and Tanasa [62], and later taken up in [47] for an SYK-type model. This interaction appears to be key for the generation of a fixed point which does not appear to feature dimensional reduction and therefore constitutes a tentative candidate for a continuum limit for three-dimensional quantum gravity. We stress that of course it requires much more than just the discovery of the fixed point to establish its relevance for three-dimensional quantum gravity; finding a fixed point without dimensional reduction is a necessary but not sufficient step in linking tensor models to a well-behaved phase of quantum gravity.

6.1. Dimensional Reduction in Tensor Models

The absence of a background geometry permits that tensor models exhibit phenomena that do not appear in local quantum fields theories. The first of these is the dynamical generation of multi-trace operators, which correspond to tensor-model vertices with a geometric interpretation as boundaries formed by disconnected pieces of geometry (such as, e.g., the two circles in the boundary of a cylinder). These multi-trace operators are however generated by connected Feynman diagrams. For instance, a matrix-connected matrix-model Feynman diagram may be dual to the triangulation of a cylinder connecting the two circles in the boundary. The corresponding interactions are thus generically generated by the flow, and are part of the quantum effective action. In particular, one finds disconnected tensor invariants with $2n$ tensors of the form $(T_{abc}T_{abc})^n$, which possess an enhanced $O(N^{l3})$-symmetry, reducing the tensor model to a vector model and producing non-Gaussian fixed points, which do not represent extended three-dimensional geometries.

In [65] we identified a mechanism that can be realized at suitable fixed points and prevents the production of multi-trace operators. It is based on the observation that the generation of $(T_{abc}T_{abc})^2$ from connected vertices requires two cyclic 4-melons with distinct preferred colors to be nonzero. Thus, the fixed points in the theory space with the enhanced $O(N') \otimes O(N'^2)$ symmetry exhibited by the cyclic melons with one preferred color do not possess non-vanishing multi-trace operators. However, this theory space exhibits dimensional reduction. Dynamical dimensional reduction at high energies is an intriguing phenomenon in several models of quantum gravity, see, e.g., [125], that are four-dimensional at large scales. In tensor models, dimensional reduction differs in that it appears to be realized at certain classes of fixed points in rank-3 and rank-4 models, such that the continuum limit is not a candidate for three- or four-dimensional quantum gravity. (Although not yet explored explicitly, the same result should be true in any rank $d > 2$.) Specifically, an enhancement of the $U(N') \otimes ... \otimes$

$U(N')$ (or $O(N') \otimes ... \otimes O(N')$) symmetry to an $U(N'^2) \otimes ... \otimes U(N')$ ($O(N'^2) \otimes ... \otimes O(N')$) symmetry goes hand in hand with an effective "fusion" of two indices into one "super-index", such that the model effectively reduces to a matrix model. This occurs at fixed points at which only cyclic melons (single-trace, multi-trace or both) of one preferred color are present. Because of the enhanced symmetry, it is always consistent to set all other interactions to zero, as one can also check by inspecting the beta functions. To fully establish the dimensional reduction, the critical exponents of the matrix model should also be reproduced. Here, the freedom in choosing r and p in Equation (11) becomes crucial: A matrix model features different canonical dimensions than a tensor model, essentially due to the reduced rank. The canonical dimensions are functions of r/p. To probe the matrix-model limit of rank-3 tensor models, one should choose $r/p = 1/2$ to obtain the canonical dimensions appropriate for a matrix model. With this choice for the scaling of the regulator, and for the prescription Equation (25), the matrix-model exponent is approximately recovered from the fixed points in tensor models.

In particular, the fixed points reported in Tables 2–4 feature dimension reduction. They were obtained for the rank-3 real model. For those which appear both in the complex and in the real model, both sets of values are shown. The tables display fixed points as well as critical exponents values obtained in the hexic truncation of the rank-3 real model. In this truncation, all $O(N') \otimes O(N') \otimes O(N')$ symmetric interactions are included in the effective average action. The resulting truncation has 21 couplings. Further details can be found in [64,65].

Table 2. This fixed point only features cyclic-melonic couplings and has one relevant direction. It features dimensional reduction to a matrix model. For the complex model, the first four critical exponents are $\theta_1 = 2.14$, $\theta_2 = -0.59$, $\theta_{3,4} = -1.26$. The slight numerical difference to the real case is due to a different choice of projection scheme. The second critical exponent of the real model misses from the universality class obtained from the complex model and is to be attributed to the additional presence of the interaction associated with $g_{4,1}^0$. Choosing $r/p = 1/2$ leads to $\theta = 1.09$ for the leading critical exponent. Using the prescription described in Section 5 the value is $\tilde{\theta} = 0.63$.

$g_{4,1}^{0\,*}$	$g_{4,1}^{2,1*}$	$g_{4,1}^{2,2*}$	$g_{4,1}^{2,3*}$	$g_{4,2}^{2\,*}$	$g_{6,1}^{0,np*}$	$g_{6,1}^{0,p*}$	$g_{6,1}^{1,i*}$	$g_{6,1}^{2,i*}$	$g_{6,1}^{3,1*}$	$g_{6,1}^{3,2*}$	$g_{6,1}^{3,3*}$	$g_{6,2}^{1\,*}$	$g_{6,2}^{3,i*}$	$g_{6,3}^{3\,*}$
0	−0.28	0	0	0	0	0	0	0	0	−0.15	0	0	0	0

θ_1	θ_2	θ_3	$\theta_{4,5}$	θ_6	$\theta_{7,8}$	θ_9	θ_{10}	$\theta_{11,12}$	$\theta_{13,14}$	$\theta_{15,16,17}$	θ_{18}	θ_{19}	$\theta_{20,21}$	η
2.19	−0.03	−0.69	−1.19	−1.68	−1.78	−2.08	−2.15	−2.18	−2.28	−2.78	−2.94	−2.98	−3.18	−0.41

Table 3. Only bubble-multi-trace interactions are present in this fixed point, i.e., interactions of the form $(T_{abc}T_{abc})^n$. It has one relevant direction and features dimensional reduction to a vector model. For the complex case, the first four critical exponents read $\theta_1 = 3.33$, $\theta_{2,3,4} = -0.83$. These were not provided in [64], but can easily be extracted from the beta functions reported in that work.

$g_{4,1}^{0\,*}$	$g_{4,1}^{2,i*}$	$g_{4,2}^{2\,*}$	$g_{6,1}^{0,np*}$	$g_{6,1}^{0,p*}$	$g_{6,1}^{1,i*}$	$g_{6,1}^{2,i*}$	$g_{6,1}^{3,i*}$	$g_{6,2}^{3,i*}$	$g_{6,2}^{1\,*}$	$g_{6,3}^{3\,*}$
0	0	−1.05	0	0	0	0	0	0	0	−2.27

θ_1	θ_2	$\theta_{3,4,5}$	$\theta_{6,7}$	$\theta_{8,9,10}$	θ_{11}	$\theta_{12,13,14,15,16,17}$	$\theta_{18,19,20}$	θ_{21}	η
3.32	−0.33	−0.83	−1.24	−1.74	−1.82	−2.24	−2.32	−3.28	−0.59

The multi-trace operators $(T_{abc}T_{abc})^n$ are invariant under the enhanced $O(N'^2) \otimes O(N')$ symmetry. Thus, one expects that there exist fixed points at which cyclic melons and multi-trace operators take non-vanishing fixed-point values. This turns out to be true and, indeed in the quartic and hexic truncations, one finds a non-Gaussian fixed point with $O(N'^2) \otimes O(N')$ symmetry and non-vanishing fixed-point values for the cyclic melons and multi-trace operators. This fixed point appears to possess two relevant directions, see Table 4, whereas the purely cyclic-melonic non-Gaussian fixed point only features one positive critical exponent in this truncation see Table 2. It should be stressed that the deviation of θ_2 from zero at the fixed point in Table 4 is smaller than the presumed systematic error of the truncation.

Table 4. This fixed point has non-vanishing melonic as well as multi-trace interactions. It features two relevant directions, the second of which has a very small critical exponent. The systematic error of the truncation is expected to be significantly larger than the deviation of θ_2 from zero. The fixed point exhibits dimensional reduction to a matrix model. For the complex case, the first four critical exponents read $\theta_1 = 2.56$, $\theta_2 = 0.44$, $\theta_{3,4} = -0.97$. The difference is to be attributed to a difference in projection scheme. Additionally, the real model features an extra critical exponent $\theta_3 = -0.66$ due to the presence of the additional interaction $g^0_{4,1}$.

$g^{0\,*}_{4,1}$	$g^{2,1*}_{4,1}$	$g^{2,(2,3)*}_{4,1}$	$g^{2\,*}_{4,2}$	$g^{0,np*}_{6,1}$	$g^{0,p*}_{6,1}$	$g^{1,i*}_{6,1}$	$g^{2,i*}_{6,1}$	$g^{3,1*}_{6,1}$	$g^{3,(2,3)*}_{6,1}$	$g^{3,1*}_{6,2}$	$g^{3,(2,3)*}_{6,2}$	$g^{1\,*}_{6,2}$	$g^{3\,*}_{6,3}$	
0	−0.27	0	−0.05	0	0	0	0	−0.13	0	−0.05	0	0	−0.01	
θ_1	θ_2	θ_3	$\theta_{4,5}$	θ_6	$\theta_{7,8}$	θ_9	θ_{10}	$\theta_{11,12}$	$\theta_{13,14}$	θ_{15}	$\theta_{16,17,18}$	θ_{19}	$\theta_{20,21}$	η
2.23	0.03	−0.66	−1.16	−1.66	−1.74	−2.04	−2.12	−2.16	−2.24	−2.73	−2.74	−3.10	−3.12	−0.42

As we have introduced the colors it is consistent to switch off the multi-trace interactions for the matrix-model limit. In matrix-model RG flows this has not been possible, as multi-trace interactions in matrix models are automatically generated from single-trace ones. Physically, this might suggest that configurations with disconnected boundaries do not have a significant impact on the path integral in two dimensions, as it appears to be possible to reach the same continuum limit both with and without the presence of multi-trace interactions.

Please note that due to the symmetry-breaking induced by the regulator, even at the single-trace cyclic-melonic fixed point, interactions with non-trivial index-dependence outside this theory space are generated and could take finite values at the fixed point. The fixed-point values of these operators are constrained by the modified $O(N')^{\otimes 3}$-Ward identity, where the only symmetry-breaking term is due to the regulator. This regulator term vanishes in the IR limit, which implies that the $O(N')^{\otimes 3}$-Ward identity turns into the constraint that all these index-dependent interactions vanish.

The analogous argument applies to all other fixed points: These fixed points will exhibit non-vanishing couplings for index-dependent vertices, but their values are constrained by the modified Ward identity. In the IR, it turns into the constraint that all couplings associated with index-dependent operators vanish. To reach this point, the initial condition for the RG flow must be chosen with an appropriate "amount" of symmetry-breaking operators, such that during the flow, the symmetry-breaking effect of the regulator compensates with that coming from the initial condition.

6.2. Candidates with Potential Relevance for Three-Dimensional Quantum Gravity

The fixed points with enhanced $O(N'^2) \otimes O(N')$ and $O(N'^3)$ symmetries appear to exhibit dimensional reduction. This might possibly be compatible with dynamical dimensional reduction in the physical UV limit, i.e., *after* the continuum limit has already been taken if these fixed points would possess a relevant direction that "inflates" additional dimensions in the IR. However, we consider this possibility unlikely, and consider it more likely that the continuum limit leads to the same topological dimension as the IR limit of the corresponding spacetime exhibits. Please note that the dimensional reduction in the spectral dimension observed in many quantum-gravity approaches is different, and does not imply that there is a reduction in the topological dimension.

A different possibility to search for quantum gravity candidate fixed points is to search for fixed points that do not possess such an enhanced symmetry. In [65] we found two possible candidates. These fixed points are isocolored, i.e., they exhibit a global symmetry under color permutation. Note that such a symmetry is not linked to dimensional reduction. In fact, the presence of cyclic-melonic interactions with all three different preferred colors is exactly what prevents the merging of two indices to one "super-index" linked to $O(N'^2) \otimes O(N')$ symmetry. Another hint about the "geometricity" of a fixed point might be the presence of the tetrahedral interaction $T_{abc}T_{ade}T_{fdc}T_{fbe}$. An isocolored fixed point at which this tetrahedral interaction takes a non-vanishing fixed-point value may describe the continuum limit of a geometric model. We stress that this is not sufficient for such a fixed point to be associated with quantum gravity. The identification as a quantum-gravity candidate can only be made

when order parameters indicate a geometric interpretation. The fixed point possesses the positive critical exponents
$$\theta_\pm = 1.35 \pm 1.56i \ldots 1.95 \pm 0.69i, \quad \theta_3 = 0.38 \ldots 0.13. \tag{30}$$
in the full hexic truncation, where the range comes from several different schemes regarding the treatment of the anomalous dimension. We stress that it should not be taken as a complete estimate of the systematic truncation error. The leading critical exponents are roughly compatible with the critical exponents found for the Einstein-Hilbert truncation in three dimensions [126] with $\theta_1 \approx 2.5$ and $\theta_2 \approx 0.8$ which are also expected to come with significant systematic errors. We caution that this comparison is subject to systematic errors on both sides. In fact, in three-dimensional continuum gravity, fixed-point searches have only been conducted in the Einstein-Hilbert truncation, not including higher-derivative operators. Therefore, it is not yet established whether there are indeed only two relevant directions, although the fact that four-derivative curvature invariants are canonically irrelevant could support such a conjecture. Accordingly, the comparison of critical exponents we perform here is to be understood as a proposal for a comparison that will become more meaningful in the future, when systematic errors are significantly reduced on both sides. Here, we only note that within the significant systematic errors that we expect these results to have, the critical exponents of the continuum and the tensor-model setting do not appear to be incompatible.

A second isocolored melonic fixed point with vanishing fixed-point value for the tetrahedral interaction was also found and discussed in the appendix of [65], but with slightly complex values for the coupling constants. The imaginary parts of the fixed-point values of the couplings exhibit a scheme-dependence that is consistent with vanishing imaginary parts of the couplings, which would make the fixed-point action real and thus physically admissible.

7. First Steps Towards Background-Independent Four-Dimensional Quantum Gravity

In this subsection, we discuss the first results obtained for rank-4 tensor models using the FRG. The purpose of our presentation is illustrative and for this reason, we restrict the analysis to a simple truncation for the effective average action. An extensive analysis employing more sophisticated truncations will be presented elsewhere.

Studying rank-4 tensor models, whose Feynman diagrams can be identified with four-dimensional triangulations, is certainly of great importance from a quantum-gravity perspective. If a suitable continuum limit can be found, they could be candidates for a description of the microscopic structure of four-dimensional quantum spacetime. While results in tensor models point towards the existence of a branched-polymer phase [127], Monte Carlo simulations indicate that causal dynamical triangulations could also give rise to extended four-dimensional geometries [11,128]. The case of Euclidean dynamical triangulations is under renewed investigation [12]. The FRG is a suitable tool to complement such simulations and discover candidates for a universal continuum limit beyond branched polymers.

We consider a complex rank-4 tensor model, i.e., we work with a random tensor T_{abcd} and its complex conjugate $\bar{T}_{abcd}$ of size N'. We focus on a model respecting the following symmetry

$$\begin{aligned} T_{a_1 a_2 a_3 a_4} &\to T'_{a_1 a_2 a_3 a_4} = U^{(1)}_{a_1 b_1} U^{(2)}_{a_2 b_2} U^{(3)}_{a_3 b_3} U^{(4)}_{a_4 b_4} T_{b_1 b_2 b_3 b_4}, \\ \bar{T}_{a_1 a_2 a_3 a_4} &\to \bar{T}'_{a_1 a_2 a_3 a_4} = \bar{U}^{(1)}_{a_1 b_1} \bar{U}^{(2)}_{a_2 b_2} \bar{U}^{(3)}_{a_3 b_3} \bar{U}^{(4)}_{a_4 b_4} \bar{T}_{b_1 b_2 b_3 b_4}, \end{aligned} \tag{31}$$

where repeated indices are summed over. The matrices $U^{(i)}_{ab}$ are unitary and therefore the model has a $U(N')^{\otimes 4}$ symmetry. Equation (31) shows that each index of the tensor transforms independently. Hence, $U(N')^{\otimes 4}$ invariance requires that the only allowed index contraction is a first index of T with a first index of $\bar{T}$, a second index of T with a second index of $\bar{T}$ and so on. Consequently, an interaction term which contains $2p$ tensors in total necessarily has p tensors T and p tensors $\bar{T}$. Invariance under Equation (31) also ensures that the indices of the tensors do not have any permutation symmetry.

A continuum limit in tensor models might fall into the universality class corresponding to the Reuter fixed point [3] (see [5,129] for recent reviews). Accordingly we bootstrap our truncation assuming a near-canonical scaling spectrum, and choose

$$\Gamma_N = \Gamma_{N,2} + \Gamma_{N,4}, \qquad (32)$$

with

$$\Gamma_{N,2} = Z_N \bar{T}_{a_1 a_2 a_3 a_4} T_{a_1 a_2 a_3 a_4}, \qquad (33)$$

and

$$\begin{aligned}
\Gamma_{N,4} &= \bar{g}_{4,1}^{2,1} \bar{T}_{a_1 a_2 a_3 a_4} T_{b_1 a_2 a_3 a_4} \bar{T}_{b_1 b_2 b_3 b_4} T_{a_1 b_2 b_3 b_4} + \bar{g}_{4,1}^{2,2} \bar{T}_{a_1 a_2 a_3 a_4} T_{a_1 b_2 a_3 a_4} \bar{T}_{b_1 b_2 b_3 b_4} T_{b_1 a_2 b_3 b_4} \\
&+ \bar{g}_{4,1}^{2,3} \bar{T}_{a_1 a_2 a_3 a_4} T_{a_1 a_2 b_3 a_4} \bar{T}_{b_1 b_2 b_3 b_4} T_{b_1 b_2 a_3 b_4} + \bar{g}_{4,1}^{2,4} \bar{T}_{a_1 a_2 a_3 a_4} T_{a_1 a_2 a_3 b_4} \bar{T}_{b_1 b_2 b_3 b_4} T_{b_1 b_2 b_3 a_4} \\
&+ \bar{g}_{4,1}^{(1,2)} \bar{T}_{a_1 a_2 a_3 a_4} T_{b_1 b_2 a_3 a_4} \bar{T}_{b_1 b_2 b_3 b_4} T_{a_1 a_2 b_3 b_4} + \bar{g}_{4,1}^{(1,3)} \bar{T}_{a_1 a_2 a_3 a_4} T_{b_1 a_2 b_3 a_4} \bar{T}_{b_1 b_2 b_3 b_4} T_{a_1 b_2 a_3 b_4} \\
&+ \bar{g}_{4,1}^{(1,4)} \bar{T}_{a_1 a_2 a_3 a_4} T_{b_1 a_2 a_3 b_4} \bar{T}_{b_1 b_2 b_3 b_4} T_{a_1 b_2 b_3 a_4} + \bar{g}_{4,2}^{2} \bar{T}_{a_1 a_2 a_3 a_4} T_{a_1 a_2 a_3 a_4} \bar{T}_{b_1 b_2 b_3 b_4} T_{b_1 b_2 b_3 b_4}.
\end{aligned} \qquad (34)$$

In Equation (33), Z_N denotes the wave-function renormalization. The interactions in Equation (34) can be represented by 4-colored graphs: For each tensor T ($\bar{T}$) we associate a white (black) vertex and a colored edge for an index. Different colors are used to indicate different index positions on the tensors. An index contraction is represented by linking a black and a white vertex by the corresponding edge. The corresponding diagrammatic representation of Equation (32) is shown in Figure 5. The notation for the couplings of different interactions encode the corresponding diagrammatics, i.e., the combinatorial structures of the interaction: The first subindex denotes the number of tensors T and $\bar{T}$, while the second one counts the number of connected components. The superindices differ for different combinatorial structures. For cyclic melons, which consist of contractions of neighboring tensors by either three lines or one line in alternating fashion, the superindices are not bracketed. The two superindices stand for the number of "submelons" and the preferred color i which is the color of the single line connecting neighboring tensors. Since this is a rank-4 model, there are four different melonic invariants, each one selecting one distinct preferred color. A symmetry-reduced theory space, the isocolored theory space is defined by a single coupling being assigned to all cyclic melons since those interactions have the same combinatorial structure and just differ by the preferred color.

A distinct combinatorial structure is indicated by bracketed superindices: The couplings $\bar{g}_{4,1}^{(1,i)}$ are associated with the *necklaces* diagrams. These interactions are such that a given white vertex is connected to a black vertex by exactly two edges. There are three such interactions due to three possible pairings of four indices into two groups of two. Each white vertex is connected to one of its neighbors by the colors $(1,i)$ in the super-index, and by the remaining two colors to its other neighbor.

Finally, the "double-trace" interaction is parameterized by the coupling $\bar{g}_{4,2}^2$: in the subindices, the number of connected components is two. The super-index represents the number of melons. At higher orders in the truncation, where the first subindex is larger than four, additional superindices must be introduced to distinguish all different combinatorial structures at fixed order in tensors and connected components.

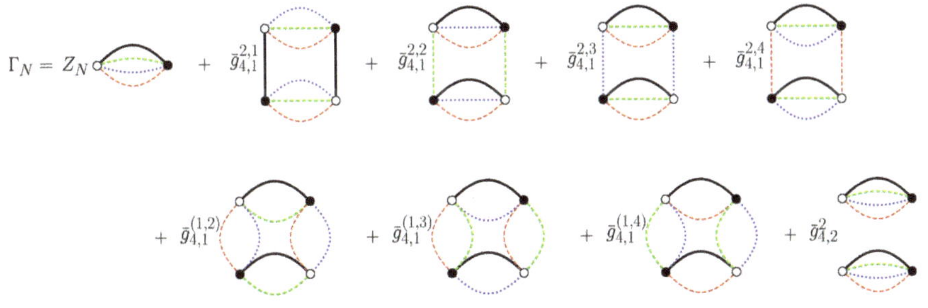

$$\Gamma_N = Z_N \bigcirc + \bar{g}_{4,1}^{2,1} \; \diagup\!\!\!\diagdown + \bar{g}_{4,1}^{2,2} \; \diagup\!\!\!\diagdown + \bar{g}_{4,1}^{2,3} \; \diagup\!\!\!\diagdown + \bar{g}_{4,1}^{2,4} \; \diagup\!\!\!\diagdown$$

$$+ \bar{g}_{4,1}^{(1,2)} \; \diagup\!\!\!\diagdown + \bar{g}_{4,1}^{(1,3)} \; \diagup\!\!\!\diagdown + \bar{g}_{4,1}^{(1,4)} \; \diagup\!\!\!\diagdown + \bar{g}_{4,2}^{2} \; \diagup\!\!\!\diagdown$$

Figure 5. Diagrammatic representation of the invariants present in our truncation

We aim at deriving the beta functions for the dimensionless couplings g_I, where

$$\bar{g}_I \equiv Z_N^2 N^{[\bar{g}_I]} g_I, \tag{35}$$

with $[\bar{g}_I]$ being the canonical dimension of the coupling $\bar{g}_I$. We will now determine the canonical dimensions, cf. Section 4. For the application of the flow equation one must choose a regulator function which acts as an "infrared" suppression term, cutting off modes with indices satisfying $a^p + b^p + c^p + d^p < N^r$, where $r, p > 0$. Our regulator choice generalizes that in [65]

$$R_N^{(r,p)}(\{a_i\},\{b_i\}) = Z_N \, \delta_{a_1 b_1} \delta_{a_2 b_2} \delta_{a_3 b_3} \delta_{a_4 b_4} \left(\frac{N^r}{\sum_{i=1}^4 a_i^p} - 1 \right) \theta\left(\frac{N^r}{\sum_{j=1}^4 a_j^p} - 1 \right). \tag{36}$$

Its scale-derivative $\partial_t = N \partial_N$ is

$$\partial_t R_N^{(r,p)}(\{a_i\},\{b_i\}) = \delta_{a_1 b_1} \delta_{a_2 b_2} \delta_{a_3 b_3} \delta_{a_4 b_4} \left[N^r \frac{\partial_t Z_N + Z_N}{\sum_{i=1}^4 a_i^p} - \partial_t Z_N \right] \theta\left(\frac{N^r}{\sum_{j=1}^4 a_j^p} - 1 \right) \tag{37}$$

$$+ Z_N \, \delta_{a_1 b_1} \delta_{a_2 b_2} \delta_{a_3 b_3} \delta_{a_4 b_4} \left(\frac{N^r}{\sum_{i=1}^4 a_i^p} - 1 \right) \cdot \delta\left(\frac{N^r}{\sum_{i=1}^4 a_i^p} - 1 \right) \cdot r \frac{N^r}{\sum_{k=1}^4 a_k^p}.$$

The term in the second line of Equation (37) does not yield a contribution to the flow of couplings of index-independent interactions, since the delta-distribution appears multiplied by its argument. With these definitions, the right-hand-side of the flow equation can be evaluated. To extract beta functions from it, suitable projections onto the monomials spanning the theory space must be used. Specifically, the distinct combinatorial structures at a given order in tensors can easily be distinguished, as the flow equation generates combinatorially different contractions on the right-hand-side. To deal with the additional index-dependence of interactions that occurs due to the symmetry-breaking induced by the regulator, we apply the prescription from [65]. Specifically, the regulator can either sit on an index forming a closed loop, or an index occurring on a tensor and antitensor. To project onto symmetry-invariant monomials only, we set indices in the regulator to zero, if they also occur on a tensor. This splits the index-trace into two parts: The contraction of tensors and their complex conjugates decouples from the regulator trace and is directly recognizable as one of the different combinatorial structures in Equation (33) or (34). The regulator trace consists of a trace over indices running through the regulator and its derivative, which can be rewritten as an integral in the large-N limit.

The resulting beta functions for the dimensionless couplings as well as the anomalous dimension $\eta \equiv -\partial_t Z_N/Z_N$ are, respectively,

$$\eta = 2\mathcal{J}_2^3(p)N^{[g_{4,1}^{2,i}]+\frac{3r}{p}}\sum_{i=1}^{4}g_{4,1}^{2,i}+4\mathcal{J}_2^2(p)N^{[g_{4,1}^{(1,i)}]+\frac{2r}{p}}\sum_{i=2}^{4}g_{4,1}^{(1,i)}+2\mathcal{J}_2^4(p)N^{[g_{4,2}^2]+\frac{4r}{p}}g_{4,2}^2, \tag{38}$$

$$\beta_{g_{4,1}^{2,i}} = \left(2\eta-[g_{4,1}^{2,i}]\right)g_{4,1}^{2,i}+4\mathcal{J}_3^3(p)N^{[g_{4,1}^{2,i}]+\frac{3r}{p}}(g_{4,1}^{2,i})^2+8\mathcal{J}_3^1(p)N^{2[g_{4,1}^{(1,i)}]-[g_{4,1}^{2,i}]+\frac{r}{p}}\sum_{j=2}^{4}\sum_{k>j}g_{4,1}^{(1,j)}g_{4,1}^{(1,k)}$$
$$+ 8\mathcal{J}_3^2(p)N^{[g_{4,1}^{(1,i)}]+\frac{2r}{p}}g_{4,1}^{2,i}\sum_{j=2}^{4}g_{4,1}^{(1,j)}, \tag{39}$$

$$\beta_{g_{4,1}^{(1,i)}} = \left(2\eta-[g_{4,1}^{(1,i)}]\right)g_{4,1}^{(1,i)}+8\mathcal{J}_3^2(p)N^{[g_{4,1}^{(1,i)}]+\frac{2r}{p}}(g_{4,1}^{(1,i)})^2+8\mathcal{J}_3^1(p)N^{[g_{4,1}^{2,i}]+\frac{r}{p}}g_{4,1}^{(1,i)}\sum_{j=1}^{4}g_{4,1}^{2,j}, \tag{40}$$

$$\beta_{g_{4,2}^2} = \left(2\eta-[g_{4,2}^2]\right)g_{4,2}^2+8\mathcal{J}_3^2(p)N^{2[g_{4,1}^{2,i}]-[g_{4,2}^2]+\frac{2r}{p}}\sum_{i=1}^{4}\sum_{j>i}g_{4,1}^{2,i}g_{4,1}^{2,j}$$
$$+ 8\mathcal{J}_3^1(p)N^{[g_{4,1}^{2,i}]+[g_{4,1}^{(1,i)}]-[g_{4,2}^2]+\frac{r}{p}}\sum_{i=1}^{4}\sum_{j=2}^{4}g_{4,1}^{2,i}g_{4,1}^{(1,j)}+8\mathcal{J}_3^3(p)N^{[g_{4,1}^{2,i}]+\frac{3r}{p}}g_{4,2}^2\sum_{i=1}^{4}g_{4,1}^{2,i}$$
$$+ 16\mathcal{J}_3^2(p)N^{[g_{4,1}^{(1,i)}]+\frac{2r}{p}}g_{4,2}^2\sum_{i=2}^{4}g_{4,1}^{(1,i)}+4\mathcal{J}_3^4(p)N^{[g_{4,2}^2]+\frac{4r}{p}}(g_{4,2}^2)^2, \tag{41}$$

where $\mathcal{J}_j^i(p)$ are threshold integrals provided in Appendix A for $p=1,2$.

The canonical dimensions for the couplings are not fixed in Equations (38)–(41). They are fixed by demanding that Equations (38)–(41) admit a $1/N$ expansion starting with a non-trivial contribution at order $(1/N)^0$. In the expression for the anomalous dimension, Equation (38), the large-N limit can be taken if the canonical dimensions satisfy the following bounds

$$\frac{3r}{p}+[g_{4,1}^{2,i}]\leq 0, \quad \frac{2r}{p}+[g_{4,1}^{(1,i)}]\leq 0, \quad \frac{4r}{p}+[g_{4,2}^2]\leq 0. \tag{42}$$

Equation (39) imposes a new constraint for the canonical dimensions, namely

$$2[g_{4,1}^{(1,i)}]-[g_{4,1}^{2,i}]+\frac{r}{p}\leq 0, \tag{43}$$

while the beta function for the necklaces (40) does not introduce any new conditions. Finally, the beta function for the double-trace couplings constrains the canonical dimensions by

$$\frac{2r}{p}+2[g_{4,1}^{2,i}]-[g_{4,2}^2]\leq 0 \quad \text{and} \quad [g_{4,1}^{2,i}]+[g_{4,1}^{(1,i)}]-[g_{4,2}^2]+\frac{r}{p}\leq 0. \tag{44}$$

From Equations (42)–(44), we obtain upper bounds for the coupling's canonical dimensions (or relations between them). Couplings decouple from the set of beta functions if their canonical dimension is chosen below the corresponding upper bounds. In this sense choosing the upper bounds as the canonical dimensions leads to the most non-trivial set of beta functions at large N. We tentatively consider a decoupling of interactions through such choices artificial, hence, we choose the upper bounds as the canonical dimension for the couplings. We start by demanding that

$$[g_{4,1}^{2,i}]=-\frac{3r}{p}. \tag{45}$$

This is exactly what one would expect based on $[g_{4,1}^{2,i}]_{\text{rank}-3}=-2r$ (for $p=1$) in the rank-3 case: The contraction of one additional index in the rank-4 case requires an additional suppression by $1/N$

(for $r/p = 1$). The first inequality of (44) implies $[\tilde{g}_{4,2}^2] \geq -4r/p$. Yet, the third inequality of (42) enforces $[\tilde{g}_{4,2}^2] \leq -4r/p$. Therefore, given the choice in Equation (45), the canonical dimension for the double-trace coupling is completely fixed, i.e.,

$$[\tilde{g}_{4,2}^2] = -\frac{4r}{p}. \tag{46}$$

This is in accordance with the expectation from the rank-3 case, as well as the reasoning that the additional "trace" of this interaction in comparison to the quartic cyclic-melonic interaction should lead to an additional suppression by $1/N$ (for $r = p$). Finally, using (45) and (46) in (42)–(44) results in $[\tilde{g}_{4,1}^{(1,i)}] \leq -2r/p$, i.e., the canonical dimension for the necklaces is not fixed uniquely. We choose

$$[\tilde{g}_{4,1}^{(1,i)}] = -\frac{2r}{p}. \tag{47}$$

The scaling dimensions are functions of the ratio r/p. Hence, if one chooses the "standard" scaling, i.e., setting the power of the infrared cutoff N equal to that of the "momentum" scale, $r = p$, the dimensions are always -3, -4 and -2 for the cyclic melons, multi-trace and necklaces interactions, respectively. For those fixed points that do not feature dimensional reduction, the choice $r = p$ is preferred based on a geometrical argument, see Section 4. Nevertheless, the threshold integrals $\mathcal{J}_j^i$ depend on those parameters in a non-trivial way which implies that different choices of (r, p) lead to different numerical coefficients in the beta functions. Thus, choosing different values r and p while keeping all canonical dimensions fixed tests the scheme/regulator dependence of our calculation.

In the large-N limit and using Equations (45)–(47) with $r = p = 1$, the system of beta functions reduces to

$$\eta = \frac{1}{20}(5-\eta)\sum_{i=1}^{4} g_{4,1}^{2,i} + \frac{1}{3}(4-\eta)\sum_{i=2}^{4} g_{4,1}^{(1,i)} + \frac{1}{90}(6-\eta)g_{4,2}^2, \tag{48}$$

which can be solved for η leading to

$$\eta = \frac{3\left(15\sum_{i=1}^{4} g_{4,1}^{2,i} + 80\sum_{i=2}^{4} g_{4,1}^{(1,i)} + 4g_{4,2}^2\right)}{180 + 9\sum_{j=1}^{4} g_{4,1}^{2,i} + 60\sum_{j=2}^{4} g_{4,1}^{(1,j)} + 2g_{4,2}^2}, \tag{49}$$

and

$$\beta_{g_{4,1}^{2,i}} = (2\eta+3)g_{4,1}^{2,i} + \frac{1}{15}(6-\eta)(g_{4,1}^{2,i})^2 + \frac{2}{3}(4-\eta)\sum_{j=2}^{4}\sum_{k>j} g_{4,1}^{(1,j)} g_{4,1}^{(1,k)} + \frac{2}{5}(5-\eta)g_{4,1}^{2,i}\sum_{j=2}^{4} g_{4,1}^{(1,j)}, \tag{50}$$

$$\beta_{g_{4,1}^{(1,i)}} = (2\eta+2)g_{4,1}^{(1,i)} + \frac{2}{5}(5-\eta)(g_{4,1}^{(1,i)})^2, \tag{51}$$

$$\beta_{g_{4,2}^2} = (2\eta+4)g_{4,2}^2 + \frac{2}{5}(5-\eta)\sum_{i=1}^{4}\sum_{j>i} g_{4,1}^{2,i} g_{4,1}^{2,j} + \frac{2}{3}(4-\eta)\sum_{i=1}^{4}\sum_{j=2}^{4} g_{4,1}^{2,i} g_{4,1}^{(1,j)}$$
$$+ \frac{2}{15}(6-\eta)g_{4,2}^2\sum_{i=1}^{4} g_{4,1}^{2,i} + \frac{4}{5}(5-\eta)g_{4,2}^2\sum_{i=2}^{4} g_{4,1}^{(1,i)} + \frac{1}{63}(7-\eta)(g_{4,2}^2)^2. \tag{52}$$

We highlight several key features of the above system: Firstly, unlike in the quartic truncation for the rank-3 complex tensor model [64], there is a class of interactions, the necklaces, which are not melonic. On the other hand, in the real rank-3 tensor model, see [65], a non-melonic interaction is present already at the quartic order. It does not contribute to the anomalous dimension at large N. As a difference to these two examples, Equations (48) and (49) show that all couplings contribute to the anomalous dimension in the complex rank-4 model, including the non-melonic (necklaces) couplings. This is the first evident structural difference between the present beta functions and the rank-3 ones [64,65].

Secondly, after choosing the canonical dimension for the melonic coupling $\bar{g}_{4,1}^{2,i}$, the canonical dimension for the double-trace coupling is fixed uniquely. Its value differs from the canonical dimension of the melonic coupling. Consequently, interactions which contain the same number of fields (tensors) as well as sums over indices (which are the analogue of an integral over momenta in ordinary quantum field theories on a background) scale with different powers at large N. This is an intrinsic property of the combinatorially non-trivial structure of the interactions in tensor models, see also [64,65,88]. We caution that if the double-trace interaction was not introduced in the present truncation, one could choose the canonical dimension for the necklaces to be the same as the canonical dimension of the melonic coupling. This would lead to the misleading conclusion that is possible to choose the same canonical dimension for all interactions with a given number of tensors.

We look for fixed points of the system of beta functions Equations (49)–(52). The strategy is the same as the one employed in [64,65]: Firstly, zeros of the beta functions are obtained in a perturbative approximation, i.e., the anomalous dimension is taken as a polynomial function of the couplings,

$$\eta_p = \frac{1}{4} \sum_{i=1}^{4} g_{4,1}^{2,i} + \frac{4}{3} \sum_{i=2}^{4} g_{4,1}^{(1,i)} + \frac{1}{15} g_{4,2}^2. \tag{53}$$

With Equation (53), the beta functions are polynomials in the couplings. Hence finding their zeros is easily achieved with computer software. Once the zeros are obtained, several criteria are applied to filter out candidates for physical fixed points. These include the regulator bound[8] $\eta_p < 1$. Furthermore, the critical exponents should stay bounded such that the bootstrap strategy for the choice of truncation is justified. Finally, we demand stability under extensions of the truncation. Given the limited nature of our investigation for the purposes of this review, the only extension is that from the perturbative form of the anomalous dimension in Equation (53) to the full expression in Equation (49).

The resulting candidates for physical universality classes can be separated into two main classes: those with enhancement of the $U(N')^{\otimes 4}$ symmetry to $U(N'^2) \otimes U(N')^{\otimes 2}$, $U(N'^3) \otimes U(N')$ or $U(N'^4)$ display dimensional reduction, i.e., the associated continuum limit would not correspond to four-dimensional geometries. In contrast, those with $U(N')^{\otimes 4}$ symmetry might be possible candidates for a suitable continuum limit which could correspond to 4d quantum gravity. The following results are quoted for the case $r = p = 1$ unless stated otherwise.

7.1. Symmetry-Enhanced Fixed Points: Dimensional Reduction in Tensor Models

The $U(N')^{\otimes 4}$ symmetric theory space contains symmetry-enhanced subspaces, such as, e.g., $U(N'^2) \otimes U(N')^{\otimes 2}$. To achieve the corresponding enhancement of symmetry, interactions which violate it must be switched off. This happens at several fixed points in our truncation[9]. The enhanced symmetry is broken if there is at least one non-vanishing interaction for each of the four colors that treats this color differently form the remaining colors. Therefore, although it appears slightly paradoxically at first glance, fixed points which are *not* invariant under color permutations typically feature a larger symmetry than $U(N')^{\otimes 4}$. The breaking of the color-permutation symmetry at the fixed point[10] allows for some interactions to vanish such that a pair, or even triple, of indices can be summarized into one "super-index". This super-index features an $U(N'^2)$ (or even $U(N'^3)$) symmetry. Consequently, the interactions which are turned on at the fixed point can be described by lower-rank tensors. This is a form of dimensional reduction, i.e., the lower-rank tensors "tessellate" lower-dimensional discrete

---

[8] As discussed in [130] and adapted to tensor models in [65], the condition that the regulator diverges at $N \to N' \to \infty$ imposes a bound on the anomalous dimension. For our regulator choice Equation (36) this is $\eta < r$.

[9] Please note that while symmetry-enhanced subspaces are invariant under the flow, this does not automatically imply that they must feature interacting fixed points; it can also be the case that the symmetry-enhanced hypersurface is a fixed surface of the flow.

[10] The color-permutation symmetry is still intact in the full theory space, as each such fixed point automatically comes with partners related by a color permutation.

geometries. For instance, at a fixed point at which two pairs of indices are summarized into two superindices, the rank-4-model reduces to a matrix model, which encodes random geometries in two dimensions.

The enhancement in symmetry and dimensional reduction entails that universality classes of lower-rank-models can be reproduced. Two comments are in order here.

Firstly, the recovery of "lower-dimensional" universality classes requires to exploit the freedom in the choice of regulator in Equation (36) such that the canonical dimensions of the interactions agree with those of the lower-rank-model. For instance, the quartic cyclic-melonic couplings have canonical dimension $-3r/p$ in the rank-4- case and -2 in the rank -3 case for $r = p = 1$. Choosing $r/p = 2/3$ for the rank-4-case leads to an agreement in the canonical dimension of the cyclic melons. Analogous choices for different fixed points will be spelled out below. We emphasize that the choice of canonical dimension of quartic interactions is grounded in geometric arguments. Therefore, for each dimensionality d, there is a unique choice of r/p for each rank n, such that the canonical scaling exponents agree with those required for an identification of the dual of the tensor model with random geometries in d dimensions. This choice appears to be $r/p = 1$ for $n = d$, but differs if $n \neq d$.

Secondly, the symmetry-enhanced fixed points are embedded in a larger theory space with symmetry-breaking directions. Therefore, additional relevant directions might exist which entail additional tuning required to reach the fixed point. Similar enlargements of universality classes are well known in statistical physics. For instance, the scaling exponents for the $O(N + M)$ Wilson-Fisher fixed point can be recovered within a $O(N) \oplus O(M)$ symmetric theory space. Yet, an additional relevant direction is associated with the additional tuning required to reach this critical point, see, e.g., [131]. We will check on a case-by-case basis whether dimensional reduction requires additional tuning, or whether it is a preferred IR endpoint of tensorial RG flows.

The set of beta functions given by Equations (49)–(52) admits the following symmetry-enhanced fixed points:

- **Cyclic-Melonic Single-trace Fixed Point:** Only one representative of the cyclic-melonic interactions $g_{4,1}^{2,i}$ is non-vanishing at this fixed point. For a given cyclic melon, e.g., $g_{4,1}^{2,1}$, the interaction can be expressed as

$$g_{4,1}^{2,1} T_{a_1 a_2 a_3 a_4} \bar{T}_{b_1 a_2 a_3 a_4} T_{b_1 b_2 b_3 b_4} \bar{T}_{a_1 b_2 b_3 b_4} \longrightarrow g_{4,1}^{2,1} T_{a_1 I} \bar{T}_{b_1 I} T_{b_1 J} \bar{T}_{a_1 J}, \tag{54}$$

where the super-index I condenses three of the initial indices and thereby enhances the symmetry of the model to $U(N') \otimes U(N'^3)$. Consequently, the fixed-point dynamics is described by a single-matrix model. The corresponding continuum limit is not associated with 4d quantum gravity but rather expected to yield the well-known pure-gravity scaling exponent in 2d. For $r = p = 1$ this fixed point features two relevant directions in our simple truncation, $\theta_1 = 3.47$ and $\theta_2 = 0.31$. Due to the systematic error associated with the truncation the present results are insufficient to establish whether the second relevant direction turns into an irrelevant one. We provide a rough estimate for a lower bound on the systematic error by exploiting the freedom in the shape function: Considering, for instance, a "spherical" cutoff function, i.e., $r = p = 2$, this fixed point also displays two relevant directions with critical exponents $\theta_1 = 3.71$ and $\theta_2 = 0.22$.

Although associated with a matrix model, the critical exponents reported are far from the exact result obtained for the pure-gravity scaling exponent in 2d, ($\theta = 0.8$). This is similar to the result obtained in the rank-3 real model in [65] and a consequence of the canonical dimensional of the cyclic-melonic coupling for $r/p = 1$. Instead setting $r = 1/3$ for $p = 1$ implies $[g_{4,1}^{2,i}] = -1$ in agreement with the canonical dimension of the quartic interaction in matrix models. In this case, the fixed point has two relevant directions with critical exponents $\theta_1 = 1.05$ and $\theta_2 = 0.11$. For the prescription for critical exponents reported in Section 5, we obtain $\tilde{\theta}_1 = 0.44$ and $\tilde{\theta}_2 = 0.11$. More sophisticated truncations are necessary to establish whether the second critical exponents are indeed positive.

- **Multi-trace-Bubble Fixed Point:** All interactions but the double-trace $g_{4,2}^2$ one vanish at this fixed point. The remaining interaction can be expressed as

$$g_{4,2}^2 \bar{T}_{a_1 a_2 a_3 a_4} T_{a_1 a_2 a_3 a_4} \bar{T}_{b_1 b_2 b_3 b_4} T_{b_1 b_2 b_3 b_4} \longrightarrow g_{4,2}^2 \bar{T}_I T_I \bar{T}_J T_J , \qquad (55)$$

where all indices are collected in one single super-index I. Such a term is characterized by an enhanced symmetry $U(N'^4)$ and it describes a vector model. This fixed point displays four relevant directions: $\theta_1 = 4.69$ and $\theta_{2,3,4} = 0.20$. The three-fold degeneracy is a consequence of an exchange-symmetry between the three directions that break the enhanced symmetry. The small absolute value of $\theta_{2,3,4}$ does not permit to determine whether there are four relevant directions in total. In fact, the same fixed-point structure is seen in the rank-3 model [65]: there, it features two relevant directions in the quartic truncation while in the hexic truncation, only one relevant direction remains, see [65].

- **Single-necklace Fixed Point:** Only one necklace interaction is non-vanishing at this fixed point. All other interactions in our truncation vanish. Beyond the truncation, only interactions respecting the corresponding enhanced symmetry are present. Due to the color-permutation symmetry in the theory space, there are three such fixed points, each characterized by a different non-vanishing necklace. If one takes, e.g., $g_{4,1}^{(1,2)}$ to be the non-vanishing necklace at the fixed point, the interaction term can be expressed as

$$g_{4,1}^{(1,2)} \bar{T}_{a_1 a_2 a_3 a_4} T_{a_1 a_2 b_3 b_4} \bar{T}_{b_1 b_2 b_3 b_4} T_{b_1 b_2 a_3 a_4} \longrightarrow g_{4,1}^{(1,2)} \bar{T}_{IJ} T_{IK} \bar{T}_{LK} T_{LJ} , \qquad (56)$$

where the two index pairs are collected in one super-index I. Therefore, the interaction features an enhanced $U(N'^2) \otimes U(N'^2)$ symmetry. Accordingly, it describes a matrix model with matrices $(\bar{T}, T)$. This symmetry enhancement is associated with an effective dimensional reduction: the continuum limit associated with this fixed point does not correspond to 4d quantum gravity, but 2d random geometries. This fixed point has one relevant direction with critical exponent $\theta = 2.27$. The naive value of the universal scaling exponent deviates strongly from the exact result $\theta = 0.8$. We attribute the difference to the fact that the canonical dimension for the necklaces couplings is -2 and not -1 as would be in the assignment of the canonical dimensions in matrix models, see [40]. By choosing $r = 1/2$ for $p = 1$, the fixed point exhibits one relevant direction with scaling exponent $\theta = 1.07$ which gets closer to the exact result. The second prescription for the universal scaling exponents yields $\tilde{\theta} = 0.42$.

As a simple check of the robustness of these results, we explore the choice $r = p = 2$. We obtain one relevant direction with critical exponent $\theta = 2.37$. Assigning dimension -1 for $p = 2$ requires $r = 1$. For this choice, the fixed point has one relevant direction with critical exponent $\theta = 1.09$. The results are qualitative and even numerically compatible with those discussed for $p = 1$, giving a first hint towards stability under different choices of scheme.

The above fixed points are characterized by a single interaction type. Symmetry-enhanced fixed points with more than one non-vanishing interaction are also possible. These include, e.g.,

- **One Cyclic-Melonic Multi-trace Fixed Point:** At this fixed point, just one cyclic-melonic interaction of a given preferred color and the double-trace interaction are non-vanishing. If one selects, e.g., the coupling $g_{4,1}^{2,1}$ to be the non-vanishing cyclic melon, the interactions at the fixed point are given by

$$g_{4,1}^{2,1} \bar{T}_{a_1 a_2 a_3 a_4} T_{b_1 a_2 a_3 a_4} \bar{T}_{b_1 b_2 b_3 b_4} T_{a_1 b_2 b_3 b_4} + g_{4,2}^2 \bar{T}_{a_1 a_2 a_3 a_4} T_{a_1 a_2 a_3 a_4} \bar{T}_{b_1 b_2 b_3 b_4} T_{b_1 b_2 b_3 b_4}$$

$$\longrightarrow g_{4,1}^{2,1} \bar{T}_{a_1 I} T_{b_1 I} \bar{T}_{b_1 J} T_{a_1 J} + g_{4,2}^2 \bar{T}_{a_1 I} T_{a_1 I} \bar{T}_{b_1 J} T_{b_1 J} , \qquad (57)$$

Three indices are condensed in one super-index I, enhancing the symmetry to $U(N') \otimes U(N'^3)$. Accordingly, the fixed point is associated with a matrix model. It features one relevant direction and the associated critical exponent is $\theta = 3.06$. In fact, for this case, one cannot reproduce both canonical scaling dimensions for matrix models. To obtain agreement for the single-trace quartic coupling, one should again choose $r/p = 1/3$. This yields a canonical dimension of -1 for the single-trace coupling, but $-4/3$ for the double-trace, whereas the corresponding dimensions in the matrix model are -1 and -2. Therefore, it is not clear whether this fixed point admits an interpretation in terms of a matrix model for random geometries.

Beyond the fixed-point candidates reported here, further zeros of the beta functions characterized by symmetry enhancement are also obtained. As particular examples, one finds zeros where one cyclic-melonic interaction together with one necklace and the multi-trace interactions are turned on. Due to the different combinatorial structures, the corresponding dynamics can be mapped to that of a rank-3 model as illustrated in Figure 6. As the cyclic melons and necklaces feature different canonical dimensions, the corresponding model would presumably be a 2-tensor model. A further zero of the system of beta functions features a single necklace and the double-trace interaction. However, in the present truncation, such zeros of the beta functions violate the regulator bound. Therefore, we tentatively discard them and do not consider them as candidates for fixed points, i.e., universal scaling regimes. More refined studies are necessary to robustly confirm this characterization.

Figure 6. Illustration of symmetry enhancement: At a fixed point which features the cyclic melon with coupling $g_{4,1}^{2,1}$ and the necklace with coupling $g_{4,1}^{(1,2)}$, two indices, represented by green and red dashed lines are collected in one red double-line which represents a super-index. At a fixed point which only features the necklace $g_{4,1}^{(1,2)}$, the index pair $(1,2)$ as well as the pair $(3,4)$ can be summarized to two superindices, entailing a reduction to a matrix model.

7.2. Candidates for Four-Dimensional Emergent Space

In this subsection, we discuss a fixed point which does not feature symmetry enhancement of the form as discussed previously and therefore cannot be mapped to a lower-rank single tensor model. Thus, it might be a potential candidate for the description of 4d quantum gravity. Of course, establishing a universality class for 4d quantum gravity requires much more than just finding a fixed point without dimensional reduction of the form discussed above. After all, the Hausdorff and spectral dimensions as well as further properties of the emergent geometry have not been studied yet. Nevertheless, the existence of a fixed point that does not admit dimensional reduction to a model of lower rank is most likely a *necessary* requirement for a universality class for 4d quantum gravity. If corroborated by further studies, our result might therefore constitute the very first step on a path towards 4d quantum gravity from tensor models.

Here, we focus on isocolored fixed points, i.e., those that display the same values for all couplings associated with different colors. In other words, we restrict the fixed points to a symmetry-enhanced subspace which explicitly realizes a discrete color-permutation symmetry in all interactions. It is still characterized by the $U(N')^{\otimes 4}$ symmetry and does not feature dimensional reduction to a model of lower rank. We conjecture that in the continuum, color-distinguishing structures should not play any role. This is based on the expectation that color is not associated with a physical property of continuum geometries and therefore only isocolored fixed points should matter. We caution that this might be a naive viewpoint, since the unequal treatment of colors could introduce more sophisticated structures. As stressed before, the identification of universality classes with actual relevance for 4d quantum gravity requires further insights into the emergent geometries. Here, we restrict ourselves to a very first mapping of different fixed-point structures in the theory space.

In the quartic truncation, one completely isocolored fixed points is found. At this fixed point, all couplings are non-vanishing. Consequently, there is no symmetry enhancement of the $U(N')^{\otimes 4}$ symmetry (apart from the discrete color-permutation symmetry) which would allow for an immediate identification of dimensional reduction. The fixed point as well as the corresponding critical exponents for $r = p = 1$ are displayed in Table 5: the isocolored fixed point features three relevant directions. A simple test of the scheme-dependence of this result can be performed by changing the regulator to $r = p = 2$. The isocolored fixed point persists and features positive critical exponents $\theta_1 = 3.41$ and $\theta_{2,3} = 0.18$ very close to the values for $r = p = 1$.

Table 5. Fixed point and critical exponents values for the isocolored fixed point with $r = p = 1$.

$g_{4,1}^{2,i}$	$g_{4,1}^{(1,j)}$	$g_{4,2}^{2}$	η	θ_1	$\theta_{2,3}$	$\theta_{4,5}$	$\theta_{6,7,8}$
−0.09	−0.07	−5.70	−0.91	3.44	0.18	$-0.40 \pm 0.07\,i$	−0.56

A subset of these relevant directions could be associated with the tuning towards the isocolored symmetry. To isolate such directions, we re-investigate the fixed point in an isocolored truncation, where the different color couplings are identified in Equation (34): $g_{4,1}^{2,1} = g_{4,1}^{2,2} = g_{4,1}^{2,3} = g_{4,1}^{2,4} = g_{4,1}^{2}$ and $g_{4,1}^{(1,2)} = g_{4,1}^{(1,3)} = g_{4,1}^{(1,4)} = g_{4,1}$. The theory space in this truncation is spanned by three couplings. For $r = p = 1$, the fixed point displays one relevant direction associated with $\theta = 3.44$. Consequently, the two extra relevant directions that appear in the non-isocolored truncation are associated with an additional tuning of couplings to achieve the color symmetry at the fixed point. As discussed above, it remains to be investigated whether color-symmetry breaking can be given any physical meaning in random geometries. Therefore, it is currently open whether one should only compare the leading relevant critical exponent $\theta_1 = 3.44$ to the critical exponents characterizing gravity in the continuum limit, i.e., the critical exponents of the Reuter fixed point, or whether one should also include $\theta_{2,3}$ in the comparison of universality classes.

The isocolored fixed point serves as a prototypical example of a fixed-point structure which does not manifest dimensional reduction at the level of the basic building blocks used to generate random geometries. This is only a necessary condition for a universality class associated with 4d quantum gravity, and the physical nature of the continuum limit associated with such a fixed point still needs to be investigated.

Going beyond the isocolored theory space, different fixed-point structures than the completely isocolored one which do not feature symmetry enhancement can be found. In particular, there are fixed points where all couplings are turned on, but for instance, not all couplings of a given combinatorial structure attain the same value. These fixed points are not color-permutation invariant. A detailed discussion of these new universality classes is beyond the scope of the present review and will be reported in a separate work.

Finally, it is worth mentioning a comparison between the results presented here and those derived for tensorial (group) field theories. As a first remark, the theory spaces of tensor models and tensorial (group) field theories are different. The reason behind that is the presence of a kinetic term in tensorial (group) field theories which is not invariant under unitary or orthogonal transformation of the tensors (in group field theories, of the generalized Fourier component of the group field). In tensor models, the regulator term is responsible for inducing such a symmetry-breaking. Therefore, symmetry-violating terms are generated along the RG flow. Nevertheless, the symmetric theory space does not feature index-dependent interactions which naturally arise in tensorial (group) field theories. In particular, this also entails a difference in the Ward-identities of the two settings. Consequently, while tensor models and tensorial GFTs share the combinatorial structure, the RG flow as well as potential fixed points differ. If the additional structure in GFTs is not relevant for the continuum limit, then the same universality class should exist in both settings. Yet, it might be the case that pure tensor models do not contain enough structure to provide a suitable continuum limit which would describe our universe. In this case, further geometric data arising from group theoretic structures should be incorporated. As long as this question is not settled, a parallel development of tensor models and (tensorial) GFTs is indicated. Furthermore, we highlight that the development of the FRG in tensor models and tensorial (group) field theories should go together since the theories have the same combinatorial structure and at least at the technical if not the conceptual level, fruitful transfers of ideas can be expected.

8. Outlook: Converging to Quantum Gravity from Different Directions

We advocate the point of view that an understanding of (key aspects of) quantum gravity can be achieved by making sense of the path integral for quantum gravity. In tensor models, the path integral is interpreted as a sum over random geometries. This sum can be tackled in a dual formulation, where rank d tensors form building blocks of d-dimensional space(time). The FRG equation is equivalent to the path integral, as it is simply a way of rewriting an integral into a differential equation that tracks the change of the integral under a change of a parameter in the integrand. This abstract setup translates into the well-known local coarse graining in quantum field theories defined on a background. We highlight that the notion of coarse graining also makes sense in a background-independent setting. In this setting, the number of degrees of freedom provides a background-independent notion of scale. In accordance with the intuition behind the a-theorem, the RG flow goes from many to few degrees of freedom. For tensor models, this corresponds to an RG flow in the tensor size N. RG fixed points play a crucial role, as they provide universality in the large N limit. Physically, this provides a phase transition in the space of couplings that leads to a continuum phase that is independent of unphysical microscopic details.

The path integral for quantum gravity is a point of convergence for a diverse set of viewpoints, e.g., [2–19,132,133]. The configuration space that is summed over in these settings typically includes a sum over (discretized) geometries. The inclusion of non-geometric configurations, e.g., [133] or summation over topologies, is one distinguishing feature of the different approaches that could be of physical relevance. Restricting to the sum over geometries, different approaches to the path integral implement this summation in mathematically distinct ways. These have diverse advantages, such as a direct access to large-scale properties of emergent geometries from lattice simulations, see, e.g., [134], a straightforward way of discovering universality classes and characterizing them by their scaling exponents in tensor models [64,65], and a direct link to phenomenological questions and the interplay between quantum gravity and matter in the continuum asymptotic safety approach [5,129] to name just a few. We advocate the point of view that such different approaches need not necessarily be considered as competitors in the race towards the goal of discovering quantum gravity. Rather, these different approaches can be viewed as different windows that allow us to view and explore distinct aspects of quantum gravity. In the best case, these are complementary, and a comprehensive understanding of quantum spacetime can emerge if key results and strengths from these diverse directions are brought together to form one coherent big picture. As in many other settings, a diversity of viewpoints can accelerate the discovery of a solution to a tough challenge—in this case, quantum gravity. Yet, a diversity of viewpoints brings a new challenge, namely the potential lack of a common language. In quantum gravity, different approaches are often formulated in mathematically very dissimilar ways, making it challenging to extract common physics. Here, we advocate that the functional RG setup could provide one option for a common language shared by different approaches. In particular, it allows one to evaluate scaling exponents linked to a universal continuum limit. These universal exponents can be compared, e.g., from continuum asymptotic safety and tensor models[11]. Once the required precision has been reached in advanced approximation of the full RG flow, such a quantitative comparison will unveil whether these approaches to the gravitational path integral encode the same physics. In the most straightforward setup for tensor models, full agreement with scaling exponents from CDTs or continuum asymptotic safety is probably not expected. This is due to the difference in configuration spaces in the respective approaches to the path integral. For instance, the gluing rules encoding causality in CDTs are expected to lead to a restriction on the allowed (multi)-tensor interactions. Following [73], the corresponding tensor model, once set up, can be explored by means of the FRG, and a characterization of the universality class is possible.

Understanding the quantum structure of spacetime is a challenging goal. We advocate that the complementarity that different approaches to the path integral for quantum gravity exhibit is a highly promising starting point. Tensor models could be helpful in this quest as they could contribute to bridging the gap between discrete numerical and analytical continuum approaches by allowing for a discrete analytical approach. We propose that background-independent functional RG techniques could potentially act as a catalyst for breakthroughs. Specifically, they allow us to discover and characterize universality classes for the continuum limit in tensor models. This could provide one of the missing links towards a background-independent understanding of quantum gravity.

Even if this hope is not realized, tensor models could constitute a stand-alone approach to the path integral for gravity. To that end, going beyond the simplest form of the large N limit, which leads to a branched-polymer phase, appears to be necessary. We highlight the potential use of the FRG in this context, as it is a highly flexible tool allowing to search for universal scaling regimes in diverse tensor-model theory spaces. In particular, setting up truncations adapted to different assumptions regarding the nature of the universality class (e.g., near-canonical vs. fully non-perturbative) could give access to different continuum limits.

Author Contributions: All authors contributed equally to all aspects of this manuscript.

Funding: This research has been funded by the DFG under grant no. Ei-1037/1.

Acknowledgments: We acknowledge helpful discussions with V. Rivasseau, S. Carrozza and J. Ben Geloun. A. E. and A. D. P. are supported by the Deutsche Forschungsgemeinschaft under grant no. Ei/1037-1. The work of T. K. was supported in part by the PAPIIT grant no. IA-103718 at the UNAM.

Conflicts of Interest: The authors declare no conflict of interest.

---

[11] We stress that if the continuum limit in the gravitational path integral can be taken, this requires asymptotic safety or asymptotic freedom to hold. In fact, to take the continuum limit in any path integral, while obtaining an interacting, and thus potentially interesting infrared limit, requires one of the two scenarios to hold. On the other hand, this does not imply that the universality class need be that of the Reuter fixed point, which is asymptotic safety of a path integral with a particular choice of configuration space. In principle, it might be that the asymptotic safety scenario is realized in mathematically and physically distinct ways in several different incarnations of the path integral for gravity. It is, therefore, important to keep in mind the distinction between asymptotic safety as a general scenario and asymptotically safe gravity defined by the Reuter fixed point.

Appendix A. Threshold Integrals

In this appendix we list the expressions for the threshold integrals that appear in the beta functions (38)–(41):

$$\begin{aligned}
\mathcal{J}_2^2(1) &= \frac{1}{12}(4-\eta), \\
\mathcal{J}_2^3(1) &= \frac{1}{40}(5-\eta), \\
\mathcal{J}_2^4(1) &= \frac{1}{180}(6-\eta), \\
\mathcal{J}_3^1(1) &= \frac{1}{12}(4-\eta), \\
\mathcal{J}_3^2(1) &= \frac{1}{20}(5-\eta), \\
\mathcal{J}_3^3(1) &= \frac{1}{60}(6-\eta), \\
\mathcal{J}_3^4(1) &= \frac{1}{252}(7-\eta), \\
\mathcal{J}_2^2(2) &= \frac{\pi}{24}(3-\eta), \\
\mathcal{J}_2^3(2) &= \frac{\pi}{70}(7-2\eta), \\
\mathcal{J}_2^4(2) &= \frac{\pi^2}{192}(4-\eta), \\
\mathcal{J}_3^1(2) &= \frac{1}{35}(7-2\eta), \\
\mathcal{J}_3^2(2) &= \frac{\pi}{48}(4-\eta), \\
\mathcal{J}_3^3(2) &= \frac{\pi}{126}(9-2\eta), \\
\mathcal{J}_3^4(2) &= \frac{\pi^2}{320}(5-\eta).
\end{aligned} \tag{A1}$$

References

1. Weinberg, S. Ultraviolet divergences in quantum theories of gravitation. In *General Relativity*; Hawking, S.W., Israel, W., Eds.; University Press: Cambridge, UK, 1979; Chapter 16.
2. Donoghue, J.F. General relativity as an effective field theory: The leading quantum corrections. *Phys. Rev. D* **1994**, *50*, 3874–3888. [CrossRef]
3. Reuter, M. Nonperturbative evolution equation for quantum gravity. *Phys. Rev. D* **1998**, *57*, 971–985. [CrossRef]
4. Reuter, M.; Saueressig, F. Quantum Einstein Gravity. *New J. Phys.* **2012**, *14*, 055022. [CrossRef]
5. Eichhorn, A. An asymptotically safe guide to quantum gravity and matter. *arXiv* **2018**, arXiv:1810.07615.
6. Feldbrugge, J.; Lehners, J.L.; Turok, N. Lorentzian Quantum Cosmology. *Phys. Rev. D* **2017**, *95*, 103508. [CrossRef]
7. Hamber, H.W. *Quantum Gravitation: The Feynman Path Integral Approach*; Springer: Berlin, Germany, 2009.
8. Perez, A. The Spin Foam Approach to Quantum Gravity. *Living Rev. Relativ.* **2013**, *16*, 3. [CrossRef] [PubMed]
9. Ambjorn, J.; Loll, R. Nonperturbative Lorentzian quantum gravity, causality and topology change. *Nucl. Phys. B* **1998**, *536*, 407–434. [CrossRef]
10. Ambjorn, J.; Jurkiewicz, J.; Loll, R. Dynamically triangulating Lorentzian quantum gravity. *Nucl. Phys. B* **2001**, *610*, 347–382. [CrossRef]
11. Ambjorn, J.; Goerlich, A.; Jurkiewicz, J.; Loll, R. Nonperturbative Quantum Gravity. *Phys. Rept.* **2012**, *519*, 127–210. [CrossRef]

12. Laiho, J.; Bassler, S.; Coumbe, D.; Du, D.; Neelakanta, J.T. Lattice Quantum Gravity and Asymptotic Safety. *Phys. Rev.* **2017**, *96*, 064015. [CrossRef]
13. Rivasseau, V. Quantum Gravity and Renormalization: The Tensor Track. *AIP Conf. Proc.* **2011**, *1444*, 18–29.
14. Rivasseau, V. The Tensor Track: An Update. In Proceedings of the 29th International Colloquium on Group-Theoretical Methods in Physics (GROUP 29), Tianjin, China, 20–26 August 2012.
15. Rivasseau, V. The Tensor Track, III. *Fortschr. Phys.* **2014**, *62*, 81–107. [CrossRef]
16. Rivasseau, V. Random Tensors and Quantum Gravity. *SIGMA* **2016**, *12*, 069. [CrossRef]
17. Hamber, H.W. Quantum Gravity on the Lattice. *Gen. Relativ. Gravit.* **2009**, *41*, 817–876. [CrossRef]
18. Freidel, L. Group field theory: An Overview. *Int. J. Theor. Phys.* **2005**, *44*, 1769–1783. [CrossRef]
19. Baratin, A.; Oriti, D. Group field theory and simplicial gravity path integrals: A model for Holst-Plebanski gravity. *Phys. Rev. D* **2012**, *85*, 044003. [CrossRef]
20. Cardy, J.L. *Scaling and Renormalization in Statistical Physics*; Cambridge University Press: Cambridge, UK, 1996.
21. Zinn-Justin, J. Quantum field theory and critical phenomena. *Int. Ser. Monogr. Phys.* **2002**, *113*, 1–1054.
22. Rosten, O.J. Fundamentals of the Exact Renormalization Group. *Phys. Rep.* **2012**, *511*, 177–272. [CrossRef]
23. Bahr, B.; Dittrich, B.; He, S. Coarse graining free theories with gauge symmetries: The linearized case. *New J. Phys.* **2011**, *13*, 045009. [CrossRef]
24. Dittrich, B.; Eckert, F.C.; Martin-Benito, M. Coarse graining methods for spin net and spin foam models. *New J. Phys.* **2012**, *14*, 035008. [CrossRef]
25. Dittrich, B. From the discrete to the continuous: Towards a cylindrically consistent dynamics. *New J. Phys.* **2012**, *14*, 123004. [CrossRef]
26. Dittrich, B.; Martín-Benito, M.; Schnetter, E. Coarse graining of spin net models: Dynamics of intertwiners. *New J. Phys.* **2013**, *15*, 103004. [CrossRef]
27. Dittrich, B.; Steinhaus, S. Time evolution as refining, coarse graining and entangling. *New J. Phys.* **2014**, *16*, 123041. [CrossRef]
28. Dittrich, B. The continuum limit of loop quantum gravity—A framework for solving the theory. In *Loop Quantum Gravity: The First 30 Years*; Ashtekar, A., Pullin, J., Eds.; World Scientific: London, UK, 2017; pp. 153–179.
29. Bahr, B. On background-independent renormalization of spin foam models. *Class. Quantum Gravity* **2017**, *34*, 075001. [CrossRef]
30. Becker, D.; Reuter, M. En route to Background Independence: Broken split-symmetry, and how to restore it with bi-metric average actions. *Ann. Phys.* **2014**, *350*, 225–301. [CrossRef]
31. Morris, T.R. Large curvature and background scale independence in single-metric approximations to asymptotic safety. *J. High Energy Phys.* **2016**, *2016*, 160. [CrossRef]
32. Dittrich, B.; Schnetter, E.; Seth, C.J.; Steinhaus, S. Coarse graining flow of spin foam intertwiners. *Phys. Rev. D* **2016**, *94*, 124050. [CrossRef]
33. Delcamp, C.; Dittrich, B. Towards a phase diagram for spin foams. *Class. Quantum Gravity* **2017**, *34*, 225006. [CrossRef]
34. Bahr, B.; Steinhaus, S. Numerical evidence for a phase transition in 4d spin foam quantum gravity. *Phys. Rev. Lett.* **2016**, *117*, 141302. [CrossRef]
35. Bahr, B.; Steinhaus, S. Hypercuboidal renormalization in spin foam quantum gravity. *Phys. Rev. D* **2017**, *95*, 126006. [CrossRef]
36. Bahr, B.; Rabuffo, G.; Steinhaus, S. Renormalization in symmetry restricted spin foam models with curvature. *arXiv* **2018**, arXiv:1804.00023.
37. Lang, T.; Liegener, K.; Thiemann, T. Hamiltonian Renormalisation I: Derivation from Osterwalder-Schrader Reconstruction. *arXiv* **2017**, arXiv:1711.05685.
38. Eichhorn, A.; Labus, P.; Pawlowski, J.M.; Reichert, M. Effective universality in quantum gravity. *SciPost Phys.* **2018**, *5*, 031. [CrossRef]
39. Brezin, E.; Zinn-Justin, J. Renormalization group approach to matrix models. *Phys. Lett. B* **1992**, *288*, 54–58. [CrossRef]
40. Eichhorn, A.; Koslowski, T. Continuum limit in matrix models for quantum gravity from the Functional Renormalization Group. *Phys. Rev. D* **2013**, *88*, 084016. [CrossRef]
41. Eichhorn, A.; Koslowski, T. Towards phase transitions between discrete and continuum quantum spacetime from the Renormalization Group. *Phys. Rev. D* **2014**, *90*, 104039. [CrossRef]

42. Gurau, R. The 1/N expansion of colored tensor models. *Ann. Henri Poincare* **2011**, *12*, 829–847. [CrossRef]
43. Gurau, R.; Ryan, J.P. Colored Tensor Models—A review. *SIGMA* **2012**, *8*, 020. [CrossRef]
44. Gurau, R. Invitation to Random Tensors. *SIGMA* **2016**, *12*, 094. [CrossRef]
45. Bonzom, V. Large N Limits in Tensor Models: Towards More Universality Classes of Colored Triangulations in Dimension $d \geq 2$. *SIGMA* **2016**, *12*, 073. [CrossRef]
46. Witten, E. An SYK-Like Model Without Disorder. *arXiv* **2016**, arXiv:1610.09758.
47. Klebanov, I.R.; Tarnopolsky, G. Uncolored random tensors, melon diagrams, and the Sachdev-Ye-Kitaev models. *Phys. Rev. D* **2017**, *95*, 046004. [CrossRef]
48. Gurau, R. The complete $1/N$ expansion of a SYK–like tensor model. *Nucl. Phys. B* **2017**, *916*, 386–401. [CrossRef]
49. Ben Geloun, J. Two and four-loop β-functions of rank 4 renormalizable tensor field theories. *Class. Quantum Gravity* **2012**, *29*, 235011. [CrossRef]
50. Ben Geloun, J. Renormalizable Models in Rank $d \geq 2$ Tensorial Group Field Theory. *Commun. Math. Phys.* **2014**, *332*, 117–188. [CrossRef]
51. Ben Geloun, J.; Livine, E.R. Some classes of renormalizable tensor models. *J. Math. Phys.* **2013**, *54*, 082303. [CrossRef]
52. Ben Geloun, J.; Rivasseau, V. A Renormalizable 4-Dimensional Tensor Field Theory. *Commun. Math. Phys.* **2013**, *318*, 69–109. [CrossRef]
53. Ben Geloun, J.; Samary, D.O. 3D Tensor Field Theory: Renormalization and One-loop β-functions. *Ann. Henri Poincare* **2013**, *14*, 1599–1642. [CrossRef]
54. Samary, D.O. Closed equations of the two-point functions for tensorial group field theory. *Class. Quantum Gravity* **2014**, *31*, 185005. [CrossRef]
55. Carrozza, S.; Oriti, D.; Rivasseau, V. Renormalization of Tensorial Group Field Theories: Abelian U(1) Models in Four Dimensions. *Commun. Math. Phys.* **2014**, *327*, 603–641. [CrossRef]
56. Carrozza, S.; Oriti, D.; Rivasseau, V. Renormalization of a SU(2) Tensorial Group Field Theory in Three Dimensions. *Commun. Math. Phys.* **2014**, *330*, 581–637. [CrossRef]
57. Jordan, S.; Loll, R. Causal Dynamical Triangulations without Preferred Foliation. *Phys. Lett. B* **2013**, *724*, 155–159. [CrossRef]
58. Gurau, R. Colored Group Field Theory. *Commun. Math. Phys.* **2011**, *304*, 69–93. [CrossRef]
59. Gurau, R.; Rivasseau, V. The 1/N expansion of colored tensor models in arbitrary dimension. *Europhys. Lett.* **2011**, *95*, 50004. [CrossRef]
60. Gurau, R. The complete 1/N expansion of colored tensor models in arbitrary dimension. *Ann. Henri Poincare* **2012**, *13*, 399–423. [CrossRef]
61. Bonzom, V.; Gurau, R.; Rivasseau, V. Random tensor models in the large N limit: Uncoloring the colored tensor models. *Phys. Rev. D* **2012**, *85*, 084037. [CrossRef]
62. Carrozza, S.; Tanasa, A. $O(N)$ Random Tensor Models. *Lett. Math. Phys.* **2016**, *106*, 1531–1559. [CrossRef]
63. Cardy, J.L. Is There a c Theorem in Four-Dimensions? *Phys. Lett. B* **1988**, *215*, 749–752. [CrossRef]
64. Eichhorn, A.; Koslowski, T. Flowing to the continuum in discrete tensor models for quantum gravity. *Ann. Inst. H. Poincare Comb. Phys. Interact.* **2018**, *5*, 173–210. [CrossRef]
65. Eichhorn, A.; Koslowski, T.; Lumma, J.; Pereira, A.D. Towards background independent quantum gravity with tensor models. *arXiv* **2018**, arXiv:1811.00814.
66. Douglas, M.R.; Shenker, S.H. Strings in Less Than One-Dimension. *Nucl. Phys. B* **1990**, *335*, 635. [CrossRef]
67. Brezin, E.; Kazakov, V.A. Exactly Solvable Field Theories of Closed Strings. *Phys. Lett. B* **1990**, *236*, 144–150. [CrossRef]
68. Gross, D.J.; Migdal, A.A. Nonperturbative Two-Dimensional Quantum Gravity. *Phys. Rev. Lett.* **1990**, *64*, 127. [CrossRef]
69. Gross, D.J.; Migdal, A.A. A Nonperturbative Treatment of Two-dimensional Quantum Gravity. *Nucl. Phys. B* **1990**, *340*, 333–365. [CrossRef]
70. Eichhorn, A. Steps towards Lorentzian quantum gravity with causal sets. *arXiv* **2019**, arXiv:1902.00391.
71. Carrozza, S. Flowing in Group Field Theory Space: A Review. *SIGMA* **2016**, *12*, 070. [CrossRef]
72. Skenderis, K. Lecture notes on holographic renormalization. *Class. Quantum Gravity* **2002**, *19*, 5849–5876. [CrossRef]
73. Benedetti, D.; Henson, J. Imposing causality on a matrix model. *Phys. Lett. B* **2009**, *678*, 222–226. [CrossRef]

74. Castro, A.; Koslowski, T. Renormalization Group Approach to the Continuum Limit of Matrix Models of Quantum Gravity with Preferred Foliation. Unpublished work, 2019.
75. Eichhorn, A. Towards coarse graining of discrete Lorentzian quantum gravity. *Class. Quantum Gravity* **2018**, *35*, 044001. [CrossRef]
76. Wetterich, C. Exact evolution equation for the effective potential. *Phys. Lett. B* **1993**, *301*, 90–94. [CrossRef]
77. Ellwanger, U. FLow equations for N point functions and bound states. *Z. Phys. C* **1994**, *62*, 503–510. [CrossRef]
78. Morris, T.R. The Exact renormalization group and approximate solutions. *Int. J. Mod. Phys. A* **1994**, *9*, 2411–2450. [CrossRef]
79. Reuter, M.; Wetterich, C. Effective average action for gauge theories and exact evolution equations. *Nucl. Phys. B* **1994**, *417*, 181–214. [CrossRef]
80. Benedetti, D.; Ben Geloun, J.; Oriti, D. Functional Renormalisation Group Approach for Tensorial Group Field Theory: A Rank-3 Model. *J. High Energy Phys.* **2015**, *2015*, 084. [CrossRef]
81. Benedetti, D.; Lahoche, V. Functional Renormalization Group Approach for Tensorial Group Field Theory: A Rank-6 Model with Closure Constraint. *Class. Quantum Gravity* **2016**, *33*, 095003. [CrossRef]
82. Ben Geloun, J.; Martini, R.; Oriti, D. Functional Renormalization Group analysis of a Tensorial Group Field Theory on $\mathbb{R}^3$. *Europhys. Lett.* **2015**, *112*, 31001. [CrossRef]
83. Ben Geloun, J.; Martini, R.; Oriti, D. Functional Renormalisation Group analysis of Tensorial Group Field Theories on $\mathbb{R}^d$. *Phys. Rev. D* **2016**, *94*, 024017. [CrossRef]
84. Lahoche, V.; Ousmane Samary, D. Functional renormalization group for the $U(1)$-T_5^6 tensorial group field theory with closure constraint. *Phys. Rev. D* **2017**, *95*, 045013. [CrossRef]
85. Carrozza, S.; Lahoche, V. Asymptotic safety in three-dimensional SU(2) Group Field Theory: evidence in the local potential approximation. *Class. Quantum Gravity* **2017**, *34*, 115004. [CrossRef]
86. Ben Geloun, J.; Koslowski, T.A. Nontrivial UV behavior of rank-4 tensor field models for quantum gravity. *arXiv* **2016**, arXiv:1606.04044.
87. Carrozza, S.; Lahoche, V.; Oriti, D. Renormalizable Group Field Theory beyond melonic diagrams: An example in rank four. *Phys. Rev. D* **2017**, *96*, 066007. [CrossRef]
88. Ben Geloun, J.; Koslowski, T.A.; Oriti, D.; Pereira, A.D. Functional Renormalization Group analysis of rank 3 tensorial group field theory: The full quartic invariant truncation. *Phys. Rev. D* **2018**, *97*, 126018. [CrossRef]
89. Lahoche, V.; Samary, D.O. Non-perturbative renormalization group beyond melonic sector: The Effective Vertex Expansion method for group fields theories. *arXiv* **2018**, arXiv:1809.00247.
90. Lahoche, V.; Samary, D.O. Ward identity violation for melonic T^4-truncation. *arXiv* **2018**, arXiv:1809.06081.
91. Krajewski, T.; Toriumi, R. Polchinski's exact renormalisation group for tensorial theories: Gaußian universality and power counting. *J. Phys. A* **2016**, *49*, 385401. [CrossRef]
92. Krajewski, T.; Toriumi, R. Exact Renormalisation Group Equations and Loop Equations for Tensor Models. *SIGMA* **2016**, *12*, 068. [CrossRef]
93. Codello, A. Polyakov Effective Action from Functional Renormalization Group Equation. *Ann. Phys.* **2010**, *325*, 1727–1738. [CrossRef]
94. Litim, D.F. Optimization of the exact renormalization group. *Phys. Lett. B* **2000**, *486*, 92–99. [CrossRef]
95. Peskin, M.E. CRITICAL POINT BEHAVIOR OF THE WILSON LOOP. *Phys. Lett. B* **1980**, *94*, 161–165. [CrossRef]
96. Gies, H. Renormalizability of gauge theories in extra dimensions. *Phys. Rev. D* **2003**, *68*, 085015. [CrossRef]
97. Morris, T.R. Renormalizable extra-dimensional models. *J. High Energy Phys.* **2005**, *2005*, 002. [CrossRef]
98. Wilson, K.G.; Fisher, M.E. Critical exponents in 3.99 dimensions. *Phys. Rev. Lett.* **1972**, *28*, 240–243. [CrossRef]
99. Gawedzki, K.; Kupiainen, A. Renormalizing the Nonrenormalizable. *Phys. Rev. Lett.* **1985**, *55*, 363–365. [CrossRef]
100. Kikukawa, Y.; Yamawaki, K. Ultraviolet Fixed Point Structure of Renormalizable Four Fermion Theory in Less Than Four-dimensions. *Phys. Lett. B* **1990**, *234*, 497. [CrossRef]
101. Falls, K.; Litim, D.F.; Nikolakopoulos, K.; Rahmede, C. A bootstrap towards asymptotic safety. *arXiv* **2013**, arXiv:1301.4191.
102. Falls, K.; Litim, D.F.; Nikolakopoulos, K.; Rahmede, C. Further evidence for asymptotic safety of quantum gravity. *Phys. Rev. D* **2016**, *93*, 104022. [CrossRef]

103. Falls, K.G.; Litim, D.F.; Schröder, J. Aspects of asymptotic safety for quantum gravity. *arXiv* **2018**, arXiv:1810.08550.
104. Eichhorn, A.; Lippoldt, S.; Pawlowski, J.M.; Reichert, M.; Schiffer, M. How perturbative is quantum gravity? *arXiv* **2018**, arXiv:1810.02828.
105. Sfondrini, A.; Koslowski, T.A. Functional Renormalization of Noncommutative Scalar Field Theory. *Int. J. Mod. Phys. A* **2011**, *26*, 4009–4051. [CrossRef]
106. Gastmans, R.; Kallosh, R.; Truffin, C. Quantum Gravity Near Two-Dimensions. *Nucl. Phys. B* **1978**, *133*, 417–434. [CrossRef]
107. Christensen, S.M.; Duff, M.J. Quantum Gravity in Two + ϵ Dimensions. *Phys. Lett. B* **1978**, *79*, 213–216. [CrossRef]
108. Kawai, H.; Ninomiya, M. Renormalization Group and Quantum Gravity. *Nucl. Phys. B* **1990**, *336*, 115–145. [CrossRef]
109. Codello, A.; Percacci, R.; Rahmede, C. Investigating the Ultraviolet Properties of Gravity with a Wilsonian Renormalization Group Equation. *Ann. Phys.* **2009**, *324*, 414–469. [CrossRef]
110. Falls, K. Renormalization of Newton's constant. *Phys. Rev. D* **2015**, *92*, 124057. [CrossRef]
111. Di Francesco, P.; Ginsparg, P.H.; Zinn-Justin, J. 2-D Gravity and random matrices. *Phys. Rep.* **1995**, *254*, 1–133. [CrossRef]
112. Ginsparg, P.H. Matrix Models of 2-d Gravity. Trieste HEP Cosmol, 1991, pp. 785–826. Available online: http://xxx.lanl.gov/abs/hep-th/9112013 (accessed on 30 January 2019).
113. Ambjorn, J. Quantization of geometry. In Proceedings of hte NATO Advanced Study Institute: Les Houches Summer School, Session 62: Fluctuating Geometries in Statistical Mechanics and Field Theory, Les Houches, France, 2 August–9 September 1994; pp. 77–195.
114. Marino, M. Les Houches lectures on matrix models and topological strings. *arXiv* **2004**, arXiv:0410165.
115. Alfaro, J.; Damgaard, P.H. The D = 1 matrix model and the renormalization group. *Phys. Lett. B* **1992**, *289*, 342–346. [CrossRef]
116. Ayala, C. Renormalization group approach to matrix models in two-dimensional quantum gravity. *Phys. Lett. B* **1993**, *311*, 55–63. [CrossRef]
117. Higuchi, S.; Itoi, C.; Sakai, N. Renormalization group approach to matrix models and vector models. *Prog. Theor. Phys. Suppl.* **1993**, *114*, 53–71. [CrossRef]
118. Higuchi, S.; Itoi, C.; Nishigaki, S.; Sakai, N. Renormalization group flow in one and two matrix models. *Nucl. Phys. B* **1995**, *434*, 283–318. [CrossRef]
119. Higuchi, S.; Itoi, C.; Nishigaki, S.; Sakai, N. Nonlinear renormalization group equation for matrix models. *Phys. Lett. B* **1993**, *318*, 63–72. [CrossRef]
120. Dasgupta, S.; Dasgupta, T. Renormalization group approach to c = 1 matrix model on a circle and D-brane decay. *arXiv* **2003**, arXiv:0310106.
121. Kazakov, V.A. The Appearance of Matter Fields from Quantum Fluctuations of 2D Gravity. *Mod. Phys. Lett. A* **1989**, *4*, 2125. [CrossRef]
122. Boettcher, I. Scaling relations and multicritical phenomena from Functional Renormalization. *Phys. Rev. E* **2015**, *91*, 062112. [CrossRef]
123. Le Guillou, J.C.; Zinn-Justin, J. Critical Exponents for the N Vector Model in Three-Dimensions from Field Theory. *Phys. Rev. Lett.* **1977**, *39*, 95–98. [CrossRef]
124. Guida, R.; Zinn-Justin, J. Critical exponents of the N vector model. *J. Phys. A* **1998**, *31*, 8103–8121. [CrossRef]
125. Carlip, S. Dimension and Dimensional Reduction in Quantum Gravity. *Class. Quant. Grav.* **2017**, *34*, 193001. [CrossRef]
126. Biemans, J.; Platania, A.; Saueressig, F. Quantum gravity on foliated spacetimes: Asymptotically safe and sound. *Phys. Rev. D* **2017**, *95*, 086013. [CrossRef]
127. Gurau, R.; Ryan, J.P. Melons are branched polymers. *Ann.s Henri Poincare* **2014**, *15*, 2085–2131. [CrossRef]
128. Ambjorn, J.; Jordan, S.; Jurkiewicz, J.; Loll, R. A Second-order phase transition in CDT. *Phys. Rev. Lett.* **2011**, *107*, 211303. [CrossRef]
129. Eichhorn, A. Status of the asymptotic safety paradigm for quantum gravity and matter. *Found. Phys.* **2018**, *48*, 1407–1429. [CrossRef]
130. Meibohm, J.; Pawlowski, J.M.; Reichert, M. Asymptotic safety of gravity-matter systems. *Phys. Rev. D* **2016**, *93*, 084035. [CrossRef]

131. Eichhorn, A.; Mesterházy, D.; Scherer, M.M. Multicritical behavior in models with two competing order parameters. *Phys. Rev. E* **2013**, *88*, 042141. [CrossRef]
132. Daum, J.E.; Reuter, M. Einstein-Cartan gravity, Asymptotic Safety, and the running Immirzi parameter. *Ann. Phys.* **2013**, *334*, 351–419. [CrossRef]
133. Surya, S. Directions in Causal Set Quantum Gravity. *arXiv* **2011**, arXiv:1103.6272.
134. Ambjorn, J.; Gorlich, A.; Jurkiewicz, J.; Loll, R. The Nonperturbative Quantum de Sitter Universe. *Phys. Rev. D* **2008**, *78*, 063544. [CrossRef]

© 2019 by the authors. Licensee MDPI, Basel, Switzerland. This article is an open access article distributed under the terms and conditions of the Creative Commons Attribution (CC BY) license (http://creativecommons.org/licenses/by/4.0/).

Review

Progress in Solving the Nonperturbative Renormalization Group for Tensorial Group Field Theory

Vincent Lahoche [1,*] and Dine Ousmane Samary [1,2,*]

1 Commissariat à l'Énergie Atomique (CEA, LIST), 8 Avenue de la Vauve, 91120 Palaiseau, France
2 Faculté des Sciences et Techniques/ICMPA-UNESCO Chair, Université d'Abomey-Calavi, Cotonou 072 BP 50, Benin
* Correspondence: vincent.lahoche@cea.fr (V.L.); dine.ousmanesamary@cipma.uac.bj (D.O.S.)

Received: 12 December 2018; Accepted: 18 March 2019; Published: 26 March 2019

Abstract: This manuscript aims at giving new advances on the functional renormalization group applied to the tensorial group field theory. It is based on the series of our three papers (Lahoche, et al., Class. Quantum Gravity 2018, 35, 19), (Lahoche, et al., Phys. Rev. D 2018, 98, 126010) and (Lahoche, et al., Nucl. Phys. B, 2019, 940, 190–213). We consider the polynomial Abelian $U(1)^d$ models without the closure constraint. More specifically, we discuss the case of the quartic melonic interaction. We present a new approach, namely the effective vertex expansion method, to solve the exact Wetterich flow equation and investigate the resulting flow equations, especially regarding the existence of non-Gaussian fixed points for their connection with phase transitions. To complete this method, we consider a non-trivial constraint arising from the Ward–Takahashi identities and discuss the disappearance of the global non-trivial fixed points taking into account this constraint. Finally, we argue in favor of an alternative scenario involving a first order phase transition into the reduced phase space given by the Ward constraint.

Keywords: nonperturbative renormalization group; quantum gravity; random geometry

1. Introduction

In seeking a theory to unify modern physics, i.e., a well-defined theory of quantum gravity, numerous contributions have been made. Despite the fact that none of them has given a complete resolution to the problem, several major advances have been observed. Of these advances, we count the very recent propositions such as loop quantum gravity [1,2], dynamical triangulation [3–5], noncommutative geometry [6,7], group field theories (GFTs) [8–12], and tensors models (TMs) [13–22]. These approaches are considered as new background independent approaches according to several theoreticians. GFTs are quantum field theories over the group manifolds and are considered as the second quantization version of loop quantum gravity [12]. These theories are characterized by the specific form of non-locality in their interactions. TMs, especially colored ones, allow one to define probability measures on simplicial pseudo-manifolds such that the tensor of rank d represents a $(d-1)$-simplex. TMs admit the large N-limit (N is the size of the tensor) dominated by the graphs called melons, thanks to the Gurau breakthrough [20–22]. The large N-limit or the leading order encodes a sum over a class of colored triangulations of the D-sphere, and its behavior is a powerful tool that allows us to understand the continuous limit of these models through, for instance, the study of critical exponents and phase transitions. TM and GFT are combined to give birth to a new class of field theories called tensorial group field theory. These class of field models enjoy renormalization and asymptotic freedom [23–38].

Using the functional renormalization group (FRG) method, it is also possible to identify the equivalent of the Wilson–Fisher fixed point for some particular cases of models.

There are several ways to introduce the FRG in field theories. The first approach is the one pioneered by Wilson, simple and intuitive and therefore yielding a powerful way to think about quantum field theories [39,40]. This method allows a smooth interpolation between the known microscopic laws IR-regime and the complicated macroscopic phenomena in physical systems' UV-regime and is constructed with the incomplete integration as the cutoff procedure. Well after, Polchinski provided a new approach called the Wilson–Polchinski FRG equation [41] to address the same question inspired by the Wilson method. This very practicable method may be integrated with an arbitrary cutoff function and expanded up to the next leading order of the derivative expansion. Despite the fact that all these approaches seem to be nonperturbative, in practice, the perturbative solution has appeared more attractive. More recently, the so-called Wetterich flow equation [42] was proposed to study the nonperturbative FRG, and this study requires approximations or truncations and numerical analysis, which is not very well controlled. The FRG equation allows determining the fixed points and probably the phase transition. These phase transitions in the case of TGFT models may help to identify the emergence of general relativity and quantum mechanics through the pre-geometrogenesis scenario [43–46]. Indeed, the way the quantum degrees of freedom are organized to shape a geometric structure that can be identified with a semi-classical space-time is one of the challenges for the GFT approach. In the geometrogenesis point of view, the standard space-time geometry is understood as an emergent property, the scenario leading to this geometric limit being assumed quite close to Bose–Einstein condensation in condensed matter physics. Evidences for this scenario were provided by FRG analysis [47–55]. In the recent works [56–58], the effective vertex expansion (EVE) method was used in the context of the FRG. This leads to the definition of a new class of equations called structure equations that help to solve the Wetterich flow equations. Taking into account the leading order contribution in the symmetric phase, the non-perturbative regime without truncation can be studied. The Ward–Takahashi (WT) identities are also derived [59–61] and become a constraint along the flow. Note that the WT-identities are universal for all field theories having a symmetry, and are not specific to TGFT. Therefore, all the fixed points must belong inside the domain of this constraint line, before being considered as an acceptable fixed point. In the case of quartic melonic TGFT models, it has been shown that the fixed point occurring from the solution of the Wetterich equation violates this constraint for any choice of the regulator function. This violation is also independent of the method used to find this fixed point, whether it is the truncation or the EVE method. This point will be discussed carefully in this note. Let us remark that most of the TGFT models previously studied in the literature were shown to admit at least a nontrivial fixed point and therefore a phase transition. The phase transitions are very useful in the likely emergence of the metric and are linked to the existence of fixed points, which becomes unavoidable in the search for models that may probably describe our universe after the geometrogenesis scenario. However, in this paper, we study the quartic T^4-TGFT models and prove that no fixed points can be found. First of all, we consider the Wilson–Polchinski renormalization group method and show the weakness of this method in the nonperturbative regime. Then, we consider the nonperturbative Wetterich flow equation from which the nonperturbative analysis can be made by an approximation on the average effective action, called truncation. The EVE method is used to get around the approximation and therefore solves the flow without truncation. The set of Ward–Takahashi identities and structure equations are derived to provide a nontrivial constraint on the reliability of the approximation schemes, i.e., the truncation and the choice of the regulator.

The paper is organized as follows: In Section 2, we recall the FGR method by Wilson–Polchinski and apply it in the context of TGFT. Despite the efficiencies of this method, we will present some questions that arise, in the search for a nonperturbative solution, and then, we will go further into the Wetterich flow equation. Section 3 is dedicated to the description of the Wetterich flow equation and the corresponding solution when the truncation method is applied. We also show that the only nontrivial fixed point, which comes from the solution of the flows, violates the Ward identities. In

Section 4, we perform new nonperturbative analysis using the so-called structure equations given, and the solutions of the flow equations are also derived. In the last Section 5, we provide a discussion and conclusion to our work.

2. Introduction to the Nonperturbative Renormalization for TGFT

FRG is a powerful ingredient to think about when it comes to quantum field theories. Generally, in every situation where the scale belongs to a range of correlated variables, the theory may be treated by the renormalization group (RG). The first conceptual framework is Wilson's version of the RG, which by Polchinski, may be applied in the case of quantum field theory. In this section, we discuss the nonperturbative renormalization group using not only the Wilson–Polchinski equation, but also the Wetterich flow equation. We discuss each method and consider the Wetterich flow equation as more suitable for the treatment of FRG applied to TGFT. Thanks to the Wilson method, the renormalization and renormalization group are understood as a coarse-graining process from a microscopic theory toward an effective long-distance theory. There are in fact different implementations of this idea, depending on the context. In the context of TGFT, we consider the pair of complex fields ϕ and $\bar\phi$, which take values of d-copies of arbitrary group G:

$$\phi, \bar\phi : G^d \to \mathbb{C}. \tag{1}$$

In a particular case, we assume that $G = U(1)$ is an Abelian compact Lie group. For the rest, we only consider the Fourier transform of the fields ϕ and $\bar\phi$ denoted by $T_{\vec p}$ and $\bar T_{\vec p}$, respectively, $\vec p \in \mathbb{Z}^d$, written as (for $\vec g \in U(1)^d, g_j = e^{i\theta_j}$):

$$\phi(\vec\theta) = \sum_{\vec p \in \mathbb{Z}^d} T_{\vec p}\, e^{i\sum_{j=1}^d \theta_j p_j}, \quad \bar\phi(\vec\theta) = \sum_{\vec p \in \mathbb{Z}^d} \bar T_{\vec p}\, e^{-i\sum_{j=1}^d \theta_j p_j}. \tag{2}$$

The description of the statistical field theory is given by the partition function $\mathcal{Z}[J, \bar J]$:

$$\mathcal{Z}[J, \bar J] = \int d\mu_C\, e^{-S_{int} + \langle J, \bar T \rangle + \langle T, \bar J \rangle}, \tag{3}$$

where S_{int} is the interaction functional action assumed to be tensor invariant, $J, \bar J$ the external currents, and $\langle J, \bar T \rangle$ a shorthand notation for:

$$\langle J, \bar T \rangle := \sum_{\vec p} J_{\vec p} \bar T_{\vec p}. \tag{4}$$

The Gaussian measure $d\mu_C$ is then fixed with the choice of the covariance C. In this paper, we adopt a Laplacian-type propagator of the form:

$$C(\vec p) = \frac{1}{\vec p^{\,2} + m^2} = \int d\mu_C\, T_{\vec p} \bar T_{\vec p}. \tag{5}$$

In order to prevent the UV divergences and suppress the high momenta contributions, the propagator (5) has to be regularized. In the usual case, the Schwinger regularization is used:

$$C_\Lambda(\vec p) = \frac{e^{-(\vec p^{\,2} + m^2)/\Lambda^2}}{\vec p^{\,2} + m^2}. \tag{6}$$

In the general case, by defining the function $\vartheta(t)$ such that the condition $|1 - \vartheta(t)| \leq C e^{-\kappa t}$ is satisfied for $C, \kappa > 0$ and $t \to +\infty$, we can write the propagator as a Laplace transform:

$$C_\Lambda(\vec p) = \int_0^{+\infty} dt\, \vartheta(t\Lambda^2)\, e^{-t(\vec p^{\,2} + m^2)}. \tag{7}$$

Then, we shall make the simplest choice $\vartheta(t) = \Theta(t-1)$, where $\Theta(t)$ is the Heaviside function, in order to recover the Schwinger regularization (6). For the rest, we keep in mind that the propagator is regularized, and the infinite limit will be given in an appropriate way. In this case, the following result in well satisfied:

Proposition 1. *Let us consider two non-normalized Gaussian measures $d\mu_C$ and $d\mu_{C'}$ whose covariances C and C' are related by $C' = C + \Delta$ and such that C, C', and Δ are assumed to be positive. Then, we get the following relation:*

$$\int d\mu_C(\bar{T}_1, T_1) d\mu_\Delta(\bar{T}_2, T_2) e^{-S_{int}(T_1+T_2,\bar{T}_1+\bar{T}_2)} = \left(\frac{\det(\Delta C)}{\det(C')}\right)^{1/2} \int d\mu_{C'}(\bar{T}, T) e^{-S_{int}(T,\bar{T})}, \quad (8)$$

where $T = T_1 + T_2$ and $\bar{T} = \bar{T}_1 + \bar{T}_2$.

Proof. The proof of this formula can simply be given using the definition of the Gaussian measure $d\mu_C$ with mean zero and covariance matrix C as:

$$d\mu_C = \det(\pi C)^{-\frac{1}{2}} e^{-\langle T, C^{-1}\bar{T}\rangle} dT\, d\bar{T}. \quad (9)$$

and the fact that:

$$\int d\mu_{C'}(T,\bar{T}) e^{-\langle J,\bar{T}\rangle - \langle T,\bar{J}\rangle} = e^{\langle J,C'\bar{J}\rangle} = e^{\langle J,C\bar{J}\rangle} e^{\langle J,\Delta\bar{J}\rangle}. \quad (10)$$

□

We introduce tensorial unitary invariants, or simply tensorial invariants. An invariant is a polynomial $P(T,\bar{T})$ in the tensor entries $T_{\vec{p}}$ and $\bar{T}_{\vec{p}}$ that is invariant under the following action of $U(N)^{\otimes d}$ (N being the size of the tensors):

$$T_{\vec{p}} \to \sum_{\vec{q}} U^{(1)}_{p_1 q_1} \cdots U^{(d)}_{p_d q_d} T_{\vec{q}}, \quad \bar{T}_{\vec{p}} \to \sum_{\vec{q}} \bar{U}^{(1)}_{p_1 q_1} \cdots \bar{U}^{(d)}_{p_d q_d} \bar{T}_{\vec{q}} \quad (11)$$

The algebra of invariant polynomials is generated by a set of polynomials labeled as bubbles. A bubble is a connected, bipartite graph, regular of degree d, whose edges must be colored with a color belonging to the set $\{1, \cdots, d\}$ and such that all d colors are incident at each vertex (and incident to exactly once). Examples of bubbles are displayed in Figure 1.

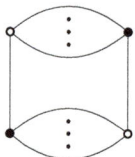

Figure 1. The four-vertex bubble for which the dots indicate multiple edges.

In this paper, we consider the quartic melonic T_5^4 model, which is proven to be renormalizable in all orders in the perturbative theory. The interaction of this model taking into account the leading order contributions (melon) is written graphically as:

$$S_{int} = \lambda_{41} \sum_{i=1}^{5} \quad \raisebox{-0.5em}{\includegraphics[height=2em]{melon}} \quad (12)$$

Note that the interaction (12) is invariant under the unitary transformations $\mathbf{U} \in U^{\otimes d}$. In contrast, this is not the case for the kinetic terms and source terms due to the non-trivial propagator and sources J and $\bar{J}$. This implies the existence of a non-trivial Ward-identity, which becomes a strong constraint and will be taking into account from the FRG point of view.

2.1. Wilson–Polchinski Equation

In this subsection, we discuss the Wilson–Polchinski RG equation and provide the corresponding solutions of the quartic melonic TGFT. For this, let us introduce a dilatation parameter $s < 1$. This parameter will be used as an evolution parameter in the integration around the UV modes. The RG idea is that if we want to describe the phenomena at scales down to s, then we should be able to use the set of variables defined at the scale s. Indeed, define the variation:

$$\Delta_{s,\Lambda}(\vec{p}) := C_\Lambda(\vec{p}) - C_{s\Lambda}(\vec{p}) \qquad (13)$$
$$= \int_0^{+\infty} dt \int_{s^2}^1 dx \frac{d}{dx} \vartheta(tx\Lambda^2) e^{-t(\vec{p}^2+m^2)}.$$

In the case where s is close to one, denoting by $D_{s,\Lambda}(\vec{p})$ the infinitesimal version of the above variation, we get:

$$\Delta_{s,\Lambda}(\vec{p}) \simeq \frac{2(1-s)}{\Lambda^2} e^{-(\vec{p}^2+m^2)/\Lambda^2} =: (1-s) D_{s,\Lambda}(\vec{p}), \qquad (14)$$

such that the partition function can be written as an integral over two fields, respectively associated with the "slow" and "rapid" modes. Starting with the partition function $\mathcal{Z}_\Lambda$ at scale Λ, we get:

$$\mathcal{Z}_\Lambda[S_{int}] := \int d\mu_{C_\Lambda}(\bar{T},T) e^{-S_{int,\Lambda}(T,\bar{T})}. \qquad (15)$$

Proposition 1 allows us to decompose $\mathcal{Z}_\Lambda[S_{int}]$ into two Gaussian integrals over two fields, $T_>$ and $T_<$, corresponding respectively to the "rapid" and "slow" modes, with covariances $\Delta_{s,\Lambda}$ and $C_{s\Lambda}$:

$$\mathcal{Z}_\Lambda[S_{int}] = \left(\frac{\det(\Delta_{s,\Lambda}C_{s\Lambda})}{\det(C_\Lambda)}\right)^{-1/2} \int d\mu_{C_{s\Lambda}}(\bar{T}_<,T_<) \int d\mu_{\Delta_{s,\Lambda}}(\bar{T}_>,T_>) e^{-S_{int}(T_<+T_>,\bar{T}_<+\bar{T}_>)}. \qquad (16)$$

Then, identify the effective action $S_{int,s\Lambda}$ at scale $s\Lambda$ as:

$$e^{-S_{int,s\Lambda}(T_<,\bar{T}_<)} := \frac{1}{\sqrt{\det \Delta_{s,\Lambda}}} \int d\mu_{\Delta_{s,\Lambda}}(\bar{T}_>,T_>) e^{-S_{int}(T_<+T_>,\bar{T}_<+\bar{T}_>)}, \qquad (17)$$

and the decomposition (16) becomes:

$$\mathcal{Z}_\Lambda = \left(\frac{\det C_{s\Lambda}}{\det C_\Lambda}\right)^{-1/2} \int d\mu_{C_{s\Lambda}}(\bar{T}_<,T_<) e^{-S_{int,s\Lambda}(T_<,\bar{T}_<)}. \qquad (18)$$

Now, for an infinitesimal step, keeping only the leading order terms in $1-s$ when s is very close to one, we find:

$$e^{-\Delta S_{int,\Lambda}(T_<,\bar{T}_<)} = 1 - \text{Tr}\left[\left(\frac{\delta^2 S_{int,\Lambda}}{\delta T \delta \bar{T}} - \frac{\delta S_{int,\Lambda}}{\delta T} \frac{\delta S_{int,\Lambda}}{\delta \bar{T}}\right) \Delta_{s,\Lambda}\right] + \mathcal{O}(1-s), \qquad (19)$$

with $\Delta S_{int,\Lambda}(T_<,\bar{T}_<) := S_{int,s\Lambda}(T_<,\bar{T}_<) - S_{int,\Lambda}(T_<,\bar{T}_<)$. At the same time, expanding the left-hand side of (18) in powers of $1-s$ and identifying the power of $1-s$ leads to:

$$\frac{dS_{int,s\Lambda}}{ds} = -\text{Tr}\left\{\left(\frac{\delta^2 S_{int,s\Lambda}}{\delta T \delta \bar{T}} - \frac{\delta S_{int,s\Lambda}}{\delta T} \frac{\delta S_{int,s\Lambda}}{\delta \bar{T}}\right) D_{s,\Lambda}\right\}. \qquad (20)$$

Graphically, this equation is given by (and is considered as the Wilson–Polchinski RG equation):

$$\frac{d}{ds}\bigcirc = \text{Tr}\left[\bigcirc - - - \bigcirc - \left(\bigcirc\right)\right]. \tag{21}$$

Note that we may consider Λ not only as a fundamental scale, but also as an arbitrary step in the flow, meaning that Equation (20) holds at each step of the flow. Physically, Equation (20) explains how the couplings are affected when the fundamental scale changes and is therefore the one pioneering ideas of the renormalization group flow firstly given by Wilson. This approach follows from a remarkably simple and intuitive idea and yields a very powerful way to think about quantum field theories. The relation (21) can be also expanded in the following result:

Proposition 2. *The set of Wilson–Polchinski renormalization group equations is given by:*

$$\frac{dV^{(n_l)}}{ds} = -\sum_{\vec{p}\vec{p}} D_{s,\Lambda,\vec{p}\vec{p}} \frac{\partial}{\partial T_{\vec{p}}}\frac{\partial}{\partial T_{\vec{p}}} V^{(n_l+1)} + \sum_{n_m=0}^{n_l-1}\sum_{\vec{p}\vec{p}} D_{s,\Lambda,\vec{p}\vec{p}} \frac{\partial V^{(n_m+1)}}{\partial T_{\vec{p}}}\frac{\partial V^{(n_l-n_m)}}{\partial T_{\vec{p}}} - n_l \eta_s V^{(n_l)}, \tag{22}$$

where $D_{s,\Lambda,\vec{p}\vec{p}} = D_{s,\Lambda}(\vec{p})\delta_{\vec{p}\vec{p}}$, $\eta_s := \frac{d}{ds}\ln Z(s)$. In this formula, we denote by n_l the number of black and white nodes in each interactions, and we consider the following expansion for $S_{int,s\Lambda}[T,\bar{T}]$:

$$S_{int,s\Lambda}[T,\bar{T}] = \sum_{n_l} \mathcal{V}^{(n_l)} = \sum_{n_l}\sum_{\{\vec{p}_i,\vec{p}_i\}} V^{(n_l)\,\vec{p}_1,\ldots,\vec{p}_l}_{\vec{p}_1,\ldots,\vec{p}_l} \prod_{i=1}^{l} T_{\vec{p}_i} \bar{T}_{\vec{p}_i}. \tag{23}$$

Proof. A pragmatic way to introduce field strength renormalization is the following. We consider a wave function $Z(s)$ and the regularized field $T = Z(s)^{\frac{1}{2}}\tilde{T}$ at the scale $s\Lambda$. A new functional $\tilde{S}_{int,s\Lambda}$ is associated to this field such as $\tilde{S}_{int,s\Lambda}[\tilde{T},\tilde{\bar{T}}] = S_{int,s\Lambda}[T,\bar{T}]$. Equation (20) is then modified into (we deleted the tildes notation):

$$\begin{aligned}\frac{dS_{int,s\Lambda}}{ds} = &-\text{Tr}\left\{\left(\frac{\delta^2 S_{int,s\Lambda}}{\delta T \delta \bar{T}} - \frac{\delta S_{int,s\Lambda}}{\delta T}\frac{\delta S_{int,s\Lambda}}{\delta \bar{T}}\right)D_{s,\Lambda}\right\} \\ &-\tfrac{1}{2}\eta_s\left[\text{Tr}\left(\frac{\delta S_{int,s\Lambda}}{\delta T}T\right) + \text{Tr}\left(\bar{T}\frac{\delta S_{int,s\Lambda}}{\delta \bar{T}}\right)\right].\end{aligned} \tag{24}$$

Then, by considering the following expansion for $S_{int,s\Lambda}[T,\bar{T}]$:

$$S_{int,s\Lambda}[T,\bar{T}] = \sum_{n_l}\mathcal{V}^{(n_l)} = \sum_{n_l}\sum_{\{\vec{p}_i,\vec{p}_i\}} V^{(n_l)\,\vec{p}_1,\ldots,\vec{p}_l}_{\vec{p}_1,\ldots,\vec{p}_l}\prod_{i=1}^{l} T_{\vec{p}_i}\bar{T}_{\vec{p}_i}, \tag{25}$$

we get the relation (22). □

The Wilson–Polchinski equation is a leading order equation in the perturbation rather than the loop expansion. Note that we can show that this equation can be turned into a Fokker–Planck equation, and therefore, it may be formally solved by a standard method. The rest of this section is devoted to a perturbative analysis of the flow equations. Before starting this computation, we have to make the approximation regime precise. We shall consider only the UV limit, which corresponds to the higher values of the scale parameter s, or to the higher momenta variables $\vec{p}$, or also for the smaller distances, and we assume that $s\Lambda$ and Λ are large. However, the analysis in the UV regime can be extended to the IR limit, which corresponds to the smaller values of the scale parameter s. More precisely, our approximation can be characterized by both $s\Lambda$ and Λ in the UV and by $s\Lambda/\Lambda$ in the

IR. At scale Λ and up to contributions of order λ_{41}^2, keeping only the melonic contribution, the action provided from (12) is assumed to be of the form:

$$S^4_{int,s\Lambda}[\bar{T},T] = \delta m^2 \sum_{\vec{p}} \bar{T}_{\vec{p}} T_{\vec{p}} + \delta Z \sum_{\vec{p}} \vec{p}^{\,2} \bar{T}_{\vec{p}} T_{\vec{p}} + \lambda_{41} \sum_{i=1}^{5} \sum_{\{\vec{p}_i,\vec{q}_i\}} W^{(i)}_{\vec{p}_1,\vec{q}_1;\vec{p}_2,\vec{q}_2} T_{\vec{p}_1} T_{\vec{p}_2} \bar{T}_{\vec{q}_1} \bar{T}_{\vec{q}_2}, \quad (26)$$

where the first two terms take into account the fact that the parameter of the Gaussian measure, the mass, and the Laplacian term can be affected by the integration of the UV modes, and these counter-terms, assumed to be of order λ_{41}, take into account these modifications. The vertex $W^{(i)}_{\vec{p}_1,\vec{q}_1;\vec{p}_2,\vec{q}_2}$ is a product of the delta function and is given by:

$$W^{(i)}_{\vec{p}_1,\vec{q}_1;\vec{p}_2,\vec{q}_2} = \delta_{p_{1i}q_{2i}} \delta_{q_{1i}p_{2i}} \prod_{j\neq i} \delta_{p_{1j}q_{1j}} \delta_{p_{2j}q_{2j}}. \quad (27)$$

Moreover, note that in this approach, the corrections to the Laplacian term are not suppressed by an effective counter-term in the action, but absorbed in the wave function renormalization. It is fixed such that all the Laplacian corrections are canceled by the η_s term in the RG equation for $\mathcal{V}^{(1)}$. We adopt the standard ansatz, namely that the generic interaction of valence n is of order $\lambda_{41}^{n/2-1}$. This allows us to organize systematically the perturbative solution, for which we shall construct the λ_{41}^2 order.

2.1.1. $\mathcal{V}^{(1)}$ at Order λ_{41}

The first corrections occur at order λ_{41} for $\mathcal{V}^{(1)}$, whose flow equation is written as:

$$\left(\frac{d}{ds} + \eta_s\right) \mathcal{V}^{(1)} = -4\lambda_{41} \sum_{\substack{\vec{p}_1,\vec{q}_1 \\ \vec{p}_2,\vec{q}_2}} D_{s\Lambda\,\vec{p}_1,\vec{p}_1} \text{Sym} W^{(i)}_{\vec{p}_1,\vec{q}_1;\vec{p}_2,\vec{q}_2} T_{\vec{p}_2} \bar{T}_{\vec{q}_2}, \quad (28)$$

where:

$$\text{Sym} W_{\vec{p}_1,\vec{q}_1;\vec{p}_2,\vec{q}_2} = W_{\vec{p}_1,\vec{q}_1;\vec{p}_2,\vec{q}_2} + W_{\vec{p}_2,\vec{q}_1;\vec{p}_1,\vec{q}_2} \quad (29)$$

and $\text{Sym}\mathcal{W} := \sum_i \text{Sym} W^{(i)}$ and $\mathcal{W} = \sum_{i=1}^{6} W^{(i)}$. The r.h.s involves two typical contributions, which are pictured graphically in Figure 2, where the contraction with $D_{s,\Lambda}$ is represented by a dotted line with a gray box.

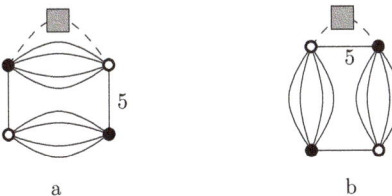

Figure 2. The two graphs contributing to the interaction $\mathcal{V}^{(1)}$ of degree two. Melonic (a) and non-melonic (b).

In the UV limit that we consider, the non-melonic contractions of the type on Figure 2b, creating only one internal face (of the color five on this figure), can be neglected in comparison to the melonic contributions of the form of Figure 2a. Retaining only the melonic contractions, Equation (28) becomes:

$$\left(\frac{d}{ds} + \eta_s\right) \mathcal{V}^{(1)} = -2\lambda_{41} \sum_{\substack{\vec{p}_1,\vec{q}_1 \\ \vec{p}_2,\vec{q}_2}} D_{s\Lambda,\vec{p}_1\vec{q}_1} W_{\vec{p}_1,\vec{q}_1;\vec{p}_2,\vec{q}_2} T_{\vec{p}_2} \bar{T}_{\vec{q}_2}, \quad (30)$$

with $D_{s,\Lambda} = dC_{s\Lambda}/ds$. Expanding this relation in powers of p_5, we generate mass and wave function corrections, and also the sub-dominant corrections, involving powers of p_5 greater than two. They correspond to the first deviation to the original form (26). Neglecting these sub-dominant contributions, we get the expansion:

$$\sum_{p_1,\ldots,p_4} \frac{2}{s^3 \Lambda^2} e^{-\frac{1}{(s\Lambda)^2}(\vec{p}^2 + m^2)} \sim 2\pi^2 s \Lambda^2 - \frac{2\pi^2}{s}(p_5^2 + m^2) + \mathcal{O}(s), \tag{31}$$

for which we only keep the leading order terms in s, and we can extract the dominant contributions to the mass and wave-function renormalization. The term in p_5^2 generates a non-local two-point interaction of the form $-\delta Z(s)\mathrm{Tr}(\bar{T}\Delta_{\vec{g}}T)$, where Δ_g is the Laplacian on $U(1)^{\times 5}$, and the first term generates a mass correction. Summing over the five colors, we find, at first order in λ_{41}:

$$\eta_s = \frac{4\pi^2 \lambda_{41}}{s}, \quad \frac{d}{ds}\delta m^2 = -4\pi^2 \lambda_{41} s \Lambda^2 + \frac{4\pi^2 \lambda_{41}}{s} m^2. \tag{32}$$

2.1.2. $\mathcal{V}^{(3)}$ and $\mathcal{V}^{(2)}$ at Order λ_{41}^2

Let us focus on the second order perturbative solution, i.e., at λ_{41}^2, in which we have to take into account the contributions of interactions of valence six, $\mathcal{V}^{(3)}$, verifying the flow equation:

$$\frac{d\mathcal{V}^{(3)\,\vec{q}_1,\vec{q}_2,\vec{q}_3}_{\vec{p}_1,\vec{p}_2,\vec{p}_3}}{ds} = 4\lambda_{41}^2 \sum_{i,j,\vec{p},\vec{q}} W^{(i)}_{\vec{p}_1,\vec{q}_1,\vec{p},\vec{q}_2} W^{(j)}_{\vec{p}_2,\vec{q}_3;\vec{p}_3,\vec{q}} D_{s,\Lambda,\vec{p}\vec{q}}, \tag{33}$$

which can be easily integrated with the initial condition $\mathcal{V}^{(3)\,\vec{q}_1,\vec{q}_2,\vec{q}_3}_{\vec{p}_1,\vec{p}_2,\vec{p}_3}(1) = 0$ as:

$$\mathcal{V}^{(3)\,\vec{q}_1,\vec{q}_2,\vec{q}_3}_{\vec{p}_1,\vec{p}_2,\vec{p}_3}(s) = -4\lambda_{41}^2 \sum_{i,j,\vec{p},\vec{q}} W^{(i)}_{\vec{p}_1,\vec{q}_1,\vec{p},\vec{q}_2} W^{(j)}_{\vec{p}_2,\vec{q}_3;\vec{p}_3,\vec{q}} (C_\Lambda - C_{s\Lambda})_{\vec{p}\vec{q}}. \tag{34}$$

As for the interaction of degree one, the structure of this effective interaction can be understood as a contraction between two bubbles, as pictured in Figure 3, where the dotted line with a gray box represents the contraction with $C_\Lambda - C_{s\Lambda}$.

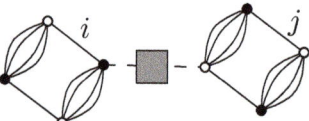

Figure 3. Typical graph contributing to the interaction of $\mathcal{V}^{(3)}$ of degree six.

Let us now build the effective coupling for the quartic melonic interaction at order λ_{41}^2, for which we shall extract only the leading behavior. From the Wilson–Polchinski flow Equation (22), it seems that the coupling evolution receives many contributions in which the first one comes from $\mathcal{V}^{(3)}$. Now, deriving two times this interaction with respect to the fields, we obtain an interaction of degree two, which can be either 1PI, when the contraction with $D_{s\Lambda}$ links two black and white nodes of two different bubbles, or one-particle reducible (1PR) if the two nodes stand on the same interaction bubble. Explicitly, we get:

$$\left[\frac{d}{ds} + 2\eta_s - 4\delta m^2 \bar{D}_{s,\Lambda}[\{p_i\},\{q_i\}]\right] \lambda_{41} W^{(i)}_{\vec{p}_2,\vec{q}_1;\vec{p}_3,\vec{q}_2} = 4\lambda_{41}^2 \sum_{\vec{p},\vec{q}\vec{p}',\vec{q}'} \left[\mathrm{S\bar{y}m}\left(W^{(i)}_{\vec{p}',\vec{q}_1;\vec{p},\vec{q}_2} W^{(i)}_{\vec{p}_2,\vec{q}';\vec{p}_3,\vec{q}}\right)\right. \\ \left. + 2\sum_j \mathrm{S\bar{y}m}\left(W^{(i)}_{\vec{p}_2,\vec{q}_1;\vec{p},\vec{q}_2} W^{(j)}_{\vec{p}',\vec{q}';\vec{p}_3,\vec{q}}\right)\right] \times (C_\Lambda - C_{s\Lambda})_{\vec{p}\vec{q}} D_{s,\Lambda,\vec{p}'\vec{q}'}, \tag{35}$$

where:

$$\widetilde{\text{Sym}}\left(\mathcal{W}^{(i)}_{\vec{p}',\vec{q}_1;\vec{p},\vec{q}_2}\mathcal{W}^{(j)}_{\vec{p}_2,\vec{q}';\vec{p}_3,\vec{q}}\right) := \mathcal{W}^{(i)}_{\vec{p}',\vec{q}_1;\vec{p},\vec{q}_2}\mathcal{W}^{(j)}_{\vec{p}_2,\vec{q}';\vec{p}_3,\vec{q}} + \mathcal{W}^{(i)}_{\vec{p}',\vec{q}_1;\vec{p}',\vec{q}_2}\mathcal{W}^{(j)}_{\vec{p}_3,\vec{q}';\vec{p}_2,\vec{q}'} \quad (36)$$

and:

$$\bar{D}_{s,\Lambda}[\{p_i\},\{q_i\}] := D_{s,\Lambda}(\vec{p}_2) + D_{s,\Lambda}(\vec{q}_1) + D_{s,\Lambda}(\vec{p}_3) + D_{s,\Lambda}(\vec{q}_2). \quad (37)$$

Equation (35) gives the exact behavior for the beta function at order λ_{41}^2, but we can easily see that it reduces to the expression of the beta function already obtained for the one-loop computation in the deep UV sector. Indeed, retaining only the melonic contributions and noting that 1PR contributions of the r.h.s are exactly canceled by the term involving the mass correction δm in the l.h.s, we get:

$$\left[\frac{d}{ds} + 2\eta_s\right]\lambda_{41}\mathcal{W}^{(i)}_{\vec{p}_2,\vec{p}_3;\vec{q}_1,\vec{q}_2} \approx 4\lambda_{41}^2 \sum_{\vec{p},\vec{q}\vec{p}',\vec{q}'} \mathcal{W}^{(i)}_{\vec{p}',\vec{p};\vec{q}_1,\vec{q}_2} \times \mathcal{W}^{(i)}_{\vec{p}_2,\vec{p}_3;\vec{q}',\vec{q}}(C_\Lambda - C_{s\Lambda})_{\vec{p}\vec{q}}D_{s,\Lambda,\vec{p}'\vec{q}'}. \quad (38)$$

The computation of the loop appearing on the r.h.s leads to:

$$\sum_{p_1,\ldots,p_4}\int_1^s ds' \frac{4}{s'^3 s^3 \Lambda^4} e^{-\left(\frac{1}{(s\Lambda)^2}+\frac{1}{(s'\Lambda)^2}\right)(\vec{p}^2+m^2)} \sim -\frac{\pi^2}{s} + \mathcal{O}(s), \quad (39)$$

from which we finally deduce that:

$$s\frac{d\lambda_{41}}{ds} = -4\pi^2\lambda_{41}^2 \quad (40)$$

which, as claimed before, is exactly the value of the one-loop beta function already obtained in the one-loop computation of the beta function.

We conclude that the main advantage of the Wilson–Polchinski equation is that it provides a very well-defined interpretation of the renormalization group flow in the space of couplings. However, except for perturbative computations, the Wilson–Polchinski equation is more adapted to mathematical and formal proofs than to non-perturbative analysis. The analysis beyond the perturbative level requires another formulation of the coarse-graining renormalization group, called *Wetterich equation*, which allows one usually to better capture the non-perturbative effects. The price to pay is an approximation scheme that is a bit more difficult to use. This non-perturbative approach to the renormalization group flow will be the subject of the next sections.

3. Wetterich Flow Equation

The Wetterich method and its incarnation into the FRG approach are a set of techniques allowing one to go beyond the difficulties coming from the Wilson–Polchinski equation, in particular in regards to tracking non-perturbative aspects. The Wetterich equation is a first order functional integro-differential equation for the effective action. The central object of the method is a continuous set of models labeled with a real parameter s running from UV scales ($s \to +\infty$) to the IR scales ($s \to -\infty$). The physical running scale e^s defined for each models what is UV and what is IR, the fluctuation with a large size with respect to the referent scale (the UV fluctuations) being integrated out. The renormalization group equation then describes how the coupling constant changes when the referent scale changes. Each model is characterized by a specific partition function $\mathcal{Z}_s$, labeled by s and defined as:

$$\mathcal{Z}_s[J,\bar{J}] := \int d\mu_C \, e^{-S_{int}(T,\bar{T})+R_s[T,\bar{T}]+\langle J,\bar{T}\rangle+\langle T,\bar{J}\rangle}. \quad (41)$$

As a result, the original model corresponds to $R_s[T,\bar{T}] = 0$, and because physically, this limit has to match with the IR limit $e^s \to 0$, we require that $R_s[T,\bar{T}]$ vanish in the same limit. The term $R_s[T,\bar{T}]$, called the *IR regulator*, plays the same role as a momentum-dependent mass term, becoming very large in the UV and vanishing in the IR. It is chosen ultra-local in the usual sense:

$$R_s[T, \bar{T}] := \sum_{\vec{p}} \bar{T}_{\vec{p}} r_s(\vec{p}) T_{\vec{p}}, \qquad (42)$$

the regulating function $r_s(\vec{p})$ being chosen to satisfy the boundary conditions in the UV/IR limit. Moreover, for s fixed, r_s aims at freezing the long-distance fluctuations, which are discarded from the functional integration. In formula: $r_s(\vec{p}) \to 0$ for $|\vec{p}|/e^s \to 0$, and $r_s(\vec{p}) \gg 1$ in the opposite limit.

The object for which we track the evolution is called *effective averaged action* Γ_s, defined as (a slightly modified version of) the Legendre transform of the standard free energy $W_s = \ln \mathcal{Z}_s$:

$$\Gamma_s[M, \bar{M}] = \langle \bar{J}, M \rangle + \langle \bar{M}, J \rangle - W_s[J, \bar{J}] - R_s[M, \bar{M}]. \qquad (43)$$

This definition ensures that Γ_s satisfies the physical boundary conditions $\Gamma_{s=\ln \Lambda} = S$, $\Gamma_{s=-\infty} = \Gamma$, where Λ denotes some fundamental UV cutoff. The fields M and $\bar{M}$ are the mean values of T and $\bar{T}$ respectively and are given by:

$$M = \frac{\partial W}{\partial \bar{J}}, \quad \bar{M} = \frac{\partial W}{\partial J} \qquad (44)$$

where $W := W_{s=-\infty}$. In general, the regulator r_s is chosen to be $r_s = Z(s) k^2 f\left(\frac{\vec{p}^2}{k^2}\right)$, $k = e^s$, and such that the boundary conditions in the UV/IR limit are well satisfied. Taking the first derivative with respect to the flow parameter s, one can deduce the Wetterich equation, describing the behavior of the effective action Γ_s when s changes [62–66]:

$$\partial_s \Gamma_s = \mathrm{Tr}\, \partial_s r_s (\Gamma_s^{(2)} + r_s)^{-1}, \qquad (45)$$

where $\Gamma_s^{(2)}$ denotes the second order partial derivative of Γ_s with respect to the mean fields M and $\bar{M}$. This equation is exact, but generally impossible to solve exactly. A large part of the FRG approach is then devoted to approximate the exact trajectory of the RG flow. In this review, we will discuss two methods, the truncation method and the effective vertex expansion method.

This section is especially devoted to the truncations. The general strategy is to cut crudely into the full theory space, projecting the flow systematically into the interior of a finite dimensional subspace. The average effective action is chosen to be of the form:

$$\Gamma_s = Z(s) \sum_{\vec{p} \in \mathbb{Z}^d} \bar{T}_{\vec{p}} (\vec{p}^2 + m^2(s)) T_{\vec{p}} + \sum_n^N \lambda_n V_n(T, \bar{T}) \qquad (46)$$

where N is finite, V_n stands for the interaction function of order n, and m^2 and λ_n are the mass and coupling constants. With this truncation and with an appropriate regulator, it is possible to solve the Wetterich flow Equation (45). In the case of quartic melonic interaction and by taking the standard modified Litim regulator [65,67,68]

$$r_s(\vec{p}) = Z(s)(e^{2s} - \vec{p}^2) \Theta(e^{2s} - \vec{p}^2) \qquad (47)$$

the Wetterich equation can be solved analytically and the phase diagram may be given [56–58]. The corresponding nontrivial fixed points can be studied taking into account the behavior of the flow around these points. Note that the validity of the fixed point requires some analysis taking into account the Ward–Takahashi identities as a new constraint along the flow line. The full violation of this constraint for quartic melonic interaction makes this class of fixed points unphysical. We discuss this point in detail in this section (for more detail, see Section 3). Taking into account only the relevant contributions for large k (in the deep UV), the flow equations are written as:

$$\begin{cases} \dot{m}^2 &= -2d\lambda I_2(0) \\ \dot{Z}(s) &= -2\lambda I'_2(q=0) \\ \dot{\lambda}_{41} &= 4\lambda_{41}^2 I_3(0) \end{cases} \quad (48)$$

with the renormalization condition:

$$m^2(s) = \Gamma_s^{(2)}(\vec{p}=\vec{0}), \quad \lambda_{41}(s) = \frac{1}{4}\Gamma_s^{(4)}(\vec{0},\vec{0},\vec{0},\vec{0}). \quad (49)$$

where:

$$I_n(q) = \sum_{\vec{p}\in\mathbb{Z}^{(d-1)}} \frac{\dot{r}_s}{(Z(s)\vec{p}^2 + Zq^2 + m^2 + r_s)^n}. \quad (50)$$

Explicitly using the integral representation of the above sum and with $d=5$, $\eta = \dot{Z}/Z$, we get:

$$I_n(0) = \frac{\pi^2 e^{6s-2ns}}{6Z(s)^{n-1}(\bar{m}^2+1)^n}(\eta+6), \quad I'_n(0) = -\frac{\pi^2 e^{4s-2ns}}{2Z(s)^{n-1}(\bar{m}^2+1)^n}(\eta+4). \quad (51)$$

In order to get an autonomous system, the standard strategy consist at extracting from the couplings the part coming from their own scaling, defining their canonical dimension. Strictly speaking, fields, couplings, and all the parameters involved in the theory are dimensionless, because there is no referent space-time, and then no referent scale. The canonical dimension emerges taking into account quantum corrections and is usually defined as the optimal scaling, with respect to the UV cut-off of the quantum corrections. Conversely, it can be defined as the scaling transformation allowing one to get an autonomous system. Note that these two points of view are note strictly equivalent, especially with respect to the choice of the initial content of the theory. For our purpose however, the two strategies provide exactly the same rescaling, and in terms of dimensionless parameter $\bar{\lambda}_{41} =: Z^2\lambda_{41}$, $\bar{m}^2 =: e^{2s}Z\bar{m}^2$, the system (48) becomes:

$$\begin{cases} \beta_m &= -(2+\eta)\bar{m}^2 - 2d\bar{\lambda}\frac{\pi^2}{(1+\bar{m}^2)^2}\left(1+\frac{\eta}{6}\right), \\ \beta_{41} &= -2\eta\bar{\lambda} + 4\bar{\lambda}^2\frac{\pi^2}{(1+\bar{m}^2)^3}\left(1+\frac{\eta}{6}\right), \end{cases} \quad (52)$$

where $\beta_m := \dot{\bar{m}}^2$, $\beta_{41} := \dot{\bar{\lambda}}$ and:

$$\eta := \frac{4\bar{\lambda}\pi^2}{(1+\bar{m}^{2a})^2 - \bar{\lambda}\pi^2}. \quad (53)$$

The solutions of the system (52) are given analytically:

$$p_{\pm} = \left(\bar{m}_{\pm}^2 = -\frac{23 \mp \sqrt{34}}{33}, \bar{\lambda}_{41,\pm} = \frac{328 \mp 8\sqrt{34}}{11979\pi^2}\right). \quad (54)$$

Numerically:

$$p_{+} = (-0.52, 0.0028), \quad p_{-} = (-0.87, 0.0036). \quad (55)$$

Apart from the fact that we have a singularity line around the point $\bar{m}^2 = -1$ in the flow Equation (48), another second singularity arises from the anomalous dimension denominator and corresponds to a line of singularity, with equation:

$$\Omega(\bar{m}, \bar{\lambda}) := (\bar{m}^2 + 1)^2 - \pi^2\bar{\lambda}_{41} = 0 \quad (56)$$

This line of singularity splits the two-dimensional phase space of the truncated theory into two connected regions characterized by the sign of the function Ω: the region I, connected to the Gaussian fixed point for $\Omega > 0$ and the region II for $\Omega < 0$. For $\Omega = 0$, the flow becomes ill defined. The existence of this singularity is a common feature for expansions around the vanishing mean field, and the region I may be viewed as the domain of validity of the expansion in the symmetric phase. Note that to ensure the positivity of the effective action, the melonic coupling must be positive, as well. Therefore, we expect that the physical region of the reduced phase space corresponds to the region $\lambda_{41} \geq 0$. From the definition of the connected region I and because of the explicit expression (53), we deduce that:

$$\eta \geq 0, \quad \text{on the symmetric phase}. \tag{57}$$

Then, only the fixed point p_+ is taken into account. In the next subsection, we will discuss the violation of the Ward identity around this fixed point p_+ and then clarify our analysis given in [56]. The phase diagram is given in Figure 4.

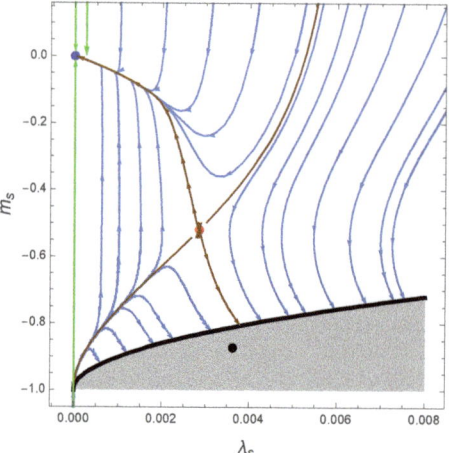

Figure 4. Renormalization group flow trajectories around the relevant fixed points obtained from a numerical integration. The Gaussian fixed point and the first non-Gaussian fixed point are respectively in blue and in red, and the last fixed point is in black. This fixed point is in the grey region bounded by the singularity line corresponding to the denominator of η. Finally, in green and brown, we draw the eigendirections around Gaussian and non-Gaussian fixed points, respectively. Note that that arrows of these fixed points are the flow oriented from IR to UV.

Convenient Search of the Ward Identities

Let $\mathcal{U} = (U_1, U_2, \cdots, U_d)$, where the $U_i \in U_\infty$ are infinite size unitary matrices in momentum representation. We define the transformation:

$$\mathcal{U}[T]_{\vec{p}} = \sum_{\vec{q}} U_{1,p_1q_1} U_{2,p_2q_2} \cdots U_{d,p_dq_d} T_{\vec{q}}, \tag{58}$$

such that the interaction term is invariant, i.e., $\mathcal{U}[S_{int}] = S_{int}$. Then, consider an infinitesimal transformation:

$$\mathcal{U} = \mathbf{I} + \vec{\epsilon}, \quad \vec{\epsilon} = \sum_i \mathbb{I}^{\otimes(i-1)} \otimes \epsilon_i \otimes \mathbb{I}^{\otimes(d-i)}, \tag{59}$$

where $\mathbb{I}$ is the identity on U_∞, $\mathbf{I} = \mathbb{I}^{\otimes d}$ the identity on $U_\infty^{\otimes d}$, and ϵ_i denotes the skew-symmetric Hermitian matrix such that $\epsilon_i = -\epsilon_i^\dagger$ and $\vec{\epsilon}_i[T]_{\vec{p}} = \epsilon_{i p_i q_i} T_{p_1, \cdots, q_i, \cdots, p_d}$. The invariance of the path integral (3) means $\vec{\epsilon}\left[\mathcal{Z}_s[J,\bar{J}]\right] = 0$, i.e.:

$$\vec{\epsilon}\left[\mathcal{Z}_s[J,\bar{J}]\right] = \int dT d\bar{T} \left[\vec{\epsilon}\left[S_{kin}\right] + \vec{\epsilon}\left[S_{int}\right] + \vec{\epsilon}\left[S_{source}\right]\right] e^{-S_s[T,\bar{T}] + \langle J,T\rangle + \langle \bar{T},\bar{J}\rangle} = 0. \tag{60}$$

Computing each term separately, we get, successively using the linearity of the operator $\vec{\epsilon}$:

$$\vec{\epsilon}[S_{int}] = 0, \tag{61}$$

$$\vec{\epsilon}[S_{source}] = -\sum_{i=1}^d \sum_{\vec{p},\vec{q}} \prod_{j \neq i} \delta_{p_j q_j} [\bar{J}_{\vec{p}} T_{\vec{q}} - \bar{T}_{\vec{p}} J_{\vec{q}}] \epsilon_{i p_i q_i}, \tag{62}$$

$$\vec{\epsilon}[S_{kin}] = \sum_{i=1}^d \sum_{\vec{p},\vec{q}} \prod_{j \neq i} \delta_{p_j q_j} \bar{T}_{\vec{p}} [C_s(\vec{p}^{\,2}) - C_s(\vec{q}^{\,2})] T_{\vec{q}}\, \epsilon_{i p_i q_i}, \tag{63}$$

where $\prod_{j \neq i} \delta_{p_j q_j} := \delta_{\vec{p}_{\perp_i} \vec{q}_{\perp_i}}$, $\vec{p}_{\perp_i} := \vec{p} \setminus \{p_i\}$, $C_s^{-1} = C_\infty^{-1} + r_s$, and $C_\infty^{-1} = Z_\infty \vec{p}^{\,2} + m_\infty^2$. Z_∞ is the renormalized wave function usually denoted by Z. We get the following result:

Proposition 3. *The Ward identity gives the relation between two- and four-point functions as:*

$$\sum_{\vec{r}_{\perp_i}, \vec{s}_{\perp_i}} \delta_{\vec{r}_{\perp_i} \vec{s}_{\perp_i}} (C_s^{-1}(\vec{r}) - C_s^{-1}(\vec{s})) \langle T_{\vec{r}} \bar{T}_{\vec{s}} T_{\vec{p}} \bar{T}_{\vec{q}} \rangle = -\delta_{\vec{p}_{\perp_i} \vec{q}_{\perp_i}} (G_s(p) - G_s(q)) \delta_{r_i s_i}, \tag{64}$$

where, defined by $\Gamma_s^{(4)}$, the 1PI four-point function, we get:

$$\langle T_{\vec{r}} \bar{T}_{\vec{s}} T_{\vec{p}} \bar{T}_{\vec{q}} \rangle = \Gamma_{s, \vec{r}\vec{s}; \vec{p}\vec{q}}^{(4)} \left(G_s(\vec{p}) G_s(\vec{q}) + \delta_{\vec{r}\vec{p}} \delta_{\vec{s}\vec{q}}\right) G_s(\vec{r}) G_s(\vec{s}) \tag{65}$$

Proof. The formal invariance of the path integral implies that the variations of these terms have to be compensated by a nontrivial variation of the source terms. Combining the expressions (60)–(63), we come to:

$$\sum_{i=1}^d \sum_{\vec{p}_{\perp_i}, \vec{q}_{\perp_i}} \delta_{\vec{p}_{\perp_i} \vec{q}_{\perp_i}} \left[\frac{\partial}{\partial J_{\vec{p}}} [C_s(\vec{p}^{\,2}) - C_s(\vec{q}^{\,2})] \frac{\partial}{\partial \bar{J}_{\vec{q}}} - \bar{J}_{\vec{p}} \frac{\partial}{\partial \bar{J}_{\vec{q}}} + J_{\vec{q}} \frac{\partial}{\partial J_{\vec{p}}} \right] e^{W_s[J,\bar{J}]} = 0, \tag{66}$$

where we have used the fact that, for all polynomial $P(T,\bar{T})$, the following identity holds:

$$\int d\mu_C\, P(T,\bar{T}) e^{\langle J,T\rangle + \langle \bar{T},\bar{J}\rangle} = \int d\mu_C\, P\left(\frac{\partial}{\partial \bar{J}}, \frac{\partial}{\partial J}\right) e^{\langle J,T\rangle + \langle \bar{T},\bar{J}\rangle}. \tag{67}$$

Equation (66) is satisfied for all i. Now, expanding each derivative, the partition function $\mathcal{Z}_s[J,\bar{J}] =: e^{W_s[J,\bar{J}]}$ of the theory defined by the action (12) verifies the following (WT identity),

$$\sum_{\vec{p}_{\perp_i}, \vec{q}_{\perp_i}} \delta_{\vec{p}_{\perp_i} \vec{q}_{\perp_i}} \left\{ [C_s(\vec{p}^{\,2}) - C_s(\vec{q}^{\,2})] \left(\frac{\partial^2 W_s}{\partial \bar{J}_{\vec{q}} \partial J_{\vec{p}}} + \bar{M}_{\vec{p}} M_{\vec{q}}\right) - \bar{J}_{\vec{p}} M_{\vec{q}} + J_{\vec{q}} \bar{M}_{\vec{p}} \right\} = 0. \tag{68}$$

The WI-identity contains some information on the relations between the Green functions. In particular, they provide a relation between four- and two-points functions, which may be translated as a relation between wave function renormalization Z and vertex renormalization Z_λ. Applying $\partial^2/\partial M_{\vec{p}}\partial \bar{M}_{\vec{q}}$ on the left-hand side of (68) and taking into account the relations:

$$\frac{\partial M_{\vec{p}}}{\partial J_{\vec{q}}} = \frac{\partial^2 W_s}{\partial \bar{J}_{\vec{p}} \partial J_{\vec{q}}} \quad \text{and} \quad \frac{\partial \Gamma_s}{\partial M_{\vec{p}}} = \bar{J}_{\vec{p}} - r_s(\vec{p})\bar{M}_{\vec{p}}, \tag{69}$$

as well as the definition $G_{s,\vec{p}\vec{q}}^{-1} := (\Gamma_s^{(2)} + r_s)_{\vec{p}\vec{q}}$, we find that:

$$\begin{aligned}\sum_{\vec{p}_{\perp i}, \vec{q}_{\perp i}} \delta_{\vec{p}_{\perp i} \vec{q}_{\perp i}} & \left[[C_s(\vec{p}^{\,2}) - C_s(\vec{q}^{\,2})] \left[\frac{\partial^2 G_{s,\vec{p}\vec{q}}}{\partial M_{\vec{r}} \partial \bar{M}_{\vec{s}}} + \delta_{\vec{p}\vec{r}} \delta_{\vec{q}\vec{s}} \right] - \Gamma^{(2)}_{s,\vec{r}\vec{p}} \delta_{\vec{s}\vec{q}} + \Gamma^{(2)}_{s,\vec{s}\vec{q}} \delta_{\vec{p}\vec{r}} \right. \\ & \left. -r_s(\vec{p}^{\,2}) \delta_{\vec{r}\vec{p}} \delta_{\vec{s}\vec{q}} + r_s(\vec{q}^{\,2}) \delta_{\vec{s}\vec{q}} \delta_{\vec{p}\vec{r}} - \Gamma^{(1,2)}_{s,\vec{r},\vec{s}\vec{p}} M_{\vec{q}} + \Gamma^{(2,1)}_{s,\vec{r}\vec{q};\vec{s}} \bar{M}_{\vec{p}} \right] = 0, \end{aligned} \tag{70}$$

and therefore, Proposition 3 is well given. □

In the deep UV, for a large-scale s, a continuous approximation for variables is suitable. Then, setting $r_1 = p_1$, $\vec{p} \to \vec{q}$, $r_1 \to s_1$, we get finally, in the deep UV, that the four- and two-point functions are related as (on both sides, $r_1 = p_1$):

$$\sum_{\vec{r}_\perp} G_s^2(\vec{r}\,) \frac{dC_s^{-1}}{dr_1^2}(\vec{r}\,) \Gamma^{(4)}_{s,\vec{r},\vec{r},\vec{p},\vec{p}} = \frac{d}{dp_1^2} \left(C_\infty^{-1}(\vec{p}) - \Gamma_s^{(2)}(\vec{p}) \right). \tag{71}$$

To give further comment on the structure of this equation, we have to specify the structure of the vertex function. To this end, we use this loop to discard the irrelevant contributions, and we keep only the melonic component of the function $\Gamma^{(4)}$, denoted by $\Gamma^{(4)}_{\text{melo}}$. In the symmetric phase, the melonic contribution $\Gamma^{(4)}_{\text{melo}}$ may be defined as the part of the function $\Gamma^{(4)}$ that decomposes as a sum of melonic diagrams in the perturbative expansion. The structure of the melonic diagrams has been extensively discussed in the literature and specifically for the approach that we propose here in [57,58]. Formally, they are defined as the graphs optimizing the power counting; and the family can be built from the recursive definition of the vacuum melonic diagrams, from the cutting of some internal edges. Among there interesting properties, these constructions imply the following statement:

Proposition 4. *Let $\mathcal{G}_N$ be a 2N-point 1PI melonic diagram built with more than one vertex for a purely-quartic melonic model. We call external vertices the vertices hooked to at least one external edge of $\mathcal{G}_N$ having:*

- *two external edges per external vertices, sharing $d-1$ external faces of length one.*
- *N external faces of the same color running through the interior of the diagram.*

As a direct consequence of Proposition 4, we expect that the melonic four-point function is decomposed as:

$$\Gamma^{(4)}_{\text{melo}} = \sum_{i=1}^d \Gamma^{(4),i}_{\text{melo}}, \tag{72}$$

the index i running from one to d corresponding to the color of the two internal faces running through the interiors of the diagrams building $\Gamma^{(4),i}_{\text{melo}}$. Moreover, the mono-colored components have the following structure:

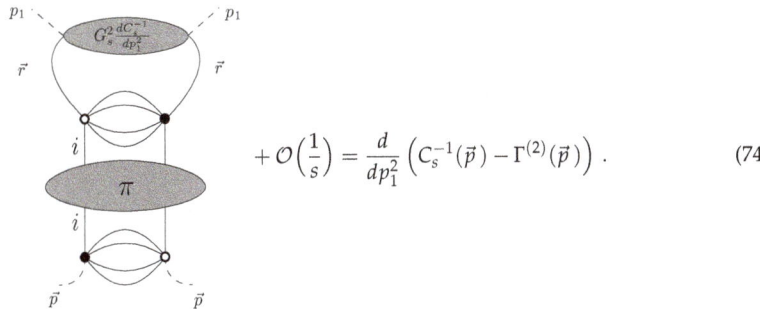
(73)

the permutation of the external momenta $\vec{p}_1$ and $\vec{p}_3$ coming from Wick's theorem: there are four ways to hook the external fields on the external vertices (two per type of field). Moreover, the simultaneous permutation of the black and white fields provides exactly the same diagram, and we count twice each configuration pictured on the previous equation. This additional factor of two is included in the definition of the matrix π, whose entries depend on the components i of the external momenta running on the boundaries of the external faces of colors i, connecting together the end vertices of the diagrams building π.

Inserting (73) into the Ward identity given from Equation (71), we get some contributions on the left-hand side, the only one relevant among them in the deep UV being, graphically:

$$+ \mathcal{O}\left(\frac{1}{s}\right) = \frac{d}{dp_1^2}\left(C_s^{-1}(\vec{p}) - \Gamma^{(2)}(\vec{p})\right). \tag{74}$$

Setting $\vec{p} = \vec{0}$ and using the definition of C_s^{-1}, as well as the definition of C_∞^{-1}, the right-hand side is reduced to $Z_{-\infty} - Z$. Moreover, the diagram on the left-hand side can be written with the following equation $Z_{-\infty}\mathcal{L}_s \pi_{00}$ such that the following equality holds:

$$Z_{-\infty}\mathcal{L}_s \pi_{00} = Z_{-\infty} - Z, \tag{75}$$

where we have defined $Z_{-\infty}\mathcal{L}_s$ as:

$$Z_{-\infty}\mathcal{L}_s := \sum_{\vec{p}\in\mathbb{Z}^d}\left(Z_{-\infty} + \frac{\partial r_s}{\partial p_1^2}(\vec{p})\right) G_s^2(\vec{p})\delta_{p_10}. \tag{76}$$

Finally, from Definition (73) we expect that $\Gamma^{(4)}_{\text{melo},\vec{0},\vec{0},\vec{0},\vec{0}} = 2\pi_{00}$, and because of the renormalization conditions (49), we must have the relation: $\pi_{00} = 2\lambda_{41}(s)$. Therefore, in the deep UV regime, the Ward identity between four- and two-point functions provides a nontrivial relation between effective coupling and wave function renormalization:

$$2Z_{-\infty}\mathcal{L}_s \lambda_{41} = Z_{-\infty} - Z. \tag{77}$$

Remark 1. *Let us give some important remarks regarding the derivation of the Ward identity* (77). *First of all, the WI is totally disconnected from the approximation used to solve the non-perturbative Wetterich Equation* (45). *The Wetterich equation and Ward identity are both functional results, deduced from the definition of the partition function, and have to be treated on the same footing. Their origins, moreover, are completely disconnected. One of them comes from the scale dependence of the model due to the regulator term; the second one comes from the symmetry violation of the action (including source terms) under the* $U(N)^d$ *group and the formal translation-invariance of the Lebesgue measure. Viewing the set* $\mathcal{Z}_s$ *having a continuous family of models, one can say that the Wetterich equation dictates how to move from* $\mathcal{Z}_s$ *to* $\mathcal{Z}_{s+\delta s}$, *whereas the WI are constraints between the observables at fixed s.*

From now on, in the hope to provide the proof that p_+ does not live in the constraint line coming from the Ward identity (77), let us give the following result, which will be proven in the next section.

Proposition 5. *Structure equation for effective coupling: In the deep UV, the effective melonic coupling is given in terms of the renormalized coupling* λ_{41}^r *and the renormalized effective loop* $\bar{\mathcal{A}}_s := \mathcal{A}_s - \mathcal{A}_{s=-\infty}$ *as:*

$$\lambda_{41}(s) = \frac{\lambda_{41}^r}{1 + 2\lambda_{41}^r \bar{\mathcal{A}}_s}, \quad \dot{\lambda}_{41} = -2\lambda_{41}^2 \dot{\mathcal{A}}_s. \tag{78}$$

where we defined the quantity $\mathcal{A}_s$ *as:* $\mathcal{A}_s := \sum_{\vec{p} \in \mathbb{Z}^{(d-1)}} G_s^2(\vec{p})$.

The constraint provided from the Ward identity, which relies on the β-functions and the anomalous dimension, is given by:

$$\mathcal{C}(\bar{\lambda}, \bar{m}^2) := \beta_{41} + \eta \bar{\lambda}_{41}\left(1 - \frac{\bar{\lambda}_{41}\pi^2}{(1+\bar{m}^2)^2}\right) - \frac{2\bar{\lambda}_{41}^2 \pi^2}{(1+\bar{m}^2)^3}\beta_m = 0 \tag{79}$$

This relation needs to be taken into account in the Wetterich flow equation and therefore in the search of fixed points. To prove this relation, let us consider the derivative of Z with respect to s using Expressions (77) and (78):

$$\dot{Z} = (Z_{-\infty} - 2\lambda_{41} Z_{-\infty}\mathcal{L}_s)\frac{\dot{\lambda}_{41}}{\lambda_{41}} - 2Z_{-\infty}\dot{\mathcal{A}}_s \lambda_{41}. \tag{80}$$

In the above relation, we have used the decomposition of $\mathcal{L}_s = \mathcal{A}_s + \Delta_s$. We remark that the Ward identity (77) can be written as $2\lambda_{41}\mathcal{L}_s = 1 - \tilde{Z}$ where $\tilde{Z} = Z/Z_{-\infty}$. Then, (80) becomes:

$$\frac{\dot{Z}}{Z} = \frac{\dot{\lambda}_{41}}{\lambda_{41}} - 2\frac{Z_{-\infty}}{Z}\dot{\Delta}_s \lambda_{41}. \tag{81}$$

We now use the dimensionless quantities $\bar{m}$, $\bar{\lambda}_{41}$, $\bar{B}_s$ such that $\Delta_s = \frac{2}{Z^2}\bar{B}_s$ and reexpressing (81) as:

$$\beta_{41} = -\eta\bar{\lambda}_{41} + 2\bar{\lambda}_{41}(-\eta\bar{B}_s + \dot{\bar{B}}_s) \tag{82}$$

where $\bar{B}_s$ and $\dot{\bar{B}}_s$ can be simply computed using the integral representation of the sum. We come to:

$$\bar{B}_s = -\frac{\pi^2}{2(1+\bar{m}^2)^2}, \quad \dot{\bar{B}}_s = \frac{\pi^2 \beta_m}{(1+\bar{m}^2)^3}, \tag{83}$$

and therefore, (79) is well given. It is time to prove that this constraint violates the existence of the fixed point p_+. Let p be a arbitrary fixed point of the theory. We get $\beta_m(p) = 0 = \beta_{41}(p) = 0$. Then, the constraint (79) implies that at the point p, we get:

$$\eta \bar{\lambda}_{41} \left(1 - \frac{\bar{\lambda}_{41} \pi^2}{(1+\bar{m}^2)^2}\right)(p) = 0. \tag{84}$$

The particular solution $\bar{\lambda}_{41} = 0$ corresponds to the Gaussian fixed point. For $\bar{\lambda}_{41} \neq 0$, we have only:

$$\eta = 0, \text{ or } \frac{\bar{\lambda}_{41} \pi^2}{(1+\bar{m}^2)^2} = 1. \tag{85}$$

It is clear that the fixed point $p_+ = (-0.55, 0.0025)$, $\eta \approx 0.7$ violates these constraints, i.e., does not satisfy the constraint Equation (85). The same conclusion can be made for all choices of the regulator; see [56]. Finally, it is possible to improve the truncation by using the so-called effective vertex expansion. In this case, the fixed point obtained by solving the flow equation also violates the Ward constraint (85). We will study this point in the next section.

4. Effective Vertex Expansion Method for the Melonic Sector

The effective vertex-expansion described in [56–58] allows establishing the structure of the Feynman graphs of our models and leads to the structure equations in the leading order sector. It can help to establish the flow equations without truncation. The Feynman graphs of the colored tensor model are $(d+1)$-colored graphs [20–22]. For the sake of completeness, we remind here about a few facts about these graphs, their representation as stranded graphs, and their uncolored version. The graphs that we consider possibly bear external edges, that is to say half-edges hooked to a unique vertex. We denote $\mathcal{G}$ a a colored graph and $\mathcal{L}(\mathcal{G})$ the set of its internal edges ($L(\mathcal{G}) = |\mathcal{L}(\mathcal{G})|$). A colored graph is said closed if it has no external edges and open otherwise. Let $\mathcal{G}$ be a $(d+1)$-colored graph and S a subset of $\{0,\ldots,d\}$. We denote $\mathcal{G}^S$ as the spanning subgraph of $\mathcal{G}$ induced by the edges of colors in S. Then, for all $0 \leq i,j \leq d$, $i \neq j$, a face of colors i,j is a connected component of $\mathcal{G}^{\{i,j\}}$. A face is open (or external) if it contains an external edge and closed (or internal) otherwise. The set of closed faces of a graph $\mathcal{G}$ is written $\mathcal{F}(\mathcal{G})$ ($F(\mathcal{G}) = |\mathcal{F}(\mathcal{G})|$). The structure of the boundary graph of $\mathcal{G}$ denoted by $\partial \mathcal{G}$ will be useful in the construction of the leading order contribution, which may be considered in the derivative expansion to compute the structure equations and therefore the flow equations.

Definition 1. *Consider $\mathcal{G}$ as a connected Feynman graph with $2N$ external edges. The boundary graph $\partial \mathcal{G}$ is obtained from $\mathcal{G}$ keeping only the external blacks and whites nodes hooked to the external edges, connected together with colored edges following the path drawn from the boundaries of the external faces in the interior of the graph $\mathcal{G}$. $\partial \mathcal{G}$ is then a tensorial invariant itself with N black (resp. whites) nodes. An illustration is given in Figure 5.*

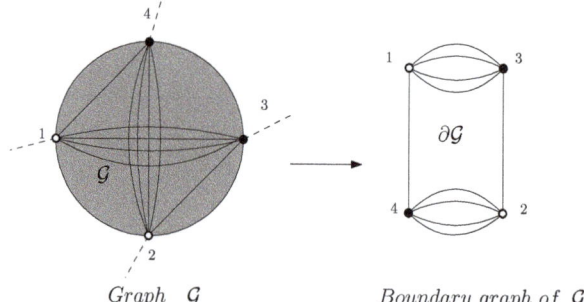

Figure 5. An opening Feynman graph with four external edges and its boundary graph. The strand in the interior of $\mathcal{G}$ represents the path following the external faces.

The power counting theorem of these models shows that the divergence degree of arbitrary Feynman graph $\mathcal{G}$ is:

$$\omega(\mathcal{G}) = -2L(\mathcal{G}) + F(\mathcal{G}) \tag{86}$$

The topological operation on the edge of the graph $\mathcal{G}$ such as contraction is studied extensively in much of the literature. We let the reader consult [20–22] and the references therein. This operation plays an important role in the power counting theorem and allows identifying the structure of the graph. It makes the connection between the divergence degree of $\mathcal{G}$ and the spanning tree denoted by $\mathcal{T}$. Let "\" be the operation of contraction, and we get the following proposition:

Proposition 6. *Under the contraction of the spanning tree edge, the number of internal faces is invariant, i.e., $F(\mathcal{G}) = F(\mathcal{G} \setminus \mathcal{T})$. The graph $\mathcal{G} \setminus \mathcal{T}$ is called the rosette.*

Note that the contraction of the edge $e \in \mathcal{L}(\mathcal{G})$, which leads to the corresponding graph $\tilde{\mathcal{G}} = \mathcal{G} \setminus \{e\}$, is such that $\omega(\mathcal{G}) = \omega(\tilde{\mathcal{G}}) - 2(V-1)$, and the divergent degree of the rosette can be easily computed, using the following formula corresponding to the contraction of the k-dipole: $\omega(\mathcal{G}) = -2L + k(L - V + 1)$. Then, the arbitrary Feynman graph $\mathcal{G}$ is melonic if its boundary graph has the elementary melon structure, i.e., the number of faces is maximal:

$$F(\mathcal{G}_{melon}) = (d-1)(L - V + 1). \tag{87}$$

The existence of the 1/N-expansion of tensors models (N denoting the size of the tensor), which provides in return a topological expansion of the partition function in terms of the generalization of the genus called the Gurau number φ, does not yield a topological expansion, but rather a combinatorial expansion in terms of the degree of the graph. For a colored closed graph $\mathcal{G}$, the degree $\varpi(\mathcal{G})$ is such that for the melon, $\varpi(\mathcal{G}_{melon}) = 0$.

4.1. Structure Equations and Compactability with Ward Identities

The structure equations are the relations between the correlation function and allow establishing a constraint between β-functions for mass, interaction couplings, and wave function renormalization. These relations are obtained in the deep UV limit (i.e., in the domain $1 \ll e^s \ll \Lambda$) without any assumption about the β-functions and without any truncation of the effective action Γ_s. The only assumption concern the choice of the initial conditions, ensuring the perturbative consistency of the full partition function. The first structure equation concerns the self energy (or 1PI two-point functions). It takes place as the *closed equation* for self energy.[1] Let us summarize this in the following proposition:

Proposition 7. *In the melonic sector, the self energy $\Sigma_s(\vec{p})$ is given by the closed equation, which takes into account the effective coupling $\lambda_{41}(s)$ as:*

$$-\Sigma_s(\vec{p}) = 2\lambda_{41}^r Z_\lambda \sum_{\vec{q}} \left(\sum_{i=1}^d \delta_{p_i q_i} \right) G_s(\vec{q}). \tag{88}$$

In the same way, in the melonic sector, the perturbative zero-momenta 1PI four-point contribution $\Gamma^{(4),i}_{s,\vec{00};\vec{00}}$ is given by:

$$\Gamma^{(4),i}_{s,\vec{00};\vec{00}} = 2\pi_{00} = \frac{4Z_\lambda \lambda_{41}^r}{1 + 2\lambda_{41}^r Z_\lambda \mathcal{A}_s}, \tag{89}$$

where $\mathcal{A}_s$ is defined as:

$$\mathcal{A}_s = \sum_{\vec{p}_\perp} [G_s(\vec{p}_\perp)]^2, \quad \vec{p}_\perp := (0, p_1, \cdots, p_d), \tag{90}$$

$G_s(\vec{p})$ being the effective propagator: $G_s^{-1}(\vec{p}) = Z_{-\infty}\vec{p}^{\,2} + m^2 + r_s(\vec{p}) - \Sigma_s(\vec{p})$. Let us recall that $Z_{-\infty}$ and m_0 are the counter-terms discarding the UV divergences of the original partition function, and the initial conditions in the UV are given such that the classical action contains only renormalizable interactions.

Proof. Concerning the proof of Relation (88), we let the reader consult [60]. Let us define $4Z_\lambda \lambda_{41}^r \Pi$ as the zero momenta melonic four-point functions made into the graphs for which two vertices may be singularized (i.e., by graphs that are at least of order two in the perturbative expansion). We have[2]:

$$2\pi_{00} =: 4Z_\lambda \lambda_{41}^r (1 + \Pi). \tag{91}$$

Because of the face connectivity of the melonic diagrams, the boundary vertices may be such that the two internal faces of the same color running on the interior of the diagrams building Π pass through of them. Then, we have the following structure:

$$-4Z_\lambda \lambda_{41}^r \Pi = \text{[diagram]} \tag{92}$$

where the grey disk is a sum of Feynman graphs. Note that it is the only configuration of the external vertices in agreement with the assumption that Π is built with the melonic diagrams. Any other configurations of the external vertices are not melonics. At the lowest order, the grey disk corresponds to propagator lines,

$$-4Z_\lambda \lambda_{41}^r \Pi^{(2)} = 8Z_\lambda^2 (\lambda_{41}^r)^2 \mathcal{A}_s|_{\lambda_{41}^r = 0} \equiv \text{[diagram]}. \tag{93}$$

---

[1] The rank of the tensors is fixed to five, and we denote it by d to clarify the proof(s).
[2] The notations are similar to the ones used for the previous proof. The context however allows excluding any confusion.

Note that the external faces have the same color. Now, we can extract the amputated component of $\bar{\Pi}$, say $\bar{\Pi}'$ (which contains at least one vertex and is irreducible by hypothesis), extracting the effective melonic propagators connected to the dotted lines linked to $\bar{\Pi}$. We get:

$$-4Z_\lambda \lambda_{41}^r \Pi = \quad\cdots\quad + \quad\cdots\quad . \tag{94}$$

At first order, $\bar{\Pi}'$ is built with a single vertex, and there is only one configuration in agreement with the melonic structure, i.e., maximizing the number of internal faces. The higher order contributions contain at least two vertices, and the argument may be repeated so that the function $\bar{\Pi}'$ appears. Finally, we deduce the closed relation:

$$\cdots = \cdots + \cdots . \tag{95}$$

This equation can be solved recursively as an infinite sum:

$$-4Z_\lambda \lambda_{41}^r \Pi = \quad \left\{ \sum_{n=1}^\infty \left(\cdots \right)^n \right\} \quad , \tag{96}$$

which can be formally solved as:

$$2\pi_{00} = 4Z_\lambda \lambda_{41}^r \left(1 - \cdots \right)^{-1} . \tag{97}$$

The loop diagram $\cdots$ may be easily computed recursively from the definition of melonic diagrams or directly using Wick theorem for a one-loop computation with the effective propagator G. The result is:

$$\cdots = -2Z_\lambda \lambda_{41}^r \mathcal{A}_s , \tag{98}$$

and the proposition is proven. □

Note that this construction can be easily checked to be compatible with the Ward identity, especially in the form of (74). Conversely, the last result may be derived directly from Equation (74) and from the closed equation for the two-point function (88) (see [58]). To prove these two results, we only assume that the classical mean field vanishes, and we deduce from our previous proof, essentially based on the assumption that the effective vertices are analytic with respect to the renormalized coupling, that the analytic domain covers what we called the symmetric phase. In the hope to extract the expression of the counter-terms at all orders and to show that the wave function renormalization and the four-point vertex renormalization are the same, we have the following result:

Proposition 8. *Choosing the following renormalization prescription:*

$$\Gamma^{(4),1}_{s=-\infty,\vec{0}\vec{0};\vec{0}\vec{0}} = 4\lambda^r_{41} \; ; \; \Gamma^{(2)}_{s=-\infty}(\vec{p}) = m_r^2 + \vec{p}^2 + \mathcal{O}(\vec{p}^2), \qquad (99)$$

where m_r^2 and λ^r_{41} are the renormalized mass and coupling constant, the counter-terms are given by:

$$Z_\lambda = \frac{1}{1 - 2\lambda^r_{41}\mathcal{A}_{s=-\infty}}, \; ; \; Z_{-\infty} = Z_\lambda \; ; \; m^2 = m_r^2 + \Sigma_{s=-\infty}(\vec{p} = 0), \qquad (100)$$

where Σ_s denotes the melonic self-energy.

Proof. From Proposition 1, we can get:

$$\Gamma^{(4),i}_{s,\vec{0}\vec{0};\vec{0}\vec{0}} = \frac{4 Z_\lambda \lambda^r_{41}}{1 + 2\lambda^r_{41} Z_\lambda \mathcal{A}_s} = \frac{4\lambda^r_{41}}{Z_\lambda^{-1} + 2\lambda^r_{41} \mathcal{A}_s}. \qquad (101)$$

Then, setting $s = -\infty$, we deduce that:

$$Z_\lambda^{-1} + 2\lambda^r_{41} \mathcal{A}_{-\infty} = 1 \to Z_\lambda = \frac{1}{1 - 2\lambda^r_{41} \mathcal{A}_{-\infty}}. \qquad (102)$$

We now concentrate our self on $Z_{-\infty}$ and m^2. Without loss of generality, the inverse of the effective propagator $\Gamma^{(2)}_s$ has the following structure:

$$\Gamma^{(2)}_{s=-\infty}(\vec{p}) = Z_{-\infty}\vec{p}^2 + m^2 - \Sigma_{s=-\infty}(\vec{p}) \qquad (103)$$

$$= Z_{-\infty}\vec{p}^2 + m^2 - \Sigma_{s=-\infty}(\vec{0}) - \vec{p}^2 \Sigma'_{s=-\infty}(\vec{0}) + \mathcal{O}(\vec{p}^2) \qquad (104)$$

$$= (Z_{-\infty} - \Sigma'_{s=-\infty}(0))\vec{p}^2 + m^2 - \Sigma_{s=-\infty}(\vec{0}) + \mathcal{O}(\vec{p}^2) \qquad (105)$$

with the notation: $\Sigma'(\vec{0}) := \partial \Sigma / \partial p_1^2(\vec{p} = \vec{0})$. Then, from the renormalization conditions, we have:

$$Z_{-\infty} - \Sigma'_{s=-\infty}(0) = 1 \; , \; m^2 - \Sigma_{s=-\infty}(\vec{0}) = m_r^2. \qquad (106)$$

Setting $s = -\infty$ in the closed equation for the two-point correlation function and by deriving with respect to p_1 for $\vec{p} = \vec{0}$, we get:

$$1 - Z_{-\infty} = -2\lambda^r_{41} Z_\lambda \mathcal{A}_{s=-\infty}. \qquad (107)$$

Using the explicit expression for Z_λ in (102), we get finally:

$$(1 - Z_{-\infty})(1 - 2\lambda^r_{41} \mathcal{A}_{s=-\infty}) = -2\lambda^r_{41} \mathcal{A}_{s=-\infty} \to Z_{-\infty} = Z_\lambda. \qquad (108)$$

□

Now, consider the mono-color four-point function $\Gamma^{(4),i}_{s,\vec{0}\vec{0};\vec{0}\vec{0}}$. If we replace Z_λ by its expression from Proposition 8, we deduce that:

$$\Gamma^{(4),i}_{s,\vec{0}\vec{0};\vec{0}\vec{0}} = \frac{4\lambda^r_{41}}{1 + 2\lambda^r_{41} \bar{\mathcal{A}}_s}, \qquad (109)$$

with the definition: $\bar{\mathcal{A}}_s := \mathcal{A}_s - \mathcal{A}_{s=-\infty}$. In other words, we have an explicit expression for the effective coupling $\lambda_{41}(s) := \frac{1}{4}\Gamma^{(4),i}_{s,\vec{0}\vec{0};\vec{0}\vec{0}}$,

$$\lambda_{41}(s) = \frac{\lambda^r_{41}}{1 + 2\lambda^r_{41} \bar{\mathcal{A}}_s}, \qquad (110)$$

from which we get:
$$\partial_s \lambda_{41}(s) = -\frac{2(\lambda_{41}^r)^2 \dot{\mathcal{A}}_s}{(1 + 2\lambda_{41}^r \Delta \mathcal{A}_s)^2} = -2\lambda_{41}^2(s)\dot{\mathcal{A}}_s. \tag{111}$$

In the above relation, we introduce the dot notation $\dot{\mathcal{A}}_s = \partial_s \mathcal{A}_s$:
$$\mathcal{A}_s = \sum_{\vec{p}_\perp} \frac{1}{[\Gamma_s^{(2)}(\vec{p}_\perp) + r_s(\vec{p}_\perp)]^2}, \quad \dot{\mathcal{A}}_s = -2 \sum_{\vec{p}_\perp} \frac{\dot{\Gamma}_s^{(2)}(\vec{p}_\perp) + \dot{r}_s(\vec{p}_\perp)}{[\Gamma_s^{(2)}(\vec{p}_\perp) + r_s(\vec{p}_\perp)]^3}. \tag{112}$$

In Proposition 8, we have investigated the relations between counter-terms, i.e., we have considered the melonic equations as Ward identities for $s = -\infty$. Far from the initial conditions, the Taylor expansion of the two-point function $\Gamma_s^{(2)}(\vec{p})$ is written as:
$$\Gamma_s^{(2)}(\vec{p}) = m_r^2 + (\Sigma_s(\vec{0}) - \Sigma_0(\vec{0})) + (Z_{-\infty} - \Sigma_s'(\vec{0}))\vec{p}^{\,2} + \mathcal{O}(\vec{p}^{\,2}). \tag{113}$$

We call the "physical" or *effective* mass parameter $m^2(s)$ the first term in the above relation:
$$m^2(s) := m_r^2 + (\Sigma_s(\vec{0}) - \Sigma_0(\vec{0})), \tag{114}$$

while the coefficient $Z_{-\infty} - \Sigma_s'(\vec{0})$ is the effective wave function renormalization and is denoted by $Z(s)$, i.e.,
$$Z(s) := Z_{-\infty} - \Sigma_s'(\vec{0}). \tag{115}$$

Now, let us consider the closed equation given in Proposition (88). By deriving with respect to p_1 and by taking $\vec{p} = \vec{0}$, we get:
$$Z - Z_{-\infty} = -2\lambda_{41}^r Z_\lambda \sum_{\vec{p}_\perp} G_s^2(\vec{p}_\perp)(Z + r_s'(\vec{p}_\perp)). \tag{116}$$

Using Equation (110), we can express $\lambda_{41}^r Z_\lambda$ in terms of the effective coupling $\lambda_{41}(s)$, and we get:
$$(Z - Z_{-\infty})(1 - 2\lambda_{41}(s)\mathcal{A}_s) = -2\lambda_{41}(s)\left(Z\mathcal{A}_s + \sum_{\vec{p}_\perp} G_s^2(\vec{p}_\perp) r_s'(\vec{p}_\perp)\right), \tag{117}$$

Then, we come to the following relation
$$Z = Z_{-\infty}(1 - 2\lambda_{41}(s)\mathcal{L}_s). \tag{118}$$

At this stage, without any confusion, let us clarify that: $Z_{-\infty}$ is the wave function counter-term, i.e., whose divergent part cancels the loop divergences and whose finite part depends on the renormalization prescription. $Z(s)$ however is fixed to one for $s = -\infty$ from our renormalization conditions.

4.2. Flow Equation from the EVE Method

There are different methods to improve the crude truncations in the FRG literature. However, their applications for TGFTs remain difficult due to the non-locality of the interactions over the group manifold on which the fields are defined. A step to go out of the truncation method was done recently in [57,58] with the effective vertex expansion (EVE) method. Basically, the strategy is to close the infinite tower of equations coming from the exact flow equation, instead of crudely truncating them. The strategy is to complete the structure Equation (78) with a structure equation for $\Gamma^{(6)}$, expressing it in terms of the marginal coupling λ and the effective propagator G_s only. In this way, the flow equations around marginal couplings are completely closed. Note that this approach crosses the first hypothesis motivating the truncation: we expect that so far from the deep UV, only the marginal interactions survive and drag the complete RG flow. Moreover, any fixed point of the autonomous set of resulting equations is automatically a fixed point for any higher effective melonic vertices built from effective quartic interactions. Finally, a strong improvement of this method with respect to the truncation method, already pointed out in [57,58], is that it allows keeping the complete momenta dependence of the effective vertex. This dependence generates a new term on the right-hand side of the equation for $\dot{Z}$, moving the critical line from its truncation's position.

Let us consider the flow equation for $\dot{\Gamma}^{(2)}$, obtained from (45), derived with respect to M and $\bar{M}$:

$$\dot{\Gamma}^{(2)}(\vec{p}) = -\sum_{\vec{q}} \Gamma^{(4)}_{\vec{p},\vec{p},\vec{q},\vec{q}} G_s^2(\vec{q}) \dot{r}_s(\vec{q}), \tag{119}$$

where we discard all the odd contributions, vanishing in the symmetric phase. Deriving on both sides with respect to p_1^2 and setting $\vec{p} = \vec{0}$, we get:

$$\dot{Z} = -\sum_{\vec{q}} \Gamma^{(4)\,\prime}_{\vec{0},\vec{0},\vec{q},\vec{q}} G_s^2(\vec{q}) \dot{r}_s(\vec{q}) - \Gamma^{(4)}_{\vec{0},\vec{0},\vec{q},\vec{q}} G_s^2(\vec{q}) \dot{r}_s(\vec{q}), \tag{120}$$

where the "prime" designates the partial derivative with respect to p_1^2. In the deep UV ($k \gg 1$), the argument used in the T^4-truncation to discard non-melonic contributions holds, and we keep only the melonic diagrams. Moreover, to capture the momentum dependence of the effective melonic vertex $\Gamma^{(4)}_{\text{melo}}$ and compute the derivative $\Gamma^{(4)\,\prime}_{\text{melo},\vec{0},\vec{0},\vec{q},\vec{q}'}$, knowledge of π_{pp} is required. It can be deduced from the same strategy as for the derivation of the structure Equation (78), up to the replacement:

$$A_s \to A_s(p) := \sum_{\vec{p} \in \mathbb{Z}^d} G_s^2(\vec{p}) \delta_{p_1 p}, \tag{121}$$

from which we get:

$$\pi_{pp} = \frac{2\lambda_{41}^r}{1 + 2\lambda_{41}^r \bar{A}_s(p)}, \quad \bar{A}_s(p) := A_s(p) - A_{-\infty}(0). \tag{122}$$

The derivative with respect to p_1^2 may be easily performed, and from the renormalization condition (49), we obtain:

$$\pi'_{00} = -4\lambda_{41}^2(s) A'_s, \tag{123}$$

while the leading order flow equation for $\dot{Z}$ becomes:

$$\dot{Z} = 4\lambda_{41}^2 A'_s(0) I_2(0) - 2\lambda_{41} I'_2(0). \tag{124}$$

As announced, a new term appears with respect to the truncated version (48), which contains a dependence on η and then moves the critical line. The flow equation for mass may be obtained from (119) setting $\vec{p} = \vec{0}$ on both sides. Finally, the flow equation for the marginal coupling λ_{41} may be obtained from Equation (45) deriving it twice with respect to each mean field M and $\bar{M}$. As explained before, it involves $\Gamma^{(6)}_{\text{melo}}$ at leading order, and to close the hierarchy, we use the marginal coupling as a driving parameter and express it in terms of $\Gamma^{(4)}_{\text{melo}}$ and $\Gamma^{(2)}_{\text{melo}}$ only. Once again, from Proposition 4, $\Gamma^{(6)}_{\text{melo}}$ have to be split into d mono-colored components $\Gamma^{(6),i}_{\text{melo}}$:

$$\Gamma^{(6)}_{\text{melo}} = \sum_{i=1}^{d} \Gamma^{(6),i}_{\text{melo}}. \tag{125}$$

The structure equation for $\Gamma^{(6),i}_{\text{melo}}$ may be deduced following the same strategy as for $\Gamma^{(4),i}_{\text{melo}}$, from Proposition 4. Starting from a vacuum diagram, a leading order four-point graph may be obtained opening successively two internal tadpole edges, both on the boundary of a common internal face. This internal face corresponds for the resulting four-point diagram to the two external faces of the same colors running through the interior of the diagram. In the same way, a leading order six-point graph may be obtained cutting another tadpole edge on this resulting graph, once again on the boundary of one of these two external faces. The reason this works is that, in this way, the number of discarded internal faces is optimal, as well as the power counting. From this construction, it is not hard to see that the zero-momenta $\Gamma^{(6),i}_{\text{melo}}$ vertex function must have the following structure (see [57,58] for more details):

$$\Gamma^{(6),i}_{\text{melo}} = (3!)^2 \left(\begin{array}{c} \text{[diagram]} \end{array} \right), \tag{126}$$

the combinatorial factor $(3!)^2$ coming from the permutation of external edges. Translating the diagram into equation and taking into account symmetry factors, we get:

$$\Gamma^{(6),i}_{\text{melo}} = 24 Z^3(s) \bar{\lambda}^3_{41}(s) e^{-2s} \bar{A}_{2s}, \tag{127}$$

with:

$$\bar{A}_{2s} := Z^{-3} e^{2s} \sum_{\vec{p} \in \mathbb{Z}^{d-1}} G_s^3(\vec{p}). \tag{128}$$

Note that this structure equation may be deduced directly from Ward identities, as pointed-out in [58,60]. The equation closing the hierarchy is then compatible with the constraint coming from unitary invariance. The flow equations involve now some new contributions depending on two sums, $\bar{A}_{2s}$ and $\bar{A}'_s$, defined without regulation function $\dot{r}_s$. However, they are both power-counting convergent in the UV, and the renormalizability theorem ensures their finiteness for all orders in the perturbation theory. For this reason, they become independent of the initial conditions at scale Λ for $\Lambda \to \infty$; and as pointed out in [58], we get, using Litim's regulator:

$$\bar{A}_{2s} = \frac{1}{2} \frac{\pi^2}{1+\bar{m}^2} \left[\frac{1}{(1+\bar{m}^2)^2} + \left(1 + \frac{1}{1+\bar{m}^2}\right) \right], \tag{129}$$

and:

$$\bar{A}'_s = \frac{1}{2} \pi^2 \frac{1}{1+\bar{m}^2} \left(1 + \frac{1}{1+\bar{m}^2}\right). \tag{130}$$

The complete flow equation for zero-momenta four-point coupling is written explicitly as:

$$\dot{\Gamma}^{(4)} = -\sum_{\vec{p}} \dot{r}_s(\vec{p}) G_s^2(\vec{p}) \left[\Gamma^{(6)}_{\vec{p},\vec{0},\vec{0},\vec{p},\vec{0},\vec{0}} - 2 \sum_{\vec{p}'} \Gamma^{(4)}_{\vec{p},\vec{0},\vec{p}',\vec{0}} G_s(\vec{p}') \Gamma^{(4)}_{\vec{p}',\vec{0},\vec{p},\vec{0}} + 2 G_s(\vec{p}) [\Gamma^{(4)}_{\vec{p},\vec{0},\vec{p},\vec{0}}]^2 \right]. \tag{131}$$

Keeping only the melonic contributions, we get finally the following autonomous system by using Litim's regulation:

$$\begin{cases} \beta_m = -(2+\eta)\bar{m}^2 - 2d\bar{\lambda}_{41} \frac{\pi^2}{(1+\bar{m}^2)^2} \left(1 + \frac{\eta}{6}\right), \\ \beta_{41} = -2\eta\bar{\lambda}_{41} + 4\bar{\lambda}^2_{41} \frac{\pi^2}{(1+\bar{m}^2)^3} \left(1 + \frac{\eta}{6}\right) \left[1 - \frac{1}{2}\pi^2 \bar{\lambda}_{41} \left(\frac{1}{(1+\bar{m}^2)^2} + \left(1 + \frac{1}{1+\bar{m}^2}\right)\right)\right]. \end{cases} \tag{132}$$

where the anomalous dimension is then given by:

$$\eta = 4\bar{\lambda}_{41}\pi^2 \frac{(1+\bar{m}^2)^2 - \frac{1}{2}\bar{\lambda}_{41}\pi^2(2+\bar{m}^2)}{(1+\bar{m}^2)^2 \Omega(\bar{\lambda}_{41},\bar{m}^2) + \frac{(2+\bar{m}^2)}{3}\bar{\lambda}_{41}^2\pi^4}. \tag{133}$$

The new anomalous dimension has two properties that distinguish it from its truncation version. First of all, as announced, the singularity line $\Omega = 0$ moves toward the $\bar{\lambda}_{41}$ axis, extending the symmetric phase domain. In fact, the improvement is optimal, the critical line being deported under the singularity line $\bar{m}^2 = -1$. In standard interpretations [57], the presence of the region II is generally assumed to come from a bad expansion of the effective average action around the vanishing mean field, becoming a spurious vacuum in this region.

However, the EVE method shows that the singularity line obtained using truncation is completely discarded taking into account the momentum dependence of the effective vertex. The second improvement comes from the fact that the anomalous dimension may be negative and vanishes on the line of equation $L(\bar{\lambda}_{41}, \bar{m}^2) = 0$, with:

$$L(\bar{\lambda}_{41}, \bar{m}^2) := (1+\bar{m}^2)^2 - \frac{1}{2}\bar{\lambda}_{41}\pi^2(2+\bar{m}^2). \tag{134}$$

Interestingly, there are now two lines in the maximally-extended region I' where physical fixed points are expected. However, numerical integrations show that the improved flow equations admit a non-Gaussian fixed point $\tilde{p}_+$, which is numerically very close to the fixed point p_+ obtained in the truncation method, i.e., $\tilde{p}_+ \approx p_+$, and then unphysical as well. Figure 6 summarize all these results.

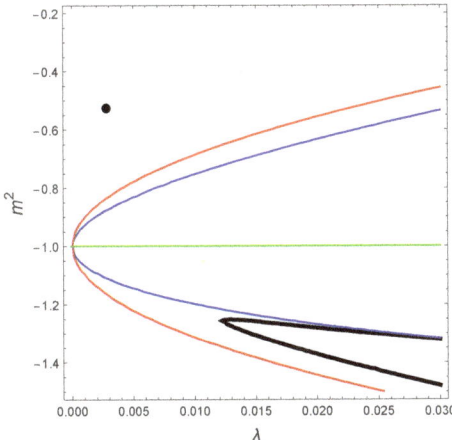

Figure 6. The relevant lines over the maximally-extended region I', bounded at the bottom with the singularity line $m^2 = -1$ (in green). The blue and red curves correspond respectively to the equations $L = 0$ and $\Omega = 0$. Moreover, the black point corresponds to the numerical non-Gaussian fixed point, so far from the two previous physical curves.

4.3. Exploration of the Physical Phase Space

In this section, we will show that the EVE method leads to an alternative first order phase transition scenario, despite the fact that the fixed point p_+ is discarded. Secondly, we also prove that this new behavior is only observed using the EVE method and cannot be obtained by implementing the usual truncation as the approximation.

(1) Despite the fact that the constraint Equation (79) is not compatible with the fixed point $p_+ = (-0.52, 0.0028)$, this is not the end of the world. The constraint $\mathcal{C} = 0$ given by Equation (79)

defines a one-dimensional subspace, say $\mathcal{E}_\mathcal{C}$, into the whole bi-dimensional phase space $(\bar{\lambda}_{41}, \bar{m}^2)$. Obviously, the Ward identity will be violated everywhere except along this one-dimensional subspace $\mathcal{E}_\mathcal{C}$; for this reason, we call the physical phase spacethis subspace.

Solving $\mathcal{C} = 0$ with respect to $\bar{m}^2$, we can extract the coupling constant $\bar{\lambda}_{41}$ as a function of the renormalized mass parameter $\bar{m}^2$. After a few hand computations, we get:

$$\bar{\lambda}_{41}^3 = 0, \quad \text{or} \quad \bar{\lambda}_{41} = -\frac{19\left(\bar{m}^2+1\right)^2}{\pi^2(4\bar{m}^2-1)} := f(\bar{m}^2). \tag{135}$$

These solutions provide only one non-trivial parametrized equation for the physical subspace $\mathcal{E}_\mathcal{C}: \bar{\lambda}_{41} = f(\bar{m}^2)$. Interestingly, it is not hard to check that the presence of the factor $(1 + \bar{m}^2)^2$ in the numerator cancels all the formal divergences occurring for $\bar{m}^2 = -1$, such that the flow becomes regular at this point. However, other divergences occur, one of them being common to each beta functions. To understand the structure of the effective flow into the physical subspace, we have to insert the solutions (135) into the flow Equation (132). However, even to do this, let us discuss the solution (135) in a few words. Because the theory is asymptotically free, we may expect that $\bar{m}^2$ and $\bar{\lambda}_{41}$ have to vanish simultaneously. What we know is that, in the vicinity of the Gaussian fixed point $\bar{m}^2 = \bar{\lambda}_{41} = 0$, the constraint $\mathcal{C} = 0$ is approximately satisfied. For instance, up to $\bar{\lambda}_{41}^3$ contributions, the Equation (79) reduces as:

$$\mathcal{C} = \beta_{41} + \eta\bar{\lambda}_{41} = 0 \tag{136}$$

which is identically satisfied by the one-loop beta equation $\beta_{41} = -\eta\bar{\lambda}_{41}$; see (132). As a result, in a small domain around $(\bar{m}^2, \bar{\lambda}_{41}) = (0,0)$, the flow behaves approximately according to the Ward constraint, but as soon as the flow leaves this region, the Ward constraint is violated, except along $\mathcal{E}_\mathcal{C}$, where it holds strictly. Note that, for $\bar{m}^2 = 0$, the value of $\bar{\lambda}_{41}$ is very large ($\bar{\lambda}_{41} \approx 1.9$) and far away from the vicinity of the Gaussian fixed point.

Now, let us move on to the solutions (135). The solution $\bar{\lambda}_{41} = 0$ corresponds to trivial flow, $\eta = 0$ and:

$$\beta_m = -2\bar{m}^2, \quad \beta_{41} = 0. \tag{137}$$

On the other hand, inserting the non-trivial solution $\bar{\lambda}_{41} = f(\bar{m}^2)$, we get:

$$\eta(\bar{m}^2) = -\frac{1026}{167 + 16\bar{m}^2}. \tag{138}$$

and:

$$\beta_m = \frac{4\bar{m}^2(173 - 8\bar{m}^2) + 760}{16\bar{m}^2 + 167}, \quad \beta_{41} = -\frac{1444(\bar{m}^2+1)(7\bar{m}^2(10\bar{m}^2-7)-137)}{\pi^2(16\bar{m}^2+167)(1-4\bar{m}^2)^2}. \tag{139}$$

As announced, the divergences at the value $\bar{m}^2 = -1$ has been discarded. However, some new divergences occur. First of all, the equation for $\mathcal{E}_\mathcal{C}$ becomes singular for the positive value $\bar{m}^2 = 1/4$. A second singularity occurs for the value $\bar{m}^2 = -\frac{167}{16} =: \bar{m}^2_{\text{div}}$, which is common for η, β_m, and β_{41}; and a third singularity occurs for $\bar{m}^2 = 1/4$ in the expression of β_{41}, which is the same as the singularity of $f(\bar{m}^2)$. We now discuss this picture. To this end, let us examine the points at which the beta function vanishes. We get:

$$\beta_m(\bar{m}_1^2) = 0 \Rightarrow \bar{m}_1^2 = \frac{1}{16}\left(173 \pm 3\sqrt{4001}\right) \tag{140}$$

$$\beta_{41}(\bar{m}_2^2) = 0 \Rightarrow \bar{m}_2^2 = \frac{1}{140}\left(49 \pm 3\sqrt{4529}\right), \quad \bar{m}_2^2 = -1. \tag{141}$$

Because $\tilde{m}_1^2 \neq \tilde{m}_2^2$, we recover our previous conclusion, in the whole theory space $(\tilde{\lambda}, \tilde{m}^2)$; no fixed point can be found using the exact FRG with the EVE method taking into account the Ward constraint. The β-function of the mass vanishes at the point $\tilde{m}_0^2 \approx -1.04$ on the projected phase space $\mathcal{E}_\mathcal{C}$. Furthermore, $\beta_m(\tilde{m}_0^2 - \epsilon) > 0$ and $\beta_m(\tilde{m}_0^2 + \epsilon) < 0$, ϵ being a small positive value, and the flow into the physical phase space changes direction at this point, pointing toward the positive mass direction for $\tilde{m}^2 > \tilde{m}_0^2$ and toward the negative mass direction for $\tilde{m}^2 < \tilde{m}_0^2$. In the last case, the flow continues on this way and reaches the singularity, where the flow becomes undefined. Both of these features are reminiscent of a first order phase transition on the physical phase space—the singularity may indicate a point at which the effective action becomes undefined, or where the expansion around the null vacuum fails to exist—the last statement having to be rigorously investigated.

The same analysis may be performed when we consider the following prescription: by extracting the mass parameter $\tilde{m}^2$ as a function of the constant $\tilde{\lambda}_{41}$: $(\tilde{m}^2 = g(\tilde{\lambda}_{41}))$ in the constraint equation and solving the β-function of the coupling. In this case, the coupling becomes the parameter, and for the point $\tilde{\lambda}_{\text{div}} = \frac{22801}{576\pi^2} \approx 4$, we get a singularity corresponding to the value $\tilde{m}_{\text{div}}^2 = -\frac{167}{16} \approx -10.43$ (see Figure 7b). Note that around $\tilde{m}_0^2$, the coupling becomes very small:

$$f(\tilde{m}_0^2) \approx 0.0007, \quad (142)$$

and we reach a new perturbative regime for small $\tilde{\lambda}_{41}$ and small $(1 + \tilde{m}^2)$.

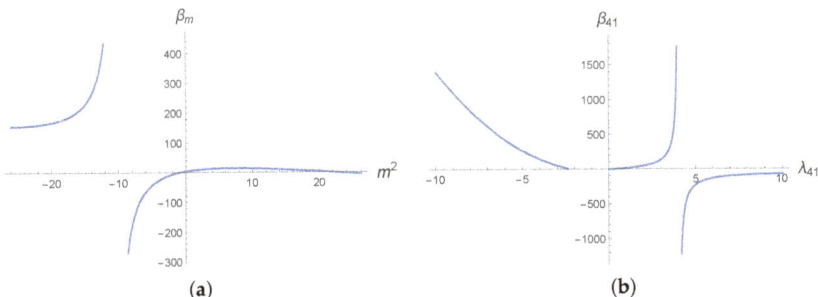

Figure 7. Plot of β_m as a function of $\tilde{m}^2$ using the constraint equation. We get the singularity at the point $\tilde{m}_{\text{div}}^2 = -\frac{167}{16}$, corresponding to the coupling value $\tilde{\lambda}_{\text{div}} = \frac{22801}{576\pi^2}$ (a). Plot of β_{41} as function of $\tilde{\lambda}_{\text{div}}$ in the parametrization $\tilde{m}^2 = \tilde{m}^2(\tilde{\lambda}_{41})$ of the physical phase space. The singularity occurs at the point $\tilde{\lambda}_{\text{div}} = \frac{22801}{576\pi^2}$ (b).

(2) When we investigated the truncation method, we did not provide such a discussion. To compare the methods, let us consider the same strategy for the phase space described with the truncation method. Solving the constraint $\mathcal{C} = 0$, we get:

$$\tilde{\lambda}_{41}^3 = 0 \text{ or } \tilde{\lambda}_{41} = \frac{11(1+\tilde{m}^2)^2}{5\pi^2}. \quad (143)$$

By replacing this solution $\tilde{\lambda}_{41}^3 = 0$ in the flow equations of mass and coupling (48), we get:

$$\beta_m = -2\tilde{m}^2, \quad \beta_{41} = 0. \quad (144)$$

Now, setting $\beta_m = 0 = \beta_{41}$, only the Gaussian fixed point $(\tilde{m}^* = 0, \tilde{\lambda}_{41}^* = 0)$ survives. Furthermore, the last solution leads to:

$$\beta_m = \frac{4}{9}(12\tilde{m}^2 + 11), \quad \beta_{41} = \frac{484(\tilde{m}^2+1)(15\tilde{m}^2+13)}{225\pi^2}. \quad (145)$$

One more time, we recover that no solutions such that $\beta_m = 0 = \beta_{41}$ exist. Moreover, we recover that β_m vanishes for a negative mass value, not so far from $\bar{m}^2 = -1$; and that the singularity at this value has been completely discarded from the solution of the Ward constraint. However, the common singularity of the beta functions as some other aspects of the previous flow equations are not reproduced in the truncation framework. The nature of the singularities, for $\bar{m}^2 = -\frac{167}{16}$ and $\bar{m}^2 = 0.25$, remains mysterious in our formalism. Obviously, they are a consequence of the improvement coming from the EVE method, and their understanding may increase our knowledge about the behavior of the TGFT renormalization group flow.

5. Conclusions

In this manuscript, we have studied with different methods the FRG applied to TGFT. First, we have derived the Wilson–Polchinski equation and given the perturbative solution. Secondly, we derived the Wetterich flow equation using the usual approximation, called truncation. The analytic solution of this equation was given. We obtained a fixed point denoted by p_+. Then, we investigated the Ward identities as a new constraint along the flow and showed that the fixed point p_+ violates this constraint. Finally, we improved the study of FRG by replacing the truncation method by the so-called EVE. The flow equation was improved, and the corresponding solution $\tilde{p}_+$ was not so far from p_+, i.e., $\tilde{p}_+ \approx p_+$. However, the Ward identities are strongly violated at this fixed point, and therefore this unique fixed point seems to be unphysical. We have also showed the importance of the EVE method in the sense that, despite the fact that the fixed point p_+ needs to be discarded, a first order phase transition exists very far from this point in the subspace $\mathcal{E}_C$ of the theory space. We have showed that this new behavior cannot be observed using the truncation as the approximation.

In this review, we focused on the EVE method for the melonic approximation, and especially on the quartic melonic just-renormalizable sector. The complete quartic sector, including all the connected quartic bubbles, has already been considered in a complementary work [57], and the conclusion about the incompatibility with nonperturbative fixed points and Ward identities holds. The graphs added to the quartic melonic ones to complete the quartic sector have been called pseudo-melons due to the similarities of their respective leading order Feynman graphs. Finally, even if we expect that some aspects of the EVE method improve the standard truncation method, some limitations have to be addressed for future works. In particular, our investigations were limited to the symmetric phase, ensuring convergence of any expansion around the vanishing classical mean field. Moreover, we have retained only the first terms in the derivative expansion of the two-point function and only considered the local potential approximation, i.e., potentials that can be expanded as an infinite sum of connected melonic (and pseudo-melonic) interactions. Finally, a rigorous investigation of the behavior of the renormalization group flow into the physical phase space has to be addressed in the continuation of current works on this topic.

Author Contributions: Conceptualization, V.L. and D.O.-S.; methodology, V.L. and D.O.-S.; software, V.L. and D.O.-S.; validation, V.L. and D.O.-S.; formal analysis, V.L. and D.O.-S.; investigation, V.L. and D.O.-S.; resources, V.L. and D.O.-S.; data curation, V.L. and D.O.-S.; writing–original draft preparation, V.L. and D.O.-S.; writing–review and editing, V.L. and D.O.-S.; visualization, V.L. and D.O.-S.

Funding: This research received no external funding.

Conflicts of Interest: The authors declare no conflict of interest.

References

1. Rovelli, C. Loop quantum gravity. *Living Rev. Relat.* **1998**, *1*, 5. [CrossRef] [PubMed]
2. Rovelli, C.; Upadhya, P. Loop quantum gravity and quanta of space: A Primer. *arXiv* **1998**, arXiv:gr-qc/9806079.
3. Ambjorn, J.; Burda, Z.; Jurkiewicz, J.; Kristjansen, C.F. Quantum gravity represented as dynamical triangulations. *Acta Phys. Pol. B* **1992**, *23*, 991–1030. [CrossRef]

4. Ambjorn, J. Quantum gravity represented as dynamical triangulations. *Class. Quantum Gravity* **1995**, *12*, 2079–2134. [CrossRef]
5. Ambjørn, J.; Görlich, A.; Jurkiewicz, J.; Loll, R. Quantum Gravity via Causal Dynamical Triangulations. In *Springer Handbook of Spacetime*; Springer: Berlin/Heidelberg, Germany, 2014; pp. 723–741.
6. Connes, A.; Lott, J. Particle Models and Noncommutative Geometry (Expanded Version). *Nucl. Phys. B Proc. Suppl.* **1991**, *18*, 29–47. [CrossRef]
7. Aastrup, J.; Grimstrup, J.M. Intersecting connes noncommutative geometry with quantum gravity. *Int. J. Mod. Phys. A* **2007**, *22*, 1589–1603. [CrossRef]
8. Oriti, D. A Quantum field theory of simplicial geometry and the emergence of spacetime. *J. Phys. Conf. Ser.* **2007**, *67*, 012052. [CrossRef]
9. de Cesare, M.; Pithis, A.G.A.; Sakellariadou, M. Cosmological implications of interacting Group Field Theory models: Cyclic Universe and accelerated expansion. *Phys. Rev. D* **2016**, *94*, 064051. [CrossRef]
10. Gielen, S.; Sindoni, L. Quantum Cosmology from Group Field Theory Condensates: A Review. *arXiv* **2016**, arXiv:1602.08104
11. Gielen, S.; Oriti, D. Cosmological perturbations from full quantum gravity. *arXiv* **2017**, arXiv:1709.01095.
12. Oriti, D.; Ryan, J.P.; Thurigen, J. Group field theories for all loop quantum gravity. *New J. Phys.* **2015**, *17*, 023042. [CrossRef]
13. Gurau, R. Colored Group Field Theory. *Commun. Math. Phys.* **2011**, *304*, 69–93. [CrossRef]
14. Rivasseau, V. Constructive Tensor Field Theory. *arXiv* **2016**, arXiv:1603.07312.
15. Rivasseau, V. Random Tensors and Quantum Gravity. *arXiv* **2016**, arXiv:1603.07278.
16. Rivasseau, V. The Tensor Theory Space. *Fortschr. Phys.* **2014**, *62*, 835–840. [CrossRef]
17. Rivasseau, V. The Tensor Track, III. *Fortschr. Phys.* **2014**, *62*, 81–107. [CrossRef]
18. Rivasseau, V. The Tensor Track, IV. *arXiv* **2016**, arXiv:1604.07860.
19. Rivasseau, V. The Tensor Track: An Update. *arXiv* **2012**, arXiv:1209.5284.
20. Gurau, R. The complete 1/N expansion of colored tensor models in arbitrary dimension. *Ann. Henri Poincare* **2012**, *13*, 399–423. [CrossRef]
21. Gurau, R. The 1/N expansion of colored tensor models. *Ann. Henri Poincare* **2011**, *12*, 829. [CrossRef]
22. Gurau, R. The 1/N Expansion of Tensor Models Beyond Perturbation Theory. *Commun. Math. Phys.* **2014**, *330*, 973–1019. [CrossRef]
23. Carrozza, S.; Oriti, D.; Rivasseau, V. Renormalization of Tensorial Group Field Theories: Abelian U(1) Models in Four Dimensions. *Commun. Math. Phys.* **2014**, *327*, 603–641. [CrossRef]
24. Carrozza, S. Tensorial methods and renormalization in Group Field Theories. *arXiv* **2013**, arXiv:1310.3736.
25. Carrozza, S.; Oriti, D.; Rivasseau, V. Renormalization of a SU(2) Tensorial Group Field Theory in Three Dimensions. *Commun. Math. Phys.* **2014**, *330*, 581–637. [CrossRef]
26. Ben Geloun, J. Renormalizable Models in Rank $d \geq 2$ Tensorial Group Field Theory. *Commun. Math. Phys.* **2014**, *332*, 117–188. [CrossRef]
27. Lahoche, V.; Oriti, D. Renormalization of a tensorial field theory on the homogeneous space SU(2)/U(1). *arXiv* **2015**, arXiv:1506.08393.
28. Lahoche, V.; Oriti, D.; Rivasseau, V. Renormalization of an Abelian Tensor Group Field Theory: Solution at Leading Order. *J. High Energy Phys.* **2015**, *2015*, 95. [CrossRef]
29. Ben Geloun, J.; Livine, E.R. Some classes of renormalizable tensor models. *J. Math. Phys.* **2013**, *54*, 082303. [CrossRef]
30. Ousmane Samary, D.; Vignes-Tourneret, F. Just Renormalizable TGFT's on $U(1)^d$ with Gauge Invariance. *Commun. Math. Phys.* **2014**, *329*, 545–578. [CrossRef]
31. Ben Geloun, J.; Ousmane Samary, D. 3D Tensor Field Theory: Renormalization and One-loop β-functions. *Ann. Henri Poincare* **2013**, *14*, 1599–1642. [CrossRef]
32. Ben Geloun, J.; Rivasseau, V. A Renormalizable 4-Dimensional Tensor Field Theory. *Commun. Math. Phys.* **2013**, *318*, 69–109. [CrossRef]
33. Ben Geloun, J.; Toriumi, R. Renormalizable Enhanced Tensor Field Theory: The quartic melonic case. *arXiv* **2017**, arXiv:1709.05141.
34. Ben Geloun, J.; Bonzom, V. Radiative corrections in the Boulatov-Ooguri tensor model: The 2-point function. *Int. J. Theor. Phys.* **2011**, *50*, 2819–2841. [CrossRef]

35. Ben Geloun, J. Two and four-loop β-functions of rank 4 renormalizable tensor field theories. *Class. Quantum Gravity* **2012**, *29*, 235011. [CrossRef]
36. Ousmane Samary, D. Beta functions of $U(1)^d$ gauge invariant just renormalizable tensor models. *Phys. Rev. D* **2013**, *88*, 105003. [CrossRef]
37. Rivasseau, V. Why are tensor field theories asymptotically free? *EPL Europhys. Lett.* **2015**, *111*, 60011. [CrossRef]
38. Carrozza, S. Discrete Renormalization Group for SU(2) Tensorial Group Field Theory. *Ann. Inst. Henri Poincaré Comb. Phys. Interact.* **2015**, *2*, 49–112. [CrossRef]
39. Wilson, K.G. Renormalization Group and Critical Phenomena. I. *Phys. Rev. B* **1971**, *4*, 3174–3183. [CrossRef]
40. Wilson, K.G. Renormalization Group and Critical Phenomena. II. *Phys. Rev. B* **1971**, *4*, 3184–3205. [CrossRef]
41. Polchinski, J. Renormalization and Effective Lagrangians. *Nucl. Phys. B* **1984**, *231*, 269–295. [CrossRef]
42. Wetterich, C. Exact evolution equation for the effective potential. *Phys. Lett. B* **1993**, *301*, 90–94. [CrossRef]
43. Oriti, D. Levels of spacetime emergence in quantum gravity. *arXiv* **2018**, arXiv:1807.04875.
44. Oriti, D. Disappearance and emergence of space and time in quantum gravity. *Stud. Hist. Philos. Sci. B* **2014**, *46*, 186–199. [CrossRef]
45. Markopoulou, F. Conserved quantities in background independent theories. *J. Phys. Conf. Ser.* **2007**, *67*, 012019. [CrossRef]
46. Wilkinson, S.A.; Greentree, A.D. Geometrogenesis under Quantum Graphity: Problems with the ripening Universe. *Phys. Rev. D* **2015**, *92*, 084007. [CrossRef]
47. Geloun, J.B.; Martini, R.; Oriti, D. Functional Renormalisation Group analysis of Tensorial Group Field Theories on $\mathbb{R}^d$. *arXiv* **2016**, arXiv:1601.08211.
48. Geloun, J.B.; Martini, R.; Oriti, D. Functional Renormalization Group analysis of a Tensorial Group Field Theory on $\mathbb{R}^3$. *Europhys. Lett.* **2015**, *112*, 31001. [CrossRef]
49. Benedetti, D.; Lahoche, V. Functional Renormalization Group Approach for Tensorial Group Field Theory: A Rank-6 Model with Closure Constraint. *arXiv* **2015**, arXiv:1508.06384.
50. Benedetti, D.; Ben Geloun, J.; Oriti, D. Functional Renormalisation Group Approach for Tensorial Group Field Theory: A Rank-3 Model. *J. High Energy Phys.* **2015**, *2015*, 84. [CrossRef]
51. Ben Geloun, J.; Koslowski, T.A.; Oriti, D.; Pereira, A.D. Functional Renormalization Group analysis of rank 3 tensorial group field theory: The full quartic invariant truncation. *Phys. Rev. D* **2018**, *97*, 126018. [CrossRef]
52. Carrozza, S.; Lahoche, V. Asymptotic safety in three-dimensional SU(2) Group Field Theory: Evidence in the local potential approximation. *Class. Quantum Gravity* **2017**, *34*, 115004. [CrossRef]
53. Lahoche, V.; Ousmane Samary, D. Functional renormalization group for the U(1)-T_5^6 tensorial group field theory with closure constraint. *Phys. Rev. D* **2017**, *95*, 045013. [CrossRef]
54. Carrozza, S.; Lahoche, V.; Oriti, D. Renormalizable Group Field Theory beyond melonic diagrams: An example in rank four. *Phys. Rev. D* **2017**, *96*, 066007. [CrossRef]
55. Ben Geloun, J. Ward–Takahashi identities for the colored Boulatov model. *J. Phys. A* **2011**, *44*, 415402. [CrossRef]
56. Lahoche, V.; Ousmane Samary, D. Ward identity violation for melonic T^4-truncation. *arXiv* **2018**, arXiv:1809.06081.
57. Lahoche, V.; Ousmane Samary, D. Nonperturbative renormalization group beyond melonic sector: The Effective Vertex Expansion method for group fields theories. *arXiv* **2018**, arXiv:1809.00247.
58. Lahoche, V.; Ousmane Samary, D. Unitary symmetry constraints on tensorial group field theory renormalization group flow. *Class. Quantum Gravity* **2018**, *35*, 195006. [CrossRef]
59. Pérez-Sánchez, C.I. The full Ward–Takahashi Identity for colored tensor models. *arXiv* **2016**, arXiv:1608.08134.
60. Ousmane Samary, D. Closed equations of the two-point functions for tensorial group field theory. *Class. Quantum Gravity* **2014**, *31*, 185005. [CrossRef]
61. Ousmane Samary, D.; Pérez-Sánchez, C.I.; Vignes-Tourneret, F.; Wulkenhaar, R. Correlation functions of a just renormalizable tensorial group field theory: The melonic approximation. *Class. Quantum Gravity* **2015**, *32*, 175012. [CrossRef]
62. Wetterich, C. Average Action and the Renormalization Group Equations. *Nucl. Phys. B* **1991**, *352*, 529–584. [CrossRef]
63. Wetterich, C. Effective average action in statistical physics and quantum field theory. *Int. J. Mod. Phys. A* **2001**, *16*, 1951–1982. [CrossRef]

64. Berges, J.; Tetradis, N.; Wetterich, C. Nonperturbative renormalization flow in quantum field theory and statistical physics. *Phys. Rep.* **2002**, *363*, 223–386. [CrossRef]
65. Delamotte, B. An Introduction to the nonperturbative renormalization group. *Lect. Notes Phys.* **2012**, *852*, 49–132.
66. Nagy, S. Lectures on renormalization and asymptotic safety. *Ann. Phys.* **2014**, *350*, 310–346. [CrossRef]
67. Litim, D.F. Optimization of the exact renormalization group. *Phys. Lett. B* **2000**, *486*, 92–99. [CrossRef]
68. Litim, D.F. Derivative expansion and renormalization group flows. *J. High Energy Phys.* **2001**, *2001*, 59. [CrossRef]

© 2019 by the authors. Licensee MDPI, Basel, Switzerland. This article is an open access article distributed under the terms and conditions of the Creative Commons Attribution (CC BY) license (http://creativecommons.org/licenses/by/4.0/).

Review

Quantum Gravity on the Computer: Impressions of a Workshop

Lisa Glaser [1,*] and Sebastian Steinhaus [2,*]

[1] Institute for Mathematics, Astrophysics and Particle Physics, Radboud University, Comeniuslaan 4, 6525 HP Nijmegen, The Netherlands
[2] Perimeter Institute for Theoretical Physics, 31 Caroline Street North, Waterloo, ON N2L 2Y5, Canada
* Correspondence: glaser@science.ru.nl (L.G.); ssteinhaus@perimeterinstitute.ca (S.S.)

Received: 29 November 2018; Accepted: 10 January 2019; Published: 18 January 2019

Abstract: Computer simulations allow us to explore non-perturbative phenomena in physics. This has the potential to help us understand quantum gravity. Finding a theory of quantum gravity is a hard problem, but, in the last several decades, many promising and intriguing approaches that utilize or might benefit from using numerical methods were developed. These approaches are based on very different ideas and assumptions, yet they face the common challenge to derive predictions and compare them to data. In March 2018, we held a workshop at the Nordic Institute for Theoretical Physics (NORDITA) in Stockholm gathering experts in many different approaches to quantum gravity for a workshop on "Quantum gravity on the computer". In this article, we try to encapsulate some of the discussions held and talks given during this workshop and combine them with our own thoughts on why and how numerical approaches will play an important role in pushing quantum gravity forward. The last section of the article is a road map providing an outlook of the field and some intentions and goalposts that were debated in the closing session of the workshop. We hope that it will help to build a strong numerical community reaching beyond single approaches to combine our efforts in the search for quantum gravity.

Keywords: quantum gravity; computer simulations; numerical methods

---

Quantum Gravity is one of the big open questions in theoretical physics. Despite recent successes in particle physics and cosmology, most notably the discovery of the Higgs boson and the direct detection of gravitational waves, we are still lacking a consistent description of physics from smallest to largest scales that reconciles gravity and the quantum nature of matter. Possible signatures and effects of quantum gravity are numerous, from singularities in the early universe and black holes to the size and origin of the cosmological constant. In addition to these fundamental issues, one might hope that future experiments could reveal other traces of quantum gravity. Hence, it is of utmost importance to push the development of quantum gravity approaches to a point where they make reliable predictions, which will allow us to verify or falsify theories.

In the last several decades, many promising non-perturbative approaches to describe space-time at the smallest scales have been developed, (causal) dynamical triangulations [1,2], causal set theory [3,4], group field theory [5,6]/tensor models [7–9], loop quantum gravity [10,11], noncommutative geometry [12], spin foam models [13,14], and others. All of these postulate discrete structures that serve as a truncation on the number of degrees of freedom and allow for well-defined non-perturbative dynamics, akin to lattice gauge theories. Previous research, in which these models are substantially simplified to be computable, has led to impressive results, e.g., the resolution of the Big Bang singularity in loop quantum cosmology as a Big Bounce [15]. However, in order to make predictions for the full theory beyond simplifications and symmetry reduced models, we have to explore their deep non-perturbative regime. The bottleneck in this is the development of numerical techniques that allow us to efficiently extract results from the models, e.g., expectation values of observables and characteristics of different phases of the theory. Encouraging developments have been made in recent years and the purpose of our workshop was to compare these across different quantum gravity approaches.

Within the last 30 years, computers have revolutionized our lives and the way science is done. While the very first physics computer simulations were 2d Ising models with 8×8 sites, the technology and its applications have evolved rapidly: today's high performance simulations can predict the gravitational waves emitted by two colliding black holes or neutron stars [16] and explain the masses of hadrons using lattice QCD [17]. These developments have slowly percolated into the quantum gravity community, and have given rise to mainly computational approaches to the problem, such as (causal) dynamical triangulations. In these approaches, the path integral for dimensions larger than two is too complicated to be tackled analytically, but numerical methods, adapted from QCD and statistical mechanics, show how a ground state with macroscopic features emerges [2].

Other approaches have followed this example: in causal set theory, Monte Carlo simulations are used to explore the space of all possible partial orders [18], which includes all geometries but also highly non-manifold like structures, and more recently to compare the prediction of a fluctuating cosmological constant to cosmological data [19]. Furthermore, in spin foam models, numerical methods are indispensable to study the dynamics of spin foams with many degrees of freedom, e.g., via the means of coarse graining/renormalization [20,21]. Moreover, calculating the fundamental spin foam amplitudes also requires numerical techniques [22].

In the workshop, we brought together experts on these approaches to discuss recent developments in quantum gravity on the computer. During the discussion, two broad clusters of topics emerged; observables that we can measure and how we can reliably measure them, and numerical methods that are efficient for the different approaches.

In this article, we would like to summarize these discussions and distill their main ideas. We hope this will serve as a record of this workshop and a reference point for the current development of the field.

In the first section of this article, we begin with a brief introduction to the various approaches discussed during the workshop. The rest of the section is split into three subsections, where we discuss subtleties in defining the theories on the computer in Section 1.2, interesting observables in Section 1.3 and numerical methods in quantum gravity in Section 1.4. In Section 2, we summarize the road map discussion of the last day and try to map goalposts and aspirations for the community. A list of participants, slides and posters can be found on the website [23].

1. Approaches, Observables and Numerical Methods

1.1. Introduction to Various Approaches to Quantum Gravity

Throughout this article, we use different theories to exemplify the issues we want to discuss; as a reminder, let us give a quick overview over frequently mentioned theories and their salient aspects. In (causal) dynamical triangulations, the path integral over geometries is regularized by introducing a triangulation; this has been explored analytically for two-dimensional geometries and through simulations in two, three and four dimensions [2]. The sum over geometries is implemented by summing over all possible triangulations, where the size of the simplices is kept fixed. In dynamical triangulations, sometimes also-called Euclidean dynamical triangulations to distinguish it from causal dynamical triangulations, these simplices are equilateral, with all edges having the same length. In causal dynamical triangulations (CDT), the time-like edges of the simplices have a different edge length, and a time-foliation of the geometries is enforced. This leads to a very different ensemble of geometries in the path integral, and in particular suppresses changes in the topology which lead to degenerate behavior in Euclidean triangulations. In both approaches, the simulations use a simplicial version of the Regge action [24] to weight the geometries.

In two dimensions, dynamical triangulations can be solved using so-called matrix models. They give probability distributions for $N \times N$ random variables—thus also-called random matrices—where the matrices are invariant under the conjugation of the unitary group. The action of these models then consists of matrix invariants, e.g., the trace of a product of three matrices. This theory can be expanded in a sum over Feynman (ribbon) diagrams, where each diagram is dual to a discrete two-dimensional surface, e.g., a triangulation if the interaction term is three-valent [25]. Tensor models were developed to explore this method in higher dimensions. Instead of integrating over random matrices and thus obtaining two dimensional surfaces, here the integrals are over higher order random tensors with an action consisting of tensor invariants, thus creating surfaces in higher dimensions [8,9].

In several ways, group field theory (GFT) [5,6] is similar to tensor models. Using the same order interaction vertices, the combinatorics of the Feynman graphs of group field theories and tensor models agree. However, in addition to the combinatorics, the Feynman diagrams carry group theoretic data encoding a discrete geometry. The fields of the theory are defined on several copies of the underlying symmetry group. Crucially, this group manifold is not related to a space-time manifold. Instead, space-time is supposed to emerge from field excitations, e.g., as a condensate [26]. Group field theories are closely related to loop quantum gravity and spin foam models, e.g., group field theories can be constructed whose Feynman diagrams are given by spin foam amplitudes [27]. As for quantum field theories, the consistency of GFTs is investigated through renormalization [28].

Spin foam models [13,14] are a path integral approach to quantum gravity sometimes also referred to as covariant loop quantum gravity. Similar to previously described approaches, spin foams regularize the gravitational path integral by introducing a discretisation, a 2-complex, which is frequently chosen to be dual to a triangulation. The discrete geometry is again encoded in group theoretic data. For a given 2-complex, the path integral is implemented by summing over this data weighted by spin foam amplitudes. A priori, there is no rule determining which 2-complex to choose for a particular calculation, and generically the results depend on this. One way to address this is by also summing over all possible 2-complexes [29], which is systematically implemented in group field theory as discussed above. Alternatively, the refinement approach [20,30] aims at consistently defining the dynamics across various 2-complexes, e.g., by relating the theories by identifying states on the boundaries of these complexes.

Among the theories discussed here, loop quantum gravity (LQG) [10,11] is the only approach aiming to canonically quantize gravity. To this end, space-time, which is assumed to be globally hyperbolic, is split into space and time. Due to diffeomorphism symmetry, the theory is totally constrained, i.e., the Hamiltonian itself is a sum of constraints, such that the dynamics amount to gauge transformations. Moreover, these constraints form the so-called hyper surface deformation algebra. The goal of LQG is to quantize this algebra of constraints via Dirac quantization. To achieve this, one defines a kinematical Hilbert space, whose states do not satisfy the constraints, and constructs suitable constraint operators and an associated operator algebra. Then, the final goal is to find the physical Hilbert space, i.e., all states annihilated by the constraints. As an alternative method to tackle this issue, spin foam models have been developed as the "covariant" theory to LQG. While the two frameworks are closely related, e.g., the boundary states of modern spin foam models are kinematical states of LQG, their connection is not completely understood [31,32].

In causal set theory (CST) space-time is reduced to a partially ordered set. The discrete events are related to each other only if they are causally connected [3]. This leads to a minimal amount of structure assumed, which is why reconstructing space-time from a causal set is a complicated problem. There are methods to recover manifold properties from a causal set, maybe the simplest is to recover time like distance between two events by counting the longest chain between them. Recovering space like distances is more complicated, but still possible [33], and we can even define a measure allowing us to identify local regions, in the sense of regions that are small compared to the curvature scale of the manifold [34]. A causal set is considered to be manifold-like if it could have, with high likelihood, arisen from a statistical, so-called sprinkling, process on a given manifold (for a good definition and an algorithm to reconstruct the embedding, see [35]).

Modern string theory describes open and closed strings in $10+1$ dimensions [36]. Since higher dimensions often lead to more trouble in computer simulations this has not extensively been explored numerically. The old, bosonic, string theory, which describes the quantization of 2D surfaces covered by strings, can be studied numerically [37]. In fact, this was one of the motivating examples for the dynamical triangulations approach. This is often called non-critical string theory, and is an example of a theory that can be solved analytically but also explored using simulations [38].

One might debate whether noncommutative geometry really offers an approach to quantum gravity, or is purely a mathematical generalization of the concept of manifolds. A compact Riemannian manifold can be expressed as an algebra of functions acting on a Hilbert space together with a Dirac operator, a so-called spectral triple. Generalizing this description to allow for noncommutative function algebras then extended the space of geometries allowed [12]. While the original examples were concerned with infinite dimensional algebras, it is also possible to construct finite matrix algebras that then converge towards continuum geometries in the limit of infinite matrix size. These are the so-called fuzzy spaces which have recently been proposed as possible states in the path integral for quantum gravity [39,40].

The asymptotic safety approach [41] hinges on Weinberg's idea [42] that quantum gravity, described as a quantum field theory, is non-perturbatively renormalizable, i.e., possesses an interacting fixed point of the renormalization group flow in the ultraviolet described by a finite amount of relevant coupling constants. In practice, this hypothesis is investigated via the functional renormalization group [43], where one integrates out short scale degrees of freedom to derive an effective theory at larger scales. Generically, this operation cannot be performed in full generality and requires truncations, e.g., only particular terms in the action, called the theory space, are considered. To check whether signs of a fixed point persist once more interactions are allowed, the theory space is consistently enlarged. Work in this theory is mostly done using analytic methods or computational algebra packages, thus not exactly qualifying it as a numerical approach. However, it can play an important role in connecting continuum to discrete theories, and thus testing predictions.

1.2. Subtleties in Defining a Theory (on the Computer)

In the past decades, we have seen tremendous progress in the definition and development of non-perturbative approaches to quantum gravity. While some of these approaches share similarities, e.g., the use of discrete structures to calculate the non-perturbative regime, they are based on very different assumptions and key ideas about what a theory of quantum gravity should be. This variety itself is an opportunity and should be embraced rather than antagonized, yet it arises due to one of the great weaknesses of quantum gravity, the lack of experimental data to guide development. However, a diverse set of approaches gives us the chance to uncover universal features across theories and to reveal the consequences of their underlying assumptions. To make the most of this chance, it is indispensable to make an effort to better understand the theories and their connections to one another.

Since we rely on numerical simulations in order to compute results, e.g., expectation values of observables, it would be ideal to know exactly how to choose the parameters of the theory, i.e., coupling constants or size of the discretisaton, to reliably and efficiently get the "right" answer. A prime example is lattice QCD [44], in which numerical methods provide accurate predictions, e.g., for the hadron spectrum [17]. Two features are crucial for its success: its direct contact to experiments and the existence of a renormalizable continuum theory. On the one hand, the renormalizability of the continuum theory, thanks to asymptotic freedom [45], makes it possible to determine the dynamics, i.e., the coupling constants, at different scales. On the other hand, experimental data fix the parameters of the theory and tells us which scale is relevant for a particular process. Naturally, this does not imply that the simulations can be straightforwardly performed, but it allows practitioners of QCD to focus their efforts on specific regions in parameter space. In his talk, Jack Laiho described in detail the challenges one faces in lattice QCD calculations, in particular with respect to fermionic degrees of freedom.

Considering their importance for the success of lattice QCD, it seems crucial to tackle the issues of renormalization, an effective continuum theory and contact to experiments in quantum gravity. Here, we understand renormalization in the Wilsonian sense [46], as a scheme to relate theories defined at different scales. Usually, one orders the degrees of freedom according to scale, then integrates out those at shorter scales to derive an effective theory on larger scales, ultimately relating a microscopic dynamics to macroscopic physics. Additionally, choice of parameters, ambiguities or the choice of discretisation in the microscopic, allegedly fundamental, theory might give rise to different continuum dynamics strongly affecting observable quantities. We would summarize these as different phases of the theory. Conversely, by exploring this phase diagram, we can identify regions of universal behaviour of the theory, unravel phase transitions and fixed points and hence check the consistency of the theory.

Finding a systematic framework that can relate theories at different scales in a background independent setting is a challenge. In her talk, Bianca Dittrich described a thoroughly studied proposal in spin foam models based on the idea to relate theories by identifying the same physical transitions on different discretisations [47,48], and thus scales, in order to find theories giving consistent answers. In particular, she emphasized that consistency is indispensable for extracting predictions from the theory, e.g., expectation values of observables. To make progress in this direction, it is worthwhile to implement approximations and simplifications in order to cover a larger part of the parameter space with given resources.

1.2.1. Relating to the Continuum

Closely related to the issue of renormalization is the question of a continuum limit or at least an effective continuum theory compatible with any particular discrete quantum gravity theory. Ideally, such a continuum theory should agree with general relativity in a suitable limit, but it might also reveal crucial deviations that experiments can search for. One possible relation discussed at the workshop, was to compare the 3-volume correlations computed in causal dynamical triangulations with an effective continuum theory. Interestingly, this can also be studied in other approaches and explored using functional renormalization group techniques [49]. However, special care is advised when comparing continuum theories and their discretisations, as relating numerical simulations to analytic solutions can give rise to new subtleties.

A particularly interesting example is the bosonic string, as pointed out by Jan Ambjørn. The bosonic string can be solved with analytic as well as numerical methods; however, these two solutions do not necessarily agree. The reason for this conundrum is an incompatibility of the renormalization procedures; the continuum theory used dimensional regularization, and hence did not generate certain terms that arose in the discrete theory. Repeating the continuum calculation using a different regularization scheme made it possible to match the continuum and discrete results [38]. This showcases how much care needs to be taken in mapping analytic and numerical results onto each other. This illustrates that "brute force" applications of known methods may not be directly applicable in the context of quantum gravity.

1.2.2. Approximations and Simplifications

Another particularly contentious issue is the use of approximations and simplifications in computer simulations. The most obvious of these is that simulated models are necessarily much smaller than the real universe. The space-time volume of our universe is about 10^{240} Planck volumes in size, which does not compare well to, for example, the size of 10^2 Planck volumes currently examined in causal set theory. Some theories do better but in general the size of the universe in current discrete approaches is of the order 10^0 to 10^5 discrete building blocks. Of course, simulating the entire universe from quantum gravity might be too ambitious, and it might suffice to simulate a small region of space-time that recovers general relativity semi classically. The current best tests of general relativity limit corrections to appear on a scale below 47 μm [50][1]. Assuming we wanted to simulate a cube of space-time of this extend in all four dimensions, we would need to simulate $\sim 10^{122}$ Planck volumes, which is still out of reach by several orders of magnitude. One might argue that it is only a question of time, and better code to improve this, but no matter how good our code will be, the size of our simulations will be limited by the need to build our computer within the universe, and out of atoms. Hence, careful reasoning and planning about how to best use our limited resources is an important part of pushing forward numerical quantum gravity.

Many current simulations, in particular those using Monte Carlo methods, use Wick rotated geometries and statistical physics methods that allow for faster convergence of the results. However, it is not clear how the theory is affected by these changes, e.g., whether the ensemble with respect to which one samples geometries is significantly altered. Moreover, effects typical for quantum superpositions might be obscured by this choice. Conversely, in some approaches, it is not clear how to define a Wick rotation in the first place. The only way to control for these factors would be to find algorithms and implementations working with oscillating amplitudes. Some methods are tensor network renormalization techniques [51], which, on the other hand, are limited by numerical cost, which increases with the complexity of the studied system. A promising future direction might be simulating quantum systems on actual quantum computers. This could avoid the problem of complex phases and make it possible to explore superpositions of states. Even disregarding these fundamental points, there are still other simplifications and limitations we need to include in our theories, and it is important to be aware of these and explore their limits.

More specifically, theory dependent examples of simplifications are the foliation in CDT, the restriction to particular geometric intertwiners in current spin foam simulations, and the 2d orders in Causal set theory. In CDT, the simulations fix space-time to be foliated into constant time slices and to have a constant topology. This limitation has proven necessary to suppress so-called "baby-universes", which have been identified as the reason that dynamical triangulations are so irregular and do not show good continuum behavior in the simplest examinations. However, this limitation has been explored and challenged: a certain rescaling of the matrix model for 2D dynamical triangulation suppresses the baby universes and leads to the same behavior as CDT [52]. In addition, in more recent work, it was shown that simulations without a strict foliation, but still conserving a time-orientability condition, lead to a good continuum behavior in two and three dimensions [53,54]. These results are expected to also hold for 4d; however, they have not been tested yet there due to technical challenges. Nevertheless, they lend some credibility to the claims that the foliation in CDT is a simplification that does not overly constrain the phase space of the model. Moreover, this foliation can be used to employ an efficient algorithm, like the transfer matrix algorithm described in Andrzej Görlich's talk—see also Section 1.4. Additional hints for this come from recent results obtained in Euclidean dynamical triangulations with an additional curvature term. These simulations show a first order phase transition, but it is conjectured that this transition ends at a critical point that could be in the same universality class as CDT [55].

Spin foams also come initially with a large theory space that is hard to explore in full generality. Indeed, calculating the fundamental amplitudes of the theory is analytically not possible and requires a lot of computational resources, even for a single building block [22,56]. Studying larger spin foams is systematically tackled in the framework of renormalization [20] described in Bianca Dittrich's talk, where effective degrees of freedom at a coarser level are defined from the full amplitude without ad hoc truncations. A suitable numerical scheme are so-called tensor network techniques [51], in which the system is rewritten as a contraction of a network of tensors, i.e., multidimensional arrays. The goal

---

[1] This number is estimated by assuming that if extra dimensions of this size can not be experimentally excluded it gives a conservative upper limit on the scale at which quantum gravity would appear.

is to approximate said network by a coarser network efficiently by locally manipulating the tensors, e.g., sorting degrees of freedom according to their relevance via a singular value decomposition. These methods are particularly useful for identifying different phases of the model, e.g., in 2D analogue spin foam models [47,57–59] and 3D lattice gauge theories [47,60], where they revealed rich phase structures and phase transitions. Benjamin Bahr presented a closely related, but less holistic ansatz suitable for studying 4D spin foams: the underlying idea is to restrict the theory space to specific geometric shapes, e.g., cuboids [61] or frusta [62], which are coarse grained by requiring agreement of expectation values of observables across discretisations. Instead of a triangulation, the combinatorics of the foam are chosen to be hypercubic such that the coarse graining procedure can be straightforwardly iterated. Integrating over all possible shapes for the polyhedra is computationally prohibitively expensive, using the simpler cuboids allowed for calculating the first 4D RG flow of (restricted) spin foam models and revealed indications for a phase transitions and a UV-attractive fixed point [21,63]. Similarly, the spectral dimension in the cuboid case showed signs of a phase transition, where one phase is characterized by a dimension of four [64]. Moreover, a candidate for a similar fixed point was also found in the frusta setting, which extends the space of allowed geometries compared to the simpler cuboids [65].

As a last example, in most of the current explorations of the dynamics in causal set theory, the path integral is restricted to only a sum over the so-called 2D orders. These are a subclass of causal sets that can always be embedded into a plane, and that are dominated by causal sets that could arise from sprinkling in $1+1$d Minkowski space. Sumati Surya told us about these and their limitations, opportunities and possible extensions in some detail. This has two practical reasons, one is that the class of 2D orders is much smaller than that of all causal sets, and hence much easier to explore on the computer. The class of all possible causal sets grows like $2^{N^2/4}$, and is dominated by the very non-manifoldlike Kleitman–Rothschild orders [66], numerically this dominance sets in for $N \gg 90$ [18], which makes it very hard to explore in computer simulations. The other reason is that the choice of 2D orders immediately answers a number of questions one needs to debate before simulating causal sets, namely those concerned with how to pick the dimension of space-time, and hence the action to use in the simulations. Furthermore, it also allows us to store the causal set in a 2D array and thus enables faster algorithms.

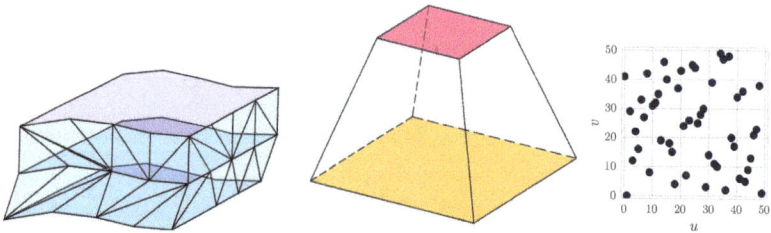

Figure 1. Examples of different simplifications used in the theories. From left to right, we see a causal triangulation with a foliation, a square frustum for spin foams and a 2D order causal set.

The three simplifications discussed above are illustrated in Figure 1.

The issue of limited numerical resources and necessary simplifications sheds a light onto the question how we can efficiently use them to reveal the properties of space-time and work towards making contact with experiments. Indeed, the latter point is certainly difficult for a theorist. Optimally, we would like to study observables that are well-defined both in discrete and continuum theories, yet connecting these to observable physical effects is usually a harder question. Thus, in order to deepen the connection between abstract quantum gravity theories and phenomenology, it is imperative that different quantum gravity approaches strive towards defining and studying the *same* observables. Then, we can unveil similarities and differences between approaches that might stimulate development and realization of experiments capable of testing multiple theories at once.

Indeed, there is great potential in studying observables in quantum gravity. In the next section, we present some proposals discussed during the workshop.

1.3. Observables

Many observables have been proposed to better understand quantum gravity. One possible consequence of quantum gravity that can arise in different ways and was discussed at length in our workshop is non-locality. However, the meaning of non-locality depends heavily on the context it is discussed in, and it is not completely understood how these notions are related. Even in classical general relativity, locality is a subtle concept. This is due to diffeomorphism invariance—the fundamental symmetry of general relativity, which encodes the independence of physics under the choice of coordinate systems. As a result, only diffeomorphism invariant quantities are physically relevant. For example, this condition severely complicates the definition of local subsystems in general relativity. Indeed, splitting systems into subsystems, e.g., to compute the entanglement entropy between them, is highly topical, yet in gravity it must be defined in a diffeomorphism invariant way. Similar to the situation in (lattice) gauge theories, this can be achieved by introducing new degrees of freedom and symmetries on the boundary separating the subsystems [67].

1.3.1. Non-Locality in Quantum Gravity

One facet of non-locality discussed at the workshop was in the context of effective quantum field theories. The essential idea put forward by Knorr and Saueressig is to define an effective continuum theory for CDT [49], where the terms and couplings in the effective field theory are chosen by comparing expectation values of the 3-volume covariance in both theories (for one specific value of parameters in CDT). The theory they define contains non-local terms, in the sense that the associated operator is a product of the field (and its derivative) evaluated at different points in space-time[2].

It remains an open question whether similar relations hold once more observables are considered or when the parameters in CDT are changed. Nevertheless, the potential implications of such non-local terms are intriguing and it will be interesting to explore whether similar effective quantum field theories can be derived from other discrete quantum gravity approaches. Non-locality can also arise in discrete theories. Spin foam models and CDT can be regarded as (at least initially) local theories, since they assign amplitudes to each building block of the triangulation, where these amplitudes only depend on the variables attached to said building block. A priori only neighbouring building blocks are "interacting" via the variables they are sharing. However, under coarse graining/renormalization, generically non-local interactions will arise involving building blocks beyond nearest neighbours.

Conversely, in causal set theory, non-locality is built into the theory from the beginning. A causal set element is connected to all causally related elements, with nearest neighbours corresponding to elements close to the light-cone. Since the light-cone in a generic space-time is non-compact, a causal set element, in an infinite causal set, would have infinitely many nearest neighbours. Additional non-locality also arises through a regularization parameter in the d'Alembertian for a scalar field [68–71]. This parameter is introduced to dampen fluctuations in the discrete theory, in effect smearing the derivative operator over several layers. This non-locality of a scalar field on a causal set, and of the causal set itself gives rise to phenomenological predictions which can be tested [72–74]. On the other hand, we do not have any current observational evidence of non-locality; hence, any non-local effects need to remain weak enough to not conflict with this. For example, when modelling the motion of a point particle through a causal set as traversing along the longest path, this introduces momentum diffusion above these stringent limits [75].

---

2 More precisely, the operator is the inverse d'Alembertian squared sandwiched by two Ricci scalars.

1.3.2. Summing over Topology

Another point of contention between different theories is the question: If we sum over different geometries, should we hold their topology fixed, or should we sum over all possible topologies? The first time this problem arose was in non-critical string theory, where the theory of strings requires a complete sum over all possible topologies of 2D geometries. If this is naively implemented, it leads to a dominance by topologies with many handles and a divergent sum [76]. However, modern models, and a suitable renormalization, make it possible to calculate the sum over topologies. In CDT, the topology of space and time individually are fixed, the simulations restrict spatial topology to either be a sphere or a torus, and, for numerical reasons, time is treated as periodic in most simulations (With the exception of the results in [77], which broadly agree with the results found using periodic time.). Causal set theory, on the other hand, does not restrict the path integral in this way, it does not even require all partial orders in the path integral to be geometries.

In loop quantum gravity and other canonical formulations of quantum gravity, the topology of space-time is usually fixed, since space-time is assumed to be globally hyperbolic in order to define the (3 + 1) split. Indeed, describing time evolution in a non-globally hyperbolic space-time is rather cumbersome. In spin foams, the issue is more subtle: any 2-complex that is compatible with given boundary data is in principle allowed. This concerns non-trivial 4-dimensional topology but also includes the possibility to change spatial topology between initial and final state. Whether one should sum over different topologies is debated in the literature [29] and depends on the interpretation of the spin foam. In the refinement approach [20], where the goal is to define a consistent theory across discretisations, one usually does not consider topology change. Since the goal is to identify the same physical process across discretisations, it is natural to fix the topology of the boundary states (If the boundary consists of several parts, e.g., initial and final state, their respective topologies can differ, but are kept fixed. Moreover, it is not clear how to embed states of differing topologies into a common discretisation.). The topology in the bulk is usually kept fixed as well, mostly for convenience. On the other hand, it is frequently argued that one should sum over all possible spin foams, i.e., all 2-complexes including all topologies. The most suitable framework to consider this are group field theories [27], in which the sum over spin foams appears as a perturbative expansion of Feynman diagrams generated by the action of the theory. Whether this theory is well-defined depends on whether it is renormalizable as a quantum field theory [28]. In the related tensor models, the sum over topologies is well defined. In particular, they can identify which topologies dominate in the perturbative expansion in the so-called large-N limit, where N is the dimension of the tensor indices [7,9].

1.3.3. Quantum Cosmology

One of the most promising routes for quantum gravity to make contact with experiments is cosmology. Quantum gravity effects may be revealed by future high precision experiments, e.g., the dynamics of the early universe might have left imprints in the cosmic microwave background. Indeed, it is an exciting prospect to see how quantum gravity can reshape our understanding of the origin of the universe, and whether it can augment, replace, or derive the current paradigm of inflation [78], which successfully explains the (almost) homogeneity, isotropy and flatness of our universe.

However, contact to this cosmological sector is difficult for non-perturbative theories of quantum gravity. While it is challenging for many approaches to define or model such a subsector of the theory, it is even more so to show how such a sector (plus fluctuations around it) could emerge dynamically. This difficulty is exemplified by the difference between loop quantum cosmology and cosmology in loop quantum gravity: in loop quantum cosmology, the system is symmetry reduced, e.g., to a homogeneous and isotropic universe, at the classical level before quantization. In loop quantum gravity, this symmetry reduction is to be implemented at the quantum level and explored in different directions. The early symmetry reduction in loop quantum cosmology simplifies calculations considerably and allows for interesting tests. For example, loop quantum cosmology (LQC) with an inflationary phase after the bounce predicts changes in the cosmic microwave background power spectrum compared to other inflationary models [79]. However, there are strong arguments that the early reduction in symmetry might remove crucial information from the theory; hence, to confirm the results of loop quantum cosmology, it is vitally important to derive symmetry reduced models from the full theory.

Antonia Zipfel gave a nice overview of the current status of the relation between LQC and LQG: In loop quantum gravity, this can be tackled directly by looking for suitably defined cosmological subsectors [80,81], e.g., by translating homogeneity and isotropy conditions on the phase space of general relativity to loop quantum gravity [82]. While this procedure is mathematically robust and relates well to the full theory, it is hard to implement in a given model and only approximately recreates the symmetry. A different idea is to study the evolution of coherent states, e.g., peaked on homogeneous and isotropic space-times [83,84]. Using these states, one can derive an effective Hamiltonian, as the expectation value of the constraint with respect to these semi-classical states. However, it is a priori not clear whether these coherent states are preserved under evolution. Another attempt to connect loop quantum gravity and loop quantum cosmology is called quantum reduced loop gravity [85], which relies on the kinematical construction of full loop quantum gravity. Then, a gauge fixing that restricts the spatial metric (and triads) to be diagonal is implemented. The symmetry reduction happens at the quantum level, where one only considers a dynamics which preserves the diagonal metric condition. Yet another perspective on the difference between imposing symmetry reduction before or after quantization is given in the context of general relativity in radial gauge [86]. In [87], the two methods are closely compared, beginning at the level of the phase space in order to identify the variables in the reduced theory with suitable phase space functions in the full theory. This analysis is continued at the quantum level, where the subsectors of the theories and the properties of operators can be compared. While a qualitative match between both theories is achieved at the kinematical level, one finds quantitative and state-dependent differences in the scaling behaviour of operators and mismatches in their commutators. This suggests that the identification of subsectors needs to be improved further.

This problem of how and where symmetry should be imposed arises in all non-perturbative approaches to quantum gravity and is dealt with in different ways. Reducing the symmetry classically and then quantizing leads to interesting toy models; however, it is important to test results obtained thus against results arising in the full non-perturbative regime. In particular, it would be fascinating if a non-perturbative path integral might give rise to a ground state that has some cosmological features. This is the case in causal dynamical triangulations, where the ground state of simulations in one phase shares some characteristics with Euclidean de Sitter space. The average volume profile of the 3-volumes, centered in time, measured in simulations assuming a spherical topology of space, matches the volume profile of Euclidean de Sitter [88]. In addition to the 3-volume, the authors also studied the covariance between 3-volumes at different time steps, which is highly peaked for the same time and drops off quickly for larger time steps. The spectral dimension in this phase of the simulations also points at 4-dimensional behavior [89]. This work has been extended to toroidal topology, where the volume profile becomes constant [90]. It can be argued that this creation of a de Sitter volume profile is a non-perturbative emergence of cosmology [91].

In group field theory, the emergence of a homogeneous state is tackled by considering condensate states [92], presented in detail by Steffen Gielen: the excitations, e.g., above a Fock vacuum, are interpreted as discrete "atoms of space-time". The heuristic idea of how a smooth, continuous space-time can emerge from this microscopic description is a hydrodynamic one. A large collection of these space-time atoms undergo a phase transition and condense similar to Bose Einstein condensates [93], such that a macroscopic, effective dynamics emerges from their collective behaviour. In the context of cosmology, one considers a gas of equilateral, uncorrelated building blocks that describe weakly interacting Bose Einstein condensates. In this setup, one can compute expectation values of observables, e.g., the volume of these building blocks. The dynamics is truncated to the classical equations of motion of the mean field of the condensate, analogous to the Gross–Pitaevskii equation of a Bose Einstein condensate. Remarkably, in this setting, the expectation values of observables satisfy effective Friedmann equations [94].

Another possible effect of non-perturbative quantum gravity on cosmology are discreteness effects. In theories where the discreteness is considered as fundamental, such as causal set theory, effects of the discreteness can lead to observable effects and explain certain phenomena. For example, the randomness inherent in the discrete causal sets can give rise to a cosmological constant of the correct order of magnitude [4]. Since this cosmological constant is no longer constant, it can vary over the age of the universe. This idea has given rise to phenomenological models that can match the standard model of cosmology and agrees with many of the observables known therein [19].

Another fascinating possibility of cosmological characteristics arising from non-perturbative dynamics was hinted at in the model system of the 2D orders in causal set theory. The closest causal set equivalent to the Hartle–Hawking wave function for the early universe is to simulate 2D orders that are fixed to begin with a single element and to end in an n element anti-chain, the closest causal set equivalent to a spatial hyper surface of fixed volume. In this model, the configurations with the highest likelihood are those that expand rapidly and are very homogeneous [95]. While this is a highly simplified model, it shows the possibility to generate features similar to those that have been observed in our universe from non-perturbative dynamics.

1.3.4. Measuring Dimension

The dimension of space-time is a familiar concept in general relativity. For each point of a d-dimensional manifold, we can find a small open region, which we can smoothly map to an open region of $\mathbb{R}^{(1,(d-1))}$ (for Lorentzian signature). As a property of the (topological) manifold, we will refer to this as the topological dimension. While this notion is intuitive in continuum gravity, it is not obvious how to define a dimension in (discrete) quantum gravity. In CDT or spin foam gravity, it is natural to regard the dimension of the fundamental building blocks as the topological dimension. However, whether this "dimension" also emerges on large scales is unclear: 4D hypercubes arranged in one long line appear one dimensional on large scales or some building blocks might be degenerate, i.e., possess vanishing 4-volume. Furthermore, in causal set theory, one cannot associate a dimension to discrete space-time events. These difficulties have motivated the definition and investigation of effective dimension measures that allow us to infer the dimension of space-time, e.g., via simulations, and potential physical consequences. Indeed, it is an important first test for any approach to quantum gravity, whether these generalized notions of dimension agree with our expectation of four space-time dimensions on large scales.

Moreover, this measured dimension may change with the scale at which space-time is probed, which is further motivation to study such observables. In general, there are several ways to define measures of dimension and all of them have different implications. One example is the Hausdorff dimension [96]. This notion of dimension can be assigned to all metric spaces, via the so-called Hausdorff measure. It is usually defined for a positive, real parameter d and considers all possible open coverings of the metric space, such that the diameter of each open subset is smaller than ϵ. The Hausdorff measure with respect to d and ϵ is then given by the infimum of the sum of all the diameters of the subsets to the power d. To find the Hausdorff dimension, we send $\epsilon \to 0$ and find the infimum d for which the Hausdorff measure vanishes, which is directly related to how quickly volumes of sets shrink with decreasing diameter. In quantum gravity, but also random geometries, this notion of dimension is frequently inferred from the exponential growth of volumes with respect to the radius. Then, the Hausdorff dimension is defined as the logarithmic derivative of the volume with respect to the radius, which can change as a function of the radius.

One definition of a scale dependent dimension prevalent in quantum gravity is the spectral dimension. After first rising to prominence in 4D simulations of CDT [89,97], it was also explored in many other theories, e.g., asymptotic safety [98], Hořava-Lifshitz gravity [99], causal set theory [100,101], loop quantum gravity [102,103], spin foams [64] and noncommutative geometry [104,105]. This dimension measure is related to studying the heat equation/a diffusion process on space-time. It crucially depends on the Laplace operator and its spectral properties. More precisely, this dimension measure is defined as the logarithmic derivative of the heat kernel. For calculations in discrete theories, the heat kernel can also be considered as the return probability of a random walker and thus calculated as an average over a sample of random walks. As a result, the spectral dimension encodes how space-time is ordered and thus might reveal interesting consequences for how matter propagates on this geometry; however, it is not obvious how to find this connection. Indeed, Giulia Gubitosi pointed out that the spectral dimension is problematic as a quantity of interest, since it cannot be measured experimentally. In most approaches, it is implemented purely on the geometry, since most computer simulations currently do not include matter. However, all currently conceived experimental measurements of space-time need test particles/test fields. Hence, to define practically observable quantities, we will need to work with matter. As an alternative, she suggested the thermal dimension, which tries to define a temperature based on the scaling of thermodynamic properties of matter [106]. This proposal is based on the dimension dependence in the Stefan Boltzmann law, describing the thermal radiation of a theory. While this is interesting in principle, and they show how it works in Hořava–Lifshitz gravity, where a preferred frame is available, the implementation for a non-perturbative theory is more challenging. Defining a temperature and other thermodynamic quantities in a background independent way can be complicated, and a nice discussion of these problems in the context of GFT is given in [107].

1.3.5. Other Observables

In addition to these larger overarching themes that were discussed at lengths and from the perspective of different theories, there were also some interesting observables discussed that are not yet explored in many theories. One such promising observable is the so-called quantum Ricci curvature [108]; the idea underlying this observable is the following: consider two points in a d-dimensional manifold with geodesic distance δ and imagine each of them to be surrounded by a sphere of radius ϵ. The points on the sphere are parametrized by a vector from the center to the sphere itself. Points on the two spheres are related by parallel transporting a vector from one sphere to the other along the geodesic connecting the centers. The average distance of points on these two spheres depends on the Ricci curvature 2-form (evaluated for the tangent vector of the geodesic connecting the centers), e.g., if the Ricci curvature is positive the average distance is smaller than δ. Since this concept is based on parallel transport, it is not straightforwardly applicable to the simplicial geometries underlying (C)DT. Instead, one considers the average distance between all points on the spheres, allowing the authors to identify the sign of curvature in constantly curved geometries. Moreover, they have tested it for 2D-(E)DT with spherical topology revealing a positively curved geometry modeled as a 5D sphere emphasizing the highly non-classical and fractal geometries in this model [109]. It will be interesting to see the behaviour of this observable in 4D CDT and whether it can be translated to other approaches of quantum gravity.

Another interesting route to explore is holography, more precisely the deep relation between a theory in the bulk and the theory on the boundary. This is most prominently represented in the continuum by the infamous AdS/CFT correspondence [110]. Naturally, it is an interesting question to ask whether these ideas can be generalized to non-perturbative approaches to quantum gravity and what the corresponding boundary theories might be. A very interesting calculation has been performed for the Ponzano–Regge model of 3D spin foams, studying the partition function and dual boundary theory of the twisted solid torus [111,112] (see also a similar calculation for linearized Regge calculus [113]). Strikingly, the results are consistent with results from perturbative quantum field theory in the continuum [114,115] and the characters of the BMS group are recovered. In addition, there have been several derivations for holographic entanglement entropy, more precisely the Ryu–Takayanagi formula for Renyi entropy, where the entropy (of the boundary theory) associated with a boundary subsystem is proportional to the minimal bulk area attached to this section of the boundary [116,117].

1.4. Numerical Methods in Quantum Gravity

Using physical intuition to develop our algorithms can lead to massive improvements in speed. At our workshop, we were introduced to two algorithms employing this, the chimera algorithm for numerical loop quantum cosmology and the transfer matrix algorithm for CDT.

1.4.1. The Chimera Algorithm

Parampreet Singh told us about the chimera algorithm developed in loop quantum cosmology. One of the key features of loop quantum cosmology is the resolution of the Big Bang singularity at the origin of the universe via a Big Bounce [15]. The vital dynamics responsible for this result is encoded in the quantum Hamiltonian constraint, which is a difference equation with uniform discretisation in volume. Indeed, for small volumes and large space-time curvature, these dynamics significantly deviate from the classical dynamics given by a Wheeler–DeWitt differential equation. However, for large space-time volume and small curvature, the quantum and classical dynamics agree very well. This is the fundamental idea underlying the chimera algorithm [118].

Difference equations, which describe the evolution in the deep quantum regime, are much more costly to compute compared to ordinary differential equations. This issue is emphasized as soon as the quantum states, whose evolution is studied, are not sharply peaked on classical configurations. Thus, the chimera algorithm introduces a hybrid lattice, where quantum evolution is only performed at small volumes and classical dynamics take over for large volumes. The intermediate region is carefully chosen for the results to match. That way, the numerical costs are drastically reduced and can be spent instead on studying the evolution of more general quantum states [118]. It will be interesting to explore whether this idea of a hybrid algorithm can be adapted in other approaches as well, e.g., spin foam models.

1.4.2. Transfer Matrix Approach

The transfer matrix approach is well known as a method to solve e.g., the Ising model in 2D. It is an analytic method based on splitting a problem into layers, e.g., time slices, calculating the dynamics of a single layer and then combining consecutive layers by convolution. This method also works to analytically solve CDT in two dimensions. In higher dimensions, CDT can only be explored using computer simulations; however, the foliated structure still makes it a prime candidate for the transfer matrix approach. Andrzej Görlich explained how this insight and a clever numerical implementation of the transfer matrix approach were used in [119]. In his algorithm, he measures the transfer matrix between slices of fixed size, such that he only has to simulate two slices of geometry, instead of the entire universe. This allows for more focused measurements, in particular improving the precision in measuring off-diagonal elements of the transition amplitude immensely. The transfer matrix approach also had another, unexpected, payoff in showing that what was before considered a single de Sitter like phase, called phase C, splits into two different phases: one that is de Sitter like, and one with alternating large and small spatial slices, called the bifurcation phase [119].

1.4.3. Markov Chain Monte Carlo Simulations

Other improvements in code are less about understanding the physical situation of the problem, and more about understanding the idiosyncrasies of a particular simulation. The most used tool to calculate a path integral using computer simulations are Markov Chain Monte Carlo (MCMC) simulations. How this algorithm is applied in their respective approaches to quantum gravity was explained by Andrzej Görlich, Jack Laiho, and Sumati Surya. In these, the ensemble of geometries is sampled with a frequency proportional to the weight of the configurations in the path integral, which makes it easy to calculate averages of observables directly from the simulations. A Markov chain is a chain in which the likelihood to transition between two states only depends on these two states. One algorithm to generate such a chain is the Metropolis–Hastings algorithm. This algorithm generates a Markov Chain by proposing a new state as a function of the old one. The function proposing these states depends on the theory used, e.g., in dynamical triangulations it is given by Pachner moves, which locally change the triangulation [120]. The probability to accept a proposed move then depends on the weight of the geometry in the path integral, given by e^{-S}, with S the action of the theory. One important feature is that a new state will always be accepted if it has a higher weight, but even states with a lower weight can still be accepted with a probability proportional to $e^{S_{old}-S_{new}}$. This makes it possible to prove that, if the moves are ergodic, the Metropolis–Hastings algorithm will find a global minimum of the action, if run sufficiently long. Unfortunately, the convergence towards this can be very slow, particularly close to phase transitions, since most proposed moves will have a very low probability of being accepted. This is known as critical slowing down and is related to the divergence of the correlation length arising there.

1.4.4. Parallel Rejection

One algorithm to overcome critical slowing down is the parallel rejection algorithm, discussed in Andrzej Görlich's talk. In general, MCMC simulations are difficult to parallelize, particular in gravity systems, since changes in the value of the action are non-local, hence proposed moves are not independent and need to be calculated sequentially. In practice, this means that most simulations are "naively parallelized" by just starting the simulations for several different points in the phase diagram, different parameter values, at the same time on different cores. Parallel rejection is an algorithm that does actual parallelization, for at least some regions of the phase diagram, where it can substantially speed up the algorithm. In regions of the parameter space where the acceptance rate of moves is particularly low, parallel rejection proposes and calculates multiple moves at the same time, on different cores. Once one of them is accepted (which can be ~1% or less of proposed moves), the geometry is updated and the parallel rejection restarted. This can drastically reduce the time in which the code remains in a given configuration in these regions of the phase diagram [55].

1.4.5. Adopting Methods from Other Fields

Another interesting option is to start using tools from other areas of science, in particular from computer science. There are many techniques that are solidly established in other fields but have not been widely adapted in numerical quantum gravity yet. For example, in QCD, the default algorithm for simulations is not the Metropolis–Hastings algorithm; instead, the algorithms in use are hybrid Monte Carlo [121] explained in Jack Laiho's talk. In these, the step of proposing a new configuration is guided by a supplementary Hamiltonian function. This Hamiltonian function is defined with respect to the probability distribution we wish to sample from and introduces fictitious momenta. While the momenta are randomly updated, a step in the configuration variables is chosen via a Metropolis algorithm with respect to the Hamiltonian equations of motion, which results in a faster convergence of results. This "Hamiltonian" is not be confused with an energy functional or the Hamiltonian constraint in gravity and serves the purpose to optimize the updating of configurations. The drawback of this method is that it requires continuous configurations, which makes it unsuitable for many proposals in quantum gravity.

Parallel tempering, discussed by Andrzej Görlich, also known as replica exchange MCMC sampling, is very useful when the configurations generated, e.g., from a Metropolis algorithm, are highly auto-correlated, that is correlated with previously generated configurations [122]. Such correlated systems may suffer from critical slowing down, in which the system is unlikely to leave said configuration via the proposed updates. To avoid this, the principle idea of parallel tempering is to start several processes with different model parameters and exchange the configurations at some point. That way, regions in configuration space that are rarely explored for certain parameters become accessible, improving the accuracy of the simulation. Often, it is proposed that the parameters only slightly vary. The probability to exchange the configurations has to satisfy the detailed balance condition. Crucially, this algorithm significantly reduces the auto-correlation time, i.e., the time it takes for configurations of the same Markov chain to become statistically independent.

In recent years, deep learning has emerged as a powerful method to analyze and search for patterns in large amounts of data. Image recognition is a particularly impressive example. Naturally, we would like to apply these methods to quantum gravity, e.g., to examine data generated in Monte Carlo simulations. In a nutshell, the idea of deep learning is to find an optimal function that quickly returns a desired output from a given large input. Deep neural networks are usually modelled to have an in- and output layer, chosen according to data and desired output. Between these layers, one implements several hidden layers, where each neuron in a hidden layer is connected to all neurons in the previous and following layer. These connections simply encode linear algebra operations on the data. Then, some of the data is used for training, i.e., these linear algebra operations get optimized to minimize a cost function. In supervised learning, where we know the desired result for a given sample, we would optimize the neural network to reproduce the already known answer. This is a particularly powerful approach when it comes to classification problems, e.g., recognizing handwriting or in quantum gravity it might help us to sort geometries with different properties. This approach bears an enormous potential, yet comes with some obvious drawbacks. Indeed, it is not obvious how to design a deep neural network that can successfully analyze a given data set. Moreover, even once we have successfully trained a neural network, it might not be obvious what the computer has learned, limiting our interpretation and understanding of the problem. Another problem is that, at least for the easiest to apply algorithms with the clearest outcomes, called supervised learning, we need to label the data set beforehand. This was beautifully demonstrated by Will Cunningham in his talk: he uses causal sets of known dimension, either $d = 2$, 3 or 4, to train a neural network to determine the dimension of the causal set. This is an interesting toy model, which demonstrates the opportunity and the challenge of machine learning at the same time. The characterization of the causal sets he obtained through the algorithm could have been done equally well using many tools that have been developed in causal set theory, e.g., the Myrrheim–Meyer dimension [123] or the interval abundance [34], which are fast and simple to use. On the other hand, these tools took time to develop and relied on our deep understanding of the problem, while the computer was obviously not aware of these and still able to solve it.

In general, quantum gravity, in particular in approaches that heavily use Monte Carlo simulations, offers many opportunities to apply machine learning. It will not always be possible to label the large data sets generated by Monte Carlo simulations and unravel all of their "hidden" information. Hence, using machine learning to search for structure within, and to possibly identify new observables, is a worthwhile endeavour.

2. Roadmap

We ended our workshop with a roadmap discussion, in which we began charting the future course of numerical quantum gravity. One outcome of this discussion is a flowchart, summarizing the discussions of the workshop and pointing towards questions for future consideration reproduced in Figure 2.

2.1. Open Science

One point of discussion which received particular emphasis in generating the roadmap was the desirability of conducting open science. In the context of computational quantum gravity, this would boil down to two points: open source code and open data.

Developing, optimizing and running code is an integral part of numerical research. In quantum gravity, currently most code is in principle available, but requires interested researchers to reach out and ask the authors for access. While this allows the authors to somewhat keep control of who has a current version of their code, it would be desirable for the development of the field to make their code open source. Open source means that the code is publicly available: anyone who is willing to improve the code or use and adapt it for their own research can do so without seeking permission of the authors.

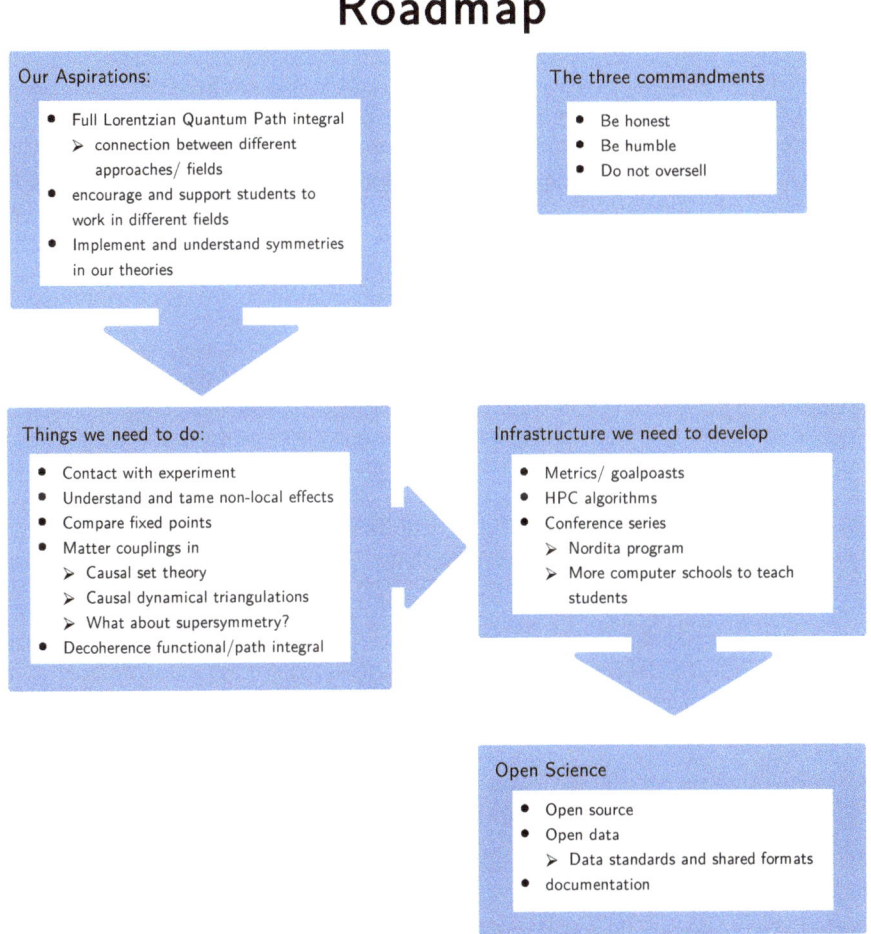

Figure 2. A slightly cleaned up version of the flowchart we created in the concluding discussion.

There exist many good solutions for storing and distributing open source code, for example the platform github (https://github.com/). This website is built around the version control software git (https://git-scm.com/). Git keeps track of any changes made to the code, and shows a history of the repository with all the changes made in various commits (In a commit, the author submits the changes made to the code to the repository.). This makes it possible to revert to previous versions and makes it straightforward to work with multiple people on the same project. Using git, anyone can download and use the code by cloning the repository. Then, they can also commit changes to the code, which must be approved by the owners of the repository.

We believe that this practice, which is standard even in closed source software development, has many advantages and its adoption by the quantum gravity community would boost the development of our numerical efforts. Indeed, authors deserve credit for their code, where the ideas and work that went into developing it are often not reflected in papers. Hence, an open source strategy makes this readily accessible, which makes it more easy for other researchers to contribute to the field and adopt ideas. Moreover, it makes research more credible and reliable, since the tools are readily available, to verify results. As a last point, open source is a good motivation to document and explain one's code, such that is usable for other people. That way, even once a researcher has left the field, their code is still available.

While there are several platforms and tools available to share and publish code, it is much more difficult to publish or exchange large amounts of data. Indeed, having public access to data generated in computer simulations is desirable for many reasons. Being able to recreate and confirm results greatly enhances the credibility of one's research. Moreover, it allows other researchers, e.g., from a different field like phenomenology, to analyze the data and use it for their own research. As a final point, large scale numerical simulations are costly and not every interested researcher has access to advanced numerical resources, e.g., in developing countries. Openly available data sets allow more people to learn and contribute to the field, e.g., students.

We envision different types of data to be uploaded, depending on the approach to quantum gravity. For approaches such as CDT and causal sets that rely on MCMC simulations, one option would be to upload the samples generated in the simulations allowing other researchers to investigate them for new patterns or calculate observables. In spin foam gravity uploading, exact values of spin foam amplitudes, which can then be readily used in other calculations, would be another straightforward example. All of the uploaded data should be reusable by other researchers and be accompanied by a documentation on how to use the data and/or contain a short program to demonstrate how to read in the data. In addition, modern efficient file formats should be used, like HDF5 or CSV, in particular for large files. On that note, uploaded files should be compressed to reduce internet traffic, and if the amount of data is particularly large, it should be split into smaller files.

Thus, at the end of the workshop, a plan was hatched to implement a quantum gravity open data repository. Together with other participants of the workshop, Benjamin Bahr, William Cunningham and Bianca Dittrich, as well as Erik Schnetter and Dustin Lang, we are actively developing the concept and realization of an open data initiative for the field of quantum gravity. The current plan is for this repository to be open to all numerical approaches, with a wiki-style website that allows authors to easily add data and link it to their papers on arXiv. Moreover, a DOI should be automatically assigned to each published dataset to make it straightforwardly citable. We are currently in the process of discussing the exact format and procuring funding for this endeavour, the working title is "The encyclopedia of quantum geometries". Any recommendations for sources of funding, or inspirations for how to set up such a project are very welcome.

2.2. Future

To keep the discussion alive, we plan to apply for funds and organize follow-up workshops and schools. Currently, the most likely schedule will be to have a school one year and then hold workshops in alternating years. For schools, we would imagine a format similar to that of "Making quantum gravity computable" at the Perimeter Institute. All participants seemed excited by the prospect of future such workshops, and there is a large number of interested parties that could not make it this year but have asked to be notified about future plans. One particularly exciting possibility would be to organize a Nordita programme on quantum gravity, embedding the workshop and the school into one longer event.

We hope to have done justice to all participants and their many brilliant contributions to this conference and hope there will be many further conferences on this exciting subject. Thus, until then, be honest, be humble and do not oversell.

Author Contributions: This article has been written and edited in close collaboration, hence it is impossible to split contributions.

Funding: We would like to thank Nordita for hosting and funding the workshop and supporting us in the organization. L.G. is supported by the People Programme (Marie Curie Actions) H2020 REA Grant No. 706349 "Renormalisation Group methods for discrete Quantum Gravity", which also supported speaker travel for the conference. This research was supported in part by the Perimeter Institute for Theoretical Physics. Research at the Perimeter Institute is supported by the Government of Canada through Innovation, Science and Economic Development Canada and by the Province of Ontario through the Ministry of Research, Innovation and Science. The conference was also supported by European Cooperation in Science and Technology (COST) Action MP1405 QSpace—'Quantum Structure of Spacetime'.

Acknowledgments: Traditionally, conferences and workshops disseminate their results through proceedings. This has the disadvantage of requiring the organizers to chase down the speakers, while it might still miss important points raised in discussions. Hence, we decided to write a recap of it ourselves, and would like to thank all participants for making the conference a huge success and giving us so much to think and write about.

Conflicts of Interest: The authors declare no conflict of interest.

References

1. Ambjorn, J.; Loll, R. Non-perturbative Lorentzian Quantum Gravity, Causality and Topology Change. *Nucl. Phys. B* **1998**, *536*, 407–434. [CrossRef]
2. Ambjørn, J.; Görlich, A.; Jurkiewicz, J.; Loll, R. Nonperturbative quantum gravity. *Phys. Rep.* **2012**, *519*, 127–210. [CrossRef]
3. Bombelli, L.; Lee, J.; Meyer, D.; Sorkin, R. Space-Time as a Causal Set. *Phys. Rev. Lett.* **1987**, *59*, 521–524. [CrossRef] [PubMed]
4. Sorkin, R.D. Causal sets: Discrete gravity. In *Lectures on Quantum Gravity*; Springer: Boston, MA, USA, 2005; pp. 305–327.
5. Freidel, L. Group field theory: An Overview. *Int. J. Theor. Phys.* **2005**, *44*, 1769–1783. [CrossRef]
6. Oriti, D. Group field theory as the microscopic description of the quantum spacetime fluid: A New perspective on the continuum in quantum gravity. *arXiv* **2007**, arXiv:0710.3276.
7. Rivasseau, V. Quantum Gravity and Renormalization: The Tensor Track. *AIP Conf. Proc.* **2011**, *1444*, 18–29.
8. Rivasseau, V. Random Tensors and Quantum Gravity. *SIGMA* **2016**, *12*, 69. [CrossRef]
9. Gurau, R. Invitation to Random Tensors. *SIGMA* **2016**, *12*, 94. [CrossRef]
10. Rovelli, C. *Quantum Gravity*; Cambridge University Press: Cambridge, UK, 2004.
11. Thiemann, T. *Modern Canonical Quantum General Relativity*; Cambridge University Press: Cambridge, UK, 2008.
12. Connes, A. *Noncommutative Geometry*; Academic Press: Cambridge, MA, USA, 1994.
13. Perez, A. The Spin Foam Approach to Quantum Gravity. *Living Rev. Relativ.* **2013**, *16*, 3. [CrossRef] [PubMed]
14. Rovelli, C.; Vidotto, F. *Covariant Loop Quantum Gravity*; Cambridge University Press: Cambrige, UK, 2014.
15. Ashtekar, A.; Pawlowski, T.; Singh, P. Quantum Nature of the Big Bang. *Phys. Rev. Lett.* **2006**, *96*, 141301. [CrossRef]
16. Pretorius, F. Evolution of Binary Black-Hole Spacetimes. *Phys. Rev. Lett.* **2005**, *95*, 121101. [CrossRef]
17. Dürr, S.; Fodor, Z.; Frison, J.; Hoelbling, C.; Hoffmann, R.; Katz, S.D.; Krieg, S.; Kurth, T.; Lellouch, L.; Lippert, T.; et al. Ab Initio Determination of Light Hadron Masses. *Science* **2008**, *322*, 1224–1227. [CrossRef] [PubMed]
18. Henson, J.; Rideout, D.P.; Sorkin, R.D.; Surya, S. Onset of the Asymptotic Regime for Finite Orders. *Exp. Math.* **2015**, *26*, 253–266. [CrossRef]
19. Zwane, N.; Afshordi, N.; Sorkin, R.D. Cosmological tests of Everpresent Λ. *Class. Quant. Grav.* **2018**, *35*, 194002. [CrossRef]
20. Dittrich, B. The continuum limit of loop quantum gravity—A framework for solving the theory. In *Loop Quantum Gravity: The First 30 Years*; Ashtekar, A., Pullin, J., Eds.; World Scientific: Singapore, 2017; pp. 153–179.
21. Bahr, B.; Steinhaus, S. Numerical evidence for a phase transition in 4d spin foam quantum gravity. *Phys. Rev. Lett.* **2016**, *117*, 141302. [CrossRef] [PubMed]
22. Donà, P.; Fanizza, M.; Sarno, G.; Speziale, S. SU(2) graph invariants, Regge actions and polytopes. *Class. Quant. Grav.* **2018**, *35*, 045011. [CrossRef]
23. Available online: nordita.org/qg2018 (accessed on 18 January 2019).
24. Regge, T. General Relativity without Coordinates. *Nuovo Cim.* **1961**, *19*, 558–571. [CrossRef]

25. Di Francesco, P.; Ginsparg, P.H.; Zinn-Justin, J. 2-D Gravity and random matrices. *Phys. Rep.* **1995**, *254*, 1–133. [CrossRef]
26. Gielen, S.; Oriti, D.; Sindoni, L. Cosmology from Group Field Theory Formalism for Quantum Gravity. *Phys. Rev. Lett.* **2013**, *111*, 031301. [CrossRef]
27. Oriti, D. *Group Field Theory and Loop Quantum Gravity*; Extended Draft Version of a Contribution to the Volume: 'Loop Quantum Gravity'; Ashtekar, A., Pullin, J., Eds.; World Scientific: Singapore, 2014.
28. Carrozza, S. Flowing in Group Field Theory Space: A Review. *SIGMA* **2016**, *12*, 70. [CrossRef]
29. Rovelli, C.; Smerlak, M. In quantum gravity, summing is refining. *Class. Quant. Grav.* **2012**, *29*, 055004. [CrossRef]
30. Bahr, B. On background-independent renormalization of spin foam models. *Class. Quant. Grav.* **2017**, *34*, 075001. [CrossRef]
31. Alesci, E.; Thiemann, T.; Zipfel, A. Linking covariant and canonical LQG: New solutions to the Euclidean Scalar Constraint. *Phys. Rev. D* **2012**, *86*, 024017. [CrossRef]
32. Thiemann, T.; Zipfel, A. Linking covariant and canonical LQG II: Spin foam projector. *Class. Quant. Grav.* **2014**, *31*, 125008. [CrossRef]
33. Eichhorn, A.; Surya, S.; Versteegen, F. Induced Spatial Geometry from Causal Structure. *arXiv* **2018**, arXiv:1809.06192.
34. Glaser, L.; Surya, S. Towards a Definition of Locality in a Manifoldlike Causal Set. *Phys. Rev. D* **2013**, *88*, 124026. [CrossRef]
35. Henson, J. Constructing an interval of Minkowski space from a causal set. *Class. Quant. Grav.* **2006**, *23*, L29–L35. [CrossRef]
36. Zwiebach, B. *A First Course in String Theory*; Cambridge University Press: Cambridge, UK, 2006.
37. Polchinski, J. String Theory. Vol. 1: An Introduction to the Bosonic String. In *Cambridge Monographs on Mathematical Physics*; Cambridge University Press: Cambridge, UK, 2007.
38. Ambjorn, J.; Makeenko, Y. String theory as a Lilliputian world. *Phys. Lett. B* **2016**, *756*, 142–146. [CrossRef]
39. Barrett, J.W.; Glaser, L. Monte Carlo simulations of random non-commutative geometries. *J. Phys. A* **2016**, *49*, 245001. [CrossRef]
40. Glaser, L. Scaling behaviour in random non-commutative geometries. *J. Phys. A* **2017**, *50*, 275201. [CrossRef]
41. Reuter, M.; Saueressig, F. Quantum Einstein Gravity. *New J. Phys.* **2012**, *14*, 055022. [CrossRef]
42. Weinberg, S. Ultraviolet Divergences in Quantum Theories of Gravitation. In *General Relativity: An Einstein Centenary Survey*; Cambridge University Press: Cambridge, UK, 1980; pp. 790–831.
43. Wetterich, C. Exact evolution equation for the effective potential. *Phys. Lett. B* **1993**, *301*, 90–94. [CrossRef]
44. Gupta, R. Introduction to lattice QCD: Course. In Proceedings of the 68th Session Summer School in Theoretical Physics "Probing the Standard Model of Particle Interactions", Les Houches, France, 28 July–5 September 1997; pp. 83–219.
45. Gross, D.J.; Wilczek, F. Ultraviolet Behavior of Nonabelian Gauge Theories. *Phys. Rev. Lett.* **1973**, *30*, 1343–1346. [CrossRef]
46. Wilson, K.G.; Kogut, J.B. The Renormalization group and the epsilon expansion. *Phys. Rep.* **1974**, *12*, 75–200. [CrossRef]
47. Dittrich, B.; Mizera, S.; Steinhaus, S. Decorated tensor network renormalization for lattice gauge theories and spin foam models. *New J. Phys.* **2016**, *18*, 053009. [CrossRef]
48. Dittrich, B.; Steinhaus, S. Time evolution as refining, coarse graining and entangling. *New J. Phys.* **2014**, *16*, 123041. [CrossRef]
49. Knorr, B.; Saueressig, F. Towards reconstructing the quantum effective action of gravity. *Phys. Rev. Lett.* **2018**, *121*, 161304. [CrossRef] [PubMed]
50. Tanabashi, M.; et al. [Particle Data Group]. Review of Particle Physics. *Phys. Rev. D* **2018**, *98*, 030001. [CrossRef]
51. Levin, M.; Nave, C.P. Tensor renormalization group approach to 2D classical lattice models. *Phys. Rev. Lett.* **2007**, *99*, 120601. [CrossRef] [PubMed]
52. Ambjorn, J.; Correia, J.; Kristjansen, C.; Loll, R. On the relation between Euclidean and Lorentzian 2D quantum gravity. *Phys. Lett. B* **2000**, *475*, 24–32. [CrossRef]
53. Jordan, S.; Loll, R. De Sitter universe from causal dynamical triangulations without preferred foliation. *Phys. Rev. D* **2013**, *88*, 044055. [CrossRef]

54. Jordan, S.; Loll, R. Causal Dynamical Triangulations without preferred foliation. *Phys. Lett. B* **2013**, *724*, 155–159. [CrossRef]
55. Laiho, J.; Bassler, S.; Coumbe, D.; Du, D.; Neelakanta, J.T. Lattice Quantum Gravity and Asymptotic Safety. *Phys. Rev. D* **2017**, *96*, 064015. [CrossRef]
56. Dona, P.; Sarno, G. Numerical methods for EPRL spin foam transition amplitudes and Lorentzian recoupling theory. *Gen. Relativ. Grav.* **2018**, *50*, 127. [CrossRef]
57. Dittrich, B.; Martín-Benito, M.; Schnetter, E. Coarse graining of spin net models: Dynamics of intertwiners. *New J. Phys.* **2013**, *15*, 103004. [CrossRef]
58. Dittrich, B.; Martin-Benito, M.; Steinhaus, S. Quantum group spin nets: Refinement limit and relation to spin foams. *Phys. Rev. D* **2014**, *90*, 024058. [CrossRef]
59. Dittrich, B.; Schnetter, E.; Seth, C.J.; Steinhaus, S. Coarse graining flow of spin foam intertwiners. *Phys. Rev. D* **2016**, *94*, 124050. [CrossRef]
60. Delcamp, C.; Dittrich, B. Towards a phase diagram for spin foams. *Class. Quant. Grav.* **2017**, *34*, 225006. [CrossRef]
61. Bahr, B.; Steinhaus, S. Investigation of the Spinfoam Path integral with Quantum Cuboid Intertwiners. *Phys. Rev. D* **2016**, *93*, 104029. [CrossRef]
62. Bahr, B.; Kloser, S.; Rabuffo, G. Towards a Cosmological subsector of Spin Foam Quantum Gravity. *Phys. Rev. D* **2017**, *96*, 086009. [CrossRef]
63. Bahr, B.; Steinhaus, S. Hypercuboidal renormalization in spin foam quantum gravity. *Phys. Rev. D* **2017**, *95*, 126006. [CrossRef]
64. Steinhaus, S.; Thürigen, J. Emergence of Spacetime in a restricted Spin-foam model. *Phys. Rev. D* **2018**, *98*, 026013. [CrossRef]
65. Bahr, B.; Rabuffo, G.; Steinhaus, S. Renormalization in symmetry restricted spin foam models with curvature. *arXiv* **2018**, arXiv:1804.00023.
66. Kleitman, D.J.; Rothschild, B.L. Asymptotic enumeration of partial orders on a finite set. *Trans. Am. Math. Soc.* **1975**, *205*, 205–220. [CrossRef]
67. Donnelly, W.; Freidel, L. Local subsystems in gauge theory and gravity. *J. High Energy Phys.* **2016**, *2016*, 102. [CrossRef]
68. Dowker, F.; Glaser, L. Causal set d'Alembertians for various dimensions. *Class. Quant. Grav.* **2013**, *30*, 195016. [CrossRef]
69. Glaser, L. A closed form expression for the causal set d'Alembertian. *Class. Quant. Grav.* **2014**, *31*, 095007. [CrossRef]
70. Aslanbeigi, S.; Saravani, M.; Sorkin, R.D. Generalized causal set d'Alembertians. *J. High Energy Phys.* **2014**, *2014*, 24. [CrossRef]
71. Sorkin, R.D. Does Locality Fail at Intermediate Length-Scales. In *Approaches to Quantum Gravity*; Oriti, D., Ed.; Cambridge University Press: Cambridge, UK, 2007; pp. 26–43.
72. Belenchia, A.; Benincasa, D.M.T.; Liberati, S. Nonlocal scalar quantum field theory from causal sets. *J. High Energy Phys.* **2015**, *2015*, 36. [CrossRef]
73. Belenchia, A.; Benincasa, D.M.; Martín-Martínez, E.; Saravani, M. Low energy signatures of nonlocal field theories. *Phys. Rev. D* **2016**, *94*, 061902. [CrossRef]
74. Belenchia, A.; Benincasa, D.M.T.; Liberati, S.; Marin, F.; Marino, F.; Ortolan, A. Tests of Quantum Gravity induced non-locality via opto-mechanical quantum oscillators. *Phys. Rev. Lett.* **2016**, *116*, 161303. [CrossRef]
75. Philpott, L.; Dowker, F.; Sorkin, R.D. Energy-momentum diffusion from spacetime discreteness. *Phys. Rev. D* **2009**, *79*, 124047. [CrossRef]
76. Ambjørn, J.; Durhuus, B. *Quantum Geometry: A Statistical Field Theory Approach*; Cambridge University Press: Cambridge, UK, 1997.
77. Cooperman, J.H.; Miller, J.M. A first look at transition amplitudes in (2 + 1)-dimensional causal dynamical triangulations. *Class. Quant. Grav.* **2014**, *31*, 035012. [CrossRef]
78. Linde, A.D. *Inflationary Cosmology*; Springer: Berlin/Heidelberg, Germany, 2008.
79. Agullo, I.; Morris, N.A. Detailed analysis of the predictions of loop quantum cosmology for the primordial power spectra. *Phys. Rev. D* **2015**, *92*, 124040. [CrossRef]

80. Fleischhack, C. Kinematical Foundations of Loop Quantum Cosmology. In Proceedings of the Conference on Quantum Mathematical Physics: A Bridge between Mathematics and Physics, Regensburg, Germany, 29 September–2 October 2014; pp. 201–232.
81. Hanusch, M. Projective Structures in Loop Quantum Cosmology. *J. Math. Anal. Appl.* **2015**, *428*, 1005–1034. [CrossRef]
82. Beetle, C.; Engle, J.S.; Hogan, M.E.; Mendonça, P. Diffeomorphism invariant cosmological sector in loop quantum gravity. *Class. Quant. Grav.* **2017**, *34*, 225009. [CrossRef]
83. Assanioussi, M.; Dapor, A.; Liegener, K.; Pawłowski, T. Emergent de Sitter Epoch of the Quantum Cosmos from Loop Quantum Cosmology. *Phys. Rev. Lett.* **2018**, *121*, 081303. [CrossRef] [PubMed]
84. Dapor, A.; Liegener, K. Cosmological Effective Hamiltonian from full Loop Quantum Gravity Dynamics. *Phys. Lett. B* **2018**, *785*, 506–510. [CrossRef]
85. Alesci, E.; Cianfrani, F. Quantum Reduced Loop Gravity and the foundation of Loop Quantum Cosmology. *Int. J. Mod. Phys. D* **2016**, *25*, 1642005. [CrossRef]
86. Bodendorfer, N.; Lewandowski, J.; Świeżewski, J. General relativity in the radial gauge: Reduced phase space and canonical structure. *Phys. Rev. D* **2015**, *92*, 084041. [CrossRef]
87. Bodendorfer, N.; Zipfel, A. On the relation between reduced quantisation and quantum reduction for spherical symmetry in loop quantum gravity. *Class. Quant. Grav.* **2016**, *33*, 155014. [CrossRef]
88. Ambjorn, J.; Jurkiewicz, J.; Loll, R. Emergence of a 4-D world from causal quantum gravity. *Phys. Rev. Lett.* **2004**, *93*, 131301. [CrossRef]
89. Ambjørn, J.; Jurkiewicz, J.; Loll, R. The Spectral Dimension of the Universe is Scale Dependent. *Phys. Rev. Lett.* **2005**, *95*, 171301. [CrossRef] [PubMed]
90. Ambjørn, J.; Gizbert-Studnicki, J.; Görlich, A.; Grosvenor, K.; Jurkiewicz, J. Four-dimensional CDT with toroidal topology. *Nucl. Phys. B* **2017**, *922*, 226–246. [CrossRef]
91. Glaser, L.; Loll, R. CDT and cosmology. *C. R. Phys.* **2017**, *18*, 265–274. [CrossRef]
92. Gielen, S.; Sindoni, L. Quantum Cosmology from Group Field Theory Condensates: A Review. *SIGMA* **2016**, *12*, 82. [CrossRef]
93. Oriti, D. The universe as a quantum gravity condensate. *C. R. Phys.* **2017**, *18*, 235–245. [CrossRef]
94. Oriti, D.; Sindoni, L.; Wilson-Ewing, E. Bouncing cosmologies from quantum gravity condensates. *Class. Quant. Grav.* **2017**, *34*, 04LT01. [CrossRef]
95. Glaser, L.; Surya, S. The Hartle–Hawking wave function in 2D causal set quantum gravity. *Class. Quant. Grav.* **2016**, *33*, 065003. [CrossRef]
96. Schleicher, D. Hausdorff Dimension, Its Properties, and Its Surprises. *Am. Math. Mon.* **2007**, *114*, 509–528. [CrossRef]
97. Benedetti, D.; Henson, J. Spectral geometry as a probe of quantum spacetime. *Phys. Rev. D* **2009**, *80*, 124036. [CrossRef]
98. Calcagni, G.; Eichhorn, A.; Saueressig, F. Probing the quantum nature of spacetime by diffusion. *Phys. Rev. D* **2013**, *87*, 124028. [CrossRef]
99. Horava, P. Spectral Dimension of the Universe in Quantum Gravity at a Lifshitz Point. *Phys. Rev. Lett.* **2009**, *102*, 161301. [CrossRef] [PubMed]
100. Eichhorn, A.; Mizera, S. Spectral dimension in causal set quantum gravity. *Class. Quant. Grav.* **2014**, *31*, 125007. [CrossRef]
101. Carlip, S. Dimensional reduction in causal set gravity. *Class. Quant. Grav.* **2015**, *32*, 232001. [CrossRef]
102. Modesto, L. Fractal Structure of Loop Quantum Gravity. *Class. Quant. Grav.* **2009**, *26*, 242002. [CrossRef]
103. Calcagni, G.; Oriti, D.; Thürigen, J. Dimensional flow in discrete quantum geometries. *Phys. Rev. D* **2015**, *91*, 084047. [CrossRef]
104. Alkofer, N.; Saueressig, F.; Zanusso, O. Spectral dimensions from the spectral action. *Phys. Rev. D* **2015**, *91*, 025025. [CrossRef]
105. Barrett, J.W.; Druce, P.J.; Glaser, L. Spectral estimators for finite non-commutative geometries. Unpublished work, 2019.
106. Amelino-Camelia, G.; Brighenti, F.; Gubitosi, G.; Santos, G. Thermal dimension of quantum spacetime. *Phys. Lett. B* **2017**, *767*, 48–52. [CrossRef]
107. Kotecha, I.; Oriti, D. Statistical Equilibrium in Quantum Gravity: Gibbs states in Group Field Theory. *New J. Phys.* **2018**, *20*, 073009. [CrossRef]

108. Klitgaard, N.; Loll, R. Introducing Quantum Ricci Curvature. *Phys. Rev. D* **2018**, *97*, 046008. [CrossRef]
109. Klitgaard, N.; Loll, R. Implementing quantum Ricci curvature. *Phys. Rev. D* **2018**, *97*, 106017. [CrossRef]
110. Maldacena, J.M. The Large N limit of superconformal field theories and supergravity. *Int. J. Theor. Phys.* **1999**, *38*, 1113–1133. [CrossRef]
111. Dittrich, B.; Goeller, C.; Livine, E.R.; Riello, A. Quasi-local holographic dualities in non-perturbative 3d quantum gravity I—Convergence of multiple approaches and examples of Ponzano–Regge statistical duals. *Nucl. Phys. B* **2019**, *938*, 807–877. [CrossRef]
112. Dittrich, B.; Goeller, C.; Livine, E.R.; Riello, A. Quasi-local holographic dualities in non-perturbative 3d quantum gravity II—From coherent quantum boundaries to BMS3 characters. *Nucl. Phys. B* **2019**, *938*, 878–934. [CrossRef]
113. Bonzom, V.; Dittrich, B. 3D holography: From discretum to continuum. *J. High Energy Phys.* **2016**, *2016*, 208. [CrossRef]
114. Barnich, G.; Gonzalez, H.A.; Maloney, A.; Oblak, B. One-loop partition function of three-dimensional flat gravity. *J. High Energy Phys.* **2015**, *2015*, 178. [CrossRef]
115. Oblak, B. Characters of the BMS Group in Three Dimensions. *Commun. Math. Phys.* **2015**, *340*, 413–432. [CrossRef]
116. Han, M.; Hung, L.Y. Loop Quantum Gravity, Exact Holographic Mapping, and Holographic Entanglement Entropy. *Phys. Rev. D* **2017**, *95*, 024011. [CrossRef]
117. Chirco, G.; Oriti, D.; Zhang, M. Group field theory and tensor networks: Towards a Ryu–Takayanagi formula in full quantum gravity. *Class. Quant. Grav.* **2018**, *35*, 115011. [CrossRef]
118. Diener, P.; Gupt, B.; Singh, P. Chimera: A hybrid approach to numerical loop quantum cosmology. *Class. Quant. Grav.* **2014**, *31*, 025013. [CrossRef]
119. Ambjørn, J.; Gizbert-Studnicki, J.; Görlich, A.; Jurkiewicz, J. The effective action in 4-dim CDT. The transfer matrix approach. *J. High Energy Phys.* **2014**, *2014*, 34. [CrossRef]
120. Pachner, U. P.L. Homeomorphic Manifolds are Equivalent by Elementary Shellings. *Eur. J. Comb.* **1991**, *12*, 129–145. [CrossRef]
121. Neal, R.M. MCMC using Hamiltonian dynamics. *arXiv* **2012**, arXiv:1206.1901.
122. Newman, M.E.J.; Barkema, G.T. *Monte Carlo Methods in Statistical Physics*; Clarendon Press: Wotton-under-Edge, UK, 1999.
123. Meyer, D.A. The Dimension of Causal Sets. Ph.D. Thesis, Massachusetts Institute of Technology, Cambridge, MA, USA, 1988.

© 2019 by the authors. Licensee MDPI, Basel, Switzerland. This article is an open access article distributed under the terms and conditions of the Creative Commons Attribution (CC BY) license (http://creativecommons.org/licenses/by/4.0/).

Review

Holographic Entanglement in Group Field Theory

Goffredo Chirco [1,2]

[1] Romanian Institute of Science and Technology (RIST), Strada Virgil Fulicea 17, 400022 Cluj-Napoca, Romania; goffredo.chirco@aei.mpg.de
[2] Max Planck Institute for Gravitational Physics, Albert Einstein Institute, Am Mühlenberg 1, 14476 Golm, Germany

Received: 5 August 2019; Accepted: 4 October 2019; Published: 9 October 2019

Abstract: This work is meant as a review summary of a series of recent results concerning the derivation of a *holographic* entanglement entropy formula for generic open spin network states in the group field theory (GFT) approach to quantum gravity. The statistical group-field computation of the Rényi entropy for a bipartite network state for a simple interacting GFT is reviewed, within a recently proposed dictionary between group field theories and random tensor networks, and with an emphasis on the problem of a consistent characterisation of the entanglement entropy in the GFT second quantisation formalism.

Keywords: holographic entanglement; quantum gravity; group field theory; random tensor networks; quantum many-body physics; quantum geometry

1. Introduction

Two potentially revolutionary ideas have inspired much work in contemporary theoretical physics. Both ideas herald from the use of the general information theoretic approach to the problem of quantum gravity. The first idea is that—*the world is holographic*—with the physics of several semi-classical systems (by which we mean systems in which matter is treated quantum-mechanically while spacetime and geometry are treated classically) entirely captured on spacetime regions of one dimension lower.

The evidence for holography is already suggestive when considering classical gravitational systems like black holes, or more general causal horizons, and the semi-classical physics of quantum fields in their vicinity. This evidence is strengthened by numerous results in the context of the AdS/CFT correspondence [1,2], where, in particular, holography as found in semi-classical gravitational systems is put in correspondence with general properties of non-gravitational, purely quantum mechanic *dual* many-body systems. Indeed, holographic features deeply characterise condensed matter physics—hence the suggestion that there may be a *purely quantum mechanic origin of holography* that may in fact underlie classical, gravitational, geometric physics as studied in the general relativistic context.

The second idea is that *geometry itself originates from entanglement*. The recent quantum information-theoretic paradigm for gravity has provided a new vision of the cosmos wherein the universe, together with its topology, its geometry and its macroscopic dynamics, arise from the entanglement between the fundamental constituents of some exotic underlying quantum system.

In this direction, in particular, along with the increasing impact of condensed matter physics in string theory and gauge/gravity duality, people have started exploring the use of tensor network (TN) algorithms [3–12] from condensed matter theory in quantum gravity [13–20], providing interesting insights on holographic duality and its generalisation in terms of geometry/entanglement correspondence [21,22]. Approaches like the AdS/MERA [10,23], where the geometry of the auxiliary tensor network decomposition of the quantum many-body vacuum state is interpreted as a representation of the dual spatial geometry, are providing an intriguing constructive framework for investigating holography beyond AdS/CFT [13–16,24].

However, if everything is quantum at its root, and this is true not only for ordinary systems living in spacetime but for *space, time and geometry themselves*, then this implies that the very holographic behaviour of the universe is the result of purely quantum properties of the microscopic constituents of spacetime. In addition, this line of research, therefore, indicates a strong need for a quantum foundation of holography.

In fact, background independence naturally leads to a quantum description of the universe in terms of fundamental quantum many-body physics of discrete and purely algebraic microscopic constituents [25–34], from which spacetime emerges only at an effective, approximate level, out of a texture of quantum correlations [35–43]. This means that the two suggestions that holography has a purely quantum origin and that geometry itself comes from entanglement are extremely natural when seen from the perspective of quantum gravity formalisms, in which spacetime and geometry are ultimately emergent notions. However, more than that, differently from the semiclassical framework, in the non-perturbative scheme, holography as detected in gravitational systems, as well as any macroscopic feature of our geometric universe, not only would result from purely quantum properties of the microscopic constituents of spacetime, but they can— *only*—be understood in this light.

This perspective is manifest in the *Group Field Theory* (GFT) formalism, a promising convergence of the insights and results from matrix models [44,45], loop quantum gravity and simplicial approaches into a background independent quantum field theory setup. The GFT approach to quantum gravity [30, 46–49] provides a very general quantum many-body formulation of the spacetime micro-structure, for instance of the spin networks and discrete quantum geometry states of Loop Quantum Gravity [25–27, 29], with a Fock space description where the quantum GFT fields create and annihilate elementary building blocks of space, interpreted as $(d-1)$-simplices in d spacetime dimensions, organised in nontrivial combinatorial tensor network structures.

As a higher order generalisation of matrix models, the GFT formalism at the same time provides a field-theoretic and inherently covariant framework for generalising the *tensor networks* approach to the holographic aspects of quantum many-body systems in condensed matter and in the AdS/CFT context. This makes GFT a very effective framework to investigate how space-time geometry, together with its holographic behaviour and macroscopic dynamics, arise from entanglement between the fundamental constituents.

In this paper, we review a series of recent results [18–20] concerning the definition of entanglement entropy in the GFT framework and the characterisation of its holographic behaviour. In particular, we focus on the definition of the notion of entanglement entropy in the second quantised formalism of GFT setting and on the set of choices which eventually lead to a holographic behaviour for the entanglement entropy.

The manuscript is organised as follows: Section 2 shortly reviews the framework of group field theory while focussing the attention of the reader on those aspects of the GFT fields that play a major role in the forthcoming derivation. Section 3 introduces the second quantisation formalism for group field theory, defines the notion of multi-particle state observable for quantum geometry in the GFT Fock space and specifies a class of GFT coherent state basis, necessary for the definition of the entanglement entropy expectation value. In Section 4, the statistical derivation of the Rényi entropy for a bipartite GFT open spin network state is reviewed, with an emphasis on the Fock space setting. The formal mapping of the expectation values of the Rényi entropy to BF theory amplitudes is described and the resulting divergence degree and scaling of the entropy analysed. Section 5 briefly comments on the results on the holographic scaling for the interacting GFT case. A brief discussion and an appendix on the notion of coherent states over-completeness close the manuscript.

2. Group Field Theory

Group field theories (GFT) are quantum field theories defined on d copies of a compact Lie group G with combinatorially non-local interactions. The dynamics of the GFT field

$$\phi : G^{\times d} \to \mathbb{C}$$

are specified by a probability measure

$$d\mu_C(\phi, \overline{\phi}) \exp\left(-S_{\text{int}}[\phi, \overline{\phi}]\right), \tag{1}$$

comprised of a *Gaussian measure* $d\mu_C$, associated with a positive covariance kernel operator C defining the propagator of the theory [45],[1] and a *perturbation* around it, given by an interaction term $S_{\text{int}}[\phi, \overline{\phi}]$, generically parametrized as

$$S_{\text{int}}[\phi, \overline{\phi}] = \sum_{\mathcal{I}} \sum_{\substack{p+q=\mathcal{I} \\ p,q \geq 0}} \int \prod_{p=1}^{p} dg'_p \, \overline{\phi}(g'_p) \prod_{q=1}^{q} dg_q \, \phi(g_q) \, \lambda_{\mathcal{I}} V_{\mathcal{I}}(g'_1, \ldots, g'_p; g_1, \ldots, g_q),$$

where $\mathcal{I}$ denotes a term in the set of elementary interactions, $V_{\mathcal{I}}$ is the specific monomial in the fields associated with the interaction, and $\lambda_{\mathcal{I}}$ is the respective coupling constant. Together with the field valence d and symmetry, the specific choice of the covariance and interaction kernels identifies the GFT model.

We are particularly concerned with three peculiar aspects of the GFT formalism, which will combine at the hearth of the following derivation. The first is that the dynamical d-valent GFT field ϕ combinatorially behaves as an infinite dimensional rank-d tensor, with indices labelled by elements of the compact Lie group G [18]. This is apparent in the combinatorially non-local structure of the interaction kernels $V_{\mathcal{I}}$, as functions on $G^{d \times |\mathcal{I}|}$ for finite sets of interactions. The kernels $V_{\mathcal{I}}$ do not impose coincidence of points in the group space $G^{\times d}$, but the whole set of the $d \times |\mathcal{I}|$ field arguments is partitioned into pairs, convoluted "strandwise" by the kernel,

$$V_{\mathcal{I}}(\{g\}_{\mathcal{I}}) = V(\{g_p\}_i \{g_q^{-1}\}_j). \tag{2}$$

Indeed, one can see GFTs as higher-rank, infinite dimensional generalizations of random matrix models [45]. For instance, if we take G as $\mathbb{Z}_N$, then group fields identically reduce to rank-d tensors,[2] where integrability with respect to the discrete Dirac measure μ is satisfied for all fields considered.

The second aspect is that specific GFT partition functions, where the group G is identified with the local gauge subgroup of gravity and the kernels properly chosen, define generating functions

---

[1] We use a vector notation for the configuration space variables and its Haar measure (We will also use the short-hand notation:

$$\overline{\varphi}_1 \cdot \varphi_2 = \int d\mathbf{g} \, \overline{\varphi}_1(\mathbf{g}) \varphi_2(\mathbf{g}),$$

for any two square-integrable functions φ_1 and φ_2 on G^d.)

$$\mathbf{g} = (g_1, \ldots, g_d) \in G^d, \qquad d\mathbf{g} = dg_1 \cdots dg_d.$$

[2] A rank-d tensor T with index cardinality N is a complex field on d copies of the cyclic group $\mathbb{Z}_N$:

$$T : \mathbb{Z}_N^{\times d} \to \mathbb{C},$$

which defines a state in $\mathcal{H}_{d,N}$ the space of

tensors with fixed rank d and index cardinality N. Neglecting the structure of the cyclic group, $\mathcal{H}_{d,N}$ is reduced to $\mathbb{C}^{N^d}$.

for the covariant quantization of Loop Quantum Gravity (LQG) in terms of spin foam models (for instance, [28]). In particular, LQG spin network states, describing three-dimensional discrete quantum geometries at the boundary of the spin foam transition amplitudes, can be expressed as expectation values of specific GFT operators. In a formalism of second quantisation for the GFT, spin network boundary states then become elements of the GFT Fock space and spin network vertices, intended as atoms of space that can be put in direct correspondence with fundamental GFT quanta that are created or annihilated by the field operators of GFT.

When G is set to correspond to the local gauge group of GR, e.g., the Lorentz group $SO(1, d-1)$ or its universal covering, or $SU(2)$ in dimension $d = 3$, the gauge symmetry leads to an invariance of the GFT action under the (right) diagonal action of G. GFT fields are constrained to satisfy a *gauge invariance condition*, defined as a global symmetry of the GFT field under simultaneous translation of its group variables:

$$\forall h \in G, \quad \phi(g_1 h, \ldots, g_d h) = \phi(g_1, \ldots, g_d). \tag{3}$$

The gauge invariance condition, or *closure constraint*, is the main dynamical ingredient of GFT models for quantum BF theory in arbitrary dimension. In $d = 3$, $SU(2)$ BF theory can be interpreted as a theory of Euclidean gravity, and therefore $SU(2)$ GFT with closure constraint provides a natural arena in which to formulate 3D Euclidean quantum gravity models. A typical example is the Boulatov model [50] (which generates Ponzano–Regge spin foam amplitudes [51]), which will constitute an important ingredient of the following derivation.

The third aspect is that, for such (simplicial) GFT models, endowed with a geometric interpretation of the dynamical fields, the very field-theoretic nature of the GFT formalism provides a powerful tool to describe quantum geometry states as a peculiar *quantum many-body* systems in the formalism of *second quantisation*. For large systems in quantum mechanics, we know that the concept of a particle fades away and is replaced by the notion of an excitation of a given mode of the field representing the particle. Similarly, in the GFT description, we expect the solid graph description of spin networks quantum geometry to fade into a dynamical net of excitations of the GFT field over a vacuum. Given the tensorial behaviour of the GFT field, such a quantum many-body description turns quantum geometry states into collective, purely combinatorial and algebraic analogues of quantum *tensor networks states*.

These three ingredients together make the GFT second quantised formalism an specially convenient setting to *quantitatively* investigate the relation between geometry and entanglement in quantum gravity, taking advantage of the most recent techniques and tools of quantum statistical mechanics, information theory and condensed matter theory. In particular, quantum tensor network algorithms provide a constructive tool to investigate the roots of the holographic behaviour of gravity at the quantum level.

3. The GFT Fock Space

In a second quantisation scheme [30], multi-particle states of the quantum GFT field $\phi(\mathbf{g})$ can be organised in a Fock space $\mathbb{F}$ generated by a Fock vacuum $|0\rangle$ and field operators

$$\hat{\varphi}(\mathbf{g}) \equiv \hat{\varphi}(g_1, \cdots, g_d), \quad \hat{\varphi}^\dagger(\mathbf{g}) \equiv \hat{\varphi}^\dagger(g_1, \cdots, g_d), \tag{4}$$

assumed to be invariant under the diagonal action of the group $\hat{\varphi}(\mathbf{g}h) \equiv \hat{\varphi}(g_1 h, \cdots, g_V h) = \hat{\varphi}(\mathbf{g})$, for $h \in G$, consistently with (3), and to obey canonical commutation relations (bosonic statistics)

$$[\hat{\varphi}(\mathbf{g}), \hat{\varphi}^\dagger(\mathbf{g}')] = \int dh \prod_1^d \delta\left(g_i h(g_i'^{-1})\right) = \mathbb{1}_G(g_i, g_i'), \tag{5}$$

$$[\hat{\varphi}(\mathbf{g}), \hat{\varphi}(\mathbf{g}')] = [\hat{\varphi}^\dagger(\mathbf{g}), \hat{\varphi}^\dagger(\mathbf{g}')] = 0.$$

In these terms, the Fock vacuum is the state with no quantum geometrical or matter degrees of freedom, satisfying $\hat{\varphi}(\mathbf{g})|0\rangle = 0$ for all arguments. A generic single-particle state $|\phi\rangle$ with wavefunction ϕ, consisting of a d-valent node with links labelled by group elements $\mathbf{g} \equiv (g_1, ..., g_d)$, is written as

$$|\phi\rangle = \int_{G^{\times d}} d\mathbf{g}\, \phi(\mathbf{g}) |\mathbf{g}\rangle, \tag{6}$$

where $d\mathbf{g} = \prod_{i=1}^{d} dg_i$ is the Haar measure, ϕ is an element of the single-particle Hilbert space $\mathbb{H} = L^2(G^{\times d})$, and $|\mathbf{g}\rangle = |g_1\rangle \times ... \times |g_d\rangle$ a basis (of Dirac distributions) in $\mathbb{H}$. For $d = 4$, this is the space of states of a quantum tetrahedron [52]. Notice then that one can think of (6) as the analogue of a quantum tensor state where each group element g corresponds to an index i variable in a continuous (∞-dim) index space $\mathbb{H}_i = L^2(G)$.[3]

The complete Fock space is given by the direct sum of n-particle sectors $\mathbb{H}^{\otimes n}$, restricted to states that are invariant under graph automorphisms of vertex relabelling in the spin network picture, in order to consider multi-particle states that only depend on the intrinsic combinatorial structure of their interaction pattern (a discrete counterpart of continuum diffeomorphisms). Therefore, one has

$$\mathbb{F} \equiv \bigoplus_{n \geq 0}^{\infty} \text{sym}[\mathbb{H}^{\otimes n}]. \tag{9}$$

The symmetry condition is consistent with the assumed *bosonic statistics* and it implies *indistinguishability* for the quanta of the quantum many-body system.

Generic GFT observables $\hat{\mathcal{O}}[\hat{\varphi}, \hat{\varphi}^\dagger]$ in the Fock space are defined in terms of a series of many-body operators expressed as a function of the field operators (or of the basic creation/annihilation operators). For instance, a (n)-body operator $\mathcal{O}_n$ acting on $\mathfrak{p}$ vertices and resulting in $\mathfrak{q}$ particles is written as [30]

$$\hat{\mathcal{O}}_n[\hat{\varphi}, \hat{\varphi}^\dagger] = \sum_{n=2}^{\infty} \sum_{\substack{p+q=n \\ p,q \geq 0}} \int \prod_{p=1}^{p} d\mathbf{g}'_p\, \hat{\varphi}^\dagger(\mathbf{g}'_p) \prod_{q=1}^{q} d\mathbf{g}_q\, \hat{\varphi}(\mathbf{g}_q)\, \mathcal{O}_n(\mathbf{g}'_1, \cdots, \mathbf{g}'_p; \mathbf{g}_1, \cdots, \mathbf{g}_q), \tag{10}$$

where $\mathcal{O}_n(\mathbf{g}'_1, \cdots, \mathbf{g}'_p; \mathbf{g}_1, \cdots, \mathbf{g}_q)$ denote the matrix elements of a corresponding first-quantized operator.

3.1. Multi-Particle State Observables

Analogous with the case of the single particle state in (6), we can think of a quantum many-body system as a collective state generated by the action of a *multi-particle group-field operator* on the Fock vacuum. We can define a **product** n-particle state, comprising n disconnected nodes, by the multiple action of the creation field operator in the group representation of the Fock space, e.g.,

$$|\mathbf{g}_1, \mathbf{g}_2, \cdots, \mathbf{g}_n\rangle = \frac{1}{\sqrt{n!}} \prod_{a=1}^{n} \hat{\varphi}^\dagger(\mathbf{g}_a) |0\rangle. \tag{11}$$

---

[3] The analogy with a tensor state is apparent again for $G = \mathbb{Z}_N$. Let $\mathcal{H}_{d,N}$ be the space of tensors with fixed rank d and index cardinality N. Neglecting the structure of the cyclic group, $\mathcal{H}_{d,N}$ is reduced to $\mathbb{C}^{N^d}$. The linear structure, the scalar product and the completeness of $\mathbb{C}^{N^d}$ establish $\mathcal{H}_{d,N}$ to be a Hilbert space. A basis of $\mathcal{H}_{d,N}$ is chosen by $|i_1, ..., i_d\rangle$, defined as:

$$\langle j_1, ...j_d | i_1, ..., i_d \rangle = \delta_{i_1, j_1} \cdot ... \cdot \delta_{i_d, j_d}. \tag{7}$$

With respect to this basis, we decompose a tensor T into its components $T_{i_1...i_d}$, which introduces an isomorphism to $\mathbb{C}^{N^d}$:

$$|T\rangle =: \sum_{i_1,...,i_d \in \mathbb{Z}_N} T_{i_1...i_d} |i_1, ..., i_d\rangle. \tag{8}$$

Because the $\hat{\varphi}^\dagger(g_a)$ commute with each other, the order of the particles does not affect the state and we have $|..., g_a, ..., g_b, ...\rangle = |..., g_b, ..., g_a, ...\rangle$. The n-particle state $|g_1, ..., g_n\rangle$ defines a multi-particle basis of the Fock space $\mathbb{F}$, with orthogonality relation given by

$$\langle g'_1, ..., g'_n | g_1, ..., g_n \rangle = \frac{1}{n!} \sum_{\pi \in S_n} \prod_{a=1}^{n} \int dh_a\, \delta^4(g'_a h_a g^{-1}_{\pi(a)}). \tag{12}$$

In addition, because of the required symmetry of the fields $\hat{\varphi}(gh) = \hat{\varphi}(g)$, $h \in G$, the n-particle state is right invariant

$$|\cdots, g_a h_a, \cdots\rangle = |\cdots, g_a, \cdots\rangle. \tag{13}$$

Finally, the resolution of identity in the Fock space can be written in terms of the n-particle state as

$$\mathbb{1}_\mathbb{F} = |0\rangle\langle 0| + \sum_{n=1}^{\infty} \int \prod_{a=1}^{n} dg_a\, |g_1, \cdots, g_n\rangle\langle g_1, \cdots, g_n|. \tag{14}$$

One can check immediately that $\mathbb{1}_\mathbb{F} |g_1, \cdots, g_n\rangle = |g_1, \cdots, g_n\rangle$.

A generic GFT **multi-particle state**, analogous to (6), will then be based on a specific configuration of n fields, characterized by a multi-particle wavefunction Ψ_n, generated by a multi-particle operator in the GFT Fock space

$$\hat{\Psi}_n[\hat{\varphi}^\dagger] = \sum_{\substack{p+q=n \\ p,q \geq 0}} \int \prod_{p=1}^{p} dg'_p\, \hat{\varphi}^\dagger(g'_p) \prod_{q=1}^{q} dg_q\, \hat{\varphi}(g_q)\, \Psi_n(g'_1, \cdots, g'_p; g_1, \cdots, g_q) \tag{15}$$

via the repeated action of the GFT field operators on the Fock space.

We are interested in the structure of quantum correlations of the multi-particle state $\Psi_n(g'_1, \cdots, g'_p; g_1, \cdots, g_q)$. As it is the case for any highly entangled quantum many-body system, disentangling the information on the quantum correlations of Ψ_n is highly nontrivial. Therefore, we focus on a special class of multi-particle operators $\hat{\Psi}_\Gamma$, where the wave-function is explicitly constructed via a pairwise contractions scheme of single node states, in correspondence with a given **network architecture** Γ. We write

$$\hat{\Psi}_\Gamma[\hat{\varphi}, \hat{\varphi}^\dagger](g_\partial) = \sum_{\substack{p+q=n \\ p,q \geq 0}} \int \prod_{p=1}^{p} dg'_p\, \hat{\varphi}^\dagger(g'_p) \prod_{q=1}^{q} dg_q\, \hat{\varphi}(g_q) \prod_{\ell \in \Gamma}^{L} dh_\ell \prod_{\ell \in \Gamma} L_\ell\left(g'_{s(\ell)} h_\ell g^{-1}_{t(\ell)}\right) \tag{16}$$

with $\hat{\varphi}^\dagger$ operators generating the nodes connected by link kernels L_ℓ to form an open network with g_∂ dangling indices, via an overall integration over $g_{\ell \in \Gamma}$. The expression of the link convolution kernel connecting the nodes pairwise is left generic at this stage, with the only requirement to preserve the overall gauge invariance of the network state.

We shall see such a class of multi-particle operators as *tensor networks* operators, where we reduce the entanglement structure of the multi-particle state to local correlations induced by the generic link kernels L, propagated non-locally via nodes.

Notice that the most generic tensor network state of the theory would involve superpositions of both network architectures (combinatorial structures) and number of particles corresponding to the same number of boundary degrees of freedom,

$$\hat{\Psi}[\hat{\varphi}, \hat{\varphi}^\dagger](g_\partial) = \sum_{\{\Gamma\}} \sum_{n=2}^{\infty} \sum_{\substack{p+q=n \\ p,q \geq 0}} \int \prod_{p=1}^{p} dg'_p\, \hat{\varphi}^\dagger(g'_p) \prod_{q=1}^{q} dg_q\, \hat{\varphi}(g_q)\, \mathcal{L}^\Gamma_{p,q}(g'_1, \cdots, g'_p; g_1, \cdots, g_q), \tag{17}$$

where, for simplicity, we indicate by $\mathcal{L}_{p,q}^{\Gamma}$ the set of pairwise link convolutions (gluing) functions associated with a given graph Γ, and some suitable symmetry quotient factor removing equivalent graph configurations is assumed.

3.2. GFT Coherent State Basis

In the context of quantum gravity, we specify the GFT formalism to the case $G = SU(2)$, the relevant local gauge subgroup of gravity, and we understand the group elements g as a generalisation of the embedded parallel transports (holonomies) of the gravitational G-connection of loop quantum gravity. The symmetric GFT d-valent fields as d-simplices (convex polyhedra, e.g., for $d = 4$, these are tetrahedra), with d number of faces labelled by dual Lie algebra-valued flux variables, become single "quanta" of twisted geometry states expressed in terms of quantum spin network basis [29].

In this setting, on the one hand, we are interested in working with states in the Fock space that can be eventually put in relation with extended macroscopic 3D geometries. To this aim, the natural choice consists of looking for a coherent state basis in $\mathbb{F}$, defined by exponential operators providing desirable coherence properties, having *macroscopic* occupation numbers for given modes controlled by the wave-function [53–56]. More concretely, this choice will allow us to compute quantum averages of many-body systems in thermal equilibrium using functional integrals over group field configurations [18–20].

The simplest class of such states is given by the single-particle (condensate) coherent states

$$|\varphi\rangle \equiv \frac{1}{\mathcal{N}_\varphi} \exp\left[\int d\mathbf{g}\, \phi(\mathbf{g}) \hat{\varphi}^\dagger(\mathbf{g})\right] |0\rangle \equiv \frac{1}{\mathcal{N}_\varphi} \sum_{n=0}^{\infty} \frac{1}{n!} \prod_a \left[\int d\mathbf{g}_a\, \phi(\mathbf{g}_a) \hat{\varphi}^\dagger(\mathbf{g}_a)\right] |0\rangle$$

$$\equiv \frac{1}{\mathcal{N}_\varphi} \sum_{n=0}^{\infty} \frac{1}{\sqrt{n!}} \int [d\mathbf{g}]^n\, \phi(\mathbf{g}_1) \times \ldots \times \phi(\mathbf{g}_n) |\mathbf{g}_1, \cdots, \mathbf{g}_n\rangle. \quad (18)$$

For the last equality, we use the definition of the n-particle state (11). $\phi(\mathbf{g})$ is the field on $\mathbb{H}$ that has the same gauge symmetry as $\hat{\varphi}^\dagger(\mathbf{g})$, namely $\phi(gh) = \phi(g)$, and $\mathcal{N}_\varphi$ is the normalization [4]

$$\mathcal{N}_\varphi^2 = \exp\left[\int d\mathbf{g}\, \overline{\phi(\mathbf{g})} \phi(\mathbf{g})\right]. \quad (22)$$

One can show that $|\varphi\rangle$ is the eigenstate of the field operator $\hat{\varphi}(\mathbf{g})$ such that

$$\hat{\varphi}(\mathbf{g})|\varphi\rangle = \phi(\mathbf{g})|\varphi\rangle. \quad (23)$$

---

[4] Define an operator $\hat{A}$ as

$$\hat{A} = \int d\mathbf{g}\, \phi(\mathbf{g}) \hat{\varphi}^\dagger(\mathbf{g}). \quad (19)$$

The commutator between $\hat{A}$ and $\hat{A}^\dagger$ is

$$[\hat{A}^\dagger, \hat{A}] = \int d\mathbf{g}\, \overline{\phi(\mathbf{g})} \phi(\mathbf{g}). \quad (20)$$

Then, $\langle \varphi | \varphi \rangle$ can be given as

$$\begin{aligned}
1 &= \langle \varphi | \varphi \rangle = \mathcal{N}_\varphi^{-2} \langle 0 | e^{\hat{A}^\dagger} e^{\hat{A}} | 0 \rangle \\
&= \mathcal{N}_\varphi^{-2} \langle 0 | e^{\hat{A}} e^{\hat{A}^\dagger} e^{[\hat{A}^\dagger, \hat{A}]} | 0 \rangle \\
&= \mathcal{N}_\varphi^{-2} \exp\left[\int d\mathbf{g}\, \overline{\phi(\mathbf{g})} \phi(\mathbf{g})\right].
\end{aligned} \quad (21)$$

In the third equality, we use the Baker–Campbell–Hausdorff formula.

Indeed, we have

$$\hat{\varphi}(\mathbf{g}) \frac{1}{n!} \prod_a^n \left[\int d\mathbf{g}_a \, \phi(\mathbf{g}_a) \hat{\varphi}^\dagger(\mathbf{g}_a) \right] \tag{24}$$

$$= \frac{1}{n!} \sum_{k=1}^n \prod_{a \neq k}^n \left[\int d\mathbf{g}_a \, \phi(\mathbf{g}_a) \hat{\varphi}^\dagger(\mathbf{g}_a) \right] \int dh \, d\mathbf{g}_k \, \phi(\mathbf{g}_k) \delta(\mathbf{g} h \mathbf{g}_k^{-1}) \tag{25}$$

$$= \frac{1}{(n-1)!} \prod_a^{n-1} \left[\int d\mathbf{g}_a \, \phi(\mathbf{g}_a) \hat{\varphi}^\dagger(\mathbf{g}_a) \right] \int dh \, \phi(\mathbf{g} h) \tag{26}$$

$$= \phi(\mathbf{g}) \frac{1}{(n-1)!} \prod_a^{n-1} \left[\int d\mathbf{g}_a \, \phi(\mathbf{g}_a) \hat{\varphi}^\dagger(\mathbf{g}_a) \right]. \tag{27}$$

In the first equality, we use the commutator (5) between $\hat{\varphi}$ and $\hat{\varphi}^\dagger$. In the last equality, we use the fact that $\phi(\mathbf{g})$ is right invariant. Thus, when $\hat{\varphi}(\mathbf{g})$ acts on $|\varphi\rangle$, it gives (23). In particular, coherent states $|\varphi\rangle$ provide an over-complete basis of the Fock space $\mathbb{F}$ (see Appendix A for details).

Via Equation (23), one immediately obtains the tensor fields $\phi(\mathbf{g})$ and $\overline{\phi(\mathbf{g})}$ in terms of expectation values of $\hat{\varphi}(\mathbf{g})$ and $\hat{\varphi}^\dagger(\mathbf{g})$ with respect to $|\varphi\rangle$

$$\langle \varphi | \hat{\varphi}(\mathbf{g}) | \varphi \rangle = \phi(\mathbf{g}), \quad \langle \varphi | \hat{\varphi}^\dagger(\mathbf{g}) | \varphi \rangle = \overline{\phi(\mathbf{g})}. \tag{28}$$

Accordingly, we can express the multi-particle state as the expectation value of a group-field network operator (16) in the 2nd quantised basis of eigenstates of the GFT quantum field operator. For the network operator $\hat{\Psi}_\Gamma[\hat{\varphi}, \hat{\varphi}^\dagger](\mathbf{g}_\partial)$ defined in Equation (16), we get

$$\langle \varphi | \hat{\Psi}_\Gamma[\hat{\varphi}, \hat{\varphi}^\dagger] | \varphi \rangle = \sum_{\substack{p+q=n \\ p,q \geq 0}} \int \prod_{p=1}^p d\mathbf{g}'_p \, \overline{\phi(\mathbf{g}'_p)} \prod_{q=1}^q d\mathbf{g}_q \, \phi(\mathbf{g}_q) \prod_{\ell \in \Gamma}^L dh_\ell \prod_{\ell \in \Gamma} L_\ell \left(\mathbf{g}_{s(\ell)} h_\ell \mathbf{g}_{t(\ell)}^{-1} \right)$$

$$= \Psi_\Gamma[\overline{\phi}, \phi](\mathbf{g}_\partial). \tag{29}$$

This is a group field tensor network state based on graph Γ with n nodes and L links. In particular, for $L_\ell \left(h_{s(\ell)} g_\ell h_{t(\ell)}^{-1} \right) = \delta \left(h_{s(\ell)} g_\ell h_{t(\ell)}^{-1} \right)$, we can think of the expectation values of the associated operators as peculiar projected entangled-pairs tensor network states (PEPS)

$$|\Psi_\Gamma\rangle \equiv \int d\mathbf{g}_\partial \, \Psi_\Gamma(\mathbf{g}_\partial) |\mathbf{g}_\partial\rangle \equiv \bigotimes_{\ell \in \Gamma} \langle L_\ell | \bigotimes_{v \in \Gamma} |\phi_v\rangle \tag{30}$$

obtained by the contraction of maximally entangled link states

$$|L_\ell\rangle = \int d\mathbf{g}_{s(\ell)} d\mathbf{g}_{t(\ell)} \, \delta \left(\mathbf{g}_{s(\ell)} h_\ell \mathbf{g}_{t(\ell)}^{-1} \right) |\mathbf{g}_{s(\ell)}\rangle \otimes |\mathbf{g}_{t(\ell)}\rangle \tag{31}$$

via tensor states $|\phi\rangle$ on some generically open graph architecture Γ.

The basis $|\mathbf{g}_\partial\rangle$ labels the uncontracted dangling indices comprising the boundary of the auxiliary tensor network representation. Differently from standard PEPS, the GFT networks are further characterised by the inherently *random* character of the tensors $|\phi\rangle$, induced by their field-theoretic statistical descritpion. In this light, in particular, we can see states like (29) as a generalisation of the random tensor network states (RTNs) recently introduced in [15], where the statistical characterisation of the standard RTN gets mapped in the momenta of the GFT partition function.[5]

---

[5] When Γ is closed, one further recognises $\Psi_\Gamma[\overline{\phi}, \phi](\mathbf{g}_{\ell \in \Gamma})$ to be equivalent to the cylindrical functions describing the quantum geometry of a closed spacial hypersurface in the kinematic Hilbert space of LQG [57].

4. Bipartite Entanglement of a GFT Network State

Given a group field tensor network state $|\Psi_\Gamma\rangle \in \mathbb{H}_\partial = \bigotimes_{\ell \in \partial} L^2_\ell[G]$, a bipartition of the boundary degrees of freedom corresponds to a factorisation of the boundary Hilbert space $\mathbb{H}_\partial$ into two subspaces $\mathbb{H}_A$ and $\mathbb{H}_B$, such that

$$\mathbb{H}_\partial = \mathbb{H}_A \otimes \mathbb{H}_B. \tag{32}$$

The entanglement of the boundary state $|\Psi_\Gamma\rangle$ across the bipartition in $\mathbb{H}_A$ and $\mathbb{H}_B$ is measured by the von Neumann entropy

$$S(A) = -\mathrm{Tr}\bar{\rho}_A \ln \bar{\rho}_A, \tag{33}$$

where

$$\bar{\rho}_A \equiv \frac{\rho_A}{\mathrm{Tr}\rho}, \quad \rho_A \equiv \mathrm{Tr}_B \rho = \mathrm{Tr}_B |\Psi_\Gamma\rangle\langle\Psi_\Gamma| \tag{34}$$

defines the (normalised) marginal on $\mathbb{H}_A$ of the GFT multi-particle density matrix,

$$\rho = |\Psi_\Gamma\rangle\langle\Psi_\Gamma| = \mathrm{Tr}\left[\bigotimes_\ell |L_\ell\rangle\langle L_\ell| \bigotimes_v |\phi_v\rangle\langle\phi_v|\right] \equiv \mathrm{Tr}\bigotimes_\ell \rho_\ell \bigotimes_v \rho_v. \tag{35}$$

A representation of ρ for the case of a simple 2-vertices graph is given in Figure 1.

It is computationally convenient to derive the von Neumann entropy as the limit of the Rényi entropy $S_N(A)$, via standard replica trick. The Rényi entropy is defined as

$$S_N(A) = \frac{1}{1-N}\ln \mathrm{Tr}\bar{\rho}_A^N = \frac{1}{1-N}\ln \frac{\mathrm{Tr}\rho_A^N}{(\mathrm{Tr}\rho)^N} \equiv \frac{1}{1-N}\ln \frac{\mathrm{Tr}(\rho^{\otimes N}\mathbb{P}_A)}{\mathrm{Tr}(\rho^{\otimes N})}, \tag{36}$$

where $\mathbb{P}_A = \mathbb{P}(\pi_A^0; n, d)$ is the 1-cycle permutation operator in S_N acting on the reduced Hilbert space $\mathbb{H}_A$,

$$\mathbb{P}(\pi_A^0; N, d) = \prod_{s=1}^N \delta_{\mu_A^{([s+1]_D)} \mu_A^{(s)}}, \tag{37}$$

and d is the dimension of the Hilbert space in the same region A (see Figure 1). Explicitly, one has

$$\mathbb{P}_A |a_1, b_1\rangle |a_2, b_2\rangle \cdots |a_N, b_N\rangle = |a_2, b_1\rangle |a_3, b_2\rangle \cdots |a_1, b_N\rangle \tag{38}$$

with $\bigotimes_i |a_i\rangle \in \mathbb{H}_A$ and $\bigotimes_i |b_i\rangle \in \mathbb{H}_B$.

The Rényi entropy $S_N(A)$ coincides with the von Neumann entropy $S(A)$ as N goes to 1, which is

$$S(A) = \lim_{N \to 1} S_N(A). \tag{39}$$

We are interested in carrying on this measurement directly in the Fock space of GFT.

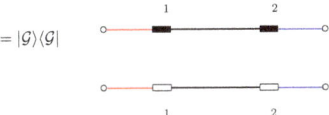

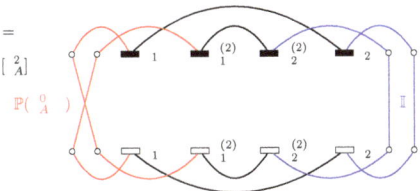

Figure 1. Graphical representation of the density matrix state for a simple bipartite GFT network state $|\mathcal{G}\rangle$, comprising two bivalent internal nodes, and trace of the $N = 2$ replica with the action of the cyclic permutation (swap) operator. Notice that the label $\bar{A}$, generally indicating the complementary marginal, corresponds to the B-labelling in the main text.

4.1. Expected Rènyi Entropy in the Fock Space

Due to the random character of the nodes $\{\phi_v\}$, induced by their dynamical GFT description, the measure of the entanglement will be necessarily given in expectation value.

Now, by construction, the Rényi entropy $S_N(A)$ is a functional of the GFT fields ϕ and $\bar{\phi}$, hence it can be promoted to an observable $S_N(A)[\hat{\varphi}, \hat{\varphi}^\dagger]$ in the GFT Fock space. We shall then derive the expectation value of the Rényi entropy in the Fock space using single-particle coherent state basis, by inserting the resolution of identity $\mathbb{1}_\mathbb{F}$ (and set $K = 0$, see Appendix A), with a normal ordering $: \cdots :$ such that all $\hat{\varphi}^\dagger$ is to the left of $\hat{\varphi}$:[6]

$$\mathbb{E}\left[S_N(A)[\hat{\varphi}, \hat{\varphi}^\dagger]\right] \equiv \frac{C}{Z_0} \text{Tr}\left(:S_N(A)[\hat{\varphi}, \hat{\varphi}^\dagger] e^{-\hat{S}[\hat{\varphi}, \hat{\varphi}^\dagger]} : \mathbb{1}_\mathbb{F}\right), \tag{41}$$

where the very GFT action $\hat{S}[\hat{\varphi}, \hat{\varphi}^\dagger]$ is constructed as a quantum many-body operator on the Fock space. By (36), the explicit form of the expectation value reads

$$\mathbb{E}[S_N(A)[\hat{\varphi}, \hat{\varphi}^\dagger]] = \frac{1}{1-N} \mathbb{E}\left[\ln \frac{\text{Tr}(\rho\,[\hat{\varphi}, \hat{\varphi}^\dagger]^{\otimes N} \mathbb{P}_A)}{\text{Tr}(\rho\,[\hat{\varphi}, \hat{\varphi}^\dagger])^N}\right]. \tag{42}$$

If we choose to work in perturbative GFT regime, such that $S[\phi, \bar{\phi}] = S_0 + \lambda S_{\text{int}}[\phi, \bar{\phi}]$, with $\lambda \ll 1$, then we deal with a polynomially perturbed generalised [45] Gaussian distribution for the random field, hence we can take advantage of the central limit theorem to get a good approximation of (42) in terms of a Taylor expansion around the mean values

$$\mathbb{E}[S_N(A)[\hat{\varphi}, \hat{\varphi}^\dagger]] \approx \frac{1}{1-N} \ln \frac{\mathbb{E}[\text{Tr}(\rho\,[\hat{\varphi}, \hat{\varphi}^\dagger]^{\otimes N} \mathbb{P}_A)]}{\mathbb{E}[\text{Tr}(\rho\,[\hat{\varphi}, \hat{\varphi}^\dagger]^{\otimes N})]}, \tag{43}$$

---

[6] The GFT vacuum amplitude in the same basis reads

$$\begin{aligned} Z_0 &= \text{CTr}\left(:e^{-\hat{S}[\hat{\varphi}, \hat{\varphi}^\dagger]}:\right) = \text{CTr}\left(:e^{-\hat{S}[\hat{\varphi}, \hat{\varphi}^\dagger]} : \mathbb{1}_\mathbb{F}\right) \\ &= \int \mathcal{D}\phi \mathcal{D}\bar{\phi}\,\delta_C[\phi - \phi_{\text{GF}}]\,e^{-S[\phi, \bar{\phi}]}. \end{aligned} \tag{40}$$

plus corrections due to fluctuations around the mean values that are suppressed whenever the dimensionality of our quantum system gets extremely large. For standard random tensor network states [15], such a *typical* regime is realised in the *large bond* limit, when the dimension of the random tensor index (bond) space $\mathcal{H}_\ell$ is extremely large, $\dim(\mathcal{H}_\ell) \gg 1$.

In the GFT setting, one has to deal with infinite dimensional bond spaces, $\mathcal{H}_\ell = L^2(G)$, which automatically set the derivation in the typicality regime of [15]. Nevertheless, link spaces are regularised via a cut-off on the group representation space, such that

$$\langle g | g' \rangle = D(\Lambda) \delta_{g,g'} \tag{44}$$

with $\delta_{g,g'}$ equal to 1 if $g = g'$ and 0, otherwise. Therefore, with $D(\Lambda)$, the dimension of the bonds (links) of the network, one can consistently assume typicality to hold in the $D(\Lambda) \gg 1$ regime.

Given (43), the expression of the N-th Rényi entropy is mapped into a ratio of averaged partition functions,

$$\frac{Z_N}{Z_0^N} \equiv \frac{\mathbb{E}\left[\mathrm{Tr}(\rho\,[\hat{\varphi}, \hat{\varphi}^\dagger]^{\otimes N} \mathbb{P}_A)\right]}{\mathbb{E}\left[\mathrm{Tr}(\rho\,[\hat{\varphi}, \hat{\varphi}^\dagger]^{\otimes N})\right]}, \tag{45}$$

where we removed the bars over Z_N, Z_0^N to simplify the notation.

Let us rewrite $\rho\,[\hat{\varphi}, \hat{\varphi}^\dagger]^{\otimes N}$ as a trace contraction of individual link and nodes density matrices, similarly to (35). We restrict for simplicity to tensor network observables (16) with n vertices and L links, and with $p = n(q = 0)$

$$\hat{\Psi}_\Gamma[\hat{\varphi}^\dagger](\mathbf{g}_\partial) = \int \prod_{\mathfrak{p}=1}^n d g_\mathfrak{p} \prod_{\mathfrak{p}=1}^n \hat{\varphi}^\dagger(\mathbf{g}_\mathfrak{p}) \prod_{\ell \in \Gamma}^L d h_\ell \prod_{\ell \in \Gamma}^L L_\ell \left(\mathbf{g}_{s(\ell)} h_\ell \mathbf{g}_{t(\ell)}^{-1} \right), \tag{46}$$

where now the $\mathfrak{p}$ label coincides with the v-label of the nodes of the graph Γ. The Nth replica of the density matrix operator describing a graph observable reads

$$\rho^{\otimes N} = \left(\int \prod_{v=1}^n dg_v dg_v' \prod_{v=1}^n \hat{\varphi}^\dagger(\mathbf{g}_v)\,\hat{\varphi}(\mathbf{g}_v') \prod_\ell^L dh_\ell\, dh_\ell' \prod_\ell^L L_\ell L_\ell' \right)^{\otimes N} \tag{47}$$

$$= \int \left(\prod_{v=1}^n dg_v dg_v' \prod_{v=1}^n \hat{\varphi}^\dagger(\mathbf{g}_v)\,\hat{\varphi}(\mathbf{g}_v') \right)^{\otimes N} \left(\prod_\ell^L dh_\ell\, dh_\ell' \prod_\ell^L L_\ell L_\ell' \right)^{\otimes N} \tag{48}$$

$$= \mathrm{Tr}\left[\left(\bigotimes_v \rho_v \right)^{\otimes N} \left(\bigotimes_\ell \rho_\ell \right)^{\otimes N} \right]. \tag{49}$$

The linearity of the trace allows for moving the expectation operator inside the integral and letting it act on the N replicas of the products of fields. We then write (45) as

$$\frac{\mathrm{Tr}\left[\bigotimes_\ell \rho_\ell^N \mathbb{E}[(\bigotimes_v \rho[\hat{\varphi}^\dagger(\mathbf{g}_v), \hat{\varphi}(\mathbf{g}_v')])_v^{\otimes N}] \mathbb{P}_A \right]}{\mathrm{Tr}\left[\bigotimes_\ell \rho_\ell^N \mathbb{E}[(\bigotimes_v \rho[\hat{\varphi}^\dagger(\mathbf{g}_v), \hat{\varphi}(\mathbf{g}_v')])_v^{\otimes N}] \right]} \tag{50}$$

and we focus on the calculation of $\mathbb{E}[\bigotimes_v \rho_v^{\otimes N}]$.

The derivation in [15,18] thereby proceeds by making two strongly simplifying assumptions. The first consists of the restriction to the case of a Gaussian (free) group field theory for describing the single node field statistics. The second assumes the expected value ϕ of the field operator at each node to be individually independently distributed (i.i.d.). The latter assumption corresponds, from a physical viewpoint, to considering a non-interacting quantum many body system. The latter condition translates in particular into a *local averaging* condition, namely

$$\mathbb{E}\left[\left(\bigotimes_v \rho_v\right)^{\otimes N}\right] \to \bigotimes_v \mathbb{E}\left[\rho_v^{\otimes N}\right] \ , \tag{51}$$

which allows for an explicit calculation of the expectation value in terms of a product of n 2N-point function of the free group field theory

$$\mathbb{E}\left[\rho_v^{\otimes N}\right] \equiv \frac{C}{Z_0}\mathrm{Tr}\left(:\rho[\hat{\varphi}^\dagger(\mathbf{g}_v),\hat{\varphi}(\mathbf{g}'_v)]^{\otimes N}\, e^{-\hat{S}_0[\hat{\varphi},\hat{\varphi}^\dagger]}:\mathbb{1}_{\mathrm{F}}\right) \tag{52}$$

$$= \frac{1}{Z_0}\int \mathcal{D}\phi \mathcal{D}\bar\phi\, \delta_C[\phi-\phi_{\mathrm{GF}}]\, \langle\varphi|:[\hat{\varphi}^\dagger(\mathbf{g}_v),\hat{\varphi}(\mathbf{g}'_v)]^{\otimes N}\, e^{-\hat{S}_0[\hat{\varphi},\hat{\varphi}^\dagger]}:|\varphi\rangle$$

$$= \frac{1}{Z_0}\int \mathcal{D}\phi \mathcal{D}\bar\phi\, \delta_C[\phi-\phi_{\mathrm{GF}}]\, (\phi,\bar\phi)^N\, e^{-S_0[\phi,\bar\phi]} \tag{53}$$

$$= \mathbb{E}_0\left[(\phi_v\bar\phi_v)^N\right] \ . \tag{54}$$

In the free case, in particular, one can evaluate the 2N-point functions at each node directly via Wick's theorem

$$\mathbb{E}_0\left[\prod_a^N \phi_v(\mathbf{g}_a)\overline{\phi_v(\mathbf{g}'_a)}\right] = C \sum_{\pi_v \in S_N}\int \prod_a^N \mathrm{d}h_a \prod_a^N \delta_v\left(h_a\mathbf{g}_a\mathbf{g}'^\dagger_{\pi(a)}\right)$$

$$= C \sum_{\pi_v \in S_N}\int \prod_a^N \mathrm{d}h_a\, \mathbb{P}_{\mathbf{h}_v}(\pi),$$

where the permutation operator acts strandwise (locally on the link spaces)

$$\mathbb{P}_{\mathbf{g}}(\pi) \equiv \prod_a^N \delta\left(h_a\mathbf{g}_a\mathbf{g}'^\dagger_{\pi(a)}\right) = \prod_{s=1}^4 \prod_a^N \delta\left(h_a g_{sa} g'^\dagger_{s\pi(a)}\right) \equiv \prod_s^4 \mathbb{P}^s_{\mathbf{g}}(\pi),$$

with $\mathbf{g}'$ independent from $\mathbf{g}$, a labelling the replica order at each node, $\mathbf{h}$ denoting the set of h_a, $a = 1, \cdots, N$.

When $h_a = \mathbb{1}$ for all a from 1 to N,

$$\mathbb{P}_\mathbb{1}(\pi) = \prod_a^N \delta\left(g_a g'^\dagger_{\pi(a)}\right) \tag{55}$$

$$= \mathbb{P}(\pi; N, D^4) = \prod_s^4 \mathbb{P}^s(\pi; N, D^4),$$

where $\mathbb{P}(\pi; N, D^4)$ and $\mathbb{P}^s(\pi; N, D^4)$ are the representations of $\pi \in S_N$ on $\mathbb{H}^{\otimes 4}$ and $\mathbb{H}$, respectively.

The averaged partition functions, Z_N and Z_0^N become

$$Z_N \approx C^{V_\Gamma} \sum_{\pi_v \in S_N} \int \prod_v dh_v \, \text{Tr}\left[\bigotimes_\ell \rho_\ell^N \bigotimes_v \mathbb{P}_{h_v}(\pi_v) \mathbb{P}(\pi_A^0; N, d)\right]$$

$$\equiv C^{V_\Gamma} \sum_{\pi_v \in S_N} \int \prod_v dh_v \, \mathcal{N}_A(h_v, \pi_v), \tag{56}$$

$$Z_0^N = C^{V_\Gamma} \sum_{\pi_v \in S_N} \int \prod_v dh_v \, \text{Tr}\left[\bigotimes_\ell \rho_\ell^N \bigotimes_v \mathbb{P}_{h_v}(\pi_v)\right]$$

$$\equiv C^{V_\Gamma} \sum_{\pi_v \in S_N} \int \prod_v dh_v \, \mathcal{N}_0(h_v, \pi_v), \tag{57}$$

respectively corresponding to summations of Feynman graphs $\mathcal{N}_A(h_v, \pi_v)$ and $\mathcal{N}_0(h_v, \pi_v)$ labelled by permutation operators $\mathbb{P}_{h_v}(\pi_v)$, at each node v, contracted with the ρ_ℓ^N densities at each link ℓ (see Figure 2). The difference between the reduced and full (density matrix) networks is encoded in the boundary condition, as Z_N is defined with $\mathbb{P}(\pi_A^0; N, d)$ on A of $\partial \Gamma$ and $\mathbb{P}(\mathbb{1}; N, d)$ on B of $\partial \Gamma$, while Z_0^N is defined with $\mathbb{P}(\mathbb{1}; N, d)$ for all boundary region $\partial \Gamma$.

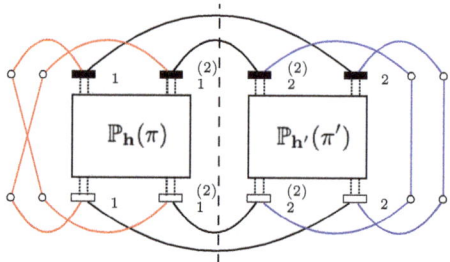

Figure 2. Free Group Field Theory propagators acting among two sets of N replicas for each node of the given graph.

It is easy to see at this stage what the impact of the i.i.d. assumption is in reducing the complexity of the propagator.[7] In this case, the permutation operator $\mathbb{P}$ acts only among the N replicas of the same node, independently taken across the graph. Clearly, $\bigotimes_v \mathbb{E}(\rho_v^N) \neq \mathbb{E}(\bigotimes_v^V \rho_v^N)$ as the permutation group S_{2N} is much smaller that S_{2NV}, reducing to $(2N!)^V$ the number of permutation patterns with respect to the $(2NV)!$ allowed patterns of the indistinguishable case. Such a strong truncation is nevertheless consistent with a restriction to the tree-like, or "melonic", sector of the Feynmann diagrams of the $2NV$-point function, which we expect to provide the leading order contribution to the divergence degree.

From a qualitative point of view, a truly general result in GFT would require to consider a fully interacting GFT and to drop the i.i.d. assumption. Moreover, it would require to leave the *tensor*

---

[7] Note that, if we keep the indistinguishability condition, the calculation changes

$$\mathbb{E}_0\left[\bigotimes_v (|\phi_v\rangle \langle \phi_v|)^{\otimes N}\right] = C \sum_{\pi \in S_{NV_\Gamma}} \mathbb{P}(\pi)$$

$$= C \sum_{\pi \in S_{NV_\Gamma}} \prod_{a=1}^{N} \prod_{n=1}^{V_\Gamma} \int dh_{na} dg_{na} dg'_{\pi(na)} \, \delta\left(h_{na} g_{na} g'_{\pi(na)}{}^{-1}\right) |g_{na}\rangle \langle g'_{\pi(na)}|, \tag{58}$$

where S_{NV_Γ} is the permutation group of NV_Γ objects, which corresponds to the permutations of NV_Γ nodes; h_{na} comes from the required gauge symmetry of the propagator; C is a constant factor.

network representation of the GFT multi-particle operator, and derive the entropy for the generic operator given in (15).

4.2. Mapping to BF Theory Partition Function

The partition functions Z_N and Z_0^N correspond to summations of two *auxiliary* networks $\mathcal{N}_A(\mathbf{h}_v, \pi_v)$ and $\mathcal{N}_0(\mathbf{h}_v, \pi_v)$. We shall proceed by giving the main ingredients of the derivation, while referring the reader to the original literature (see e.g., [18]) for a fully detailed description of the combinatorics of the calculation.

First of all, in opening the expressions for Z_N and Z_0^N, we shall notice that the action of $\mathbb{P}_{\mathbf{h}_v}(\pi_v)$ at each node is decoupled among the incident legs. Due to the strandwise action of the propagator, the value of the networks $\mathcal{N}_A(\mathbf{h}_v, \pi_v)$ and $\mathcal{N}_0(\mathbf{h}_v, \pi_v)$ can be written as factorised products over internal (e.g., trivial propagators acting in the picture below) and boundary links

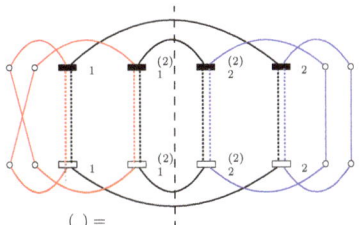

$$\mathcal{N}_A(\mathbf{h}_v, \pi_v) = \prod_{e \in \Gamma} \mathcal{L}_e(\pi_v, \pi_{v'}; \mathbf{h}_v, \mathbf{h}_{v'}) \prod_{e \in A} \mathcal{L}_e(\pi_v, \pi_A^0; \mathbf{h}_v) \prod_{e \in B} \mathcal{L}_e(\pi_v, \mathbb{1}; \mathbf{h}_v), \qquad (59)$$

$$\mathcal{N}_0(\mathbf{h}_v, \pi_v) = \prod_{e \in \Gamma} \mathcal{L}_e(\pi_v, \pi_{v'}; \mathbf{h}_v, \mathbf{h}_{v'}) \prod_{e \in \partial \Gamma} \mathcal{L}_e(\pi_v, \mathbb{1}; \mathbf{h}_v). \qquad (60)$$

On the internal (dotted) links, $\mathcal{L}(\pi, \pi', \mathbf{h}, \mathbf{h}')$ can be written as a trace of a modified representation of a permutation group element $\omega \equiv (\pi')^{-1} \pi$, such that

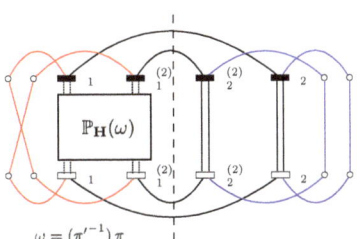

$$\mathcal{L}(\pi, \pi'; \mathbf{h}, \mathbf{h}') = \text{Tr}\left[\mathbb{P}_{\mathbf{h}}(\pi) \rho_e^N \mathbb{P}_{\mathbf{h}'}(\pi)\right] = \text{Tr}\left[\mathbb{P}_{\mathbf{h}}\left((\pi')^{-1} \pi\right)\right] \equiv \text{Tr}\left[\mathbb{P}_{\mathbf{h}}(\omega)\right], \qquad (61)$$

where

$$\mathbf{h} = \left\{ H_a \mid H_a \equiv \left(h'_{\omega(a)}\right)^\dagger h_a, \; \forall a = 1, \cdots, N \right\}.$$

In particular, any element $\omega \in S_N$ can be expressed as the product of disjoint cycles $\mathcal{C}_i$

$$\omega \equiv \prod_i^{\chi(\omega)} \mathcal{C}_i, \qquad (62)$$

leading to a specific simple form for the individual link contributions

$$\mathcal{L}(\pi, \pi'; \mathbf{g}, \mathbf{g}') = \text{Tr}\left[\mathbb{P}_{\mathbf{g}}(\varpi)\right] = \prod_i \text{Tr}\left[\mathbb{P}_{\mathbf{g}}(\mathcal{C}_i)\right] \tag{63}$$

$$= \prod_i \int \prod_{k=1}^{r_i} dg_{a_k^i} \, \delta\left(H_{a_k^i} g_{a_k^i} g_{a_{[k]r_i+1}^i}^\dagger\right) = \prod_i^{\chi(\varpi)} \delta\left(\overleftarrow{\prod_{k=1}^{r_i} H_{a_k^i}}\right),$$

which is expressed as the product of the traces of the individual cycles $\mathcal{C}_i$. Indeed, one can eventually realise [18,20] that the integral of the pattern networks $\mathcal{N}(\mathbf{h}_v, \boldsymbol{\pi}_v)$ on the gauge holonomies are equivalent to the amplitudes of a three-dimensional topological BF field theory [50], with given boundary condition. Such amplitudes are discretized on a specific 2-complex comprised by the N replicas of the networks, with each different pattern $\mathbb{P}$ corresponding to a different 2-complex. The simple form of the various functions entering the calculation of the entropy directly follows from the specific approximations used in the calculation of expectation values, namely the choice to neglect the GFT interactions and to consider tensor-network like observables with simple delta functions associated with the links' kernels.

In the local averaging setting, the expectation values of the partition functions therefore reduce to contractions of single node $2N$-point functions of the GFT model, expanded in series of Feynman amplitudes and corresponding to a local BF gravity spin-foams (see Figure 3). In particular, if one specifies to a 3-valent GFT field theory, with simplicial four-field interaction kernels, each domain amplitude will correspond to a Boulatov (or 3-d BF theory) spinfoam amplitude, whose semi-classical limit coincides with Ponzano–Regge gravity [50].

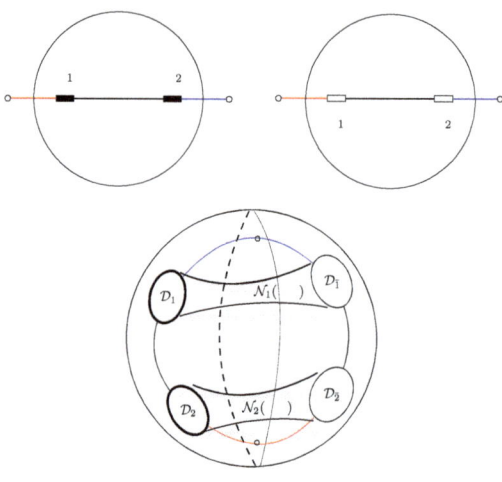

Figure 3. (Above) Simple open graph associated with a boundary tensor state (and its conjugate). (Down) Cartoon picture of the expectation value of the partition function defined by the trace of the boundary tensor state density matrix. The N replicas of the nodes of the graph define domain regions $\mathcal{D}$, connected among them via (replicas of) the graph links. The boundary links are contracted among the conjugate graphs via a global trace. The expectation value generates a set of independent spin-foam channels $\mathcal{N}$ among conjugate domains (GFT $2N$-point functions), due to the i.i.d. assumption considered. Indistinguishability would instead merge the different spin-foams into a single one, raising the degree of correlations of the boundary degrees of freedom. Nevertheless, the local approximation seems to convey the leading order contribution of the quantum entanglement of the bipartite system.

4.3. Entanglement Scaling and Divergence Degree

Given the expression for the Rényi entropy, $e^{(1-N)S_N} = Z_N/Z_0^N$, finding the scaling of the entanglement entropy amounts to identifying the most divergent terms of the partition functions Z_N and Z_0^N, which corresponds to the divergence degree of the BF theory amplitudes discretized on a lattice [58–62]. The recipe given in [18] consists of a combination of coarse graining and combinatorics.

Once given global boundary conditions for $\mathcal{N}_0$, $\pi = \mathbb{1}$ and $h = \mathbb{1}$, one can coarse-grain the boundary of $\mathcal{N}_0$ into a single node with $\pi = \mathbb{1}$ and $h = \mathbb{1}$. Accordingly, the boundary of $\mathcal{N}_A$ gets coarse-grained into two nodes, one of which corresponds to A with $\pi = \mathcal{C}_0$, $h = \mathbb{1}$ and the other to B with $\pi = \mathbb{1}$ and $h = \mathbb{1}$. The corresponding closed graphs are denoted as Γ_0 and Γ_{AB}. Once we take the expectation value on the N replicas at each node, then a given pattern $\mathbb{P}(\pi_v)$ divides Γ_0 and Γ_{AB} into regions (set of nodes), each one coloured with permutation group π_m and N integrals over h_m (see Figure 4 below).

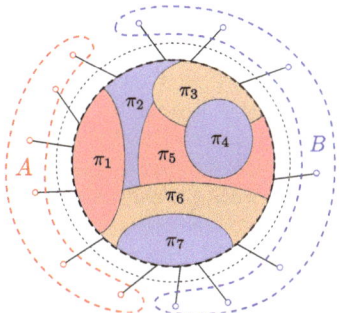

Figure 4. Example of a permutation pattern $\mathbb{P}(\pi_v)$ dividing Γ_{AB} into regions (set of nodes), each one colored with permutation group element π and N integrals over holonomies **h**.

In full analogy with the S_N Ising model [15], links connecting different regions identify boundaries which can be interpreted as domain walls. We are interested in finding the scaling behaviour of the Rényi entropy in the large bond regime $D(\Lambda) \gg 1$, classically corresponding to the long-range ordered phase for such an Ising-like model, where the entropy of a boundary region is known to be directly related to the energy of the domain wall between domains of the order parameter [15].

Different local regions R, with uniform boundary conditions, can be further coarse-grained via gauge invariance to single block nodes. For a given region, the degree of divergence is counted by the number of links in the region, minus the number of trivialised links on a maximal spanning tree T_R

$$\#_R = (\#_{e\in R} - \#_{T_R})N. \qquad (64)$$

One can then generally show that the pattern with no domain walls, Γ_0, corresponding to the ordered phase in which all nodes are assigned with the same permutation group, has the highest degree of divergence

$$\#_0 = (\#_{e\in \Gamma_0} - \#_{T_{\Gamma_0}})N, \qquad (65)$$

where $\#_{e\in \Gamma_0}$ is the number of links in the Γ_0 graph.

For the bipartite graph Γ_{AB}, defined by the assignment of different boundary conditions for A and B, patterns with a single domain wall have higher divergence degree than the multi-domain walls configurations. The coarse-grained graph contains only two block nodes and one finds [18]

$$\#_{\Gamma_{AB}(\pi_m)} \leq \#_{AB} = \#_0 + (1-N)\min(\#_{e\in \partial AB}). \qquad (66)$$

The leading order divergence terms of the amplitudes are given by

$$Z_0^N = C^{V_\Gamma}[D(\Lambda)]^{\#_0},$$
$$Z_N = C^{V_\Gamma}[D(\Lambda)]^{\#_0+(1-N)\min(\#_{e\in\partial_{AB}})}\left[1+\mathcal{O}(D(\Lambda)^{-1})\right],$$

and the Nth order Rényi entropy S_N eventually reads

$$e^{(1-N)S_N} = \frac{Z_N}{Z_0^N} = [D(\Lambda)]^{(1-N)\min(\#_{e\in\partial_{AB}})}\left[1+\mathcal{O}(D(\Lambda)^{-1})\right]. \tag{67}$$

When N goes to 1, the leading term of the entanglement entropy S_{EE} is given by

$$S_{EE} = \min(\#_{e\in\partial_{AB}})\ln D(\Lambda), \tag{68}$$

where $D(\Lambda)$ with $\Lambda \gg 1$ reads as a regularisation of each BF bubble divergence $\delta(\mathbb{1})$.

The result, within a different formal setting, reproduces the universal behaviour typical of random tensor network states [15]. However, differently from the case of a standard random tensor network, the considered GFT states carry an inherent geometric characterisation. The graph Γ is dual to a 2D simplicial complex. Each node is dual to a triangle and each link is dual to an edge of this complex, and the specific GFT model endows the simplicial complex with dynamical geometric data. In particular, the proportionality of the entropy to the cardinality of the minimal domain wall $\sigma_{\min} \equiv \min(\#_{e\in\partial_{AB}})$ has a clear geometric interpretation, which becomes apparent in passing from a group element to a spin representation description of the dynamical fields, in the sense of discrete geometry. In this case, indeed, we have

$$\text{Area}(\sigma_{\min}) = \sum_{e\in\sigma_{\min}} \ell_e(j_e) = \langle \ell_{j_e}\rangle |\sigma_{\min}|, \tag{69}$$

where the length of each edge ℓ_j, in any given eigenstate of the length operator [26], is a function of the irreducible representation j_e associated with it, and to the dual link. Therefore, the cardinality of the minimal global domain wall can be interpreted as the 1D-area (length) of a dual discrete minimal one-dimensional path and we can write $|\sigma_{\min}| = \text{Area}(\sigma_{\min})/\langle\ell_{j_e}\rangle$. [20].

Equation (68) can then be understood as the discrete tensor network analogue of the Ryu–Takayanagi formula in the context of group field theory [21], if we consider the path integral averaging over the open network Γ as a simplified model of a bulk/boundary (spinfoam/network state) duality [18].

5. Holographic Scaling for Interacting with GFT

A complete geometric characterisation of the result requires working with a fully interacting GFT model. Indeed, it is the simplicial character of the interaction kernels that actually provides a connection between GFT and quantum gravity spin-foams. In this sense, both consistency and robustness of the result in (68) require exploring the possible modifications to the RT formula induced by group field interactions.

At the perturbative level, this amounts to calculating the first order correction of $\mathbb{E}[S_N(A)]$ in the GFT coupling constant λ, namely $\mathbb{E}[\text{Tr}(\rho^{\otimes N}\mathbb{P}_A)]$ and $\mathbb{E}[\text{Tr}(\rho^{\otimes N})]$, with

$$\mathbb{E}[\text{Tr}(\rho^{\otimes N}\mathbb{P}_A)] = \mathbb{E}_0[\text{Tr}(\rho^{\otimes N}\mathbb{P}_A)] + \lambda\mathbb{E}_0\left[S_{\text{int}}[\phi,\overline{\phi}]\text{Tr}(\rho^{\otimes N}\mathbb{P}_A)\right], \tag{70}$$
$$\mathbb{E}[\text{Tr}(\rho^{\otimes N})] = \mathbb{E}_0[\text{Tr}(\rho^{\otimes N})] + \lambda\mathbb{E}_0\left[S_{\text{int}}[\phi,\overline{\phi}]\text{Tr}(\rho^{\otimes N})\right].$$

This is a combinatorially highly non-trivial problem, as the interaction processes correspond to further stranded diagrams that contribute to the expectation value of $Z_{A/0}^{(N)}$ (see Figure 5). The behaviour of the amplitudes for rank-3 fields and Boulatov (simplicial) four fields interaction kernel $\mathcal{V}^{\text{sym}}(\mathbf{g}^{(1)}\mathbf{g}^{(2)}\underline{\mathbf{g}}^{(1)}\underline{\mathbf{g}}^{(2)})$

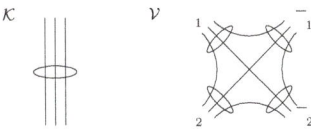

$$\int \prod_{\ell=1}^{4} dh_\ell\, \delta(h_1 g_1^{(1)}, h_3 \underline{g}_1^{(1)}) \delta(h_1 g_2^{(1)}, h_4 \underline{g}_2^{(2)}) \delta(h_1 g_3^{(1)}, h_2 g_3^{(2)})$$
$$\delta(h_2 g_1^{(2)}, h_4 \underline{g}_1^{(2)}) \delta(h_2 g_2^{(2)}, h_3 \underline{g}_2^{(1)}) \delta(h_3 g_3^{(1)}, h_4 \underline{g}_3^{(2)})$$

is studied in [20]. Therein, the authors provide a series of theorems aimed at constraining the complexity of the combinatorial pattern. In particular, it was shown that patterns with a single interaction happening between two incoming and two outgoing fields of the same network node v leads at most to the same number of divergences as in a maximal case of the free theory, where the interaction at v is replaced by a free propagation [20].

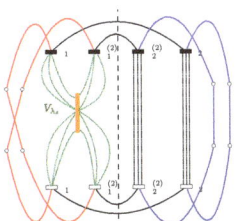 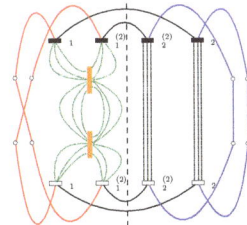

Figure 5. Examples of insertions of Boulatov 4-fields interaction vertices $\mathcal{V}$ in the amplitude among replicas of the individual node of the graph. Multiple of interaction vertices generate new faces in the amplitudes.

Interestingly, the situation changes if one drops the assumption of gauge invariance (closure) for the GFT field. In this case, the expected Nth Rényi entanglement entropy is estimated to be

$$\mathbb{E}_{\text{n-sym}}[S_N(\rho_A)] \approx [\ln D(\Lambda) - \lambda N] \min(\#_{e \in \partial_{AB}}). \tag{71}$$

The linear order corrections do modify the asymptotical scaling of the Rènyi entanglement entropy with the area of a minimal surface. In particular, the results in [20] show that the proportionality factor is corrected by an additive term linear in the perturbed group field theory coupling constant λ. However, no additive leading order correction to the area scaling entropy formula emerges from the analysis. A more systematic analysis of such dynamical regularisation and its relation with the emergence of an effective gravitational coupling is an open interesting issue for future work.

6. Discussion and Conclusions

Group field theories realise a kinematical description of quantum space geometry in terms of discrete, pre-geometric degrees of freedom of combinatorial and algebraic nature, described in terms of spin-network states. In loop quantum gravity (LQG), such theories play the role of *auxiliary* field theories, whose partition functions, for appropriate choices of the kernels, provide generating functionals for the LQG spinfoams: a covariant path integral realisation of spacetime as a transition amplitude between boundary spin network states. More generally, GFTs provide a versatile field-theoretic tool to study the very emergence of space-time quantum geometry via path integral techniques and a quantum many-body approach associated with their second quantised formalism [57,63]. Moreover, as higher order generalisations of matrix models, group field theories can be put in direct formal relation with tensor network algorithms [10–12,17,18] in condensed matter theory.

In this review, we considered a setting *similar* to that of a gauge field theory on a lattice, which in the background independent quantum gravity context consists of nets of fundamental quanta of space that admit an interpretation as quantized fuzzy geometries.

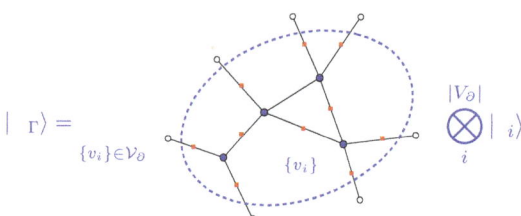

The entanglement structure of the wave-function for such collections of quanta is encoded into a ∞-rank tensor of coefficients, which we interpret here as an *open* GFT network state. In [18–20], such an open GFT spin network is interpreted as a *tensor network representation* of some quantum geometry wave-function, with physical indices corresponding to boundary degrees of freedom of the "auxiliary" GFT network state, realised by a spin network.

In this setting, the degree of entanglement of a generic quantum region of space can be measured *holographically*, in terms of the entanglement entropy of the bipartite auxiliary spin network. Such entanglement is directly related to the topology of the internal network, which, differently from many similar derivations in quantum gravity, is only partially fixed by the choice of the auxiliary network architecture. Indeed, due to the random character of the nodes, expectation values imply a dynamical characterisation of the internal graph topology, which end up being directly specified by the choice of the GFT model. Therefore, different GFT models a priori induce different architecture for the internal network, with higher order interactions corresponding to a higher degree of connectivity of the graph.

The GFT formalism, along the lines first proposed by [15], allows for an explicit computation of the entanglement entropy, which in [18–20] is realised within a set of simplifying assumptions:

1. indistinguishability and i.i.d.: the symmetry of the group field naturally selects a bosonic quantum many-body characterisation for the quanta comprising our network states. This translates into indistinguishability for the nodes comprising the network. Furthermore, along with the results in [15], in the statistical evaluation of the averaged Rényi entropy, 'tensor' fields at the nodes are individually independently distributed. This leads to a sensible simplification of the auxiliary bulk averages, reduced to a set of *local* amplitudes defined among conjugate domains of replicas of the fields at each node. Such i.i.d. assumption is a natural choice in [15], where random tensor networks are associated with maximally mixed states. In our setting, accordingly, the i.i.d. assumption presupposes actual *absence of dynamical interaction* among the quanta sitting at the nodes of the graph. Effectively, quanta couples through the network only via local adjacent link entanglement. Such a non-interacting quantum many-body assumption is very strong. Nevertheless, it is consistent with the simplifying choice of dealing with *product* coherent states in the adopted 2nd quantised formalism. On the other hand, the i.i.d. assumption embodies some a priori knowledge of the position of the individual nodes (their replicas and conjugates) on the graph, hence somehow violating indistinguishability.

2. maximally entangled bonds: the class of GFT tensor networks considered is further radically characterised by link states given by maximally entangled pair states, associated with gluing kernels realised in terms of bivalent intertwiners with $\delta\left(g_{s(\ell)}h_\ell g_{t(\ell)}^{-1}\right)$ coefficients. In principle, the choice of the link state can be generalised to a more general kernel, while still being expressed in terms of the mutual information among the half link states [14].
3. propagator holonomies are set to $h_\ell = 1$ for all link $\ell \in \Gamma$: this assumption makes the state $|\Psi_\Gamma\rangle$ lying in the flat vacuum state recently considered in the context of Loop Quantum Gravity [64].
4. finite link space (bond) dimension D: while taking a large leg space dimension $D \gg 1$ limit, we always deal with a finite-dimensional restriction of the leg spaces $L^2[G,\mu]$, obtained through the introduction of a sharp cut-off Λ in the group representation, such that $\delta(g) = D(\Lambda)$, for $g \in G$. More radically, one could regularise the divergences via "box" normalization of $\delta(g) \in L^2[G,\mu]$ by using quantum groups. Interestingly, such a quántum deformation can be related to the cosmological constant Λ in the semi-classical regime of the spinfoam formalism (see e.g., [65,66]).
5. tensor network setting: we work with regular networks of 3-valent nodes. The GFT interaction kernels adopted correspond to the ones defining the Boulatov model for 3D gravity. The coupling λ of the interaction term is always assumed to be much smaller than 1, allowing for a perturbative expansion of the expectation value of the Rényi entropy in λ.

As a result, the set of assumptions above induces a structure of correlations that is essentially local in its *leading* contribution to the entropy, even when the interaction among fields at different nodes is considered [18,20]. This suggests that the emergence of the dominant holographic behaviour for the entropy is intertwined with the universality features of the large bond regime. In the presence of (weak) interactions, such area scaling behaviour is shown to remain a solid feature of the *quantum typical regime* [20]. [8]

Many aspects can be tuned to further test such a holographic behaviour in the specific aproach. At the level of the free theory, it would be interesting to look at changes in the area scaling due to minimal modification of the very free propagator, via more general heat-kernel techniques, possibly expressed in terms of a different choice of mutual information among the half link states [14]. For higher order perturbations, where the diagrams are dominated by the bulk structure, induced by the creation of new interaction vertices [20], we may expect significant changes in the area scaling behaviour, even if the local averaging scheme induced by the i.i.d. assumption is preserved. Outside the typicality regime, it is natural to expect that fluctuations in the annealed average (43) become more and more relevant and the area scaling behaviour is eventually lost. A detailed study in this sense requires strong computational efforts, necessary to explore the properties and combinatorics of the dynamically induced bulk auxiliary networks.

More generally, concerning the potential of the GFT formalism in the study of holographic entanglement in quantum geometry, much still remain to be understood. We showed how the formalism of *second quantization* for GFT, and the use of a coherent state basis, allow for a new approach to the study of entanglement in quantum geometry. Standard quantum mechanical correlations among links and nodes of the quantum many-body network state are replaced by correlations among *excitations* of modes of the tensor field representing the particle. This is likely the right formalism for understanding the diverse roles played by entanglement in quantum gravity, from the local entanglement among single quanta responsible for the connectivity of a quantum geometry configuration, to the local and non-local entanglement among modes of the GFT field, intended as elementary excitations induced by the *collective* behaviour of the underlying many-body quantum system. This would effectively allow for maximally disregarding the auxiliary spin network structures and work in full indistinguishability with bosonic GFT quanta of geometry in the Fock space.

---

[8] A possibly deep relation between such universal regime and the universality proper of tensor models in the large D regime (see e.g., [67]) has been pointed out and discussed in [18].

However, harnessing entanglement in the 2nd quantisation formalism is still a fundamental open problem already in condensed matter and quantum information theory [68–70]. In our derivation, the second quantisation scheme was in fact only partially realised. Indistinguishability would have actually prevented us from defining an AB-factorisation of the Hilbert space. In this regard, we proceeded by selecting a single graph Hilbert space within the full Fock space, hence coming back to a first quantisation formalism. Indistinguishability was further violated by the local averaging approach, in Section 4.1, induced by the i.i.d. assumption. A GFT derivation that makes full use of the mode aspect of second quantization is therefore substantially left for future work. Success in this sense would be evocative of (and consistent with) the general perspective that sees continuum spacetime and geometry as emergent from the collective, quantum many-body description of the fundamental GFT degrees of freedom, the same perspective motivating a large part of the literature and in particular the one concerned with GFT renormalization (both perturbative and non-perturbative) [71–82].

Most recent work in this direction focussed on entanglement among GFT modes induced by interaction, starting from a generalization of the Bogoliubov description of a weakly interacting Bose gas to the GFT framework [83]. Analogous insights on the inherently dynamic character of entanglement among GFT modes in second quantization appear in the series of recent results proposing a general procedure for constructing states that describe macroscopic, spatially homogeneous universes as Bose–Einstein condensates (see e.g., [53]). Strong results in this sense would provide a new quantitative tool to unravel the relation between entanglement, holography and emergence of spacetime geometry in quantum gravity.

For instance, an interesting direct correspondence between black holes and Bose–Einstein condensates (BECs) of gravitons at the point of maximal packing was recently proposed in [84], though within a standard *effective field theory framework*. In this case, the physics of maximally-packed gravitational systems is identified with the general behaviour of BECs at the critical point of the quantum phase transition, while collective nearly gapless excitations of the quantum condensate are shown to define the holographic degrees of freedom responsible for the known semiclassical holographic properties of black holes (BH) [84]. In this exemplary case, GFTs would provide a unique *non-perturbative* formalism to investigate the foundations of black hole entropy [85] and holography as a general quantum phenomenon of nature.

Funding: This research received no external funding.

Acknowledgments: It is a pleasure to acknowledge Daniele Oriti, Mingyi Zhang and Alex Goeßmann for their precious contribution in the exploration of such a fascinating topic. I am particularly grateful to Hermann Nicolai and to the Albert Einstein Institute in Potsdam, for the support received during the preparation of the manuscript.

Conflicts of Interest: The author declare no conflict of interest.

Appendix A. Coherent States Over-Completeness

Consider two coherent state $|\varphi\rangle$ and $|\varphi'\rangle$. One can prove that they are not orthogonal

$$\langle \varphi | \varphi' \rangle = \frac{1}{\mathcal{N}_\varphi \mathcal{N}_{\varphi'}} \exp\left[\int d\mathbf{g}\, \overline{\varphi(\mathbf{g})} \varphi'(\mathbf{g}) \right]. \tag{A1}$$

The state $|\varphi\rangle$ can be decomposed by the n-particle state basis $|\cdots, g_a, \cdots\rangle$ as

$$|\varphi\rangle = |0\rangle\langle 0|\varphi\rangle + \sum_{n=1}^{\infty} \int \prod_{a=1}^{n} d g_a\, |g_1, \cdots, g_n\rangle \langle g_1, \cdots, g_n | \varphi \rangle, \tag{A2}$$

where $\langle g_1, \cdots, g_n | \varphi \rangle$ is given as

$$\langle g_1, \cdots, g_n | \varphi \rangle = \frac{1}{\mathcal{N}_\varphi} \frac{1}{\sqrt{n!}} \prod_a^n \left[\int dg'_a \, \varphi(g'_a) \right] \langle g_1, \cdots, g_n | g'_1, \cdots, g'_n \rangle$$

$$= \frac{1}{\mathcal{N}_\varphi} \frac{1}{(n!)^{3/2}} \sum_{\pi \in S_n} \prod_a^n \left[\int dg'_a \, dh_a \, \varphi(g'_a) \delta(g_{\pi(a)} h_a (g'_a)^\dagger) \right]$$

$$= \frac{1}{\mathcal{N}_\varphi} \frac{1}{\sqrt{n!}} \prod_a^n \varphi(g_a). \tag{A3}$$

In order to obtain the resolution of identity in terms of $|\varphi\rangle$, let us first introduce the gauge fixed field φ_{GF}

$$\varphi_{GF}(\mathbf{g}) \equiv \varphi(\mathbf{g} h_1^{-1}) = \varphi(\mathbb{1}, h_2 h_1^{-1}, h_3 h_1^{-1}, h_4 h_1^{-1}) \equiv \varphi([\mathbf{g}]). \tag{A4}$$

Then, we have the identities

$$\int \mathcal{D}\varphi \mathcal{D}\overline{\varphi} \, \delta_{\mathbb{C}}[\varphi - \varphi_{GF}] \, e^{-K \int d\mathbf{g} \, \overline{\varphi(\mathbf{g})} \varphi(\mathbf{g})} \langle g_1, \cdots, g_n | \varphi \rangle \langle \varphi | g'_1, \cdots, g'_n \rangle$$

$$= \frac{1}{n!} \int \mathcal{D}\varphi \mathcal{D}\overline{\varphi} \, \delta_{\mathbb{C}}[\varphi - \varphi_{GF}] \, e^{(-K-1) \int d\mathbf{g} \, \overline{\varphi(\mathbf{g})} \varphi(\mathbf{g})} \prod_a^n \varphi(g_a) \overline{\varphi(g'_a)}$$

$$= \frac{1}{n!} \int \mathcal{D}\varphi_{GF} \mathcal{D}\overline{\varphi_{GF}} \, e^{(-K-1) \int d\mathbf{g} \, \overline{\varphi_{GF}(\mathbf{g})} \varphi_{GF}(\mathbf{g})} \prod_a^n \varphi_{GF}(g_a) \overline{\varphi_{GF}(g'_a)}$$

$$= C \frac{1}{n!} \sum_{\pi \in S_n} \prod_{a=1}^n \delta^3([g'_a][g^\dagger_{\pi(a)}])$$

$$= C \frac{1}{n!} \sum_{\pi \in S_n} \prod_{a=1}^n \int dh_a \, \delta^4(g'_a h_a g^\dagger_{\pi(a)})$$

$$= C \langle g_1, \cdots, g_n | g'_1, \cdots, g'_n \rangle, \tag{A5}$$

where K is a parameter which we assume to be real, and C is a constant number

$$C \equiv \int \mathcal{D}\varphi \mathcal{D}\overline{\varphi} \, \delta_{\mathbb{C}}[\varphi - \varphi_{GF}] \, \exp\left[(-K-1) \int d\mathbf{g} \, \overline{\varphi(\mathbf{g})} \varphi(\mathbf{g})\right]. \tag{A6}$$

For the third equality, we use the Wick's theorem, and, for the fourth equality, we reintroduce the gauge symmetry such that the arguments of φ are equal in weight.

The resolution of identity in the Fock space $\mathbb{F}$ is in terms of $|\varphi\rangle$

$$\mathbb{1}_{\mathbb{F}} = C^{-1} \int \mathcal{D}\varphi \mathcal{D}\overline{\varphi} \, \delta_{\mathbb{C}}[\varphi - \varphi_{GF}] \, e^{-K \int d\mathbf{g} \, \overline{\varphi(\mathbf{g})} \varphi(\mathbf{g})} \, |\varphi\rangle \langle \varphi|. \tag{A7}$$

References

1. Maldacena, J.M. The Large N Limit of Superconformal Field Theories and Supergravity. *Adv. Theor. Math. Phys.* **1998**, *2*, 231–252. [CrossRef]
2. Aharony, O.; Gubser, S.S.; Maldacena, J.; Ooguri, H.; Oz, Y. Large N Field Theories, String Theory and Gravity. *Phys. Rep.* **2000**, *323*, 183–386. [CrossRef]
3. Orus, R. A Practical Introduction to Tensor Networks: Matrix Product States and Projected Entangled Pair States. *Ann. Phys.* **2014**, *349*, 117–158. [CrossRef]
4. Bridgeman, J.C.; Chubb, C.T. Hand-waving and Interpretive Dance: An Introductory Course on Tensor Networks. *J. Phys. A Math. Theor.* **2017**, *50*, 223001. [CrossRef]
5. Wen, X.-G. *Quantum Field Theory of Many-Body Systems: From the Origin of Sound to an Origin of Light and Electrons*; Oxford University Press: New York, NY, USA, 2004.
6. Wen, X.-G. Zoo of quantum-topological phases of matter. *Rev. Mod. Phys.* **2017**, *89*, 41004. [CrossRef]

7. Cirac, J.I.; Verstraete, F. Renormalization and tensor product states in spin chains and lattices. *J. Phys. A Math. Theor.* **2009**, *42*, 504004. [CrossRef]
8. Verstraete, F.; Murg, V.; Cirac, J.I. Matrix product states, projected entangled pair states, and variational renormalization group methods for quantum spin systems. *Adv. Phys.* **2008**, *57*, 143–224. [CrossRef]
9. Augusiak, R.; Cucchietti, F.M.; Lewenstein, M. Modern theories of many-particle systems in condensed matter physics. *Lect. Not. Phys.* **2012**, *843*, 245–294.
10. Vidal, G. Class of quantum many-body states that can be efficiently simulated. *Phys. Rev. Lett.* **2008**, *101*, 110501. [CrossRef]
11. Singh, S.; Pfeifer, R.N.; Vidal, G. Tensor network decompositions in the presence of a global symmetry. *Phys. Rev. A* **2010**, *82*, 050301. [CrossRef]
12. Evenbly, G.; Vidal, G. Tensor network states and geometry. *J. Stat. Phys.* **2011**, *145*, 891–918. [CrossRef]
13. Pastawski, F.; Yoshida, B.; Harlow, D.; Preskill, J. Holographic quantum error-correcting codes: Toy models for the bulk/boundary correspondence. *J. High Energy Phys.* **2015**, *2015*, 149. [CrossRef]
14. Cao, C.; Carroll, S.M.; Michalakis, S. Space from Hilbert space: recovering geometry from bulk entanglement. *Phys. Rev. D* **2017**, *95*, 024031. [CrossRef]
15. Hayden, P.; Nezami, S.; Qi, X.L.; Thomas, N.; Walter, M.; Yang, Z. Holographic duality from random tensor networks. *J. High Energy Phys.* **2016**, *2016*, 9. [CrossRef]
16. Miyaji, M.; Takayanagi, T.; Watanabe, K. From path integrals to tensor networks for the AdS/CFT correspondence. *Phys. Rev. D* **2017**, *95*, 066004. [CrossRef]
17. Han, M.; Hung, L.Y. Loop quantum gravity, exact holographic mapping, and holographic entanglement entropy. *Phys. Rev. D* **2017**, *95*, 024011. [CrossRef]
18. Chirco, G.; Oriti, D.; Zhang, M. Group field theory and tensor networks: towards a Ryu-Takayanagi formula in full quantum gravity. *Class. Quant. Grav.* **2018**, *35*, 115011. [CrossRef]
19. Chirco, G.; Oriti, D.; Zhang, M. Ryu-Takayanagi formula for symmetric random tensor networks. *Phys. Rev. D* **2018**, *97*, 126002. [CrossRef]
20. Chirco, G.; Goeßmann, A.; Oriti, D.; Zhang, M. Group Field Theory and Holographic Tensor Networks: Dynamical Corrections to the Ryu-Takayanagi formula. *arXiv* **2019**, arXiv:1903.07344.
21. Ryu, S.; Takayanagi, T. Holographic derivation of entanglement entropy from the anti–de sitter space/conformal field theory correspondence. *Phys. Rev. Lett.* **2006**, *96*, 181602. [CrossRef]
22. Van Raamsdonk, M. Building up spacetime with quantum entanglement. *Gen. Relativ. Gravit.* **2010**, *42*, 2323–2329. [CrossRef]
23. Swingle, B. Entanglement renormalization and holography. *Phys. Rev. D* **2012**, *86*, 065007. [CrossRef]
24. Bonzom, V.; Dittrich, B. 3D holography: From discretum to continuum. *J. High Energy Phys.* **2016**, *2016*, 208. [CrossRef]
25. Ashtekar, A.; Lewandowski, J. Background independent quantum gravity: A status report. *Class. Quant. Grav.* **2004**, *21*, R53. [CrossRef]
26. Rovelli, C. *Quantum Gravity*; Cambridge University Press: London, UK, 2004.
27. Thiemann, T. *Modern Canonical Quantum General Relativity*; Cambridge University Press: London, UK, 2008.
28. Perez, A. The spin-foam approach to quantum gravity. *Living Rev. Relativ.* **2013**, *16*, 3. [CrossRef] [PubMed]
29. Rovelli, C.; Vidotto, F. *Covariant Loop Quantum Gravity: An Elementary Introduction to Quantum Gravity and Spinfoam Theory*; Cambridge University Press: London, UK, 2014.
30. Oriti, D. Group Field Theory and Loop Quantum Gravity. *arXiv* **2014**, arXiv:1408.7112.
31. Koslowski, T.A. Dynamical Quantum Geometry (DQG Programme). *arXiv* **2007**, arXiv:0709.3465.
32. Konopka, T.; Markopoulou, F.; Smolin, L. Quantum Graphity. *arXiv* **2006**, arXiv:hep-th/0611197.
33. Rivasseau, V. Quantum gravity and renormalization: The tensor track. *AIP Conf. Proc.* **2012**, *1444*, 18–29.
34. Steinacker, H. Emergent geometry and gravity from matrix models: an introduction. *Class. Quant. Grav.* **2010**, *27*, 133001. [CrossRef]
35. Bianchi, E.; Myers, R.C. On the architecture of spacetime geometry. *Class. Quant. Grav.* **2014**, *31*, 214002. [CrossRef]
36. Chirco, G.; Rovelli, C.; Ruggiero, P. Thermally correlated states in loop quantum gravity. *Class. Quant. Grav.* **2015**, *32*, 035011. [CrossRef]
37. Hamma, A.; Hung, L.Y.; Marciano, A.; Zhang, M. Area law from loop quantum gravity. *Phys. Rev. D* **2018**, *97*, 064040. [CrossRef]

38. Bianchi, E.; Hackl, L.; Yokomizo, N. Entanglement entropy of squeezed vacua on a lattice. *Phys. Rev. D* **2015**, *92*, 085045. [CrossRef]
39. Livine, E.R. Intertwiner entanglement on spin networks. *Phys. Rev. D* **2018**, *97*, 026009. [CrossRef]
40. Chirco, G.; Mele, F.M.; Oriti, D.; Vitale, P. Fisher metric, geometric entanglement, and spin networks. *Phys. Rev. D* **2018**, *97*, 046015. [CrossRef]
41. Livine, E.R.; Terno, D.R. Quantum black holes: Entropy and entanglement on the horizon. *Nucl. Phys. B* **2006**, *741*, 131–161. [CrossRef]
42. Livine, E.R.; Terno, D.R. Reconstructing quantum geometry from quantum information: Area renormalisation, coarse-graining and entanglement on spin networks. *arXiv* **2006**, arXiv:gr-qc/0603008.
43. Donnelly, W. Entanglement entropy in loop quantum gravity. *Phys. Rev. D* **2018**, *77*, 104006. [CrossRef]
44. Gurau, R. Invitation to random tensors. *arXiv* **2016**, arXiv:1609.06439.
45. Rivasseau, V. Random tensors and quantum gravity. *arXiv* **2016**, arXiv:1603.07278.
46. Baratin, A.; Oriti, D. Ten questions on Group Field Theory (and their tentative answers). *J. Phys. Conf. Ser.* **2012**, *360*, 012002. [CrossRef]
47. Oriti, D. The microscopic dynamics of quantum space as a group field theory. *arXiv* **2011**, arXiv:1110.5606.
48. Oriti, D. The group field theory approach to quantum gravity: Some recent results. *arXiv* **2009**, arXiv:0912.2441.
49. Gurau, R.; Ryan, J.P. Colored Tensor Models—A Review. *SIGMA* **2012**, *8*, 020. [CrossRef]
50. Boulatov, D.V. A model of three-dimensional lattice gravity. *Mod. Phys. Lett. A* **1992**, *7*, 1629–1646. [CrossRef]
51. Rovelli, C. Basis of the Ponzano-Regge Turaev-Viro-Ooguri quantum-gravity model is the loop representation basis. *Phys. Rev. D* **1993**, *48*, 2702. [CrossRef] [PubMed]
52. Baez, J.C.; Barrett, J.W. The quantum tetrahedron in 3 and 4 dimensions. *Adv. Theor. Math. Phys.* **1999**, *3*, 815–850. [CrossRef]
53. Gielen, S.; Oriti, D.; Sindoni, L. Homogeneous cosmologies as group field theory condensates. *J. High Energy Phys.* **2014**, *2014*, 13. [CrossRef]
54. Oriti, D.; Pranzetti, D.; Ryan, J.P.; Sindoni, L. Generalized quantum gravity condensates for homogeneous geometries and cosmology. *Class. Quant. Grav.* **2015**, *23*, 235016. [CrossRef]
55. Oriti, D.; Pereira, R.; Sindoni, L. Coherent states in quantum gravity: A construction based on the flux representation of loop quantum gravity. *J. Phys. A Math. Theor.* **2012**, *45*, 244004. [CrossRef]
56. Oriti, D.; Pereira, R.; Sindoni, L. Coherent states for quantum gravity: Toward collective variables. *Class. Quant. Grav.* **2012**, *29*, 135002. [CrossRef]
57. Oriti, D. Group field theory as the second quantization of loop quantum gravity. *Class. Quant. Grav.* **2016**, *33*, 085005. [CrossRef]
58. Freidel, L.; Gurau, R.; Oriti, D. Group field theory renormalization in the 3D case: Power counting of divergences. *Phys. Rev. D* **2009**, *80*, 044007. [CrossRef]
59. Magnen, J.; Noui, K.; Rivasseau, V.; Smerlak, M. Scaling behavior of three-dimensional group field theory. *Class. Quant. Grav.* **2009**, *26*, 185012. [CrossRef]
60. Bonzom, V.; Smerlak, M. Bubble divergences from cellular cohomology. *Lett. Math. Phys.* **2010**, *93*, 295–305. [CrossRef]
61. Bonzom, V.; Smerlak, M. Bubble divergences from twisted cohomology. *Commun. Math. Phys.* **2012**, *312*, 399–426. [CrossRef]
62. Bonzom, V.; Smerlak, M. Bubble divergences: sorting out topology from cell structure. *Annales Henri Poincare* **2012**, *13*, 185–208. [CrossRef]
63. Oriti, D. Disappearance and emergence of space and time in quantum gravity. *Stud. Hist. Phil. Sci.* **2014** *46*, 186–199. [CrossRef]
64. Dittrich, B.; Geiller, M. A new vacuum for loop quantum gravity. *Class. Quant. Grav.* **2015**, *11*, 112001. [CrossRef]
65. Han, M. Cosmological constant in loop quantum gravity vertex amplitude. *Phys. Rev. D* **2011**, *84*, 064010. [CrossRef]
66. Major, S.; Smolin, L. Quantum deformation of quantum gravity. *Nucl. Phys. B* **1996**, *473*, 267–290. [CrossRef]
67. Gurau, R. Universality for random tensors. *Annales de l'IHP Probabilités et Statistiques* **2014**, *50*, 1474–1525. [CrossRef]

68. Vedral, V. Entanglement in the second quantization formalism. *Central Eur. J. Phys.* **2003**, *1*, 289–306. [CrossRef]
69. Benatti, F.; Floreanini, R.; Titimbo, K. Entanglement of identical particles. *Open Syst. Inf. Dyn.* **2014**, *21*, 1440003. [CrossRef]
70. Simon, C. Natural entanglement in Bose-Einstein condensates. *Phys. Rev. A* **2002**, *66*, 052323. [CrossRef]
71. Dittrich, B.; Mizera, S.; Steinhaus, S. Decorated tensor network renormalization for lattice gauge theories and spin foam models. *New J. Phys.* **2016**, *18*, 053009. [CrossRef]
72. Delcamp, C.; Dittrich, B. Towards a phase diagram for spin foams. *Class. Quant. Grav.* **2016**, *34*, 225006. [CrossRef]
73. Dittrich, B.; Eckert, F.C.; Martin-Benito, M. Coarse graining methods for spin net and spin foam models. *New J. Phys.* **2012**, *14*, 035008. [CrossRef]
74. Dittrich, B.; Schnetter, E.; Seth, C.J.; Steinhaus, S. Coarse graining flow of spin foam intertwiners. *Phys. Rev. D* **2016**, *94*, 124050. [CrossRef]
75. Carrozza, S.; Oriti, D.; Rivasseau, V. Renormalization of a SU (2) tensorial group field theory in three dimensions. *Commun. Math. Phys.* **2014**, *330*, 581–637. [CrossRef]
76. Benedetti, D.; Geloun, J. B.; Oriti, D. Functional renormalisation group approach for tensorial group field theory: a rank-3 model. *J. High Energy Phys.* **2015**, *2015*, 84. [CrossRef]
77. Carrozza, S.; Lahoche, V. Asymptotic safety in three-dimensional SU (2) Group Field Theory: evidence in the local potential approximation. *Class. Quant. Grav.* **2016**, *34*, 115004. [CrossRef]
78. Carrozza, S. Flowing in group field theory space: A review. *arXiv* **2016**, arXiv:1603.01902.
79. Geloun, J.B.; Martini, R.; Oriti, D. Functional renormalization group analysis of tensorial group field theories on $\mathbb{R}^d$. *Phys. Rev. D* **2016**, *94*, 024017. [CrossRef]
80. Lahoche, V.; Oriti, D. Renormalization of a tensorial field theory on the homogeneous space SU (2)/U (1). *J. Phys. A Math. Theor.* **2017**, *50*, 025201. [CrossRef]
81. Bahr, B.; Dittrich, B.; Hellmann, F.; Kaminski, W. Holonomy spin foam models: definition and coarse graining. *Phys. Rev. D* **2013**, *87*, 044048. [CrossRef]
82. Bahr, B.; Steinhaus, S. Numerical evidence for a phase transition in 4d spin-foam quantum gravity. *Phys. Rev. Lett.* **2016**, *117*, 141302. [CrossRef] [PubMed]
83. Bielli, D.; Chirco, G.; Thiam, G. Entanglement of Bogoliubov Modes in Quantum Gravity Condensates. in preparation.
84. Dvali, G.; Gomez, C. Black Hole's quantum N-portrait. *Fortschritte der Physik* **2013**, *61*, 742–767. [CrossRef]
85. Oriti, D.; Pranzetti, D.; Sindoni, L. Horizon entropy from quantum gravity condensates. *Phys. Rev. Lett.* **2016**, *116*, 211301. [CrossRef]

© 2019 by the author. Licensee MDPI, Basel, Switzerland. This article is an open access article distributed under the terms and conditions of the Creative Commons Attribution (CC BY) license (http://creativecommons.org/licenses/by/4.0/).

Article

Switching Internal Times and a New Perspective on the 'Wave Function of the Universe'

Philipp A. Höhn [1,2]

[1] Institute for Quantum Optics and Quantum Information, Austrian Academy of Sciences, Boltzmanngasse 3, 1090 Vienna, Austria; p.hoehn@univie.ac.at
[2] Vienna Center for Quantum Science and Technology (VCQ), Faculty of Physics, University of Vienna, Boltzmanngasse 5, 1090 Vienna, Austria

Received: 7 March 2019; Accepted: 7 May 2019; Published: 14 May 2019

Abstract: Despite its importance in general relativity, a quantum notion of general covariance has not yet been established in quantum gravity and cosmology, where, given the a priori absence of coordinates, it is necessary to replace classical frames with dynamical quantum reference systems. As such, quantum general covariance bears on the ability to consistently switch between the descriptions of the same physics relative to arbitrary choices of quantum reference system. Recently, a systematic approach for such switches has been developed. It links the descriptions relative to different choices of quantum reference system, identified as the correspondingly reduced quantum theories, via the reference-system-neutral Dirac quantization, in analogy to coordinate changes on a manifold. In this work, we apply this method to a simple cosmological model to demonstrate how to consistently switch between different internal time choices in quantum cosmology. We substantiate the argument that the conjunction of Dirac and reduced quantized versions of the theory defines a complete relational quantum theory that not only admits a quantum general covariance, but, we argue, also suggests a new perspective on the 'wave function of the universe'. It assumes the role of a perspective-neutral global state, without immediate physical interpretation that, however, encodes all the descriptions of the universe relative to all possible choices of reference system at once and constitutes the crucial link between these internal perspectives. While, for simplicity, we use the Wheeler-DeWitt formulation, the method and arguments might be also adaptable to loop quantum cosmology.

Keywords: quantum relational dynamics; switching relational clocks; quantum symmetry reduction; quantum cosmology; quantum general covariance; Dirac and reduced quantization; Hamiltonian constraint; wave function of the universe

1. Introduction

General covariance is a celebrated feature of general relativity. It asserts that all the laws of physics are the same in all reference frames and independent of coordinates. It not only permits us to describe the physics from arbitrary choices of reference frame, but also to switch between the different descriptions at will. General covariance is the origin of the diffeomorphism invariance of the theory and thereby leads to profound conceptual consequences [1]: Physical systems are neither localized nor evolve with respect to a background spacetime, but relative to one another. General covariance thus already implies classically that coordinates are not a fundamental concept in physics. While they are practical for any concrete calculations of the physics *in* a given spacetime, already classically, one could, instead, use dynamical degrees of freedom as reference systems relative to which to describe the physics, incl. the dynamics *of* spacetime [1–9].

In quantum cosmology and quantum gravity the situation becomes more extreme: Since one does not quantize spacetime and its matter content relative to a background, coordinate systems are a priori absent altogether. Consequently, it becomes a necessity to employ dynamical degrees of freedom as *quantum* reference systems relative to which to describe the physics [1–4,9–37]. Ordinary coordinate systems are only expected to be reconstructed from such reference systems in a semiclassical, large-scale limit.

A question that has so far received little attention in quantum gravity and cosmology is how to establish a quantum notion of general covariance, despite its fundamental importance to the theory supposed to be quantized. A reason is perhaps the absence of coordinates and the (attempted) outright diffeomorphism invariance in quantum gravity. However, already classically, general covariance is less about coordinates and, operationally, primarily about linking the descriptions relative to different reference frames. Similarly, given the absence of coordinates, quantum general covariance can only refer to the ability to consistently switch between the descriptions of the same physics relative to arbitrary choices of quantum reference system. This includes both spatial and temporal reference systems.

As an initial step, we shall address this question in the context of a simple isotropic and homogeneous quantum cosmological model in this article, exploiting a novel framework for quantum reference systems [38–41] and building up on the earlier works [20–22]. As such, we will here not be concerned with spatial reference systems [38,39,41], but only internal times to which one usually resorts for defining temporal localization in quantum cosmology [9–12,16–37].

The use of different choices of internal times in parametrized systems and cosmological models has been considered, e.g., in [11,30,42–44], but no explicit switches between the different choices were constructed. Instead, the so-called *multiple choice problem* associated with the problem of time was diagnosed [11,12]. This is the purported problem that generically there are no distinguished internal time choices and that different choices of internal times would lead to unitarily inequivalent quantum theories. Switching between different internal time choices was only later studied in a semiclassical approach [20–22,45] and, for a restricted set of choices, at the level of reduced quantization [46–49]. Nevertheless, the meaning of quantum general covariance remained elusive.

One of our aims here will be to begin clarifying both technically and conceptually what quantum general covariance is, at least in the simplified context of quantum cosmology. The method and concepts, however, extend, at least in principle, to full quantum gravity. To this end, I will invoke a recent unifying approach to switching quantum reference systems in both quantum foundations and gravity [38–40]. This approach blends operational quantum reference frame methods [41], aiming at quantum covariance too, with the ideas underlying the semiclassical clock switches in [20–22,45] and conceptual arguments concerning the 'wave function of the universe' and how to accommodate different frame perspectives in it [50]. In particular, in [40] it was already shown that it provides a systematic method for switching between different choices of relational quantum clocks and this will be exploited below.

The key feature of the method in [38–40] is that it identifies a consistent quantum reduction procedure that maps the Dirac quantized theory to the various reduced quantized versions of it relative to different choices of quantum reference systems. It identifies the physical Hilbert space of the Dirac quantization as a reference-system-neutral quantum super structure and the various reduced quantum theories as the physics described relative to the corresponding choice of reference system. In analogy to a coordinate change on a manifold, one can then switch between different choices of quantum reference system by inverting a given quantum reduction map and concatenating it with the forward reduction map associated with the new choice of reference system. Just as with coordinate changes, this will not always work globally, but this, I argue, is the structure defining quantum general covariance in a canonical formulation [38–40]. In particular, a *complete* relational quantum theory, admitting quantum general covariance, is the conjunction of its Dirac and various reduced quantized versions, just like the classical theory contains both the constraint surface and reduced phase spaces [40]. By linking the various (generally unitarily inequivalent) reduced quantizations, the multiple choice problem becomes a *multiple choice feature* of the complete relational quantum theory [40], just like general covariance is a feature of general relativity.

In this work, I will apply this method to the simple flat Friedman-Robertson-Walker (FRW) universe filled with a massless, homogeneous scalar field and show how to consistently switch between choosing either the scale factor or the field as an internal time in both the classical and quantum theory and how the different descriptions are explicitly linked. This model has become a fairly standard example in Wheeler-DeWitt type quantum cosmology [30,31,37] and loop quantum cosmology [25,26,34,36] and has recently even been reconstructed from a full quantum gravity theory [27,28]. Our discussion will be of relevance to each of these approaches, although loop quantization related subtleties need be taken into account before this framework can be directly applied to the latter two approaches (see later comments).

In the conclusions, I will use this explicit construction to argue more generally that quantum general covariance also entails a novel perspective on the 'wave function of the universe'. It provides the handle to relate quantum states of subsystems 'as seen' by other subsystems to 'the wave function of the universe', linking frame-dependent and frame-independent descriptions of the physics and thereby suggesting a new interpretation of states in quantum cosmology. In particular, I propose to view the 'wave function of the universe' as a perspective-neutral global state that does not admit an immediate physical interpretation, but that encodes all the descriptions of the universe relative to all possible choices of reference system at once and constitutes the crucial link between all these internal perspectives. This will substantiate (and partially amend) an earlier proposal for interpreting the 'wave function of the universe' and rendering it compatible with operationally significant relative states [50] (see also the earlier discussion in [51]).

2. The Flat FRW Model with Massless Scalar Field

Consider an isotropic and homogeneous FRW universe, filled with a homogeneous scalar field $\phi(t)$ and described by a metric $ds^2 = -dt^2 + a^2(t)(dr^2/(1-kr^2) + r^2 d\Omega^2)$, where $a(t)$ is the scale factor and $k = -1, 0, +1$ characterize open, flat and closed universes, respectively. For quantization later, it will be convenient to rather choose $\alpha := \ln a$, so that $(\alpha, \phi) \in \mathbb{R}^2$ will be our configuration variables. This choice also simplifies the form of the Hamiltonian constraint, generating the dynamics, and yields[1] $C_H = p_\phi^2 - p_\alpha^2 - 4k \exp(4\alpha) + 4m^2 \phi^2 \exp(6\alpha)$, where m is the mass of the field, e.g., see [25,26,30–37]. For illustrative purposes, we shall henceforth set the mass and curvature to zero, $m = k = 0$, such that the Hamiltonian constraint takes a particularly simple Klein-Gordon form

$$C_H = p_\phi^2 - p_\alpha^2 \approx 0, \qquad (1)$$

where $\approx$ denotes a weak equality [52,53]. Hence, we can equivalently interpret the dynamics as either a flat FRW model with massless scalar field, or as a relativistic particle in 1 + 1 dimensions.

To understand the quantum internal time switches, it is necessary to first carefully revisit the classical model.

2.1. Classical Relational Dynamics and Internal Time Switches

It is clear that p_ϕ, p_α are *dependent* constants of motion and thus Dirac observables, as are

$$\Lambda = p_\alpha \phi + p_\phi \alpha, \qquad L = p_\phi \phi + p_\alpha \alpha. \qquad (2)$$

We have not yet selected a temporal reference with respect to which to interpret the dynamics. The constraint surface $\mathcal{C}$ defined by (1) encodes all possible internal time choices at once, as reflected

---

[1] In fact, we have included a choice of lapse function $N = e^{3\alpha}$.

also in the redundancy of its description, and constitutes an internal-time-neutral super structure [40] (see also [38,39]). As such, $\mathcal{C}$ itself does not admit the interpretation as the physics described relative to a reference system; it is also not a phase space, but a pre-symplectic manifold.

Using Λ or L, we can construct *relational* Dirac observables [1,3,5–9,14–22,40] in various ways. For simplicity, we choose α, ϕ as internal times, exploiting that they are globally monotonic. For compactness of notation, denote by e and t the evolving and clock configuration degree of freedom, respectively, which are either $e = \phi$ and $t = \alpha$, or vice versa. The relational observable describing the evolution of e with respect to t can be easily constructed by evaluating the right-hand side of $\Lambda = p_\alpha \phi + p_\phi \alpha$ along the trajectories generated by C_H (with flow parameter s) and noting that Λ is a constant of motion, producing

$$e(\tau) := e(s)\big|_{t(s)=\tau} = \frac{1}{p_t}\left(\Lambda - p_e \tau\right) = -\frac{p_e}{p_t}(\tau - t) + e. \qquad (3)$$

(The situation is completely symmetric in α and ϕ.) This parameter family of Dirac observables gives the value of e when the clock t reads τ. We would have to carefully regularize the inverse powers of p_t in the subsequent reduced phase spaces and quantum theory. While this can be done [40], it will be convenient to make a variable change in the evolving degrees of freedom to avoid these complications. Instead of the canonical pair (e, p_e), we will henceforth look at the evolution of the *affine* pair $(E := e\, p_e, p_e)$, satisfying $\{E, p_e\} = p_e$, with respect to t. This amounts to evaluating L instead of Λ and yields

$$E(\tau) := E(s)\big|_{t(s)=\tau} \approx L - p_t\, \tau = -p_t\,(\tau - t) + E, \qquad p_e(\tau) := p_e(s)\big|_{t(s)=\tau} = p_e, \qquad (4)$$

so that we have no singular behavior to worry about.

We wish to remove the redundant clock degrees of freedom from among the dynamical variables through reduction [40]. To this end, it will be convenient to factorize (1),

$$C_H = s_t\, C_+^t\, C_-^t, \qquad C_\pm^t := p_t \pm h_e, \qquad h_e := |p_e|, \qquad s_t := \begin{cases} +1, & t = \phi, \\ -1, & t = \alpha. \end{cases} \qquad (5)$$

h_e will assume the role of a Hamiltonian. We have the following situation:

(i) On $\mathcal{C}_\pm^t \subset \mathcal{C}$, defined by $C_\pm^t = 0$ and $p_t \neq 0$, we have

$$\frac{d\cdot}{ds} = \{\cdot, C_H\} \approx \mp 2 s_t\, h_e\, \{\cdot, C_\pm^t\}, \qquad (6)$$

so that $C_\pm^t$ generates the dynamics on $\mathcal{C}_\pm^t$. Since $h_e > 0$, the flows generated by C_+^t and C_-^t are opposite to and aligned with that of C_H, respectively, for $t = \phi$ and aligned with and opposite to that of C_H, respectively, for $t = \alpha$. That is, ϕ runs 'backward' on $\mathcal{C}_+^\phi$ and 'forward' on $\mathcal{C}_-^\phi$, while α expands on $\mathcal{C}_+^\alpha$ and contracts on $\mathcal{C}_-^\alpha$. Backward/expanding and forward/contracting will correspond to positive and negative frequency solutions, respectively, in the quantum theory.

(ii) The set $p_\alpha = p_\phi = 0$ is the shared boundary between $\mathcal{C}_+^\phi$ and $\mathcal{C}_-^\phi$, as well as between $\mathcal{C}_+^\alpha$ and $\mathcal{C}_-^\alpha$. Notice that orbits with $p_\alpha = p_\phi = 0$ are just points in $\mathcal{C}$ so that the latter is stratified by gauge orbits of different dimension. Since $dC_H = 0$ for $p_\alpha = p_\phi = 0$, no gauge-fixing surface can pierce every such gauge orbit once and only once.

The situation is summarized in Figure 1 for convenience.

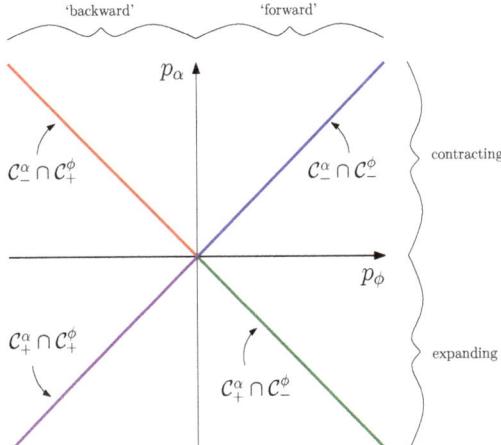

Figure 1. Schematic representation of the constraint surface $\mathcal{C}$, defined by (1), as a 'light cone' in momentum space. Its four components have the following physical interpretation. Red: contracting universe, but ϕ runs 'backward'. Blue: contracting universe and ϕ runs 'forward'. Green: expanding universe and ϕ runs 'forward'. Purple: expanding universe, but ϕ runs 'backward'. At the intersection point (the origin) the dynamics is static.

On $\mathcal{C}^t_\pm$ we can thus use $C^t_\pm$ as evolution generators and their *relational* dynamics is equivalent to that of C_H. Indeed, on $\mathcal{C}^t_\pm$ we find ($s_\pm$ denotes the flow parameter of $C^t_\pm$):

$$E_\pm(\tau) := E(s_\pm)\big|_{t(s_\pm)=\tau} = \pm|p_e|(\tau - t) + E, \qquad p_{e\pm}(\tau) := p_e(s_\pm)\big|_{t(s_\pm)=\tau} = p_e, \qquad (7)$$

which is (4) after solving (1). The relational Dirac observables being gauge-invariant extensions of gauge-restricted quantities [5,6,40,53], we can now gauge fix the clock to, e.g., $t = 0$ and evaluate (7) on this surface $\mathcal{G}_{t=0}$ without loss of dynamical information. This will produce two separate reduced phase spaces $\mathcal{P}^{e(t)}_\pm \simeq \bar{\mathcal{C}}^t_\pm \cap \mathcal{G}_{t=0}$ for positive/negative frequency modes, where $\bar{\mathcal{C}}^t_\pm$ is $\mathcal{C}^t_\pm$ including its boundary $p_\alpha = p_\phi = 0$. Of course, due to (ii) these gauge-fixed reduced phase spaces will miss all point-like orbits with $p_\alpha = p_\phi = 0$ and $t \neq 0$ and so their union does *not* coincide with the space of orbits $\mathcal{C}/\sim$, where $\sim$ identifies points in the same orbit. We comment on this shortly.

The Dirac bracket for any functions F, G on $\mathcal{C}^t_\pm$ reads

$$\{F, G\}_{D_\pm} = \{F, G\} - \{F, C^t_\pm\}\{t, G\} + \{F, t\}\{G, C^t_\pm\}. \qquad (8)$$

All Dirac brackets involving the redundant clock variables (t, p_t) vanish, which can thus be removed. Furthermore, the affine bracket

$$\{E, p_e\}_{D_\pm} \equiv \{E, p_e\} = p_e \qquad (9)$$

is well-defined everywhere. By contrast, the canonical $\{e, p_e\}_{D_\pm}$ is undefined for $p_e = 0$. Hence, we take $\mathcal{P}^{e(t)}_\pm$ to be fundamentally defined through the affine algebra (9). Then we could *define* $e := E/p_e$ on $\mathcal{P}^{e(t)}_\pm$, yielding a *derived* canonical relation $\{e, p_e\}_{D_\pm} = 1$.

On $\mathcal{P}^{e(t)}_\pm$ the relational observables (7) become

$$E_\pm(\tau) = \pm|p_e|\tau + E, \qquad p_{e\pm}(\tau) = p_e, \qquad (10)$$

and satisfy the following equations of motion

$$\frac{dE_\pm}{d\tau} = \pm|p_e| = \{E_\pm, \pm h_e\}_{D_\pm} \qquad \frac{dp_{e\pm}}{d\tau} = 0 = \{p_{e\pm}, \pm h_e\}_{D_\pm}, \qquad (11)$$

which are thus generated by the physical Hamiltonian $\pm h_e$.

This can now genuinely be interpreted as the evolution described relative to the clock t, which, being the reference system, has become dynamically redundant and an evolution *parameter* τ (see also [40]). Notice also that the measure-zero set of ignored orbits that distinguishes (the union of) $\mathcal{P}_\pm^{e(t)}$ from the space of orbits is redundant for relational dynamics. Indeed, the ignored orbits correspond to the static point-like orbits with $p_\alpha = p_\phi = 0$ and $t \neq 0$, where (E, p_e) are directly independent observables. However, all their information is already encoded in (10): for $p_e = 0$, $E_\pm(\tau) = E$ does not depend on τ, which, however, runs over all possible values of t. It is thus physically justified to work with the gauge-fixed reduced phase space $\mathcal{P}_\pm^{e(t)}$ rather than the abstract reduced phase space $\mathcal{C}/\sim$. We will also see that the relation between the Dirac and reduced quantized theories is consistent with this observation.

Next, we interchange the roles of e and t, i.e., we switch to using e as the clock and t as an evolving variable [40]. The corresponding map between the corresponding reduced phase spaces $\mathcal{P}_\pm^{e(t)}$ and $\mathcal{P}_\pm^{t(e)}$ involves the gauge transformation generated by C_H which maps $\mathcal{C} \cap \mathcal{G}_{t=0}$ to, e.g., $\mathcal{C} \cap \mathcal{G}_{e=0}$. Solving the equations of motion generated by C_H, one easily finds that one has to flow a parameter distance $s = s_t t_0/2p_t$ in $\mathcal{C}$, where t_0 is the clock value prior to the transformation. Dropping the redundant variables, this yields the following maps[2]

$$S_{t_+ \to e_\pm} : \mathcal{P}_+^{e(t)} \to \mathcal{P}_\pm^{t(e)}, \qquad (E, p_e) \mapsto (T = E, p_t = -|p_e|),$$
$$S_{t_- \to e_\pm} : \mathcal{P}_-^{e(t)} \to \mathcal{P}_\pm^{t(e)}, \qquad (E, p_e) \mapsto (T = E, p_t = +|p_e|), \qquad (12)$$

where $T = t\, p_t$ is the evolving affine variable after the clock switch. Notice that gauge transformations preserve $\mathcal{C}_i^\alpha \cap \mathcal{C}_j^\phi, i,j = +,-$, i.e., the four quadrants of Figure 1. Hence, e.g., $S_{t_+ \to e_\pm}$ maps the $p_e < 0$ and $p_e > 0$ halves of $\mathcal{P}_+^{e(t)}$ onto the $p_t < 0$ halves of $\mathcal{P}_+^{t(e)}$ and $\mathcal{P}_-^{t(e)}$, respectively, etc. (For example, $S_{\alpha_+ \to \phi_-}$ switches from the description of the 'expanding-forward' sector (green quadrant in Figure 1) relative to α to its description relative to ϕ.) Respecting this, one obtains a 'continuous' relational evolution, despite the clock switch: Using (10) and setting $\tau_e^i = E_+(\tau_t^f)/p_e =: e_+(\tau_t^f)$ as the initial value of the new clock e after the clock switch, where τ_t^f was the final value of the old clock t prior to it, one consistently finds[3]

$$T_\pm(\tau_e^i) = p_t\, \tau_t^f, \qquad \text{on } \mathcal{P}_\pm^{t(e)}. \qquad (13)$$

We shall see the quantum analog of this later. We emphasize that due to the intermediate gauge transformation the clock switch proceeds via the internal-time-neutral $\mathcal{C}$ [40].

2.2. Reduced Quantization Relative to a Choice of Internal Time

We proceed by quantizing the gauge-fixed reduced phase spaces $\mathcal{P}_\pm^{e(t)}$ of this model universe. Subsequently, we will link the various reduced quantum theories *via* the internal-time-neutral Dirac quantized theory. For simplicity, we resort to the Wheeler-DeWitt formulation in the Dirac procedure, but we note that the loop quantization of this FRW model can be cast into a very similar form (modulo observables) [34]. There is thus good hope that the below framework for switching internal times can be adapted to loop quantum cosmology. Before doing so, however, one must overcome loop quantization related subtleties, which I briefly comment on in the conclusions.

---

[2] For more details of this procedure in a different model, see [40].
[3] Please note that generally $T_\pm(\tau_e^i) \neq E_+(\tau_t^f)$, despite the form (12).

Since we will encounter several Hilbert spaces and transformations along the way, we summarize the various classical and quantum reduction steps and their relation in Figure 2 for guidance.

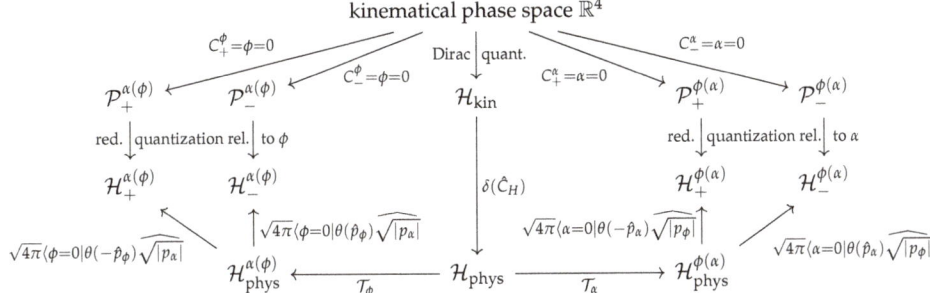

Figure 2. Diagrammatic overview of the relation between Dirac and the four reduced quantizations. In a nutshell, the physical Hilbert space is mapped to any of the four (positive or negative frequency) reduced Hilbert spaces by first trivializing the Hamiltonian constraint via $\mathcal{T}_\phi$ or $\mathcal{T}_\alpha$ to the corresponding choice of internal time variable and subsequently projecting onto the classical internal time gauge-fixing condition. (Details in the main text.)

Recall that $\mathcal{P}_\pm^{e(t)}$ is defined through the affine algebra (9). However, it turns out to be equivalent to quantize these phase spaces in either the affine or standard canonical method. We promote the Dirac bracket $\{.,.\}_{D_\pm}$ to a commutator $[.,.]$ and (e, p_e) to conjugate or (E, p_e) to affinely related operators on a Hilbert space $\mathcal{H}_\pm^{e(t)} := L^2(\mathbb{R})$. In the canonical momentum representation, we represent states as

$$|\psi\rangle_\pm^{e(t)} = \int_{-\infty}^{+\infty} dp_e \, \psi_\pm^{e(t)}(p_e) \, |p_e\rangle_e , \qquad (14)$$

the inner product as

$$\langle\psi|\chi\rangle_\pm^{e(t)} = \int_{-\infty}^{+\infty} dp_e \, [\psi_\pm^{e(t)}(p_e)]^* \, \chi_\pm^{e(t)}(p_e) , \qquad (15)$$

$\hat{p}_e$ as a multiplication operator and the configuration observables as[4]

$$\hat{e} \, \psi_\pm^{e(t)}(p_e) = i \partial_{p_e} \psi_\pm^{e(t)}(p_e), \qquad \hat{E} \, \psi_\pm^{e(t)}(p_e) = i \left(p_e \partial_{p_e} + \frac{1}{2} \right) \psi_\pm^{e(t)}(p_e). \qquad (16)$$

These are self-adjoint and for states with $\lim_{p_e \to \pm\infty} \sqrt{|p_e|} \, \psi_\pm^{e(t)}(p_e) = 0$ we can equivalently work with $\hat{e}$ or $\hat{E}$ which also satisfies $\hat{E} = \frac{1}{2}(\hat{e}\hat{p}_e + \hat{p}_e\hat{e})$.[5] The evolving observables (10) become

$$\hat{E}_\pm(\tau) = \pm|\hat{p}_e|\, \tau + \hat{E}, \qquad \hat{p}_{e\pm}(\tau) = \hat{p}_e , \qquad (17)$$

---

[4] We set $\hbar = 1$.

[5] In the affine momentum representation, states are represented as $|\psi\rangle_\pm^{e(t)} = \int_{-\infty}^{+\infty} \frac{dp_e}{|p_e|} \, \tilde{\psi}_\pm^{e(t)}(p_e) \, |p_e\rangle_{\text{aff}}$, where $\tilde{\psi}_\pm^{e(t)} = \sqrt{|p_e|}\, \psi_\pm^{e(t)}$ and $\langle p_e | p_e' \rangle_{\text{aff}} = |p_e|\, \delta(p_e - p_e')$. The inner product then reads $\langle\psi|\chi\rangle_\pm^{e(t)} = \int_{-\infty}^{+\infty} \frac{dp_e}{|p_e|} \, [\tilde{\psi}_\pm^{e(t)}(p_e)]^* \tilde{\chi}_\pm^{e(t)}(p_e)$ and the configuration observables are represented as $\hat{E}\, \tilde{\psi}_\pm^{e(t)} = i p_e \partial_{p_e} \tilde{\psi}_\pm^{e(t)}$ and $\hat{e}\, \tilde{\psi}_\pm^{e(t)} = (i\partial_{p_e} - \frac{i}{2p_e}) \tilde{\psi}_\pm^{e(t)}$. It is easy to check that this affine representation is equivalent to the canonical one above.

and satisfy the Heisenberg equations with Hamiltonian $\hat{H} = \pm \hat{h}_e = \pm |\hat{p}_e|$ on $\mathcal{H}_\pm^{e(t)}$

$$\frac{d\hat{E}_\pm}{d\tau} = \pm |\hat{p}_e| = -i[\hat{E}_\pm, \hat{H}] \qquad \frac{d\hat{p}_{e\pm}}{d\tau} = 0 = -i[\hat{p}_{e\pm}, \hat{H}]. \qquad (18)$$

2.3. The Internal-Time-Neutral Dirac Quantization

We continue with Dirac quantization (see Figure 2), promoting (α, p_α) and (ϕ, p_ϕ) to conjugate operators on a kinematical Hilbert space $\mathcal{H}_{\text{kin}} := L^2(\mathbb{R}^2)$. The solutions to the quantum constraint

$$\hat{C}_H |\psi\rangle_{\text{phys}} = (\hat{p}_\phi^2 - \hat{p}_\alpha^2) |\psi\rangle_{\text{phys}} \stackrel{!}{=} 0. \qquad (19)$$

will define the physical Hilbert space $\mathcal{H}_{\text{phys}}$. Using *group averaging* [23,38–40,54–57], $|\psi\rangle_{\text{phys}} = \delta(\hat{C}_H)|\psi\rangle_{\text{kin}}$, and working in momentum representation with kinematical wave functions $\psi_{\text{kin}}(p_\phi, p_\alpha)$, we find physical states to be of the form

$$|\psi\rangle_{\text{phys}} = \int_{-\infty}^{+\infty} \frac{dp_e}{2|p_e|} \left[\psi_{\text{kin}}^{e(t)}(-|p_e|, p_e) \,|-|p_e|\rangle_t |p_e\rangle_e + \psi_{\text{kin}}^{e(t)}(|p_e|, p_e) \,||p_e|\rangle_t |p_e\rangle_e \right] \qquad (20)$$

and the physical inner product as

$$\langle \psi | \chi \rangle_{\text{phys}} = \int_{-\infty}^{+\infty} \frac{dp_e}{2|p_e|} \left[(\psi_{\text{kin}}^{e(t)}(-|p_e|, p_e))^* \chi_{\text{kin}}^{e(t)}(-|p_e|, p_e) + (\psi_{\text{kin}}^{e(t)}(|p_e|, p_e))^* \chi_{\text{kin}}^{e(t)}(|p_e|, p_e) \right], \qquad (21)$$

where for compactness of notation we have set

$$\psi_{\text{kin}}^{e(t)}(p_t = \mp|p_e|, p_e) := \begin{cases} \psi_{\text{kin}}(\mp|p_\alpha|, p_\alpha), & t = \phi, \; e = \alpha, \\ \psi_{\text{kin}}(p_\phi, \mp|p_\phi|), & t = \alpha, \; e = \phi. \end{cases} \qquad (22)$$

The situation is completely symmetric in α and ϕ and we will exploit this for the internal time switches. For interpretation it is useful to note that the position representation of the states reads

$$\psi_{\text{phys}}^\pm(e, t) = \int_{-\infty}^{+\infty} \frac{dp_e}{4\pi |p_e|} e^{i(\mp|p_e| t + p_e e)} \psi_{\text{kin}}^{e(t)}(\mp|p_e|, p_e), \qquad (23)$$

where $\psi_{\text{phys}}^\pm$ are the positive/negative frequency solutions of the Klein-Gordon equation. It is easy to convince oneself that

$$\langle \psi | \chi \rangle_{\text{phys}} = 2\pi \left[\left(\psi_{\text{phys}}^+, \chi_{\text{phys}}^+ \right)_{\text{KG}} - \left(\psi_{\text{phys}}^-, \chi_{\text{phys}}^- \right)_{\text{KG}} \right], \qquad (24)$$

where $(\psi, \chi)_{\text{KG}} = i \int de \, (\psi^* \partial_t \chi - (\partial_t \psi^*) \chi)$ is the usual Klein-Gordon inner product in which positive and negative frequency solutions are orthogonal (see also [54]). Physical states and inner product thus decompose into a sum of positive and negative frequency modes. It follows from Figure 1 that for $e = \alpha$ and $t = \phi$ positive/negative frequency solutions correspond to classical backward/forward evolution in ϕ. Conversely, for $e = \phi$ and $t = \alpha$, positive/negative frequency solutions correspond to evolving relative to an expanding/contracting α. It is standard (and usually justified) to ignore the negative frequency solutions [25,26,34]; here we shall not do that as they will be interesting when switching internal times. In particular, it is easy to convince oneself, using (21) and Figure 1, that both the positive and negative frequency part of a physical state for $e = \alpha$ overlap with both the positive and negative frequency part of the same physical state associated with $e = \phi$.

Choosing a symmetric ordering, the relational Dirac observables (4) are quantized as

$$\hat{E}(\tau) = -\hat{p}_t \tau + \frac{1}{2}\left(\hat{p}_t \hat{t} + \hat{t} \hat{p}_t + \hat{e} \hat{p}_e + \hat{p}_e \hat{e} \right) + i = -\hat{p}_t \tau + \hat{t} \hat{p}_t + \hat{p}_e \hat{e} + i, \qquad \hat{p}_e(\tau) = \hat{p}_e \qquad (25)$$

and commute with $\hat{\mathcal{C}}_H$; however, $\hat{E}(\tau)$ *only* does so on $\mathcal{H}_{phys}$. This is also the reason for the $+i$ term, which ensures that $\hat{E}(\tau)$ is Hermitian with respect to (21) and ultimately self-adjoint on $\mathcal{H}_{phys}$, see Appendix A.

In analogy to the classical $\mathcal{C}$, I propose to conceive of $\mathcal{H}_{phys}$ as the internal-time-neutral quantum structure [40]. In the Dirac quantized theory, we have not yet chosen a temporal reference system with respect to which to interpret the dynamics. This is reflected in the redundancy of the representation of states (20), inner product (21) and relational observables (25); we have not yet decided whether $t = \alpha$ or ϕ and we could have selected a different internal time altogether. Just as with $\mathcal{C}$, $\mathcal{H}_{phys}$ encodes all internal clock choices at once and it features no Heisenberg evolution equations for relational observables.

2.4. Quantum Reduction: From Dirac to Reduced Quantization

Next, we perform the quantum reduction procedure that maps the Dirac to the various reduced quantized theories [38–40] and ultimately permits us to switch internal times also in the quantum theory. In analogy to the classical case, it proceeds as follows (see Figure 2): (i) choose an internal time; (ii) trivialize the constraint to the internal time to render it redundant; (iii) project onto the classical gauge-fixing conditions, corresponding to the choice of internal time, to remove the redundancy.

We define the *trivialization map*

$$\mathcal{T}_t := \mathcal{T}_{t+} + \mathcal{T}_{t-}, \qquad \mathcal{T}_{t\pm} := \exp\left(\pm i \hat{t}\,(\hat{h}_e - \epsilon)\right) \theta(\mp \hat{p}_t), \tag{26}$$

where $\theta(0) = \frac{1}{2}$. The theta function separates positive and negative frequency modes and the transformation is akin to the time evolution map in t time, except that the latter appears as an operator. In consequence, $\mathcal{T}_t$ does *not* commute with $\hat{\mathcal{C}}_H$ and maps $\mathcal{H}_{phys}$ to a new Hilbert space $\mathcal{H}_{phys}^{e(t)} := \mathcal{T}_t(\mathcal{H}_{phys})$. Using the tools of [40], one can check that its inverse $\mathcal{T}_t^{-1} : \mathcal{H}_{phys}^{e(t)} \to \mathcal{H}_{phys}$ is given by

$$\mathcal{T}_t^{-1} := \mathcal{T}_{t+}^{-1} + \mathcal{T}_{t-}^{-1}, \qquad \mathcal{T}_{t\pm}^{-1} := \exp\left(\mp i \hat{t}\,(\hat{h}_e - \epsilon)\right) \theta(\mp \hat{p}_t). \tag{27}$$

and satisfies $\mathcal{T}_t^{-1} \mathcal{T}_t = \theta(-\hat{p}_t) + \theta(\hat{p}_t) = \mathbb{1}$ only on $\mathcal{H}_{phys}$ and *only* for $\epsilon > 0$. The role of the parameter ϵ is thus to render (26) invertible.

The key property of (26) is that it trivializes $\hat{\mathcal{C}}_H$ to the clock variables. More precisely,

$$\mathcal{T}_{t\pm}\,\hat{\mathcal{C}}_\pm^t\,\mathcal{T}_{t\pm}^{-1} = (\hat{p}_t \pm \epsilon)\,\theta(\mp \hat{p}_t), \qquad \mathcal{T}_{t\mp}\,\hat{\mathcal{C}}_\pm^t\,\mathcal{T}_{t\mp}^{-1} = (\hat{p}_t \pm 2\hat{h}_e \mp \epsilon)\,\theta(\pm \hat{p}_t), \tag{28}$$

and so $\mathcal{T}_{t\pm}$ trivializes $\hat{\mathcal{C}}_\pm^t$ from (5) in the positive/negative frequency sector such that it *only* acts on the clock variables upon transformation. Together

$$\mathcal{T}_q\,\hat{\mathcal{C}}_H\,\mathcal{T}_q^{-1} = s_t\left(\hat{p}_t - 2\hat{h}_e + \epsilon\right)(\hat{p}_t + \epsilon)\,\theta(-\hat{p}_t) + s_t\left(\hat{p}_t + 2\hat{h}_e - \epsilon\right)(\hat{p}_t - \epsilon)\,\theta(\hat{p}_t). \tag{29}$$

It is thus not surprising to find the states of $\mathcal{H}_{phys}^{e(t)}$ in the form

$$|\psi\rangle_{phys}^{e(t)} := \mathcal{T}_t|\psi\rangle_{phys} = \int_{-\infty}^{+\infty} \frac{dp_e}{2|p_e|}\left[\psi_{kin}^{e(t)}(-|p_e|,p_e)\,|-\epsilon\rangle_t|p_e\rangle_e + \psi_{kin}^{e(t)}(|p_e|,p_e)\,|\epsilon\rangle_t|p_e\rangle_e\right]. \tag{30}$$

Hence, apart from distinguishing the positive/negative frequency sectors, the clock-slot of the state has become redundant. It is easy to convince oneself that $\mathcal{T}_t$ constitutes an isometry from $\mathcal{H}_{phys}$ to $\mathcal{H}_{phys}^{e(t)}$.

After a straightforward calculation one finds that the relational Dirac observables (25) transform as follows to $\mathcal{H}_{\text{phys}}^{e(t)}$:

$$T_t \hat{E}(\tau) T_t^{-1} = (|\hat{p}_e| \tau + \hat{p}_e \hat{e} + i) \theta(-\hat{p}_t) + (-|\hat{p}_e| \tau + \hat{p}_e \hat{e} + i) \theta(\hat{p}_t),$$
$$T_t \hat{p}_e(\tau) T_t^{-1} = \hat{p}_e \theta(-\hat{p}_t) + \hat{p}_e \theta(\hat{p}_t). \tag{31}$$

On the respective positive/negative frequency sectors, these almost coincide with the reduced evolving observables (17) on $\mathcal{H}_\pm^{e(t)}$.

To complete the quantum reduction to $\mathcal{H}_\pm^{e(t)}$ we note the following:

$$\langle \psi | \chi \rangle_{\text{phys}} \equiv \frac{1}{2} \langle \psi | \chi \rangle_+^{e(t)} + \frac{1}{2} \langle \psi | \chi \rangle_-^{e(t)}, \tag{32}$$

where $\langle \psi | \chi \rangle_\pm^{e(t)}$ is the inner product (15), provided

$$\psi_\pm^{e(t)}(p_e) := \frac{\psi_{\text{kin}}^{e(t)}(\mp |p_e|, p_e)}{\sqrt{|p_e|}}. \tag{33}$$

The reduced state is thereby essentially the Newton-Wigner wave function associated with the positive/negative frequency solutions of the constraint (19). However, there is a small difference: usually, one restricts to positive frequency solutions in which case the Newton-Wigner wave function involves an additional factor $1/\sqrt{2}$ [58]. This would here imply $\langle \psi | \chi \rangle_{\text{phys}} \equiv \langle \psi | \chi \rangle_+^{e(t)}$. While this could be done, here we shall not discard negative frequency modes as they are also physically interesting, in particular when switching internal times in cosmology, see Figure 1 (e.g., we would be discarding forward evolution in ϕ). Therefore, we keep the normalization as in (33), so that positive and negative frequency modes can be simultaneously normalized.

It is now easy to see that with an additional transformation for the measure

$$\widehat{\sqrt{|p_e|}} T_t |\psi\rangle_{\text{phys}} = \frac{1}{2} |-\epsilon\rangle_t |\psi\rangle_+^{e(t)} + \frac{1}{2} |+\epsilon\rangle_t |\psi\rangle_-^{e(t)}, \tag{34}$$

we can identify $|\psi\rangle_\pm^{e(t)}$ with the reduced states (14) on $\mathcal{H}_\pm^{e(t)}$. We also recover the reduced evolving observables (17) in the corresponding sectors (here the $+i$ term in (25) is crucial)

$$\widehat{\sqrt{|p_e|}} T_t \hat{E}(\tau) T_t^{-1} (\widehat{\sqrt{|p_e|}})^{-1} = (|\hat{p}_e| \tau + \hat{E}) \theta(-\hat{p}_t) + (-|\hat{p}_e| \tau + \hat{E}) \theta(\hat{p}_t)$$
$$= \hat{E}_+(\tau) \theta(-\hat{p}_t) + \hat{E}_-(\tau) \theta(\hat{p}_t), \tag{35}$$
$$\widehat{\sqrt{|p_e|}} T_t \hat{p}_e(\tau) T_t^{-1} (\widehat{\sqrt{|p_e|}})^{-1} = \hat{p}_e \theta(-\hat{p}_t) + \hat{p}_e \theta(\hat{p}_t).$$

Projecting onto the classical gauge-fixing conditions $t = 0$, in some analogy to the Page-Wootters construction [59], removes the redundant clock-slot and finally yields the states of the reduced theory

$$|\psi\rangle_\pm^{e(t)} = 2\sqrt{2\pi} \, {}_t\langle t = 0 | \theta(\mp \hat{p}_t) \widehat{\sqrt{|p_e|}} T_t |\psi\rangle_{\text{phys}}. \tag{36}$$

This projection is compatible with the observables and the inner product. Its image is the Heisenberg picture on $\mathcal{H}_\pm^{e(t)}$; e.g., (36) can be interpreted as an initial state at $t = 0$. This completes the quantum reduction from the Dirac quantized theory to the reduced one relative to internal time t, see Figure 2.

2.5. Quantum Internal Time Switches

This quantum reduction procedure now enables us to switch from the relational quantum dynamics relative to t to that relative to e [40]. Just as with the classical case, we can thus interchange the roles of t and e and the following is the quantum analog of it. In analogy to a coordinate change on a manifold, we have to invert the quantum reduction map associated with t and concatenate it with that associated with e. This will map from the reduced Hilbert spaces $\mathcal{H}_\pm^{e(t)}$ via the internal-time-neutral $\mathcal{H}_{\text{phys}}$ to $\mathcal{H}_\pm^{t(e)}$:

$$\hat{S}_{t_+ \to e_\pm} : \mathcal{H}_+^{e(t)} \to \mathcal{H}_\pm^{t(e)} \tag{37}$$
$$\hat{S}_{t_+ \to e_\pm} := 2\sqrt{2\pi}\,_e\langle e=0|\,\theta(\mp\hat{p}_e)\,\widehat{\sqrt{|\hat{p}_t|}}\,\mathcal{T}_{e\pm}\,\mathcal{T}_{t+}^{-1}\,(\widehat{\sqrt{|p_e|}})^{-1}\,\tfrac{|p_t=-\epsilon\rangle_t}{2}\otimes,$$

where $\mathcal{T}_e$ is identical to (26), except that t and e are interchanged. Here, $|p_t=-\epsilon\rangle_t\otimes$ mean tensoring the input state $|\psi\rangle_+^{e(t)}$ with this factor, which amounts to restoring gauge invariance as $|p_t=-\epsilon\rangle_t = 1/\sqrt{2\pi}\int dt\,\exp(-it\epsilon)|t\rangle_t$ averages over the classical gauge-fixing conditions $t=\text{const}$. Similarly, for the negative frequency modes, we have

$$\hat{S}_{t_- \to e_\pm} : \mathcal{H}_-^{e(t)} \to \mathcal{H}_\pm^{t(e)} \tag{38}$$
$$\hat{S}_{t_- \to e_\pm} := 2\sqrt{2\pi}\,_e\langle e=0|\,\theta(\mp\hat{p}_e)\,\widehat{\sqrt{|\hat{p}_t|}}\,\mathcal{T}_{e\pm}\,\mathcal{T}_{t-}^{-1}\,(\widehat{\sqrt{|p_e|}})^{-1}\,\tfrac{|p_t=+\epsilon\rangle_t}{2}\otimes,$$

It is clear that just as in the classical case, the internal time switches must preserve the four quadrants of Figure 1. Indeed, in Appendix B we show that

$$\hat{S}_{t_+ \to e_\pm}|\psi\rangle_+^{e(t)} = \theta(-\hat{p}_t)|\psi\rangle_\pm^{t(e)}, \qquad \hat{S}_{t_- \to e_\pm}|\psi\rangle_-^{e(t)} = \theta(\hat{p}_t)|\psi\rangle_\pm^{t(e)}, \tag{39}$$

where the reduced states on the left- and right-hand sides of the equations correspond via (33) to the same physical state. We also demonstrate in Appendix B that the complicated expressions (37) and (38) vastly simplify, being equivalent to

$$\hat{S}_{t_+ \to e_\pm} \equiv \mathcal{P}_{e_\pm \to t_+}\,\theta(\mp\hat{p}_e), \qquad \hat{S}_{t_- \to e_\pm} \equiv \mathcal{P}_{e_\pm \to t_-}\,\theta(\mp\hat{p}_e), \tag{40}$$

where we have introduced the clock-switch operators

$$\mathcal{P}_{e_\pm \to t_+}|p_e\rangle_e := |-|p_e|\rangle_t, \qquad \mathcal{P}_{e_\pm \to t_-}|p_e\rangle_e := ||p_e|\rangle_t \tag{41}$$

in close analogy to [40] and the parity-swap operator of [38,41].

The quantum clock-switch procedure can be summarized in a commutative diagram:

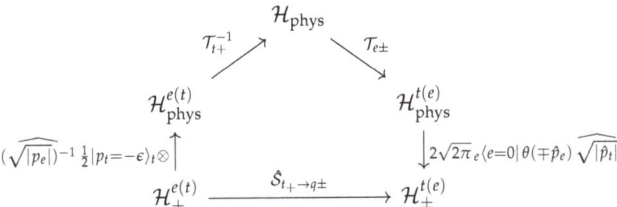

and analogously for (38). Notice that the quantum clock switch thereby has the structure $\varphi_e \circ \varphi_t^{-1}$ of a coordinate transformation, where the internal-time-neutral $\mathcal{H}_{\text{phys}}$ assumes the role of the 'manifold'. This is the appropriate structure for a quantum notion of general covariance that pertains to switching between the descriptions of the physics relative to different quantum reference systems, supporting the arguments in [38–40].

The inverse clock switch from e to t is due to the symmetry of the problem the same as above, except that one must interchange the e and t labels everywhere. It is now easy to check how the elementary observables transform from $\mathcal{H}_\pm^{e(t)}$ to $\mathcal{H}_\pm^{t(e)}$:

$$\begin{aligned}\hat{S}_{t_\pm \to e_+} \hat{E} \, \hat{S}_{e_+ \to t_\pm} &= \hat{T}\, \theta(\mp \hat{p}_t), & \hat{S}_{t_\pm \to e_+} \hat{p}_e \, \hat{S}_{e_+ \to t_\pm} &= \pm \hat{p}_t \, \theta(\mp \hat{p}_t), \\ \hat{S}_{t_\pm \to e_-} \hat{E} \, \hat{S}_{e_- \to t_\pm} &= \hat{T}\, \theta(\mp \hat{p}_t), & \hat{S}_{t_\pm \to e_-} \hat{p}_e \, \hat{S}_{e_- \to t_\pm} &= \mp \hat{p}_t \, \theta(\mp \hat{p}_t).\end{aligned} \qquad (42)$$

Notice that the image of $\hat{S}_{e_+ \to t_\pm}$ is $\theta(-\hat{p}_e)\left(\mathcal{H}_\pm^{e(t)}\right)$ so that in the first line one can set $\hat{p}_e = -|\hat{p}_e|$. Similarly, in the second line one can set $\hat{p}_e = |\hat{p}_e|$. Then it is obvious that the transformations (42) are exactly the quantum version of the classical maps between the corresponding reduced phase spaces in (12), which have been obtained through gauge transformations. While there are no gauge transformations in the quantum theory (except on $\mathcal{H}_{\text{kin}}$) [38–40], this is their quantum analog.

These relations permit us to transform the reduced relational observables (17) from $\mathcal{H}_\pm^{e(t)}$ to $\mathcal{H}_\pm^{t(e)}$:

$$\begin{aligned}\hat{S}_{t_\pm \to e_+} \hat{E}_\pm(\tau_t) \, \hat{S}_{e_+ \to t_\pm} &= (\pm |\hat{p}_t|\, \tau_t + \hat{T})\, \theta(\mp \hat{p}_t), \\ \hat{S}_{t_\pm \to e_-} \hat{E}_\pm(\tau_t) \, \hat{S}_{e_- \to t_\pm} &= (\pm |\hat{p}_t|\, \tau_t + \hat{T})\, \theta(\mp \hat{p}_t)\end{aligned} \qquad (43)$$

($\hat{p}_{e_\pm}(\tau_t)$ is already transformed in (42)). The right-hand side is *not* $\hat{T}_\pm(\tau_e)$, despite looking like it, due to the appearance of τ_t, which runs over the values of t, rather than τ_e, which runs over the values of e. Instead, it is the representation of $\hat{E}_\pm(\tau_t)$ on $\mathcal{H}_\pm^{t(e)}$ and could be used to set initial values τ_e^i for e after the clock switch.

In contrast to the classical case, there does not seem to be a unique procedure, given that $\hat{E}_\pm$ is now an operator. However, in analogy to the classical case, we can define the initial reading τ_e^i of the new clock e in terms of expectation values, e.g.:

$$\tau_e^i := \frac{\langle \hat{E}_\pm(\tau_t^f) \rangle_\pm^{e(t)}}{\langle \hat{p}_e \rangle_\pm^{e(t)}}. \qquad (44)$$

Indeed, we prove in Appendix C that this leads to exactly the classical 'continuity' relation (13) in terms of expectation values

$$\left\langle \hat{T}_\pm(\tau_e^i) \right\rangle_\pm^{t(e)} = \tau_t^f \, \langle \hat{p}_t \rangle_\pm^{t(e)} \qquad \text{on } \mathcal{H}_\pm^{t(e)}, \qquad (45)$$

so that one also finds a continuous quantum relational evolution, despite the intermediate clock switch.

2.6. Illustration in Concrete States

Let us briefly illustrate this internal time switch for example states. We pick semiclassical kinematical states, built according to the recipe for elliptic coherent states in [60] (and adapt the normalization):

$$\psi_{\text{kin}}(p_\phi, p_\alpha) = \sqrt{\frac{2}{\Gamma(n)}} (p_\phi + i\, p_\alpha)^n \exp\left(-\frac{p_\alpha^2 + p_\phi^2}{2}\right). \qquad (46)$$

For concreteness, we restrict to the green quadrant in Figure 1, where we have $p_\alpha = -p_\phi \leq 0$ and so an expanding universe with forward evolution in ϕ. Using the Newton-Wigner type identification (33), this gives semiclassical reduced negative and positive frequency wave functions on $\theta(-\hat{p}_\alpha)(\mathcal{H}_-^{\alpha(\phi)})$ and $\theta(\hat{p}_\phi)(\mathcal{H}_+^{\phi(\alpha)})$, respectively,

$$\psi_-^{\alpha(\phi)}(p_\alpha) = \sqrt{\frac{2}{\Gamma(n)\,|p_\alpha|}} p_\alpha^n (i-1)^n \exp\left(-p_\alpha^2\right), \qquad \psi_+^{\phi(\alpha)}(p_\phi) = \sqrt{\frac{2}{\Gamma(n)\,p_\phi}} p_\phi^n (i-1)^n \exp\left(-p_\phi^2\right).$$

For visualization, we provide plots of their probability distributions in Figure 3.

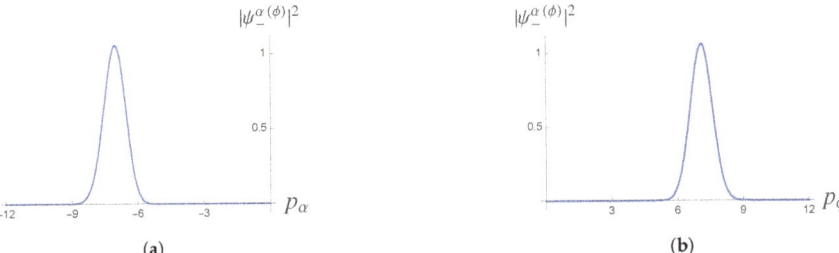

(a)   (b)

Figure 3. Reduced probability distributions coming from the same physical state (defined through (20) and (46) and here $n = 100$), but described relative to the choices of (**a**) ϕ and (**b**) α as internal times in the 'expanding-forward' (green) quadrant of Figure 1. Recall that in the reduced theory the usual modulus square of the wave function is the probability distribution, see (15). Due to the symmetry of the model in α and ϕ, reduced probability distributions will always behave symmetrically.

One easily finds that $\langle \hat{A} \rangle_-^{\alpha(\phi)} = \langle \hat{\Phi} \rangle_+^{\phi(\alpha)} = 0$, where $\hat{A}, \hat{\Phi}$ are the reduced quantizations (16) of $A = \alpha p_\alpha$ and $\Phi = \phi p_\phi$, and

$$\left\langle \hat{A}_-(\tau_\phi) \right\rangle_-^{\alpha(\phi)} = \tau_\phi \langle \hat{p}_\alpha \rangle_-^{\alpha(\phi)} = -\tau_\phi \frac{\Gamma(n+\frac{1}{2})}{\sqrt{2}\,\Gamma(n)},$$
$$\left\langle \hat{\Phi}_+(\tau_\alpha) \right\rangle_+^{\phi(\alpha)} = \tau_\alpha \langle \hat{p}_\phi \rangle_+^{\phi(\alpha)} = +\tau_\alpha \frac{\Gamma(n+\frac{1}{2})}{\sqrt{2}\,\Gamma(n)}.$$
(47)

Suppose we evolve first in ϕ and then switch to α time. Then invoking (44) immediately yields

$$\tau_\alpha^i = \tau_\phi^f \quad \Rightarrow \quad \left\langle \hat{\Phi}_+(\tau_\alpha^i) \right\rangle_+^{\phi(\alpha)} = \tau_\phi^f \langle \hat{p}_\phi \rangle_+^{\phi(\alpha)}.$$
(48)

This simple switch from ϕ to α time is illustrated in Figure 4.

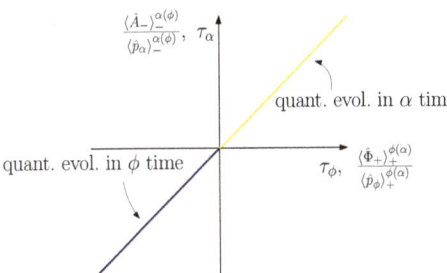

Figure 4. Illustration of the quantum relational evolution given in (47) and (48) for an internal time switch from ϕ to α at $\tau_\phi^f = \tau_\alpha^i = 0$. The blue branch corresponds to the evolution of $\left\langle \hat{A}_-(\tau_\phi) \right\rangle_-^{\alpha(\phi)} / \langle \hat{p}_\alpha \rangle_-^{\alpha(\phi)}$ in τ_ϕ, while the golden branch depicts the evolution of $\left\langle \hat{\Phi}_+(\tau_\alpha) \right\rangle_+^{\phi(\alpha)} / \langle \hat{p}_\phi \rangle_+^{\phi(\alpha)}$ in τ_α. Together they trace out a continuous classical trajectory, describing an expanding universe.

3. Perspective on the 'Wave Function of the Universe'

We have illustrated in a very simple quantum cosmological model, namely the flat FRW universe with massless scalar field, how to consistently switch between the quantum relational dynamics relative to the scale factor and that relative to the field used as internal times. In particular, just as with the classical case, the quantum relational evolution is *continuous*, despite the intermediate internal time switch, and no information gets lost. This extends the quantum clock-switch method of [40] (see also [38,39,41] for spatial reference systems) to the relativistic case and offers a full Hilbert space alternative to the semiclassical effective approach of [20–22].

Owing to the symmetry of the model in ϕ and α, the internal time switches are particularly simple here and the relational dynamics in a given physical (i.e., internal-time-neutral) state looks essentially 'the same' relative to these two possible choices (up to relabeling the evolving variables). This will no longer be the case in models which are not symmetric relative to different internal time choices, e.g., see [40], and especially not in the presence of the so-called global problem of time [11–14,18–22,32,61,62], which arises, e.g., for interactions between evolving and internal time degrees of freedom [16,17,22,63]. However, our method is general and applies to generic models if one suitably takes into account the Gribov problem and the fact that a description relative to a choice of reference system, just like a coordinate choice, will generally not be globally valid [20–22,38–40].

Indeed, the internal time switch proceeds in complete analogy to coordinate changes $\varphi_t \circ \varphi_{t'}^{-1}$ on a manifold [38,40]: it inverts the quantum reduction map relative to one time choice, mapping the corresponding reduced quantized theory back into the internal-time-neutral physical Hilbert space of the Dirac quantization and subsequently applies the quantum reduction map to the reduced quantization relative to the other internal time choice. The same compositional structure appears for changes of spatial quantum reference systems [38,39]. This permits us to interpret the physical Hilbert space of the Dirac quantized theory as encoding the 'perspective-neutral' (i.e., reference-system-neutral) physics [38–40,50] and the quantum reduction maps as defining 'quantum coordinate' descriptions of these physics relative to a choice of quantum reference system. This is precisely the structure that one would expect for establishing a genuine quantum notion of general covariance, which refers to the ability to consistently switch, within one theory, between arbitrary choices of quantum reference systems, each of which can be used as a vantage point to describe the physics of the remaining degrees of freedom.

Accordingly, in line with our earlier discussion in [38–40], we thus propose to define a *complete* relational quantum theory, admitting a quantum general covariance, as the *conjunction* of the quantum-reference-system-neutral Dirac quantized theory and the multitude of reduced quantum theories associated with the different choices of quantum reference system. Just like the classical theory contains a (perspective-neutral) constraint surface *and* the multitude of reduced phase spaces, together comprising a complete classical description, the complete quantum theory contains their corresponding quantum structures, as illustrated here, and this is a complete quantum description. Specifically, we propose this conjunction to overcome the so-called multiple choice facet of the problem of time [11,12] (the arguments of which could also be applied to spatial reference systems) and to turn it into a multiple choice *feature* of the complete relational quantum theory [40].

For simplicity, we have illustrated the novel procedure using the Wheeler-DeWitt approach in the Dirac quantization; however, the loop quantization of the simple model of this article can actually be formulated in the same physical Hilbert space [34]. It is thus suggestive that the present framework for switching internal times can be extended to loop quantum cosmology as well. To this end, however, at least two loop quantization related subtleties need be suitably taken into account. For instance, loop quantization leads to superselection sectors in the geometric degrees of freedom [26]. Presumably, the framework should be applied per superselection sector, although subtleties remain to be checked as different sectors might have different physical properties. Secondly, loop quantization leads to a deformation of gauge covariance as embodied in the constraint algebra [64–66]. While this should not be a problem for homogeneous cosmological models, it arises as an additional challenge when attempting to extend the framework to a loop quantization of inhomogeneous models where non-trivial diffeomorphism constraints arise. This would be relevant, e.g., when studying relational dynamics in the context of loop quantum cosmology modifications of the 'no-boundary' proposal [67].

Our proposal also entails a novel perspective on the 'wave function of the universe', i.e., the global quantum state for the universe as a whole, which appears ubiquitously and in various interpretational guises in quantum cosmology [24–26,31,35,68–72]. It is usually taken to be a solution to the

Wheeler-DeWitt equation, in the present article (19), and thereby a physical state of the Dirac quantized theory. The proposal here suggests viewing the 'wave function of the universe' as a perspective-neutral global state that thereby does *not* admit an immediate physical interpretation; it is not the description of the universe relative to any physical reference system. Instead, while in the simple model here we have only illustrated it for two possible choices, it contains the information about *all* the relative states at once, i.e., all the descriptions relative to all possible choices of quantum reference system, and provides the crucial linking structure between all these relative descriptions. In fact, it is these relative reduced states that admit the immediate physical interpretation and should be taken as relevant for observational and operational predictions (although the 'wave function of the universe' encodes that information too).

This offers a consistent link between operationally significant subsystem structures in quantum cosmology and gravity, relative to a choice of quantum reference system, and a perspective-neutral (in particular, observer-independent) global state that contains all degrees of freedom [50]. Specifically, this also suggests a novel perspective on the notorious problem of how to interpret the probabilities defined by the 'wave function of the universe'. While the global probability density defined by it through the physical inner product (here (21)) does not admit an immediate operational interpretation, the 'wave function of the universe' gives rise to all the relative states through quantum reduction, and these *do* admit an immediate physical interpretation. Indeed, the relative states admit a physically relevant reduced probability distribution (here via (15)), and the quantum reduction always implies their inner product through the inner product of the corresponding physical states (here see (32) and [38–40] for further examples of the method). However, crucially, the two kinds of probability distributions live on different spaces: the 'wave function of the universe' technically defines an abstract probability distribution over *all* the degrees of freedom of the universe, while the relative states define a probability distribution over all degrees of freedom of the universe, except those of the associated reference system. As such, the latter admits the interpretation as the probability distribution 'seen' by that reference system.

Note that the proposal here is general and not specific to any detailed interpretation of quantum theory and its probabilities. There is no obvious reason it should conflict with any of many-worlds, relational, QBism, consistent histories, Copenhagen, or realist interpretations. In particular, it is worthwhile to point out that it might actually reconcile relational and informational state interpretations [51,73–80] with the global 'wave function of the universe'. While the details depend on the specific interpretation, relational interpretations take a state to be defined relative to an agent, or, more generally, reference system, and this state is taken to be the observer's 'catalog of knowledge' about the observed system. One can then argue [50,51] that such interpretations deny a global operationally meaningful quantum state as the self-reference problem [81,82] impedes a given observer or reference system to infer the global state of the entire universe (incl. itself) from its interactions with the rest. Accordingly, relative to any subsystem, one can assign a 'catalog of knowledge' about the rest of the universe but, without external observer or reference frame, there can then be no global, operationally meaningful 'catalog of knowledge' about the entire universe at once (see also related discussions in [83–86]). In the proposal of this article, the global 'wave function of the universe' indeed does not admit an immediate operational interpretation as an informational state, yet it links all the different relational reference system perspectives on the universe consistently [50], something that was missing, e.g., in the discussion of [51,73,74,79,80,83–85].

Specifically, this might reconcile the seemingly subjective relational states (an observer's 'catalog of knowledge') with the objective 'wave function of the universe'. Being a physical system too, the subjective degrees of beliefs, i.e., 'catalogs of knowledge' of any observer about states of other systems should be encoded in physical degrees of freedom of this observer. However, the 'wave function of the universe'—as a perspective-neutral global state—encodes *all* physical degrees of freedom of the universe and thus 'knows', in particular, what information any observing system has in its memory. Hence, while the relative states may be interpreted as subjective 'catalogs of knowledge' of the observing systems, the 'wave function of the universe', as proposed here, contains all these 'catalogs of knowledge' at once and would actually consistently and objectively link them. However, to manifest this more specific interpretation, one would have to clarify how state collapses occur in the relative descriptions from measurement interactions at the perspective-neutral level, i.e., one has to revisit the measurement problem (and specifically the Wigner friend paradox [73,79,87–89]), but now with a complete relational quantum theory at hand, as proposed here, which contains both a perspective-neutral description and all the individual perspectives, a structure that was not available before.

Finally, it can be shown in simple examples that quantum correlations will generally depend on the choice of quantum reference system [38,41]. This immediately raises some interesting questions since both quantum reference systems and quantum correlations appear ubiquitously in quantum cosmology. For example, given the phenomenological importance of CMB correlations and propagators, does the quantum frame dependence of correlations, which surely must be expected in quantum cosmology too, have *any* observational significance? This question could be studied, e.g., in Bianchi models with inhomogeneous perturbations and the tools for these investigations are now, in principle, available.

Funding: The project leading to this publication has initially received funding from the European Union's Horizon 2020 research and innovation programme under the Marie Sklodowska-Curie grant agreement No 657661. The author also acknowledges support through a Vienna Center for Quantum Science and Technology Fellowship.

Acknowledgments: I thank Bianca Dittrich, Steffen Gielen and Merce Martín-Benito for discussion.

Conflicts of Interest: The author declares no conflict of interest.

Appendix A. Hermiticity of the Relational Observable $\hat{E}(\tau)$ on $\mathcal{H}_{phys}$

We prove the claim that $\hat{E}(\tau)$ as given in (25) is Hermitian with respect to the physical inner product. To this end, it suffices to consider the symmetric quantization of the Dirac observable L in (2) on $\mathcal{H}_{kin}$

$$\hat{L} = \frac{1}{2}(\hat{p}_\phi \hat{\phi} + \hat{\phi} \hat{p}_\phi + \hat{p}_\alpha \hat{\alpha} + \hat{\alpha} \hat{p}_\alpha). \tag{A1}$$

This is a Hermitian and, in particular, self-adjoint operator on $\mathcal{H}_{kin}$. However, it fails to be Hermitian with respect to the physical inner product. To see this, note that

$$[\hat{C}_H, \hat{L}] = -2i\,\hat{C}_H. \tag{A2}$$

Hence, $\hat{L}$ commutes with the constraint *only* on $\mathcal{H}_{phys}$. The physical inner product (21) comes from group averaging [23,40,54–57] and is given by

$$\langle \psi | \chi \rangle_{phys} := \langle \psi_{kin} | \delta(\hat{C}_H) | \chi_{kin} \rangle, \tag{A3}$$

where $\langle \cdot | \cdot \rangle$ is the standard inner product on $\mathcal{H}_{kin}$ and

$$\delta(\hat{C}_H) = \frac{1}{2\pi} \int_{-\infty}^{+\infty} ds\, e^{is\hat{C}_H}. \tag{A4}$$

Using (A2) one easily finds $[\hat{C}_H^n, \hat{L}] = -2i\,n\,\hat{C}_H^n$ and thereby

$$\begin{aligned}
[\delta(\hat{C}_H), \hat{L}] &= \tfrac{1}{\pi} \int ds\, s\, \hat{C}_H\, e^{is\hat{C}_H} = -2i\, \tfrac{d}{dx} \tfrac{1}{2\pi} \int ds\, e^{ixs\hat{C}_H}\Big|_{x=1} \\
&= -2i\, \tfrac{d}{dx} \delta(x\,\hat{C}_H)\Big|_{x=1} = -2i\, \tfrac{d}{dx} |x|^{-1}\Big|_{x=1} \delta(\hat{C}_H) = 2i\,\delta(\hat{C}_H).
\end{aligned} \tag{A5}$$

From this result it is clear that $\hat{L}$ is *not* Hermitian with respect to (A3). However, using (A5), we have

$$\langle\psi_{\text{kin}}|\,(\hat{L}+i)\,\delta(\hat{C}_H)\,|\chi_{\text{kin}}\rangle = \langle\psi_{\text{kin}}|\,\delta(\hat{C}_H)\,(\hat{L}-i)\,|\chi_{\text{kin}}\rangle = \langle((\hat{L}+i)\,\delta(\hat{C}_H)\,\psi_{\text{kin}}|\chi_{\text{kin}}\rangle, \qquad (A6)$$

where in the last step we have made use of the fact that both $\hat{L}$ and $\delta(\hat{C}_H)$ are symmetric on $\mathcal{H}_{\text{kin}}$. Consequently, $\hat{L}+i$ is Hermitian with respect to the physical inner product and, in turn, also $\hat{E}(\tau)$ in (25). This operator can also be densely defined and is thus essentially self-adjoint.

Appendix B. Changes of Internal Times in the Quantum Theory

We begin by proving the left equation in (39). Recall that

$$\hat{S}_{t_+\to e_\pm} := 2\sqrt{2\pi}\,_e\langle e=0|\,\theta(\mp\hat{p}_e)\,\widehat{\sqrt{|\hat{p}_t|}}\,\mathcal{T}_{e\pm}\,\mathcal{T}_{t+}^{-1}\,(\sqrt{|p_e|})^{-1}\,\frac{|p_t=-\epsilon\rangle_t}{2}\otimes\,. \qquad (A7)$$

We make use of the definition of the reduced positive and negative frequency wave functions (33) and

$$\begin{aligned}\psi_{\text{kin}}^{t(e)}\!\left(p_e=\mp|p_t|,|p_t|\right) &= \psi_{\text{kin}}^{e(t)}\!\left(|p_e|,\mp|p_e|\right),\\ \psi_{\text{kin}}^{t(e)}\!\left(p_e=\mp|p_t|,-|p_t|\right) &= \psi_{\text{kin}}^{e(t)}\!\left(-|p_e|,\mp|p_e|\right),\end{aligned} \qquad (A8)$$

which is implied by (22). Then,

$$\begin{aligned}\hat{S}_{t_+\to e_\pm}|\psi\rangle_+^{e(t)} &= \hat{S}_{t_+\to e_\pm}\int_{-\infty}^{\infty}dp_e\,\frac{\psi_{\text{kin}}^{e(t)}(-|p_e|,p_e)}{\sqrt{|p_e|}}|p_e\rangle_e\\ &= 2\sqrt{2\pi}\,_e\langle e=0|\,\theta(\mp\hat{p}_e)\,\widehat{\sqrt{|\hat{p}_t|}}\,\mathcal{T}_{e\pm}\int_{-\infty}^{\infty}\frac{dp_e}{2|p_e|}\,\psi_{\text{kin}}^{e(t)}(-|p_e|,p_e)\,|-|p_e|\rangle_t|p_e\rangle_e\\ &= 2\sqrt{2\pi}\,_e\langle e=0|\,\theta(\mp\hat{p}_e)\,\widehat{\sqrt{|\hat{p}_t|}}\,\mathcal{T}_{e\pm}\\ &\quad\times\int_{-\infty}^{0}\frac{dp_t}{2|p_t|}\left[\psi_{\text{kin}}^{t(e)}(-|p_t|,p_t)\,|-|p_t|\rangle_e|p_t\rangle_t+\psi_{\text{kin}}^{t(e)}(|p_t|,p_t)\,||p_t|\rangle_e|p_t\rangle_t\right]\\ &= \sqrt{2\pi}\,_e\langle e=0|\,\theta(\mp\hat{p}_e)\int_{-\infty}^{0}dp_t\left[\psi_+^{t(e)}(p_t)\,|-\epsilon\rangle_e|p_t\rangle_t+\psi_-^{t(e)}(p_t)\,|+\epsilon\rangle_e|p_t\rangle_t\right]\\ &= \theta(-\hat{p}_t)\,|\psi\rangle_\pm^{t(e)}.\end{aligned}$$

From the second to the third line, we have performed a variable change $p_e=p_t$ for $p_e<0$ and $p_e=-p_t$ for $p_e>0$ and used (A8).

To prove that this transformation is equivalent to $\mathcal{P}_{e_\pm\to t_+}\,\theta(\mp\hat{p}_e)$, as claimed in (40), where $\mathcal{P}_{e_\pm\to t_+}$ is defined in (41), write

$$|\psi\rangle_+^{e(t)} = \int_{-\infty}^{0}\frac{dp_e}{\sqrt{|p_e|}}\,\psi_{\text{kin}}^{e(t)}(-|p_e|,p_e)\,|p_e\rangle_e+\int_{0}^{+\infty}\frac{dp_e}{\sqrt{|p_e|}}\,\psi_{\text{kin}}^{e(t)}(-|p_e|,p_e)\,|p_e\rangle_e,$$

perform a variable transformation $p_e=p_t$ in the left and $p_e=-p_t$ in the right integral and invoke (A8) and the definition (41).

The right equations in (39) and (40) are shown in complete analogy.

Appendix C. Continuity of the Quantum Relational Dynamics during a Switch

We briefly prove the continuity of the quantum relational dynamics, as expressed in (45), notwithstanding the intermediate internal time switch. For concreteness, we restrict our attention to one quadrant of Figure 1, e.g., either the green or red quadrant where $p_\alpha = -p_\phi$. Notice first that (40) implies

$$\theta(-\hat{p}_e)(\mathcal{H}_-^{e(t)}) \xrightleftharpoons[\hat{\mathcal{S}}_{e_+ \to t_-}]{\hat{\mathcal{S}}_{t_- \to e_+}} \theta(+\hat{p}_t)(\mathcal{H}_+^{t(e)}). \tag{A9}$$

Clearly, using (17), we have

$$\langle \hat{E}_-(\tau_t)\rangle_-^{e(t)} = -\tau_t \langle|\hat{p}_e|\rangle_-^{e(t)} + \langle \hat{E}\rangle_-^{e(t)} = \tau_t \langle\hat{p}_e\rangle_-^{e(t)} + \langle \hat{E}\rangle_-^{e(t)} \quad \text{on } \theta(-\hat{p}_e)(\mathcal{H}_-^{e(t)}),$$

$$\langle \hat{T}_+(\tau_e)\rangle_+^{t(e)} = +\tau_e \langle|\hat{p}_t|\rangle_+^{t(e)} + \langle \hat{T}\rangle_+^{t(e)} = \tau_e \langle\hat{p}_t\rangle_+^{t(e)} + \langle \hat{T}\rangle_+^{t(e)} \quad \text{on } \theta(+\hat{p}_t)(\mathcal{H}_+^{t(e)}).$$

Setting now the initial value of the new clock e, as in (44), to

$$\tau_e^i := \frac{\langle \hat{E}_-(\tau_t^f)\rangle_-^{e(t)}}{\langle \hat{p}_e\rangle_-^{e(t)}} = \tau_t^f + \frac{\langle \hat{E}\rangle_-^{e(t)}}{\langle \hat{p}_e\rangle_-^{e(t)}}, \tag{A10}$$

we find

$$\langle \hat{T}_+(\tau_e^i)\rangle_+^{t(e)} = \tau_t^f \langle \hat{p}_t\rangle_+^{t(e)} + \frac{\langle \hat{E}\rangle_-^{e(t)}}{\langle \hat{p}_e\rangle_-^{e(t)}} \langle \hat{p}_t\rangle_+^{t(e)} + \langle \hat{T}\rangle_+^{t(e)}. \tag{A11}$$

Now we invoke (42) and, in particular,

$$\hat{\mathcal{S}}_{t_- \to e_+} \hat{E}\, \hat{\mathcal{S}}_{e_+ \to t_-} = \hat{T}\,\theta(+\hat{p}_t), \qquad \hat{\mathcal{S}}_{t_- \to e_+} \hat{p}_e\, \hat{\mathcal{S}}_{e_+ \to t_-} = -\hat{p}_t\,\theta(+\hat{p}_t). \tag{A12}$$

Using (15) and (39), this implies

$$\frac{\langle \hat{E}\rangle_-^{e(t)}}{\langle \hat{p}_e\rangle_-^{e(t)}} = -\frac{\langle \hat{T}\rangle_+^{t(e)}}{\langle \hat{p}_t\rangle_+^{t(e)}} \tag{A13}$$

and thereby

$$\langle \hat{T}_+(\tau_e^i)\rangle_+^{t(e)} = \tau_t^f \langle \hat{p}_t\rangle_+^{t(e)}, \tag{A14}$$

as claimed. The proof for the other quadrants is completely analogous.

References

1. Rovelli, C. *Quantum Gravity*; Cambridge University Press: Cambridge, UK, 2004.
2. Rovelli, C. What is observable in classical and quantum gravity? *Class. Quant. Grav.* **1991**, *8*, 297–316. [CrossRef]
3. Rovelli, C. Quantum reference systems. *Class. Quant. Grav.* **1991**, *8*, 317–332. [CrossRef]
4. Brown, J.D.; Kuchař, K.V. Dust as a standard of space and time in canonical quantum gravity. *Phys. Rev. D* **1995**, *51*, 5600–5629.
5. Dittrich, B. Partial and complete observables for Hamiltonian constrained systems. *Gen. Rel. Grav.* **2007**, *39*, 1891–1927. [CrossRef]
6. Dittrich, B. Partial and complete observables for canonical General Relativity. *Class. Quant. Grav.* **2006**, *23*, 6155–6184.
7. Dittrich, B.; Tambornino, J. A perturbative approach to Dirac observables and their space-time algebra. *Class. Quant. Grav.* **2007**, *24*, 757–784. [CrossRef]

8. Dittrichm, B.; Tambornino, J. Gauge invariant perturbations around symmetry reduced sectors of general relativity: Applications to cosmology. *Class. Quant. Grav.* **2007**, *24*, 4543–4586. [CrossRef]
9. Tambornino, J. Relational observables in gravity: A review. *Symmetry Integr. Geom.* **2012**, *8*, 17–30.
10. DeWitt, B.S. Quantum theory of gravity. I. The canonical theory. *Phys. Rev.* **1967**, *160*, 1113–1148. [CrossRef]
11. Kuchař, K. Time and interpretations of quantum gravity. *Int. J. Mod. Phys. Proc. Suppl. D* **2011**, *20*, 3–86. [CrossRef]
12. Isham, C. Canonical quantum gravity and the problem of time. In *Integrable Systems, Quantum Groups, and Quantum Field Theories*; Kluwer Academic Publishers: Dordrecht, The Netherlands, 1993; pp. 157–287.
13. Anderson, E. *The Problem of Time*; Springer International Publishing: Cham, Switzerland, 2017; Volume 190.
14. Rovelli, C. Quantum mechanics without time: A model. *Phys. Rev. D* **1990**, *42*, 2638–2646. [CrossRef]
15. Rovelli, C. Time in quantum gravity: Physics beyond the Schrödinger regime. *Phys. Rev. D* **1991**, *43*, 442–456. [CrossRef]
16. Marolf, D. Almost ideal clocks in quantum cosmology: A brief derivation of time. *Class. Quant. Grav.* **1995**, *12*, 2469–2486. [CrossRef]
17. Marolf, D. Solving the problem of time in minisuperspace: Measurement of Dirac observables. *Phys. Rev.* **1995**, *12*, 2469–2486.
18. Dittrich, B.; Höhn, P.A.; Koslowski, T.A.; Nelson, M.I. Can chaos be observed in quantum gravity? *Phys. Lett. B* **2017**, *769*, 554–560. [CrossRef]
19. Dittrich, B.; Höhn, P.A.; Koslowski, T.A.; Nelson, M.I. Chaos, Dirac observables and constraint quantization. *arXiv* **2015**, arXiv:1508.01947.
20. Bojowald, M.; Höhn, P.A.; Tsobanjan, A. An Effective approach to the problem of time. *Class. Quant. Grav.* **2011**, *28*, 035006. [CrossRef]
21. Bojowald, M.; Höhn, P.A.; Tsobanjan, A. Effective approach to the problem of time: General features and examples. *Phys. Rev. D* **2011**, *83*, 125023. [CrossRef]
22. Höhn, P.A.; Kubalova, E.; Tsobanjan, A. Effective relational dynamics of a nonintegrable cosmological model. *Phys. Rev. D* **2012**, *86*, 065014. [CrossRef]
23. Thiemann, T. *Modern Canonical Quantum General Relativity*; Cambridge University Press: Cambridge, UK, 2007.
24. Bojowald, M. *Canonical Gravity and Applications: Cosmology, Black Holes and Quantum Gravity*; Cambridge University Press: Cambridge, UK, 2011.
25. Ashtekar, A.; Singh, P. Loop Quantum Cosmology: A status report. *Class. Quant. Grav.* **2011**, *28*, 213001. [CrossRef]
26. Banerjee, K.; Calcagni, G.; Martín-Benito, M. Introduction to Loop Quantum Cosmology. *Symmetry Integr. Geom.* **2012**, *8*, 016. [CrossRef]
27. Oriti, D.; Sindoni, L.; Wilson-Ewing, E. Emergent Friedmann dynamics with a quantum bounce from quantum gravity condensates. *Class. Quant. Grav.* **2016**, *33*, 224001. [CrossRef]
28. Gielen, S. Emergence of a low spin phase in group field theory condensates. *Class. Quant. Grav.* **2016**, *33*, 224002. [CrossRef]
29. Gielen, S. Group field theory and its cosmology in a matter reference frame. *Universe* **2018**, *4*, 103. [CrossRef]
30. Blyth, W.; Isham, C. Quantization of a Friedmann universe filled with a scalar field. *Phys. Rev. D* **1975**, *11*, 768–778. [CrossRef]
31. Hawking, S. Quantum cosmology. In *Relativity, Groups and Topology II, Les Houches Summer School, 1983*; DeWitt, B., Stora, R., Eds.; North Holland: Amsterdam, The Netherlands, 1984; p. 333.
32. Hájíček, P. Origin of nonunitarity in quantum gravity. *Phys. Rev. D* **1986**, *34*, 1040. [CrossRef]
33. Kiefer, C. Wave packets in minisuperspace. *Phys. Rev. D* **1988**, *38*, 1761. [CrossRef]
34. Ashtekar, A.; Corichi, A.; Singh, P. Robustness of key features of loop quantum cosmology. *Phys. Rev. D* **2008**, *77*, 024046. [CrossRef]
35. Bojowald, M. Quantum cosmology. *Lect. Notes Phys.* **2011**, *835*, 1–308.
36. Ashtekar, A.; Pawlowski, T.; Singh, P. Quantum Nature of the Big Bang: An Analytical and Numerical Investigation. I. *Phys. Rev. D* **2006**, *73*, 124038. [CrossRef]
37. Kamenshchik, A.Y.; Tronconi, A.; Vardanyan, T.; Venturi, G. Time in quantum theory, the Wheeler-DeWitt equation and the Born-Oppenheimer approximation. *arXiv* **2018**, arXiv:1809.08083.

38. Vanrietvelde, A.; Höhn, P.A.; Giacomini, F.; Castro-Ruiz, E. A change of perspective: Switching quantum reference frames via a perspective-neutral framework. *arXiv* **2018**, arXiv:1809.00556.
39. Vanrietvelde, A.; Höhn, P.A.; Giacomini, F. Switching quantum reference frames in the N-body problem and the absence of global relational perspectives. *arXiv* **2018**, arXiv:1809.05093.
40. Höhn, P.A.; Vanrietvelde, A. How to switch between relational quantum clocks. *arXiv* **2018**, arXiv:1810.04153.
41. Giacomini, F.; Castro-Ruiz, E.; Brukner, Č. Quantum mechanics and the covariance of physical laws in quantum reference frames. *Nat. Commun.* **2019**, *10*, 494. [CrossRef] [PubMed]
42. Hartle, J.B. Time and time functions in parametrized nonrelativistic quantum mechanics. *Class. Quant. Grav.* **1996**, *13*, 361–376. [CrossRef]
43. Hajicek, P. Choice of gauge in quantum gravity. *Nucl. Phys. Proc. Suppl.* **2000**, *80*, 1213.
44. Gambini, R.; Porto, R.A. Relational time in generally covariant quantum systems: Four models. *Phys. Rev. D* **2001**, *63*, 105014. [CrossRef]
45. Bojowald, M.; Halnon, T. Time in quantum cosmology. *Phys. Rev. D* **2018**, *98*, 066001. [CrossRef]
46. Malkiewicz, P. Multiple choices of time in quantum cosmology. *Class. Quant. Grav.* **2015**, *32*, 135004. [CrossRef]
47. Malkiewicz, P. What is Dynamics in Quantum Gravity?. *Class. Quant. Grav.* **2017**, *34*, 205001. [CrossRef]
48. Malkiewicz, P. Clocks and dynamics in quantum models of gravity. *Class. Quant. Grav.* **2017**, *34*, 145012. [CrossRef]
49. Malkiewicz, P.; Miroszewski, A. Internal clock formulation of quantum mechanics. *Phys. Rev. D* **2017**, *96*, 046003. [CrossRef]
50. Höhn, P.A. Reflections on the information paradigm in quantum and gravitational physics. *J. Phys. Conf. Ser.* **2017**, *880*, 012014. [CrossRef]
51. Höhn, P.A. Toolbox for reconstructing quantum theory from rules on information acquisition. *Quantum* **2017**, *1*, 38. [CrossRef]
52. Dirac, P.A. *Lectures on Quantum Mechanics*; Yeshiva University Press: New York, NY, USA, 1964.
53. Henneaux, M.; Teitelboim, C. *Quantization of Gauge Systems*; Princeton University Press: Princeton, NJ, USA, 1992.
54. Hartle, J.B.; Marolf, D. Comparing formulations of generalized quantum mechanics for reparametrization—Invariant systems. *Phys. Rev. D* **1997**, *56*, 6247–6257. [CrossRef]
55. Marolf, D. Refined algebraic quantization: Systems with a single constraint. *arXiv* **1995**, arXiv:gr-qc/9508015.
56. Marolf, D. Group averaging and refined algebraic quantization: Where are we now? *arXiv* **2000**, arXiv:gr-qc/0011112.
57. Ashtekar, A.; Lewandowski, J.; Marolf, D.; Mourao, J.; Thiemann, T. Quantization of diffeomorphism invariant theories of connections with local degrees of freedom. *J. Math. Phys.* **1995**, *36*, 6456–6493. [CrossRef]
58. Haag, R. *Local Quantum Physics: Fields, Pparticles, Algebras*; Springer Science & Business Media: Berlin/Heidelberg, Germany, 2012.
59. Page, D.N.; Wootters, W.K. Evolution without evolution: Dynamics described by stationary observables. *Phys. Rev. D* **1983**, *27*, 2885. [CrossRef]
60. Pollet, J.; Méplan, O.; Gignoux, C. Elliptic eigenstates for the quantum harmonic oscillator. *J. Phys. A Math. Gen.* **1995**, *28*, 7287. [CrossRef]
61. Hájíček, P. Group quantization of parametrized systems. I. Time levels. *J. Math. Phys.* **1995**, *36*, 4612–4638. [CrossRef]
62. Hájíček, P. Time evolution and observables in constrained systems. *Class. Quant. Grav.* **1996**, *13*, 1353–1376. [CrossRef]
63. Giddings, S.B.; Marolf, D.; Hartle, J.B. Observables in effective gravity. *Phys. Rev. D* **2006**, *74*, 064018. [CrossRef]
64. Bojowald, M.; Paily, G.M. Deformed General Relativity. *Phys. Rev. D* **2013**, *87*, 044044. [CrossRef]
65. Bojowald, M.; Brahma, S.; Reyes, J.D. Covariance in models of loop quantum gravity: Spherical symmetry. *Phys. Rev. D* **2015**, *92*, 045043. [CrossRef]
66. Bojowald, M.; Brahma, S. Covariance in models of loop quantum gravity: Gowdy systems. *Phys. Rev. D* **2015**, *92*, 065002. [CrossRef]
67. Bojowald, M.; Brahma, S. Loops rescue the no-boundary proposal. *Phys. Rev. Lett.* **2018**, *121*, 201301. [CrossRef] [PubMed]

68. Everett, H., III. The Theory of the Universal Wave Function. Ph.D. Thesis, Princeton University, Princeton, NJ, USA, 1956. Available online: http://www-tc.pbs.org/wgbh/nova/manyworlds/pdf/dissertation.pdf (accessed on 10 May 2019).
69. Everett, H. Relative state formulation of quantum mechanics. *Rev. Mod. Phys.* **1957**, *29*, 454–462. [CrossRef]
70. Hartle, J.; Hawking, S. Wave function of the universe. *Phys. Rev. D* **1983**, *28*, 2960–2975. [CrossRef]
71. Hawking, S. The quantum state of the universe. *Nucl. Phys. B* **1984**, *239*, 257. [CrossRef]
72. Page, D.N. Hawking's wave function for the universe. In *Quantum Concepts in Space and Time*; Penrose, R., Isham, C., Eds.; Clarendon Press: Oxford, UK, 1986; p. 274.
73. Rovelli, C. Relational quantum mechanics. *Int. J. Theor. Phys.* **1996**, *35*, 1637–1678. [CrossRef]
74. Rovelli, C. Space is blue and birds fly through it. *Philos. Trans. R. Soc. A* **2018**, *376*, 20170312. [CrossRef] [PubMed]
75. Höhn, P.A.; Wever, C.S.P. Quantum theory from questions. *Phys. Rev. A* **2017**, *95*, 012102. [CrossRef]
76. Höhn, P.A. Quantum theory from rules on information acquisition. *Entropy* **2017**, *19*, 98. [CrossRef]
77. Martin-Dussaud, P.; Rovelli, C.; Zalamea, F. The notion of locality in relational quantum mechanics. *Found. Phys.* **2019**, *49*, 96–106
78. Koberinski, A.; Müller, M.P. *Quantum Theory as a Principle Theory: Insights from an Information-Theoretic Reconstruction*; Cambridge University Press: Cambridge, UK, 2018; p. 257.
79. Brukner, Č. On the quantum measurement problem. In *Quantum [Un]Speakables II*; Springer: Berlin/Heidelberg, Germany, 2017; pp. 95–117.
80. Fuchs, C.A.; Stacey, B.C. QBist Quantum Mechanics: Quantum Theory as a Hero's Handbook. *arXiv* **2016**, arXiv:1612.07308
81. Breuer, T. The impossibility of accurate state self-measurements. *Philos. Sci.* **1995**, *62*, 197–214. [CrossRef]
82. Dalla Chiara, M.L. Logical self reference, set theoretical paradoxes and the measurement problem in quantum mechanics. *J. Philos. Logic* **1977**, *6*, 331–347. [CrossRef]
83. Crane, L. Clock and category: Is quantum gravity algebraic?. *J. Math. Phys.* **1995**, *36*, 6180–6193. [CrossRef]
84. Markopoulou, F. Planck scale models of the universe. *arXiv* **2002**, arXiv:gr-qc/0210086.
85. Markopoulou, F. New directions in background independent quantum gravity. *arXiv* **2007**, arXiv:gr-qc/0703097.
86. Hackl, L.F.; Neiman, Y. Horizon complementarity in elliptic de Sitter space. *Phys. Rev. D* **2015**, *91*, 044016
87. Wigner, E.P. Remarks on the mind-body question. In *Philosophical Reflections and Syntheses*; Springer: Berlin/Heidelberg, Germany, 1995; pp. 247–260.
88. Deutsch, D. Quantum theory as a universal physical theory. *Int. J. Theor. Phys.* **1985**, *24*, 1–41. [CrossRef]
89. Frauchiger, D.; Renner, R. Quantum theory cannot consistently describe the use of itself. *Nat. Commun.* **2018**, *9*, 3711. [CrossRef] [PubMed]

© 2019 by the author. Licensee MDPI, Basel, Switzerland. This article is an open access article distributed under the terms and conditions of the Creative Commons Attribution (CC BY) license (http://creativecommons.org/licenses/by/4.0/).

Article
Thermal Quantum Spacetime

Isha Kotecha [1,2]

[1] Max Planck Institute for Gravitational Physics (Albert Einstein Institute), Am Mühlenberg 1, 14476 Potsdam-Golm, Germany; isha.kotecha@aei.mpg.de
[2] Institut für Physik, Humboldt-Universität zu Berlin, Newtonstraße 15, 12489 Berlin, Germany

Received: 2 July 2019; Accepted: 7 August 2019; Published: 12 August 2019

Abstract: The intersection of thermodynamics, quantum theory and gravity has revealed many profound insights, all the while posing new puzzles. In this article, we discuss an extension of equilibrium statistical mechanics and thermodynamics potentially compatible with a key feature of general relativity, background independence; and we subsequently use it in a candidate quantum gravity system, thus providing a preliminary formulation of a thermal quantum spacetime. Specifically, we emphasise an information-theoretic characterisation of generalised Gibbs equilibrium that is shown to be particularly suited to background independent settings, and in which the status of entropy is elevated to being more fundamental than energy. We also shed light on its intimate connections with the thermal time hypothesis. Based on this, we outline a framework for statistical mechanics of quantum gravity degrees of freedom of combinatorial and algebraic type, and apply it in several examples. In particular, we provide a quantum statistical basis for the origin of covariant group field theories, shown to arise as effective statistical field theories of the underlying quanta of space in a certain class of generalised Gibbs states.

Keywords: background independence; generalised statistical equilibrium; quantum gravity; entropy

1. Introduction

Background independence is a hallmark of general relativity that has revolutionised our conception of space and time. The picture of physical reality it paints is that of an impartial dynamical interplay between matter and gravitational fields. Spacetime is no longer a passive stage on which matter performs; it is an equally active performer in itself. Coordinates are gauge, thus losing their physical status of non-relativistic settings. In particular, the notion of time is modified drastically. It is no longer an absolute, global, external parameter uniquely encoding the full dynamics. It is instead a gauge parameter associated with a Hamiltonian constraint.

On the other hand, the well-established fields of quantum statistical mechanics and thermodynamics have been of immense use in the physical sciences. From early applications to heat engines and study of gases, to modern day uses in condensed matter systems and quantum optics, these powerful frameworks have greatly expanded our knowledge of physical systems. However, a complete extension of them to a background independent setting, such as for a gravitational field, remains an open issue [1–3]. The biggest challenge is the absence of an absolute notion of time, and thus of energy, which is essential to any statistical and thermodynamical consideration. This issue is particularly exacerbated in the context of defining statistical equilibrium, for the natural reason that the standard concepts of equilibrium and time are tightly linked. In other words, the constrained dynamics of a background independent system lacks a non-vanishing Hamiltonian in general, which makes formulating (equilibrium) statistical mechanics and thermodynamics, an especially thorny problem. This is a foundational issue, and tackling it is important and interesting in its own right, and even more so because it could provide deep insights into the nature of (quantum) gravitational systems. This paper is devoted to addressing precisely these points.

The importance of addressing these issues is further intensified in light of the deep interplay between thermodynamics, gravity and the quantum theory, first uncovered for black holes. The laws of black hole mechanics [4] were a glimpse into a curious intermingling of thermodynamics and classical gravity, even if originally only at a formal level of analogy. The discovery of black hole entropy and radiation [5–7] further brought quantum mechanics into the mix. This directly led to a multitude of new conceptual insights along with many puzzling questions which continue to be investigated still after decades. The content of the discovery, namely that a black hole must be assigned physical entropy and that it scales with the area of its horizon in Planck units, has birthed several distinct lines of thoughts, in turn leading to different (even if related) lines of investigations, such as thermodynamics of gravity, analogue gravity and holography. Moreover, early attempts at understanding the physical origin of this entropy [8] made evident the relevance of quantum entanglement, thus also contributing to the current prolific interest in fascinating connections between quantum information theory and gravitational physics.

This discovery further hinted at a quantum microstructure underlying a classical spacetime. This perspective is shared, to varying degrees of details, by various approaches to quantum gravity such as loop quantum gravity (and related spin foams and group field theories), string theory and AdS/CFT, simplicial gravity and causal set theory to name a few. Specifically within discrete non-perturbative approaches, spacetime is replaced by more fundamental entities that are discrete, quantum, and pre-geometric in the sense that no notion of smooth metric geometry and spacetime manifold exists yet. The collective dynamics of such quanta of geometry, governed by some theory of quantum gravity is then hypothesised to give rise to an emergent spacetime, corresponding to certain phases of the full theory. This would essentially entail identifying suitable procedures to extract a classical continuum from a quantum discretuum, and to reconstruct general relativistic gravitational dynamics coupled with matter (likely with quantum corrections). This emergence in quantum gravity is akin to that in condensed matter systems in which also coarse-grained macroscopic (thermodynamic) properties of the physical systems are extracted from the microscopic (statistical and) dynamical theories of the constituent atoms. In this sense our universe can be understood as an unusual condensed matter system, brought into the existing smooth geometric form by a phase transition of a quantum gravity system of pre-geometric 'atoms' of space; in particular, as a condensate [9].

This brings our motivations full circle, and to the core of this article: to illustrate, the potential of and preliminary evidence for, a rewarding exchange between a background independent generalisation of statistical mechanics and discrete quantum gravity; and show that ideas from the former are vital to investigate statistical mechanics and thermodynamics of quantum gravity, and that its considerations in the latter could in turn provide valuable insights into the former.

These are the two facets of interest to us here. In Section 2, we discuss a potential background independent extension of equilibrium statistical mechanics, giving a succinct yet complete discussion of past works in Section 2.1.1, and subsequently focussing on a new thermodynamical characterisation for background independent equilibrium in Section 2.1.2, which is based on a constrained maximisation of information entropy. In Section 2.2, we detail further crucial properties of this characterisation, while placing it within a bigger context of the issue of background independent statistical equilibrium, also in comparison with the previous proposals. Section 2.3 is more exploratory, remarking on exciting new connections between the thermodynamical characterisation and the thermal time hypothesis, wherein information entropy and observer dependence are seen to play instrumental roles. In Section 2.4, we discuss several aspects of a generalised thermodynamics based on the generalised equilibrium statistical mechanics derived above, including statements of the zeroth and first laws. Section 3 is devoted to statistical mechanical considerations of candidate quantum gravity degrees of freedom of combinatorial and algebraic type. After clarifying the framework for many-body mechanics of such atoms of space in Section 3.1, we give an overview of examples in Section 3.2, thus illustrating the applicability of the generalised statistical framework in quantum gravity. The one case for which we give a slightly more detailed account is that of deriving a generic covariant group field theory as an effective statistical field theory starting from a particular class of quantum Gibbs states of the underlying microscopic system. Finally, we conclude and offer some outlook.

2. Background Independent Equilibrium Statistical Mechanics

Covariant statistical mechanics [1–3] broadly aims at addressing the foundational issue of defining a suitable statistical framework for constrained systems. This issue, especially in the context of gravity, was brought to the fore in a seminal work [1], and developed subsequently in [2,3,10,11]. Valuable insights from these studies on spacetime relativistic systems [1–3,11–13] have also formed the conceptual backbone of first applications to discrete quantum gravity [14–16]. In this section, we present extensions of equilibrium statistical mechanics to background independent[1] systems, laying out different proposals for a generalised statistical equilibrium, but emphasising on one in particular, and based on which further aspects of a generalised thermodynamics are considered. The aim here is thus to address the fundamental problem of formulating these frameworks in settings where the conspicuous absence of time and energy is particularly tricky.

Section 2.1 discusses background independent characterisations of equilibrium Gibbs states, of the general form $e^{-\sum_a \beta_a \mathcal{O}_a}$. In Section 2.1.1, we touch upon various proposals for equilibrium put forward in past studies on spacetime covariant systems [1,3,11,17,18]. From Section 2.1.2 onwards, we focus on Jaynes' information-theoretic characterisation [19,20] for equilibrium. This was first suggested as a viable proposal for background independent equilibrium, and illustrated with an explicit example in the context of quantum gravity, in [14]. Using the terminology of [14], we call this a 'thermodynamical' characterisation of equilibrium, to contrast with the customary Kubo–Martin–Schwinger (KMS) [21] 'dynamical' characterisation[2].

We devote Section 2.2 to discussing various aspects of the thermodynamical characterisation, including highlighting many of its favourable features, also compared to the other proposals. In fact, we point out how this characterisation can comfortably accommodate the other proposals for Gibbs equilibrium.

Further, as is evident shortly, the thermodynamical characterisation hints at the idea that entropy is a central player, which has been a recurring theme across modern theoretical physics. In Section 2.3, we present a tentative discussion on some of these aspects. In particular, we notice compelling new relations between the thermodynamical characterisation and the thermal time hypothesis, which further seem to hint at intriguing relations between entropy, observer dependence and thermodynamical time. We further propose to use the thermodynamical characterisation as a constructive criterion of choice for the thermal time hypothesis.

Finally, in Section 2.4, we define the basic thermodynamic quantities which can be derived immediately from a generalised equilibrium state, without requiring any additional physical and/or interpretational inputs. We clarify the issue of extracting a single common temperature for the full system from a set of several of them, and end with the zeroth and first laws of a generalised thermodynamics.

---

[1] In the original works mentioned above, the framework is usually referred to as covariant or general relativistic statistical mechanics. However, we choose to call it background independent statistical mechanics as our applications to quantum gravity are evident of the fact that the main ideas and structures are general enough to be used in radically background independent systems devoid of any spacetime manifold or associated geometric structures.

[2] For a more detailed discussion of the comparison between these two characterisations, we refer the reader to [14]. The main idea is that the various proposals for generalised Gibbs equilibrium can be divided into these two categories. Which characterisation one chooses to use in a given situation depends on the information/description of the system that one has at hand. For instance, if the description includes a one-parameter flow of physical interest, then using the dynamical characterisation, i.e., satisfying the KMS condition with respect to it, will define equilibrium with respect to it. The procedures defining these two categories can thus be seen as 'recipes' for constructing a Gibbs state, and which one is more suitable depends on our knowledge of the system.

2.1. Generalised Equilibrium

Equilibrium states are a cornerstone of statistical mechanics, which in turn is the theoretical basis for thermodynamics. They are vital in the description of macroscopic systems with a large number of microscopic constituents. In particular, Gibbs states $e^{-\beta E}$ have a vast applicability across a broad range of fields such as condensed matter physics, quantum information and tensor networks, and (quantum) gravity, to name a few. They are special, being the unique class of states in finite systems satisfying the KMS condition[3]. Furthermore, usual coarse-graining techniques also rely on the definition of Gibbs measures. In treatments closer to hydrodynamics, one often considers the full (non-equilibrium) system as being composed of many interacting subsystems, each considered to be at local equilibrium. While in the context of renormalisation group flow treatments, each phase at a given scale, for a given set of coupling parameters is also naturally understood to be at equilibrium, each described by (an inequivalent) Gibbs measure.

Given this physical interest in Gibbs states, the question then is how to define them for background independent systems. The following are different proposals, all relying on different principles originating in standard non-relativistic statistical mechanics, extended to a relativistic setting.

2.1.1. Past Proposals

The first proposal [1,12] was based on the idea of statistical independence of arbitrary (small, but macroscopic) subsystems of the full system. The notion of equilibrium is taken to be characterised by the factorisation property of the state, $\rho_{12} = \rho_1 \rho_2$, for any two subsystems 1 and 2; and the full system is at equilibrium if any one of its subsystems satisfies this property with all the rest. We notice that the property of statistical independence is related to an assumption of weak interactions [22].

This same dilute gas assumption is integral also to the Boltzmann method of statistical mechanics. It characterises equilibrium as the most probable distribution, that is one with maximum entropy[4]. This method is used in [11] to study a gas of constrained particles[5].

The work in [3] puts forward a physical characterisation for an equilibrium state. The suggestion is that ρ (itself a well-defined state on the physical, reduced state space) is said to be a physical Gibbs state if its modular Hamiltonian $h = -\ln \rho$, is a well-defined function on the physical state space; and, is such that there exists a (local) clock function $T(\mathbf{x})$ on the extended state space (with its conjugate momentum $p_T(\mathbf{x})$), such that the (pull-back) of h is proportional to (the negative of) p_T. Importantly, when this is the case the modular flow ('thermal time', see Section 2.2) is a geometric (foliation) flow in spacetime, in which sense ρ is said to be 'physical'. Notice that the built-in strategy here is to define KMS equilibrium in a deparameterised system (thus it is an example of using the dynamical characterisation), since it basically identifies a state's modular Hamiltonian with a (local) clock Hamiltonian on the base spacetime manifold.

Another strategy [17] is based on the use of the ergodic principle and introduction of clock subsystems to define (clock) time averages. Again, this characterisation, similar to a couple of the previous ones, relies on the validity of a postulate, even if traditionally a fundamental one.

Finally, the proposal of [18] interestingly characterises equilibrium by a vanishing information flow between interacting histories. The notion of information used is that of Shannon (entropy), $I = \ln N$, where N is the number of microstates traversed in a given history during interaction.

---

[3] The algebraic KMS condition [21] is well known to provide a comprehensive characterisation of statistical equilibrium in systems of arbitrary sizes, as long as there exists a well-defined one-parameter dynamical group of automorphisms of the system. This latter point, of the required existence of a preferred time evolution of the system, is exactly the missing ingredient in our case, thus limiting its applicability.

[4] Even though this method relies on maximising the entropy similar to the thermodynamical characterisation, it is more restrictive than the latter, as is made clear in Section 2.2.

[5] We remark that except for this one work, all other studies in spacetime covariant statistical mechanics are carried out from the Gibbs ensemble point of view.

Equilibrium between two histories 1 and 2 is encoded in a vanishing information flow, $\delta I = I_2 - I_1 = 0$. This characterisation of equilibrium is evidently information-theoretic, even if relying on an assumption of interactions. Moreover, it is much closer to our thermodynamical characterisation, because the condition of vanishing δI is nothing but an optimisation of information entropy.

These different proposals, along with the thermal time hypothesis [1,2], have led to some remarkable results, such as recovering the Tolman–Ehrenfest effect [13,23], relativistic Jüttner distribution [23] and Unruh effect [24]. However, they all assume the validity of one or more principles, postulates or assumptions about the system. Moreover, none (at least presently) seems to be general enough, as with the proposal below, to be implemented in a full quantum gravity setup, while also accommodating within it the rest of the proposals.

2.1.2. Thermodynamical Characterisation

This brings us to the proposal of characterising a generalised Gibbs state based on a constrained maximisation of information (Shannon or von Neumann) entropy [14–16], along the lines advocated by Jaynes [19,20] purely from the perspective of evidential statistical inference. Jaynes' approach is fundamentally different from other more traditional ones of statistical physics, as is the thermodynamical characterisation, compared with the others outlined above, which is exemplified in the following. It is thus a new proposal for background independent equilibrium [14,25], which has the potential of incorporating also the others as special cases, from the point of view of constructing a Gibbs state.

Consider a macroscopic system with a large number of constituent microscopic degrees of freedom. Our (partial) knowledge of its macrostate is given in terms of a finite set of averages $\{\langle \mathcal{O}_a \rangle = U_a\}$ of the observables we have access to. Jaynes suggests that a fitting probability estimate (which, once known, will allow us to infer also the other observable properties of the system) is not only one that is compatible with the given observations, but also that which is least-biased in the sense of not assuming any more information about the system than what we actually have at hand (namely, $\{U_a\}$). In other words, given a limited knowledge of the system (which is always the case in practice for any macroscopic system), the least-biased probability distribution compatible with the given data should be preferred. As shown below, this turns out to be a Gibbs distribution with the general form $e^{-\sum_a \beta_a \mathcal{O}_a}$.

Let Γ be a finite-dimensional phase space (be it extended or reduced), and on it consider a finite set of smooth real-valued functions $\mathcal{O}_a$. Denote by ρ a smooth statistical density (real-valued, positive and normalised function) on Γ, to be determined. Then, the prior on the macrostate gives a finite number of constraints,

$$\langle \mathcal{O}_a \rangle_\rho = \int_\Gamma d\lambda \, \rho \, \mathcal{O}_a = U_a \tag{1}$$

where $d\lambda$ is a Liouville measure on Γ, and the integrals are taken to be well-defined. Further, ρ has an associated Shannon entropy

$$S[\rho] = -\langle \ln \rho \rangle_\rho . \tag{2}$$

By understanding S to be a measure of uncertainty quantifying our ignorance about the details of the system, the corresponding bias is minimised (compatibly with the prior data) by maximising S (under the set of constraints in Equation (1), plus the normalisation condition for ρ) [19]. The method of Lagrange multipliers then gives a generalised Gibbs distribution of the form,

$$\rho_{\{\beta_a\}} = \frac{1}{Z_{\{\beta_a\}}} e^{-\sum_a \beta_a \mathcal{O}_a} \tag{3}$$

where the partition function $Z_{\{\beta_a\}}$ encodes all thermodynamic properties in principle, and is assumed to be convergent. This can be done analogously for a quantum system [20], giving a Gibbs density operator on a representation Hilbert space

$$\hat{\rho}_{\{\beta_a\}} = \frac{1}{Z_{\{\beta_a\}}} e^{-\sum_a \beta_a \hat{\mathcal{O}}_a} \,. \tag{4}$$

A generalised Gibbs state can thus be defined, characterised fully by a finite set of observables of interest $\mathcal{O}_a$, and their conjugate generalised "inverse temperatures" β_a, which have entered formally as Lagrange multipliers. Given this class of equilibrium states, it should be evident that some thermodynamic quantities (e.g., generalised "energies" U_a) can be identified immediately. Aspects of a generalised thermodynamics are discussed in Section 2.4.

Finally, we note that the role of entropy is shown to be instrumental in defining (local[6]) equilibrium states: "...thus entropy becomes the primitive concept with which we work, more fundamental even than energy..." [19]. It is also interesting to notice that Bekenstein's arguments [6] can be observed to be influenced by Jaynes' information-theoretic insights surrounding entropy, and these same insights have now guided us in the issue of background independent statistical equilibrium.

2.2. Remarks

1. There are two key features of this characterisation. The first is the use of evidential (or epistemic, or Bayesian) probabilities, thus taking into account the given evidence $\{U_a\}$; and second is a preference for the least-biased (or most "honest") distribution out of all the different ones compatible with the given evidence. It is not enough to arbitrarily choose any that is compatible with the prior data. An aware observer must also take into account their own ignorance, or lack of knowledge honestly, by maximising the information entropy.

2. This notion of equilibrium is inherently observer-dependent because of its use of the macrostate thermodynamic description of the system, which in itself is observer-dependent due to having to choose a coarse-graining, that is the set of macroscopic observables.

3. Given a generalised Gibbs state, the question arises as to which flow it is stationary with respect to. Any density distribution or operator satisfies the KMS condition (which implies stationarity) with respect to its own modular flow. In fact, by the Tomita–Takesaki theorem [21], any faithful algebraic state over a von Neumann algebra is KMS with respect to its own one-parameter modular (Tomita) flow.[7] Given this, then $\rho_{\{\beta_a\}}$ is clearly KMS with respect to the flow $X_\rho \sim \partial/\partial t$ (or $\hat{U}_\rho(t) \sim e^{iht}$) generated by its modular Hamiltonian $h = \sum_a \beta_a \mathcal{O}_a$. In particular, $\rho_{\{\beta_a\}}$ is not stationary with respect to the individual flows X_a generated by $\mathcal{O}_a$, unless they satisfy $[X_a, X_{a'}] = 0$ for all a, a' [15]. In fact, this last property shows that the proposal in [1,12] based on statistical independence (that is, $[X_{\rho_1}, X_{\rho_2}] = 0$) can be understood as a special case of this one, when the state is defined for a pair of observables that are defined on mutually exclusive subspaces of the state space. In this case, their respective flows will automatically commute and the state will be said to satisfy statistical independence.

4. To be clear, the use of the "most probable" characterisation for equilibrium is not new in itself. It was used by Boltzmann in the late 19th century, and utilised (also within a Boltzmann interpretation of statistical mechanics) in a constrained system in [11]. The fact that equilibrium configurations maximise the system's entropy is also not new: it was well known already in the time of Gibbs[8]. The novelty here is: in the revival of Jaynes' perspective, of deriving

---

[6] Local, in the sense of being observer-dependent (see Section 2.2).
[7] This is also the main ingredient of the thermal time hypothesis [1,2], which we return to below.
[8] However, as Jaynes points out in [19], these properties were relegated to side remarks in the past, not really considered to be fundamental to the theory or to the justifications for the methods of statistical mechanics.

equilibrium statistical mechanics in terms of evidential probabilities, solely as a problem of statistical inference without depending on the validity of any further conjectures, physical assumptions or interpretations; and in the suggestion that it is general enough to apply to genuinely background independent systems, including quantum gravity. Below, we list some of these more valuable features.

- The procedure is versatile, being applicable to a wide array of cases (both classical and quantum), relying only on a sufficiently well-defined mathematical description in terms of a state space, along with a set of observables with dynamically constant averages U_a defining a suitable macrostate of the system[9].
- Evidently, this manner of defining equilibrium statistical mechanics (and from it, thermodynamics) does not lend any fundamental status to energy, nor does it rely on selecting a single, special (energy) observable out of the full set $\{\mathcal{O}_a\}$. It can thus be crucial in settings where concepts of time and energy are dubious at the least, or not defined at all as in non-perturbative quantum gravity.
- It has a technical advantage of not needing any (one-parameter) symmetry (sub-)groups of the system to be defined a priori, unlike the dynamical characterisation based on the standard KMS condition.
- It is independent of any additional physical assumptions, hypotheses or principles that are common to standard statistical physics, and, in the present context, to the other proposals of generalised equilibrium recalled in Section 2.1. Some examples of these extra ingredients (not required in the thermodynamical characterisation) that we have already encountered are ergodicity, weak interactions, statistical independence, and often a combination of them.
- It is independent of any physical interpretations attached (or not!) to the quantities and setup involved. This further amplifies its appeal for use in quantum gravity where the geometrical (and physical) meanings of the quantities involved may not necessarily be clear from the start.
- One of the main features (which helps accommodate the other proposals as special cases of this one) is the generality in the choice of observables $\mathcal{O}_a$ allowed naturally by this characterisation. In principle, they need only be mathematically well-defined in the given description of the system (regardless of whether it is kinematic i.e., working at the extended state space level, or dynamic, i.e., working with the physical state space), satisfying convexity properties so that the resultant Gibbs state is normalisable. More standard choices include a Hamiltonian in a non-relativistic system, a clock Hamiltonian in a deparameterised system [3,14], and generators of kinematic symmetries such as rotations, or more generally of one-parameter subgroups of Lie group actions [26,27]. Some of the more unconventional choices include geometric observables such as volume [14,28], (component functions of the) momentum map associated with geometric closure of classical polyhedra [15,16], half-link gluing (or face-sharing) constraints of discrete gravity [15], a projector in group field theory [15,29], and generic gauge-invariant observables (not necessarily symmetry generators) [11]. We refer to [14] for a more detailed discussion.

In Section 3.2, we outline some examples of using this characterisation in quantum gravity, while a detailed investigation of its consequences in particular for covariant systems on a spacetime manifold is left to future studies.

---

[9] In fact, in hindsight, we could already have anticipated a possible equilibrium description in terms of these constants, whose existence is assumed from the start.

2.3. Relation to Thermal Time Hypothesis

This section outlines a couple of new intriguing connections between the thermodynamical characterisation and the thermal time hypothesis, which we think are worthwhile to be explored further. Thermal time hypothesis [1,2] states that the (geometric) modular flow of the (physical, equilibrium) statistical state that an observer happens to be in is the time that they experience. It thus argues for a thermodynamical origin of time [30].

What is this state? Pragmatically, the state of a macroscopic system is that which an observer is able to observe and assigns to the system. It is not an absolute property since one can never know everything there is to know about the system. In other words, the state that the observer "happens to be in" is the state that they are able to detect. This leads us to suggest that the thermodynamical characterisation can provide a suitable *criterion of choice* for the thermal time hypothesis.

What we mean by this is the following. Consider a macroscopic system that is observed to be in a particular macrostate in terms of a set of (constant) observable averages. The thermodynamical characterisation then provides the least biased choice for the underlying (equilibrium) statistical state. Given this state then, the thermal time hypothesis would imply that the (physical) time experienced by this observer is the (geometric) modular flow of the observed state.

Jaynes [19,20] turned the usual logic of statistical mechanics upside-down to stress on entropy and the observed macrostate as the starting point, to define equilibrium statistical mechanics in its entirety *from* it (and importantly, a further background independent generalisation, as shown above). Rovelli [1], later with Connes [2], turned the usual logic of the definition of time upside-down to stress on the choice of a statistical state as the starting point to identify a suitable time flow *from* it. The suggestion here is to merge the two and get an operational way of implementing the thermal time hypothesis.

It is interesting to see that the crucial property of observer-dependence of relativistic time arises as a natural consequence of our suggestion, directly because of the observer-dependence of any state defined using the thermodynamical characterisation. This way, thermodynamical time is intrinsically "perspectival" [31] or "anthropomorphic" [32].

To be clear, this criterion of choice will not single out a preferred state, by the very fact that it is inherently observer-dependent. It is thus compatible with the basic philosophy of the thermal time hypothesis, namely that there is no preferred physical time.

Presently the above suggestion is rather conjectural, and certainly much work remains to be done to understand it better, and explore its potential consequences for physical systems. Here, it may be helpful to realise that the thermal time hypothesis can be sensed to be intimately related with (special and general) relativistic systems, and so might the thermodynamical characterisation when considered in this context. Thus, for instance, Rindler spacetime or stationary black holes might offer suitable settings to begin investigating these aspects in more detail.

The second connection that we observe is much less direct, and is via information entropy. The generator of the thermal time flow [1], $-\ln\rho$, can immediately be observed to be related to Shannon entropy in Equation (2). Moreover, in the general algebraic (quantum) field theoretic setting, the generator is the log of the modular operator Δ of von Neumann algebra theory [2]. A modification of it, the relative modular operator, is known to be an algebraic measure of relative entropy [33], which in fact has seen a recent revival in the context of quantum information and gravity. This is a remarkable feature in our opinion, which compels us to look for deeper insights it may have to offer, in further studies.

2.4. Generalised Thermodynamic Potentials, Zeroth and First Laws

Traditional thermodynamics is the study of energy and entropy exchanges. However, what is a suitable generalisation of it for background independent systems? This, as with the question of a generalised equilibrium statistical mechanics which we have considered until now, is still open. In the following, we offer some insights gained from preceding discussions, including identifying certain thermodynamic potentials, and generalised zeroth and first laws.

Thermodynamic potentials are vital, particularly in characterising the different phases of the system. The most important one is the partition function $Z_{\{\beta_a\}}$, or equivalently the free energy

$$\Phi(\{\beta_a\}) := -\ln Z_{\{\beta_a\}}. \tag{5}$$

It encodes complete information about the system from which other thermodynamic quantities can be derived in principle. Notice that the standard definition of a free energy F comes with an additional factor of a (single, global) temperature, that is we normally have $\Phi = \beta F$. However, for now, Φ is the more suitable quantity to define and not F since we do not (yet) have a single common temperature for the full system. We return to this point below.

Next is the thermodynamic entropy (which by use of the thermodynamical characterisation has been identified with information entropy), which is straightforwardly

$$S(\{U_a\}) = \sum_a \beta_a U_a - \Phi \tag{6}$$

for generalised Gibbs states of the form in Equation (3). Notice again the lack of a single β scaling the whole equation at this more general level of equilibrium.

By varying S such that the variations dU_a and $\langle d\mathcal{O}_a \rangle$ are independent [19], a set of generalised heats can be defined

$$dS = \sum_a \beta_a (dU_a - \langle d\mathcal{O}_a \rangle) =: \sum_a \beta_a \, dQ_a \tag{7}$$

and, from it (at least part of the[10]) work done on the system dW_a [15], can be identified

$$dW_a := \langle d\mathcal{O}_a \rangle = \frac{1}{\beta_a} \int_\Gamma d\lambda \, \frac{\delta \Phi}{\delta \mathcal{O}_a} \, d\mathcal{O}_a \,. \tag{8}$$

From the setup of the thermodynamical characterisation presented in Section 2.1.2, we can immediately identify U_a as generalised "energies". Jaynes' procedure allows these quantities to *democratically* play the role of generalised energies. None had to be selected as being *the* energy in order to define equilibrium. This a priori democratic status of the several conserved quantities can be broken most easily by preferring one over the others. In turn, if its modular flow can be associated with a physical evolution parameter (relational or not), then this observable can play the role of a dynamical Hamiltonian.

Thermodynamic conjugates to these energies are several generalised inverse temperatures β_a. By construction, each β_a is the periodicity in the flow of $\mathcal{O}_a$, in addition to being the Lagrange multiplier for the ath constraint in Equation (1). Moreover, these same constraints can determine β_a, by inverting the equations

$$\frac{\partial \Phi}{\partial \beta_a} = U_a; \tag{9}$$

or equivalently from

$$\frac{\partial S}{\partial U_a} = \beta_a \,. \tag{10}$$

---

[10] By this we mean that the term $\langle d\mathcal{O}_a \rangle$, based on the *same* observables defining the generalised energies U_a, can be seen as reflecting some work done on the system. However, naturally, we do not expect or claim that this is all the work that is/can be performed on the system by external agencies. In other words, there could be other work contributions, in addition to the terms dW_a. A better understanding of work terms in this background independent setup, will also contribute to a better understanding of the generalised first law presented below.

In general, $\{\beta_a\}$ is a multi-variable inverse temperature. In the special case when $\mathcal{O}_a$ are component functions of a dual vector, then $\vec{\beta} \equiv (\beta_a)$ is a vector-valued temperature. For example, this is the case when $\vec{\mathcal{O}} \equiv \{\mathcal{O}_a\}$ are dual Lie algebra-valued momentum maps associated with Hamiltonian actions of Lie groups, as introduced by Souriau [26,27], and appearing in the context of classical polyhedra in [15].

As shown above, a generalised equilibrium is characterised by several inverse temperatures, but an identification of a single common temperature for the full system is of obvious interest. This can be done as follows [12,15]. A state of the form in Equation (3), with modular Hamiltonian

$$h = \sum_a \beta_a \mathcal{O}_a \tag{11}$$

generates a modular flow (with respect to which it is at equilibrium), parameterised by

$$t = \sum_a \frac{t_a}{\beta_a} \tag{12}$$

where t_a are the flow parameters of $\mathcal{O}_a$. The strategy now is to reparameterise the same trajectory by a rescaling of t,

$$\tau := t/\beta \tag{13}$$

for a real-valued β. It is clear that τ parameterises the modular flow of a rescaled modular hamiltonian $\tilde{h} = \beta h$, associated with the state

$$\tilde{\rho}_\beta = \frac{1}{\tilde{Z}_\beta} e^{-\tilde{h}} = \frac{1}{\tilde{Z}_\beta} e^{-\beta h} \tag{14}$$

characterised now by a single inverse temperature β.

In fact, this state can be understood as satisfying the thermodynamical characterisation for a single constraint

$$\langle h \rangle = \text{constant} \tag{15}$$

instead of several of them as in Equation (1). Clearly, this rescaling is not a trivial move. It corresponds to the case of a weaker, single constraint which by nature corresponds to a different physical situation wherein there is exchange of information between the different observables (so that they can thermalise to a single β). This can happen for instance when one observable is special (e.g., the Hamiltonian) and the rest are functionally related to it (e.g., the volume or number of particles). Whether such a determination of a single temperature can be brought about by a more physically meaningful technique is left to future work. Having said that, it will not change the general layout of the two cases as outlined above.

One immediate consequence of extracting a single β is regarding the free energy, which can now be written in the familiar form as

$$\Phi = \beta F. \tag{16}$$

This is most directly seen from the expression for the entropy,

$$\tilde{S} = -\langle \ln \tilde{\rho}_\beta \rangle_{\tilde{\rho}_\beta} = \beta \sum_a \beta_a \tilde{U}_a + \ln \tilde{Z} \quad \Leftrightarrow \quad \tilde{F} = \tilde{U} - \beta^{-1} \tilde{S} \tag{17}$$

where $\tilde{U} = \sum_a \beta_a \tilde{U}_a$ is a total energy, and tildes mean that the quantities are associated with the state $\tilde{\rho}_\beta$. Notice that the above equation clearly identifies a single conjugate variable to entropy, the temperature β^{-1}.

It is important to remark that, in the above method to get a single β, we still do not need to choose a special observable, say $\mathcal{O}'$, out of the given set of $\mathcal{O}_a$. If one were to do this, i.e., select $\mathcal{O}'$ as a dynamical energy (so that by extension the other $\mathcal{O}_a$ are functions of this one), then by standard arguments, the rest of the Lagrange multipliers will be proportional to β', which in turn would then be the common inverse temperature for the full system. The point is that this latter instance is a special case of the former.

We end this section with zeroth and first laws of generalised thermodynamics. The crux of the zeroth law is a definition of equilibrium. Standard statement refers to a thermalisation resulting in a single temperature being shared by any two systems in thermal contact. This can be extended by the statement that at equilibrium, all inverse temperatures β_a are equalised. This is in exact analogy with all intensive thermodynamic parameters, such as the chemical potential, being equal at equilibrium.

The standard first law is basically a statement about conservation of energy. In the generalised equilibrium case corresponding to the set of individual constraints in Equation (1), the first law is satisfied ath-energy-wise,

$$dU_a = dQ_a + dW_a. \tag{18}$$

The fact that the law holds a-energy-wise is not surprising because the separate constraints in Equation (1) for each a mean that observables $\mathcal{O}_a$ do not exchange any information amongst themselves. If they did, then their Lagrange multipliers would no longer be mutually independent and we would automatically reduce to the special case of having a single β after thermalisation.

On the other hand, for the case with a single β, variation of the entropy in Equation (17) gives

$$d\tilde{S} = \beta \sum_a \beta_a (dU_a - \langle d\mathcal{O}_a \rangle) =: \beta d\tilde{Q} \tag{19}$$

giving a first law with a more familiar form, in terms of total energy, total heat and total work variations

$$d\tilde{U} = d\tilde{Q} + d\tilde{W}. \tag{20}$$

As before, in the even more special case where β is conjugate to a single preferred energy, then this reduces to the traditional first law. We leave the verification of the second law for the generalised entropy to future work. Further, the quantities introduced above and the consequences of this setup also need to be investigated in greater detail.

3. Equilibrium Statistical Mechanics in Quantum Gravity

Emergence of spacetime is the outstanding open problem in quantum gravity that is being addressed from several directions. One such is based on modelling quantum spacetime as a many-body system [34], which further complements the view of a classical spacetime as an effective macroscopic thermodynamic system. This formal suggestion allows one to treat extended regions of quantum spacetime as built out of discrete building blocks whose dynamics is dictated by non-local, combinatorial and algebraic mechanical models. Based on this mechanics, a formal statistical mechanics of the quanta of space can be studied [14,15]. Statistical mixtures of quantum gravity states are better suited to describe generic boundary configurations with a large number of quanta. This is in the sense that given a region of space with certain known macroscopic properties, a more reasonable modelling of its underlying quantum gravity description would be in in terms of a mixed state rather than a pure state, essentially because we cannot hope to know precisely all microscopic details to prefer one particular microstate. A simple example is having a region with a fixed spatial volume and wanting to estimate the underlying quantum gravity (statistical) state [11,14].

In addition to the issue of emergence, investigating the statistical mechanics and thermodynamics of quantum gravity systems would be expected to contribute towards untangling the puzzling knot between thermodynamics, gravity and the quantum theory, especially when applied to more physical settings, such as cosmology [28].

In the rest of this article, we use results from the previous sections to outline a framework for equilibrium statistical mechanics for candidate quanta of geometry (along the lines presented in [14,15], but generalising further to a richer combinatorics based on [35]), and within it give an overview of some concrete examples. In particular, we show that a group field theory can be understood as an effective statistical field theory derived from a coarse-graining of a generalised Gibbs configuration of the underlying quanta. In addition to providing an explicit quantum statistical basis for group field theories, it further reinforces their status as being field theories for quanta of geometry [36–39]. As expected, we see that even though the many-body viewpoint makes certain techniques available that are almost analogous to standard treatments, there are several non-trivialities such as that of background independence, and physical (possible pre-geometric and effective geometric) interpretations of the statistical and thermodynamic quantities involved.

3.1. Framework

The candidate atoms of space considered here are geometric (quantum) d-polyhedra (with d faces), or equivalently open d-valent nodes with its half-links dressed by the appropriate algebraic data [40]. This choice is motivated strongly by loop quantum gravity [41], spin foam [42], group field theory [36–39] and lattice quantum gravity [43] approaches in the context of 4d models. Extended discrete space and spacetime can be built out of these fundamental atoms or "particles", via kinematical compositions (or boundary gluings) and dynamical interactions (or bulk bondings), respectively. In this sense, the perspective innate to a many-body quantum spacetime is a constructive one, which is naturally also extended to the statistical mechanics based on this mechanics.

Two types of data specify a mechanical model, combinatorial and algebraic. States and processes of a model are supported on combinatorial structures, here abstract[11] graphs and 2-complexes, respectively; and algebraic dressings of these structures adds discrete geometric information. Thus, different choices of combinatorics and algebraic data gives different mechanical models. For instance, the simplest spin foam models (and their associated group field theories) for 4d gravity are based on: boundary combinatorics based on a 4-valent node (or a tetrahedron), bulk combinatorics based on a 4-simplex interaction vertex, and algebraic (or group representation) data of $SU(2)$ labelling the boundary 4-valent graphs and bulk 2-complexes.

Clearly, this is not the only choice, in fact far from it. The vast richness of possible combinatorics, compatible with our constructive point of view, is comprehensively illustrated in [35][12]. The various choices for variables to label the discrete structures with (so that they may encode a suitable notion of discrete geometry, which notion depending exactly on the variables chosen and constraints imposed on them) have been an important subject of study, starting all the way from Regge [45–50]. Accommodation of these various different choices is yet another appeal of the constructive many-body viewpoint and this framework. After clarifying further some of these aspects in the following, we choose to work with simplicial combinatorics and $SU(2)$ holonomy-flux data for the subsequent examples.

3.1.1. Atoms of Quantum Space and Kinematics

In the following, we make use of some of the combinatorial structures defined in [35]. However we are content with introducing them in a more intuitive manner, and not recalling the rigorous definitions as that would not be particularly valuable for the present discussion. The interested reader can refer to [35] for details[13].

---

[11] Thus, not necessarily embedded into any continuum spatial manifold.
[12] In fact, the work in [35] is phrased in a language closer to the group field theory approach, but the structures are general enough to apply elsewhere, such as in spin foams, as evidenced in [44].
[13] For clarity, we note that the terminology used here is slightly different from that in [35]. Specifically, the dictionary between here ↔ there is: combinatorial atom or particle ↔ boundary patch; interaction/bulk vertex ↔ spin foam atom; boundary node ↔ boundary multivalent vertex $\bar{v}$; link or full link ↔ boundary edge connecting two multivalent vertices $\bar{v}_1, \bar{v}_2$; half-link ↔ boundary edge connecting a multivalent vertex $\bar{v}$ and a bivalent vertex $\hat{v}$. This minor difference is mainly due to a minor difference in the purpose for the same combinatorial structures. Here, we are in a setup where the accessible states are boundary states, for which a statistical mechanics is defined; and the case of interacting dynamics is considered as defining a suitable (amplitude) functional over the the boundary state space. On the other hand, the perspective in [35] is more in a spin foam constructive setting, so that modelling the 2-complexes as built out of fundamental spin foam atoms is more natural there.

The primary objects of interest to us are boundary patches, which we take as the combinatorial atoms of space. To put simply, a boundary patch is the most basic unit of a boundary graph, in the sense that the set of all boundary patches generates the set of all connected bisected boundary graphs. A bisected boundary graph is simply a directed boundary graph with each of its full links bisected into a pair of half-links, glued at the bivalent nodes (see Figure 1). Different kinds of atoms of space are then the different, inequivalent boundary patches (dressed further with suitable data), and the choice of combinatorics basically boils down to a choice of the set of admissible boundary patches. Moreover, a model with multiple inequivalent boundary patches can be treated akin to a statistical system with multiple species of atoms.

The most general types of boundary graphs are those with nodes of arbitrary valence, and including loops. A common and natural restriction is to consider loopless structures, as they can be associated with combinatorial polyhedral complexes [35]. As the name suggests, loopless boundary patches are those with no loops, i.e., each half-link is bounded on one end by a unique bivalent node (and on the other by the common, multivalent central node). A loopless patch is thus uniquely specified by the number of incident half-links (or equivalently, by the number of bivalent nodes bounding the central node). A d-patch, with d number of incident half-links, is simply a d-valent node. Importantly for us, it is the combinatorial atom that supports (quantum) geometric states of a d-polyhedron [40,51,52]. A further common restriction is to consider graphs with nodes of a single, fixed valence, that is to consider d-regular loopless structures.

Let us take an example. Consider the boundary graph of a 4-simplex as shown in Figure 1. The fundamental atom or boundary patch is a 4-valent node. This graph can be constructed starting from five open 4-valent nodes (denoted $m, n, ..., q$), and gluing the half-links, or equivalently the faces of the dual tetrahedra, pair-wise, with the non-local combinatorics of a complete graph on five 4-valent nodes. The result is ten bisected full links, bounded by five nodes. It is important to note here that a key ingredient of constructing extended boundary states from the atoms are precisely the half-link gluing, or face-sharing conditions on the algebraic data decorating the patches. For instance, in the case of standard loop quantum gravity holonomy-flux variables of $T^*(SU(2))$, the face-sharing gluing constraints are area matching [48], thus lending a notion of discrete classical twisted geometry to the graph. This is much weaker than a Regge geometry, which could have been obtained for the same variables if instead the so-called shape-matching conditions [47] are imposed on the pair-wise gluing of faces/half-links. Thus, kinematic composition (boundary gluings) that creates boundary states depends on two crucial ingredients, the combinatorial structure of the resultant boundary graph, and face-sharing gluing conditions on the algebraic data.

From here on, we restrict ourselves to a single boundary patch for simplicity, a (gauge-invariant) 4-valent node dressed with $SU(2)$ data, i.e., a quantised tetrahedron [40,51]. However, it should be clear from the brief discussion above (and the extensive study in [35]) that a direct generalisation of the present (statistical) mechanical framework is possible also for these more enhanced combinatorial structures.

The phase space of a single classical tetrahedron, encoding both intrinsic and extrinsic degrees of freedom (along with an arbitrary orientation in $\mathbb{R}^3$) is

$$\Gamma = T^*(SU(2)^4/SU(2)) \tag{21}$$

where the quotient by $SU(2)$ imposes geometric closure of the tetrahedron. The choice of domain space is basically the choice of algebraic data. For instance, in Euclidean 4d settings a more apt choice would be the group $Spin(4)$, and $SL(2,\mathbb{C})$ for Lorentzian settings. Then, states of a system of N tetrahedra belong to $\Gamma_N = \Gamma^{\times N}$, and observables would be smooth (real-valued) functions defined on Γ_N [14,15].

The quantum counterparts are,

$$\mathcal{H} = L^2(SU(2)^4/SU(2)) \tag{22}$$

for the single-particle Hilbert space, and $\mathcal{H}_N = \mathcal{H}^{\otimes N}$ for an N-particle system. In the quantum setting, we can go a step further and construct a Fock space based on the above single-particle Hilbert space,

$$\mathcal{H}_F = \bigoplus_{N \geq 0} \text{sym}\, \mathcal{H}_N \qquad (23)$$

where the symmetrisation of N-particle spaces implements a choice of bosonic statistics for the quanta, mirroring the graph automorphism of node exchanges. One choice for the algebra of operators on $\mathcal{H}_F$ is the von Neumann algebra of bounded linear operators. A more common choice though is the larger *-algebra generated by ladder operators $\hat{\varphi}, \hat{\varphi}^\dagger$, which generate the full $\mathcal{H}_F$ by acting on a cyclic Fock vacuum, and satisfy a commutation relations algebra

$$[\hat{\varphi}(\vec{g}_1), \hat{\varphi}^\dagger(\vec{g}_2)] = \int_{SU(2)} dh \prod_{I=1}^{4} \delta(g_{1I} h g_{2I}^{-1}) \qquad (24)$$

where $\vec{g} \equiv (g_I) \in SU(2)^4$ and the integral on the right ensures $SU(2)$ gauge invariance. In fact, this is the Fock representation of an algebraic bosonic group field theory defined by a Weyl algebra [14,29,53].

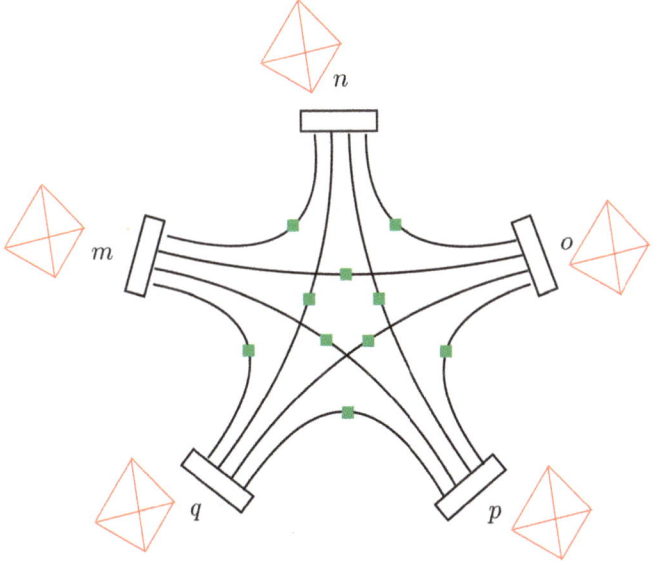

Figure 1. Bisected boundary graph of a 4-simplex, as a result of non-local pair-wise gluing of half-links. Each full link is bounded by two 4-valent nodes (denoted here by $m, n, ...$), and bisected by one bivalent node (shown here in green).

3.1.2. Interacting Quantum Spacetime and Dynamics

Coming now to dynamics, the key ingredients here are the specifications of propagators and admissible interaction vertices, including both their combinatorics, and functional dependences on the algebraic data, i.e., their amplitudes.

The combinatorics of propagators and interaction vertices can be packaged neatly within two maps defined in [35], the bonding map and the bulk map, respectively. A bonding map is defined between two bondable boundary patches. Two patches are bondable if they have the same number of nodes and links. Then, a bonding map between two bondable patches identifies each of their nodes and links, under the compatibility condition that if a bounding bivalent node in one patch is identified with a particular one in another, then their respective half-links (attaching them to their respective central nodes) are also identified with each other. Thus, a bonding map basically bonds two bulk vertices via (parts of) their boundary graphs to form a process (with boundary). This is simply a bulk edge, or propagator.

The set of interaction vertices can themselves be defined by a bulk map. This map augments the set of constituent elements (multivalent nodes, bivalent nodes, and half-links connecting the two) of *any* bisected boundary graph, by one new vertex (the bulk vertex), a set of links joining each of the original boundary nodes to this vertex, and a set of two-dimensional faces bounded by a triple of the bulk vertex, a multivalent boundary node and a bivalent boundary node. The resulting structure is an interaction vertex with the given boundary graph[14]. The complete dynamics is then given by the chosen combinatorics, supplemented with amplitude functions that reflect the dependence on the algebraic data.

The interaction vertices can in fact be described by vertex operators on the Fock space in terms of the ladder operators. An example vertex operator, corresponding to the 4-simplex boundary graph shown in Figure 1, is

$$\hat{V}_{4sim} = \int_{SU(2)^{20}} [dg] \, \hat{\varphi}^\dagger(\vec{g}_1) \hat{\varphi}^\dagger(\vec{g}_2) V_{4sim}(\vec{g}_1, ..., \vec{g}_5) \hat{\varphi}(\vec{g}_3) \hat{\varphi}(\vec{g}_4) \hat{\varphi}(\vec{g}_5) \qquad (25)$$

where the interaction kernel $V_{4sim} = V_{4sim}(\{g_{ij}g_{ji}^{-1}\}_{i<j})$ (for $i,j = 1, ..., 5$) encodes the combinatorics of the boundary graph. There are of course other vertex operators associated with the *same* graph (that is with the same kernel), but including different combinations of creation and annihilation operators[15].

Thus, a definition of kinematics entails: defining the state space, which includes specifying the combinatorics (choosing the set of allowed boundary patches, which generate the admissible boundary graphs), and the algebraic data (choosing variables to characterise the discrete geometric states supported on the boundary graphs); and defining the algebra of observables acting on the state space. A definition of dynamics entails: specifying the propagator and bulk vertex combinatorics and amplitudes. Together, they specify the many-body mechanics.

3.1.3. Generalised Equilibrium States

Outlined below is a generalised equilibrium statistical mechanics for these systems [14,15], along the lines laid out in Section 2. For a system of many classical tetrahedra (in general, polyhedra), a statistical state ρ_N can be formally defined on the state space Γ_N. If it satisfies the thermodynamical characterisation with respect to a set of functions on Γ_N, then it will be an equilibrium state. Further, a configuration with a varying number of tetrahedra can be described by a grand-canonical type state [15] of the form

$$Z = \sum_{N \geq 0} e^{\mu N} Z_N \qquad (26)$$

where $Z_N = \int_{\Gamma_N} d\lambda \, \rho_N$, and μ is a chemical potential. Similarly, for a system of many quantum tetrahedra, a generic statistical state $\hat{\rho}$ is a density operator on $\mathcal{H}_F$; and generalised equilibrium states with a varying number of quanta are

$$Z = \text{Tr}_{\mathcal{H}_F}(e^{-\sum_a \beta_a \hat{\mathcal{O}}_a + \mu \hat{N}}) \qquad (27)$$

---

[14] An interesting aspect is that the bulk map is one-to-one, so that for every distinct bisected boundary graph, there is a unique interaction vertex which can be defined from it.
[15] This would generically be true for any second quantised operator [29].

where $\hat{N} = \int d\vec{g}\, \hat{\varphi}^\dagger(\vec{g})\hat{\varphi}(\vec{g})$ is the number operator on $\mathcal{H}_F$. Operators of natural interest here are the ones encoding the dynamics, i.e. vertex (and kinetic) operators (see Section 3.2 below). Such grand-canonical type boundary states are important because one would expect quantum gravity dynamics to not be number conserving in general [15,29]. In addition, naturally, in both cases, what the precise content of equilibrium is depends crucially on which observables $\mathcal{O}_a$ are used to define the state. As pointed out in Section 2.2, and exemplified in the cases below in Section 3.2, there are many choices and types of observables one could consider in principle. Which ones are the relevant ones in a given situation is in fact a crucial part of the problem.

3.2. Applications

We briefly sketch below some examples of applying the above framework.

A couple of examples for a classical system are studied in [15]. In the process of applying the thermodynamical characterisation, these cases introduce a statistical, effective manner of imposing a given (set of) first class constraint(s), that is $\langle C \rangle = 0$, instead of the exact, strong way $C = 0$. In one case, the condition of closure of a classical d-polyhedron is relaxed in this statistical manner, while in the other the boundary gluing constraints amongst the polyhedral atoms of space are relaxed in this way to describe fluctuating twisted geometries. Brief summaries of these follow.

In the first example, starting from the extended state space $\Gamma_{ex} = \times_I S^2_{A_I}$ of intrinsic geometries of a d-polyhedron with face areas $\{A_I\}_{I=1,\ldots,d}$, closure is implemented via the following $\mathfrak{su}(2)^*$-valued function on Γ_{ex},

$$J = \sum_{I=1}^{d} x_I \tag{28}$$

which is the momentum map associated with the diagonal action of $SU(2)$. Satisfying closure exactly is to have $J = 0$. Then, applying the thermodynamic characterisation to the scalar component functions of J, that is requiring $\langle J_a \rangle = 0$ ($a = 1, 2, 3$), gives a Gibbs distribution on Γ_{ex} of the form $e^{-\sum_a \beta_a J_a}$ with a vector-valued temperature $(\beta_a) \in \mathfrak{su}(2)$. Thus, we have a thermal state for a classical polyhedron that is fluctuating in terms of its closure, with the fluctuations controlled by the parameter β. In fact, this state generalises Souriau's Gibbs states [26,27] to the case of Lie group (Hamiltonian) actions associated with first class constraints.

In the other example, the set of half-link gluing (or face-sharing) conditions for a boundary graph are statistically relaxed. It is known that an oriented (closed) boundary graph γ, with M nodes and L links, labelled with $(g, x) \in T^*(SU(2))$ variables admits a notion of discrete (closed) twisted geometry [48]. Twisted geometries are a generalisation of the more rigid Regge geometries, wherein the shapes of the shared faces are left arbitrary and only their areas are constrained to match. From the present constructive many-body viewpoint, one can understand these states instead as a result of satisfying a set of $SU(2)$- and $\mathfrak{su}(2)^*$-valued gluing conditions (denoted, respectively, by $\{C\}$ and $\{D\}$) on an initially disconnected system of several labelled open nodes. That is, starting from a system of M number of labelled open nodes, one ends up with a twisted geometric configuration if the set of gluing constraints on the holonomy and flux variables corresponding to a given γ, $\{C_{\ell,a}(g_{n\ell}g_{m\ell}^{-1}) = 0, D_{\ell,a}(x_{n\ell} - x_{m\ell}) = 0\}_\gamma$, are satisfied strongly (component-wise). Here, $\ell = 1, 2, \ldots, L$ labels a full link, $a = 1, 2, 3$ is $SU(2)$ component index, and subscripts $n\ell$ refer to the half-link (belonging to the full link) ℓ of node n. We can then choose instead to impose these constraints weakly by requiring only its statistical averages in a state to vanish. This gives a γ-dependent state on Γ_M, written formally as

$$\rho_{\{\gamma,\alpha,\beta\}} \propto e^{-\sum_\ell \sum_a \alpha_{\ell,a} C_{\ell,a} + \beta_{\ell,a} D_{\ell,a}} \equiv e^{-G_\gamma(\alpha,\beta)} \tag{29}$$

where $\alpha, \beta \in \mathbb{R}^{3L}$ are generalised inverse temperatures characterising this fluctuating twisted geometric configuration. In fact, one can generalise this state to a probabilistic superposition of such internally fluctuating twisted geometries for an N particle system (thus, defined on Γ_N), which includes contributions from distinct graphs, each composed of a possibly variable number of nodes M. A state of this kind can formally be written as,

$$\rho_N = \frac{1}{Z_N(M_{\max}, \lambda_\gamma, \alpha, \beta)} e^{-\sum_{M=2}^{M_{\max}} \sum_{\{\gamma\}_M} \frac{1}{\text{Aut}(\gamma)} \lambda_\gamma \sum_{i_1 \neq \ldots \neq i_M=1}^{N} G_\gamma(\vec{g}_{i_1}, \vec{x}_{i_1}, \ldots, \vec{g}_{i_M}, \vec{x}_{i_M}; \alpha, \beta)} \tag{30}$$

where i is the particle index, and $M_{\max} \leq N$. The value of $M_{\max}$ and the set $\{\gamma\}_M$ for a fixed M are model-building choices. The first sum over M includes contributions from all admissible (depending on the model, determined by $M_{\max}$) different M-particle subgroups of the full N particle system, with the gluing combinatorics of various different boundary graphs with M nodes. The second sum is a sum over all admissible boundary graphs γ, with a given fixed number of nodes M. Furthermore, the third sum takes into account all M-particle subgroup gluings (according to a given fixed γ) of the full N particle system. We note that the state in Equation (30) is a further generalisation of that presented in [15]; specifically, the latter is a special case of the former for the case of a single term $M = M_{\max} = N$ in the first sum. Further allowing for the system size to vary, that is considering a variable N, gives the most general configuration, with a set of coupling parameters linked directly to the underlying microscopic model,

$$Z(M_{\max}, \lambda_\gamma, \alpha, \beta) = \sum_{N \geq 0} e^{\mu N} Z_N(M_{\max}, \lambda_\gamma, \alpha, \beta). \tag{31}$$

A physically more interesting example is considered in [14], which defines a thermal state with respect to a spatial volume operator,

$$\hat{\rho} = \frac{1}{Z} e^{-\beta \hat{V}} \tag{32}$$

where $\hat{V} = \int d\vec{g}\, v(\vec{g}) \hat{\varphi}^\dagger(\vec{g}) \hat{\varphi}(\vec{g})$ is a positive, self-adjoint operator on $\mathcal{H}_F$, and the state is a well-defined density operator on the same. In fact, with a grand-canonical extension of it, this system can be shown to naturally support Bose–Einstein condensation to a low-spin phase [14]. Clearly, this state encodes thermal fluctuations in the volume observable, which is especially an important one in the context of cosmology. In fact, the rapidly developing field of condensate cosmology [54] for atoms of space of the kind considered here, is based on modelling the underlying system as a condensate, and subsequently extracting effective physics from it. These are certainly crucial steps in the direction of obtaining cosmological physics from quantum gravity [9]. It is equally crucial to enrich further the microscopic quantum gravity description itself, and extract effective physics for these new cases. One such important case is to consider thermal fluctuations of the gravitational field at early times, during which our universe is expected to be in a quantum gravity regime. That is, to consider *thermal* quantum gravity condensates using the frameworks laid out in this article (as opposed to the zero temperature condensates that have been used till now), and subsequently derive effective physics from them. This case would then directly reflect thermal fluctuations of gravity as being of a proper quantum gravity origin. This is investigated in [28].

We end this section by making a direct link to the definition of group field theories using the above framework. Group field theories (GFT) [37–39] are non-local field theories defined over (copies of) a Lie group. Most widely studied (Euclidean) models are for real or complex scalar fields, over copies of $SU(2)$, $Spin(4)$ or $SO(4)$. For instance, a complex scalar GFT over $SU(2)$ is defined by a partition function of the following general form,

$$Z_{\text{GFT}} = \int [D\mu(\varphi, \bar{\varphi})]\, e^{-S_{\text{GFT}}[\varphi, \bar{\varphi}]} \tag{33}$$

where μ is a functional measure which in general is ill-defined, and S_{GFT} is the GFT action of the form (for commonly encountered models),

$$S_{GFT} = \int_G dg_1 \int_G dg_2 \, K(g_1, g_2) \bar{\varphi}(g_1) \varphi(g_2) + \int_G dg_1 \int_G dg_2 \cdots V(g_1, g_2, \ldots) f(\varphi, \bar{\varphi}) \qquad (34)$$

where $g \in G$, and the kernel V is generically non-local, which convolutes the arguments of several φ and $\bar{\varphi}$ fields (written here in terms of a single function f). It defines the interaction vertex of the dynamics by enforcing the combinatorics of its corresponding (unique, via the inverse of the bulk map) boundary graph.

Z_{GFT} defines the covariant dynamics of the GFT model encoded in S_{GFT}. Below we outline a way to derive such covariant dynamics from a suitable quantum statistical equilibrium description of a system of quanta of space defined previously in Section 3.1. The following technique of using field coherent states is the same as in [15,29], but with the crucial difference that here we do not claim to define, or aim to achieve any correspondence (even if formal) between a canonical dynamics (in terms of a projector operator) and a covariant dynamics (in terms of a functional integral). Here we simply show a quantum statistical basis for the covariant dynamics of a GFT, and in the process, reinterpret the standard form of the GFT partition function in Equation (33) as that of an effective statistical field theory arising from a coarse-graining and further approximations of the underlying statistical quantum gravity system.

We saw in Section 3.1 that the dynamics of the polyhedral atoms of space is encoded in the choices of propagators and interaction vertices, which can be written in terms of kinetic and vertex operators in the Fock description. In our present considerations with a single type of atom ($SU(2)$-labelled 4-valent node), let us then consider the following generic kinetic and vertex operators,

$$\mathbb{K} = \int_{SU(2)^8} [dg] \, \hat{\varphi}^\dagger(\vec{g}_1) K(\vec{g}_1, \vec{g}_2) \hat{\varphi}(\vec{g}_2) \quad , \quad \mathbb{V} = \int_{SU(2)^{4N}} [dg] \, V_\gamma(\vec{g}_1, \ldots, \vec{g}_N) \hat{f}(\hat{\varphi}, \hat{\varphi}^\dagger) \qquad (35)$$

where $N > 2$ is the number of 4-valent nodes in the boundary graph γ, and $\hat{f}$ is a function of the ladder operators with all terms of a single degree N. For example, when $N = 3$, this function could be $\hat{f} = \lambda_1 \hat{\varphi} \hat{\varphi} \hat{\varphi}^\dagger + \lambda_2 \hat{\varphi}^\dagger \hat{\varphi} \hat{\varphi}^\dagger$. As shown above, in principle, a generic model can include several distinct vertex operators. Even though what we have considered here is the simple of case of having only one, the argument can be extended directly to the general case.

Operators $\mathbb{K}$ and $\mathbb{V}$ have well-defined actions on the Fock space $\mathcal{H}_F$. Using the thermodynamical characterisation then, we can consider the formal constraints[16] $\langle \mathbb{K} \rangle =$ constant and $\langle \mathbb{V} \rangle =$ constant, to write down a generalised Gibbs state on $\mathcal{H}_F$,

$$\hat{\rho}_{\{\beta_a\}} = \frac{1}{Z_{\{\beta_a\}}} e^{-\beta_1 \mathbb{K} - \beta_2 \mathbb{V}} \qquad (36)$$

where $a = 1, 2$ and the partition function[17] is,

$$Z_{\{\beta_a\}} = \text{Tr}_{\mathcal{H}_F}\left(e^{-\beta_1 \mathbb{K} - \beta_2 \mathbb{V}}\right) . \qquad (37)$$

An effective field theory can then be extracted from the above by using a basis of coherent states on $\mathcal{H}_F$ [15,29,55]. Field coherent states give a continuous representation on $\mathcal{H}_F$ where the parameter labelling each state is a wave (test) function [55]. For the Fock description mentioned in Section 3.1, the coherent states are

$$|\psi\rangle = e^{\hat{\varphi}^\dagger(\psi) - \hat{\varphi}(\psi)} |0\rangle \qquad (38)$$

---

[16] A proper interpretation of these constraints is left for future work.
[17] This partition function will in general be ill-defined as expected. One reason is the operator norm unboundedness of the ladder operators.

where $|0\rangle$ is the Fock vacuum (satisfying $\hat{\varphi}(\vec{g})|0\rangle = 0$ for all $\vec{g}$), $\hat{\varphi}(\psi) = \int_{SU(2)^4} \bar{\psi}\hat{\varphi}$ and its adjoint are smeared operators, and $\psi \in \mathcal{H}$. The set of all such states provides an over-complete basis for $\mathcal{H}_F$. The most useful property of these states is that they are eigenstates of the annihilation operator,

$$\hat{\varphi}(\vec{g})|\psi\rangle = \psi(\vec{g})|\psi\rangle. \tag{39}$$

The trace in the partition function in Equation (37) can then be evaluated in this basis,

$$Z_{\{\beta_a\}} = \int [D\mu(\psi,\bar{\psi})] \langle\psi| e^{-\beta_1 \hat{\mathbb{K}} - \beta_2 \hat{\mathbb{V}}} |\psi\rangle \tag{40}$$

where μ here is the coherent state measure [55]. The integrand can be treated and simplified along the lines presented in [15] (to which we refer for details), to get an effective partition function,

$$Z_0 = \int [D\mu(\psi,\bar{\psi})] \, e^{-\beta_1 K[\bar{\psi},\psi] - \beta_2 V[\bar{\psi},\psi]} = Z_{\{\beta_a\}} - Z_{\mathcal{O}(\hbar)} \tag{41}$$

where subscript 0 indicates that we have neglected higher order terms, collected inside $Z_{\mathcal{O}(\hbar)}$, resulting from normal orderings of the exponent in $Z_{\{\beta_a\}}$, and the functions in the exponent are $K = \langle\psi| : \hat{\mathbb{K}} : |\psi\rangle$ and $V = \langle\psi| : \hat{\mathbb{V}} : |\psi\rangle$. It is then evident that Z_0 has the precise form of a generic GFT partition function. It thus *defines* a group field theory as an effective statistical field theory, that is

$$Z_{\text{GFT}} := Z_0. \tag{42}$$

From this perspective, it is clear that the generalised inverse temperatures (which are basically the intensive parameters conjugate to the energies in the generalised thermodynamics setting of Section 2.4) *are* the coupling parameters defining the effective model, thus characterising the phases of the emergent statistical group field theory, as would be expected. Moreover, from this purely statistical standpoint, we can understand the GFT action more appropriately as Landau–Ginzburg free energy (or effective "Hamiltonian', in the sense that it encodes the effective dynamics), instead of a Euclidean action, which might imply having Wick rotated a Lorentzian measure, even in an absence of any such notions as is the case presently. Lastly, deriving in this way the covariant definition of a group field theory, based entirely on the framework presented in Section 3.1, strengthens the statement that a group field theory is a field theory of combinatorial and algebraic quanta of space [38,39].

4. Conclusions and Outlook

We have presented an extension of equilibrium statistical mechanics for background independent systems, based on a collection of results and insights from old and new studies. While various proposals for a background independent notion of statistical equilibrium have been summarised, one in particular, based on the constrained maximisation of information entropy, has been stressed upon. We have argued in favour of its potential by highlighting its many unique and valuable features. We have remarked on interesting new connections with the thermal time hypothesis, in particular suggesting to use this particular characterisation of equilibrium as a criterion of choice for the application of the hypothesis. Subsequently, aspects of a generalised framework for thermodynamics have been investigated, including defining the essential thermodynamic potentials, and discussing generalised zeroth and first laws.

We have then considered the statistical mechanics of a candidate quantum gravity system, composed of many atoms of space. The choice of (possibly different types of) these quanta is inspired directly from boundary structures in loop quantum gravity, spin foam and group field theory approaches. They are combinatorial building blocks (or boundary patches) of graphs, labelled with suitable algebraic data encoding discrete geometric information, with their constrained many-body dynamics dictated by bulk bondings between interaction vertices and amplitude functions.

Generic statistical states can then be defined on a many-body state space, and generalised Gibbs states can be defined using the thermodynamical characterisation [14]. Finally, we have given an overview of applications in quantum gravity [14–16,28]. In particular, we have derived the covariant definition of group field theories as a coarse-graining using coherent states of a class of generalised Gibbs states of the underlying system with respect to dynamics-encoding kinetic and vertex operators; and in this way reinterpreted the GFT partition function as an effective statistical field theory partition function, extracted from an underlying statistical quantum gravity system.

More investigations along these directions will certainly be worthwhile. For example, the thermodynamical characterisation could be applied in a spacetime setting, such as for stationary black holes with respect to the mass, charge and angular momentum observables, to explore further its physical implications. The black hole setting could also help unfold how the selection of a single preferred temperature can occur starting from a generalised Gibbs measure. Moreover, it could offer insights into relations with the thermal time hypothesis, and help better understand some of our more intuitive reasonings presented in Section 2.3, and similarly for generalised thermodynamics. It requires further development, particularly for the first and second laws. For instance, in the first law as presented above, the additional possible work contributions need to be identified and understood, particularly in the context of background independence. For these, and other thermodynamical aspects, we could benefit from Souriau's generalisation of Lie group thermodynamics [26,27].

There are many avenues to explore also in the context of statistical mechanics and thermodynamics of quantum gravity. In the former, for example, it would be interesting to study potential black hole quantum gravity states [56]. In general, it is important to be able to identify suitable observables to characterise an equilibrium state of physically relevant cases. On the cosmological side, for instance, those phases of the complete quantum gravity system which admit a cosmological interpretation will be expected to have certain symmetries whose associated generators could then be suitable candidates for the generalised energies. Another interesting cosmological aspect to consider is that of inhomogeneities induced by early time volume thermal fluctuations of quantum gravity origin, possibly from an application of the volume Gibbs state [14] (or a suitable modification of it) recalled above. The latter aspect of investigating thermodynamics of quantum gravity would certainly benefit from confrontation with studies on thermodynamics of spacetime in semiclassical settings. We may also need to consider explicitly the quantum nature of the degrees of freedom, and use insights from the field of quantum thermodynamics [57], which itself has fascinating links to quantum information [58].

Funding: This research received no external funding.

Acknowledgments: Many thanks are due to Daniele Oriti for valuable discussions and comments on the manuscript. Special thanks are due also to Goffredo Chirco and Mehdi Assanioussi for insightful discussions. The generous hospitality of the Visiting Graduate Fellowship program at Perimeter Institute, where part of this work was carried out, is gratefully acknowledged. Research at Perimeter Institute is supported in part by the Government of Canada through the Department of Innovation, Science and Economic Development Canada and by the Province of Ontario through the Ministry of Economic Development, Job Creation and Trade.

Conflicts of Interest: The author declares no conflict of interest.

References

1. Rovelli, C. Statistical mechanics of gravity and the thermodynamical origin of time. *Class. Quantum Gravity* **1993**, *10*, 1549–1566. [CrossRef]
2. Connes, A.; Rovelli, C. Von Neumann algebra automorphisms and time thermodynamics relation in general covariant quantum theories. *Class. Quantum Gravity* **1994**, *11*, 2899–2918. [CrossRef]
3. Rovelli, C. General relativistic statistical mechanics. *Phys. Rev.* **2013**, *87*, 084055. [CrossRef]
4. Bardeen, J.M.; Carter, B.; Hawking, S.W. The Four laws of black hole mechanics. *Commun. Math. Phys.* **1973**, *31*, 161–170. [CrossRef]
5. Bekenstein, J.D. Black holes and the second law. *Lett. Nuovo Cimento* **1972**, *4*, 737–740. [CrossRef]
6. Bekenstein, J.D. Black holes and entropy. *Phys. Rev.* **1973**, *7*, 2333–2346. [CrossRef]

7. Hawking, S.W. Particle Creation by Black Holes. *Commun. Math. Phys.* **1975**, *43*, 199–220. [CrossRef]
8. Bombelli, L.; Koul, R.K.; Lee, J.; Sorkin, R.D. A Quantum Source of Entropy for Black Holes. *Phys. Rev. D* **1986**, *34*, 373–383. [CrossRef]
9. Oriti, D. The universe as a quantum gravity condensate. *C. R. Phys.* **2017**, *18*, 235–245. [CrossRef]
10. Rovelli, C. The Statistical state of the universe. *Class. Quantum Gravity* **1993**, *10*, 1567. [CrossRef]
11. Montesinos, M.; Rovelli, C. Statistical mechanics of generally covariant quantum theories: A Boltzmann-like approach. *Class. Quantum Gravity* **2001**, *18*, 555–569. [CrossRef]
12. Chirco, G.; Haggard, H.M.; Rovelli, C. Coupling and thermal equilibrium in general-covariant systems. *Phys. Rev.* **2013**, *88*, 084027. [CrossRef]
13. Rovelli, C.; Smerlak, M. Thermal time and the Tolman-Ehrenfest effect: Temperature as the "speed of time". *Class. Quantum Gravity* **2011**, *28*, 075007. [CrossRef]
14. Kotecha, I.; Oriti, D. Statistical Equilibrium in Quantum Gravity: Gibbs states in Group Field Theory. *New J. Phys.* **2018**, *20*, 073009. [CrossRef]
15. Chirco, G.; Kotecha, I.; Oriti, D. Statistical equilibrium of tetrahedra from maximum entropy principle. *Phys. Rev.* **2019**, *99*, 086011. [CrossRef]
16. Chirco, G.; Kotecha, I. Generalized Gibbs Ensembles in Discrete Quantum Gravity. In *Geometric Science of Information 2019*; Nielsen, F., Barbaresco, F., Eds.; Springer: Cham, Switzerland, 2019.
17. Chirco, G.; Josset, T.; Rovelli, C. Statistical mechanics of reparametrization-invariant systems. It takes three to tango. *Class. Quantum Gravity* **2016**, *33*, 045005. [CrossRef]
18. Haggard, H.M.; Rovelli, C. Death and resurrection of the zeroth principle of thermodynamics. *Phys. Rev.* **2013**, *87*, 084001.
19. Jaynes, E.T. Information Theory and Statistical Mechanics. *Phys. Rev.* **1957**, *106*, 620–630. [CrossRef]
20. Jaynes, E.T. Information Theory and Statistical Mechanics. II. *Phys. Rev.* **1957**, *108*, 171–190. [CrossRef]
21. Bratteli, O.; Robinson, D.W. *Operator Algebras and Quantum Statistical Mechanics—I, II*; Springer: Berlin/Heidelberg, Germany, 1987.
22. Landau, L.D.; Lifshitz, E.M. *Statistical Physics, Part 1*; Volume 5 of Course of Theoretical Physics; Butterworth-Heinemann: Oxford, UK, 1980.
23. Chirco, G.; Josset, T. Statistical mechanics of covariant systems with multi-fingered time. *arXiv* **2016**, arXiv:1606.04444.
24. Martinetti, P.; Rovelli, C. Diamonds's temperature: Unruh effect for bounded trajectories and thermal time hypothesis. *Class. Quantum Gravity* **2003**, *20*, 4919–4932. [CrossRef]
25. Haggard, H.M. Gibbsing spacetime: A group field theory approach to equilibrium in quantum gravity. *New J. Phys.* **2018**, *20*, 071001. [CrossRef]
26. Souriau, J.-M. *Structure des Systemes Dynamiques*; Dunod: Paris, France, 1969.
27. Marle, C.-M. From tools in symplectic and poisson geometry to J.-M. Souriau's theories of statistical mechanics and thermodynamics. *Entropy* **2016**, *18*, 370. [CrossRef]
28. Assanioussi, M.; Kotecha, I. Thermal quantum gravity condensates and group field theory cosmology. In progress.
29. Oriti, D. Group field theory as the 2nd quantization of Loop Quantum Gravity. *Class. Quantum Gravity* **2016**, *33*, 085005. [CrossRef]
30. Rovelli, C. "Forget time". *Found. Phys.* **2011**, *41*, 1475–1490. [CrossRef]
31. Rovelli, C. Is Time's Arrow Perspectival? In *Simon Saunders*; Silk, J., Barrow, J.D., Chamcham, K., Eds.; The Philosophy of Cosmology: Cambridge, UK, 2017; pp. 285–296.
32. Jaynes, E.T. The Gibbs paradox. In *Maximum Entropy and Bayesian Methods*; Smith, C.R., Erickson, G.J., Neudorfer, P.O., Eds.; Springer: Dordrecht, The Netherlands, 1992; pp. 1–21.
33. Araki, H. Relative Entropy of States of Von Neumann Algebras. *Publ. Res. Inst. Math. Sci. Kyoto* **1976**, *1976*, 809–833.
34. Oriti, D. Spacetime as a quantum many-body system. *arXiv* **2017**, arXiv:1710.02807.
35. Oriti, D.; Ryan, J.P.; Thurigen, J. Group field theories for all loop quantum gravity. *New J. Phys.* **2015**, *17*, 023042. [CrossRef]
36. Michael, P. Reisenberger and Carlo Rovelli. Space-time as a Feynman diagram: The Connection formulation. *Class. Quantum Gravity* **2001**, *18*, 121–140.
37. Freidel, L. Group field theory: An Overview. *Int. J. Theory Phys.* **2005**, *44*, 1769–1783. [CrossRef]

38. Oriti, D. The group field theory approach to quantum gravity. In *Approaches to Quantum Gravity: Toward a New Understanding of Space, Time and Matter*; Oriti, D., Ed.; Cambridge University Press: Cambridge, UK, 2009.
39. Oriti, D. The microscopic dynamics of quantum space as a group field theory. *arXiv* **2011**, arXiv:1110.5606.
40. Bianchi, E.; Dona, P.; Speziale, S. Polyhedra in loop quantum gravity. *Phys. Rev. D* **2011**, *83*, 044035. [CrossRef]
41. Bodendorfer, N. An elementary introduction to loop quantum gravity. *arXiv* **2016**, arXiv:1607.05129.
42. Perez, A. The Spin Foam Approach to Quantum Gravity. *Living Rev. Relativ.* **2013**, *16*, 3. [CrossRef]
43. Hamber, H.W. Quantum Gravity on the Lattice. *Gen. Relativ. Gravit.* **2009**, *41*, 817–876. [CrossRef]
44. Finocchiaro, M.; Oriti, D. Spin foam models and the Duflo map. *arXiv* **2018**, arXiv:1812.03550.
45. Regge, T. General relativity without coordinates. *Nuovo Cimento* **1961**, *19*, 558–571. [CrossRef]
46. Regge, T.; Williams, R.M. Discrete structures in gravity. *J. Math. Phys.* **2000**, *41*, 3964–3984. [CrossRef]
47. Dittrich, B.; Speziale, S. Area-angle variables for general relativity. *New J. Phys.* **2008**. [CrossRef]
48. Freidel, L.; Speziale, S. Twisted geometries: A geometric parametrisation of SU(2) phase space. *Phys. Rev. D* **2010**, *82*, 084040. [CrossRef]
49. Rovelli, C.; Speziale, S. On the geometry of loop quantum gravity on a graph. *Phys. Rev.* **2010**, *82*, 044018. [CrossRef]
50. Dittrich, B.; Ryan, J.P. Phase space descriptions for simplicial 4d geometries. *Class. Quantum Gravity* **2011**, *28*, 065006. [CrossRef]
51. Barbieri, A. Quantum tetrahedra and simplicial spin networks. *Nucl. Phys.* **1998**, *518*, 714–728. [CrossRef]
52. Baez, J.C.L.; Barrett, J.W. The Quantum tetrahedron in three-dimensions and four-dimensions. *Adv. Theor. Math. Phys.* **1999**, *3*, 815–850. [CrossRef]
53. Kegeles, A.; Oriti, D.; Tomlin, C. Inequivalent coherent state representations in group field theory. *Class. Quantum Gravity* **2018**, *35*, 125011. [CrossRef]
54. Pithis, A.G.; Sakellariadou, M. Group field theory condensate cosmology: An appetizer. *Universe* **2019**, *5*, 147. [CrossRef]
55. Klauder, J.; Skagerstam, B. *Coherent States*; World Scientific: Singapore, 1985.
56. Oriti, D.; Pranzetti, D.; Sindoni, L. Black Holes as Quantum Gravity Condensates. *Phys. Rev.* **2018**, *97*, 066017. [CrossRef]
57. Vinjanampathy, S.; Anders, J. Quantum thermodynamics. *Contemp. Phys.* **2016**, *57*, 545–579. [CrossRef]
58. Goold, J.; Huber, M.; Riera, A.; del Rio, L.; Skrzypczyk, P. The role of quantum information in thermodynamics—A topical review. *J. Phys. A Math. Theor.* **2016**, *49*, 143001. [CrossRef]

© 2019 by the author. Licensee MDPI, Basel, Switzerland. This article is an open access article distributed under the terms and conditions of the Creative Commons Attribution (CC BY) license (http://creativecommons.org/licenses/by/4.0/).

Article

Spin Foam Vertex Amplitudes on Quantum Computer—Preliminary Results

Jakub Mielczarek [1,2]

[1] CPT, Aix-Marseille Université, Université de Toulon, CNRS, F-13288 Marseille, France; jakub.mielczarek@uj.edu.pl
[2] Institute of Physics, Jagiellonian University, Łojasiewicza 11, 30-348 Cracow, Poland

Received: 16 April 2019; Accepted: 24 July 2019; Published: 26 July 2019

Abstract: Vertex amplitudes are elementary contributions to the transition amplitudes in the spin foam models of quantum gravity. The purpose of this article is to make the first step towards computing vertex amplitudes with the use of quantum algorithms. In our studies we are focused on a vertex amplitude of 3+1 D gravity, associated with a pentagram spin network. Furthermore, all spin labels of the spin network are assumed to be equal $j = 1/2$, which is crucial for the introduction of the intertwiner qubits. A procedure of determining modulus squares of vertex amplitudes on universal quantum computers is proposed. Utility of the approach is tested with the use of: IBM's *ibmqx4* 5-qubit quantum computer, simulator of quantum computer provided by the same company and QX quantum computer simulator. Finally, values of the vertex probability are determined employing both the QX and the IBM simulators with 20-qubit quantum register and compared with analytical predictions.

Keywords: Spin networks; vertex amplitudes; quantum computing

1. Introduction

The basic objective of theories of quantum gravity is to calculate transition amplitudes between configurations of the gravitational field. The most straightforward approach to the problem is provided by the Feynman's path integral

$$\langle \Psi_f | \Psi_i \rangle = \int D[g] D[\phi] e^{\frac{i}{\hbar}(S_G + S_\phi)}, \tag{1}$$

where S_G and S_ϕ are the gravitational and matter actions respectively. While the formula (1) is easy to write, it is not very practical for the case of continuous gravitational field, characterized by infinite number of degrees of freedom. One of the approaches to determine (1) utilizes discretization of the gravitational field associated with some cut-off scale. The expectation is that continuous limit of such discretized theory can be recovered at the second order phase transition [1,2]. The essential step in this challenge is to generate different discrete space-time configurations (triangulations) contributing to the path integral (1). In Causal Dynamical Triangulations (CDT) [1] which is one of the approaches to the problem, Markov chain of elementary moves is used to explore different triangulations between initial and final state. In practice, the Markov chain is implemented after performing Wick rotation in Equation (1). In the last 20 years, the procedure has been extensively studied running computer simulations [3]. However, in 1+1 D case analytical methods of generating allowed triangulations are also available. In particular, it has been shown that Feynman diagrams of auxiliary random matrix theories generate graphs dual to the triangulations [4]. An advantage of the method is that in the large N (color) limit of such theories, symmetry factors associated with given triangulations can be recovered [5].

Another path to the problem of determining (1) is provided by the Loop Quantum Gravity (LQG) [6,7] approach to the Planck scale physics. Here, discreteness of space is not due to the applied by hand cut-off but is a consequence of the procedure of quantization. Accordingly, the spatial configuration of the gravitational is encoded in the so-called *spin network* states [8]. In consequence, the transition amplitude (1) is calculated between two spin network states. The geometric structures (2-complexes) representing the path integral are called *Spin Foams* [9,10]. The elementary processes contributing to the spin foam amplitudes are associated with vertices of the spin foams and are called *vertex amplitudes* [11,12]. The employed terminology of spin networks and spin foams is clarified in Figure 1 in an example of 2+1 D gravity.

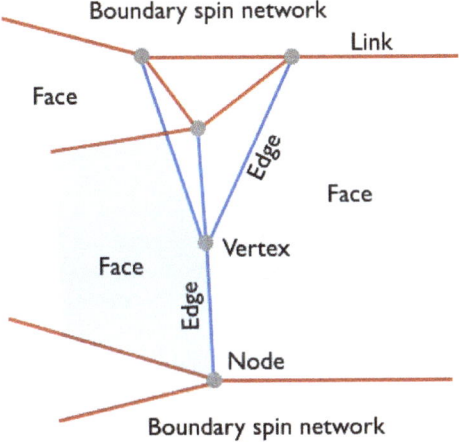

Figure 1. Pictorial representation of a simple spin foam associated with 2+1 D gravity. The plot has been inspired by Figure 4.2 in Ref. [7].

In Figure 1 two boundary spin networks with 3-valent nodes are shown. Such nodes are dual to two-dimensional triangles. In this article we will focus on the 3+1 D case in which the spin network nodes (related with non-vanishing volumes) are 4-valent. The nodes are dual to tetrahedra (3-simplex). In the example presented in Figure 1, the four edges of the 2-complex meet at the vertex. However, in the 3+1 D case the valence of the vertex is higher and equal to 5. For the purpose of this article it is crucial to note that such 5-valent vertices can be enclosed by a boundary represented by a spin network containing five nodes. Each of the nodes is placed on one of the five edges entering the vertex. The boundary has topology of a three-sphere, S^3. This is higher dimensional extension of the 2+1 D case, where a vertex can be enclosed by the two-sphere, S^2.

In analogy to the random matrix theories in case of the 2D triangulations, the spin foams (2-complexes) can also be obtained as Feynman diagrams of some auxiliary field theory. Namely, the so-called Group Field Theories (GFTs) have been introduced to generate structure of vertices and edges associated with spin foams [13–15]. In particular, the 3+1 D theory with 5-valent vertices requires GFT with five-order interaction terms, known as Ooguri's model [16]. There has recently been great progress in the field of GFTs with many interesting results (see e.g., [17,18]).

The aim of this article is to investigate a possibility of employing universal quantum computers to compute vertex amplitudes of 3+1 D spin foams. The idea has been suggested in Ref. [19], however, it is not investigated there. Here, we make the first attempt to materialize this concept. In our studies, we consider a special case of spin networks with spin labels corresponding to fundamental representations of the $SU(2)$ group, for which *intertwiner qubits* [19–21] can be introduced. The qubits will be implemented on IBM Q 5-qubit quantum computer (*ibmqx4*) as well as with the use of quantum computer simulator provided by the same company [22]. In the case of real 5-qubit quantum computer the qubits are physically realized as superconducting circuits [23] operating at millikelvin temperatures. Furthermore, the QX quantum computer simulator [24], available on the Quantum Inspire [25] platform, will be employed.

The studies contribute to our broader research program focused on exploring the possibility of simulating Planck scale physics with the use of quantum computers. The research is in the spirit of the original Feynman's idea [26] of performing the so-called *exact simulations* of quantum systems with the use of quantum information processing devices. In our previous articles [21,27] we have preliminary explored the possibility of utilizing Adiabatic Quantum Computers [28] to simulate quantum gravitational systems. Here, we are making the first steps towards the application of Universal Quantum Computers [29,30].

2. Intertwiner Qubit

The basic question a skeptic can ask is why it is worth considering quantum computers to study Planck scale physics at all? Can we not just do it employing classical supercomputers as in the case of CDT approach to quantum gravity? Let me answer these questions by giving two arguments. The first concerns the huge dimensionality of a Hilbert space for many-body quantum system. For a single spin-1/2 (qubit) Hilbert space $\mathcal{H}_{1/2} = \text{span}\{|0\rangle, |1\rangle\}$ the dimension is equal to 2. However, considering N such spins (qubits) the resulting Hilbert space is a tensor product of N copies of the qubit Hilbert space. The dimension of such space grows exponentially with N:

$$\dim(\underbrace{\mathcal{H}_{1/2} \otimes \mathcal{H}_{1/2} \otimes \cdots \otimes \mathcal{H}_{1/2}}_{N}) = 2^N. \tag{2}$$

This exponential behavior is the main obstacle behind simulating quantum systems on classical computers. With the present most powerful classical supercomputers we can simulate quantum systems with $N = 64$ at most [31]. The difficulty is due to the fact that quantum operators acting on 2^N dimensional Hilbert space are represented by $2^N \times 2^N$ matrices. Operating with such matrices for $N > 50$ is challenging to the currently available supercomputers. On the other hand, such companies as IBM or Rigetti Computing are developing quantum chips with $N > 100$ and certain topologies of couplings between the qubits. The possibility of simulating quantum systems which are unattainable to classical supercomputers may, therefore, emerge in the coming decade leading to the so-called *quantum supremacy* [32]. See Appendix A for more detailed discussion of the state of the art of the quantum computing technologies and prospects for the near future. The second argument concerns *quantum speed-up* leading to reduction of computational complexity of some classical problems. Such possibility is provided by certain quantum algorithms (e.g., Deutsch, Grover, Shor, ...) thanks to the so-called *quantum parallelism*. For more information on quantum algorithms please see Appendix B, where elementary introduction to quantum computing can be found.

Taking the above arguments into account we are convinced that it is justified to explore the possibility of simulating quantum gravitational physics on quantum computers. The fundamental question is, however, whether gravitational degrees of freedom can be expressed with qubits, which are used in the current implementations of quantum computers[1]. Fortunately, it has recently been shown that at least in Loop Quantum Gravity approach to quantum gravity notion of qubit degrees of freedom can be introduced and is associated with the intertwiner space of a certain class of spin networks (see Refs. [19–21,27]).

Let us briefly explain it. Namely, nodes of the spin networks are where Hilbert spaces associated with the links meet. The gauge invariance (enforced by the Gauss constraint) implies that the total spin at the node has to be equal zero. The 4-valent nodes are of special interest since they are associated

---

[1] In general, quantum variables associated with higher dimensional Hilbert spaces may be considered.

with the non-vanishing eigenvalues of the volume operator (see e.g., Ref. [7]). As already mentioned in the introduction, in the picture of discrete geometry, the 4-valent nodes are dual to tetrahedra. The class of spin networks that we are focused on here are those with links of the spin networks labelled by fundamental representations of the $SU(2)$ group (i.e., the spin labels are equal $j = 1/2$) and the nodes are 4-valent. For such spin networks the Hilbert spaces at the nodes are given by the following tensor products:

$$\mathcal{H}_{1/2} \otimes \mathcal{H}_{1/2} \otimes \mathcal{H}_{1/2} \otimes \mathcal{H}_{1/2} = 2\mathcal{H}_0 \oplus 3\mathcal{H}_1 \oplus \mathcal{H}_2. \tag{3}$$

The Gauss constraint implies that only singlet configurations ($\mathcal{H}_0$) are allowed. Because there are two copies of the spin-zero configurations in the tensor product (3), the so-called intertwiner Hilbert space is two-dimensional:

$$\dim \mathrm{Inv}(\mathcal{H}_{1/2} \otimes \mathcal{H}_{1/2} \otimes \mathcal{H}_{1/2} \otimes \mathcal{H}_{1/2}) = 2. \tag{4}$$

We associate the two-dimensional invariant subspace with the *intertwiner qubit* $|\mathcal{I}\rangle \in \mathcal{H}_0 \oplus \mathcal{H}_0$. The 4-valent node (at which the intertwiner qubit is defined) together with the entering links is dual to the tetrahedron in a way shown in Figure 2.

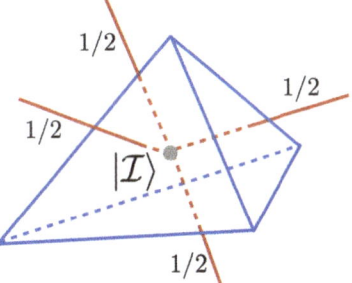

Figure 2. A single 4-valent node together with the entering links with spin labels $j = 1/2$. The intertwiner qubit $|\mathcal{I}\rangle$ is a degree of freedom defined at the node. The node is dual to the tetrahedron (3-symplex) as represented in the picture.

The two basis states of the intertwiner qubit $|\mathcal{I}\rangle$ are basically the two singlets we can obtain for a system of four spins $1/2$. The basis states can be expressed composing familiar singlets and triplet states for two spin-1/2 particles:

$$|S\rangle = \frac{1}{\sqrt{2}}(|01\rangle - |10\rangle), \tag{5}$$

$$|T_+\rangle = |00\rangle, \tag{6}$$

$$|T_0\rangle = \frac{1}{\sqrt{2}}(|01\rangle + |10\rangle), \tag{7}$$

$$|T_-\rangle = |11\rangle. \tag{8}$$

Namely, in the s-channel (which is one of the possible superpositions) the intertwiner qubit basis states can be expressed as follows:

$$|0_s\rangle = |S\rangle \otimes |S\rangle, \tag{9}$$

$$|1_s\rangle = \frac{1}{\sqrt{3}}(|T_+\rangle \otimes |T_-\rangle + |T_-\rangle \otimes |T_+\rangle - |T_0\rangle \otimes |T_0\rangle). \tag{10}$$

The $|0_s\rangle$ state is simply a tensor product of two singlets for two spin-1/2 particles, while the state $|1_s\rangle$ does not have such a simple product structure. The states $|0_s\rangle$ and $|1_s\rangle$ form an orthonormal basis of the intertwiner qubit. Worth stressing is that other bases being linear superpositions of $|0_s\rangle$ and $|1_s\rangle$ might be considered. In particular, the eigenbasis of the volume operator turns out to be useful (see Ref. [27]). Here we stick to the s-channel basis $\{|0_s\rangle, |1_s\rangle\}$ in which a general intertwiner state (neglecting the total phase) can be expressed as

$$|\mathcal{I}\rangle = \cos(\theta/2)|0_s\rangle + e^{i\phi}\sin(\theta/2)|1_s\rangle, \tag{11}$$

where $\theta \in [0, \pi]$ and $\phi \in [0, 2\pi)$ are angles parametrizing the Bloch sphere.

In the context of quantum computations it is crucial to define a quantum algorithm (a unitary operation $\hat{U}_\mathcal{I}$ acting on the input state) which will allow us to create the intertwiner state (11) from the input state $|0000\rangle$, i.e.,

$$|\mathcal{I}\rangle = \hat{U}_\mathcal{I}|0000\rangle. \tag{12}$$

The general construction of the operator $\hat{U}_\mathcal{I}$ can be performed applying the procedure introduced in Ref. [33], and will be discussed in a sequel to this article [34]. Here, for the purpose of illustration of the method of computing vertex amplitude we will focus on the special case of the intertwiner states being the first basis state: $|\mathcal{I}\rangle = |0_s\rangle = |S\rangle \otimes |S\rangle$. The contributing two-particle singlet states can easily be generated as a sequence of elementary gates used to construct quantum circuits (see also Appendix B):

$$|S\rangle = \widehat{\text{CNOT}}(\hat{H} \otimes \hat{I})(\hat{X} \otimes \hat{X})|00\rangle. \tag{13}$$

Here, the $\hat{X}$ is the so-called *bit-flip* (NOT) operator (Pauli σ_x matrix) which transforms $|0\rangle$ into $|1\rangle$ and $|1\rangle$ into $|0\rangle$ (i.e., $\hat{X}|0\rangle = |1\rangle$ and $\hat{X}|1\rangle = |0\rangle$). The $\hat{H}$ is the Hadamard operator defined as $\hat{H}|0\rangle = \frac{1}{\sqrt{2}}(|0\rangle + |1\rangle)$ and $\hat{H}|1\rangle = \frac{1}{\sqrt{2}}(|0\rangle - |1\rangle)$. Finally, the $\widehat{\text{CNOT}}$ is the Controlled NOT 2-qubit gate defined as $\widehat{\text{CNOT}}(|a\rangle \otimes |b\rangle) = |a\rangle \otimes |a \oplus b\rangle$, where $a, b \in \{0, 1\}$ and $\oplus$ is the XOR (exclusive or) logical operation, such that $0 \oplus 0 = 0$, $0 \oplus 1 = 1$, $1 \oplus 0 = 1$ and $1 \oplus 1 = 0$. In consequence, the $|0_s\rangle$ basis state can be expressed as follows:

$$|0_s\rangle = (\widehat{\text{CNOT}} \otimes \widehat{\text{CNOT}})(\hat{H} \otimes \hat{I} \otimes \hat{H} \otimes \hat{I})(\hat{X} \otimes \hat{X} \otimes \hat{X} \otimes \hat{X})|0000\rangle$$
$$= \frac{1}{2}(|0101\rangle + |1010\rangle - |0110\rangle - |1001\rangle). \tag{14}$$

In Figure 3 a quantum circuit generating (and measuring) the intertwiner state $|0_s\rangle$ has been presented.

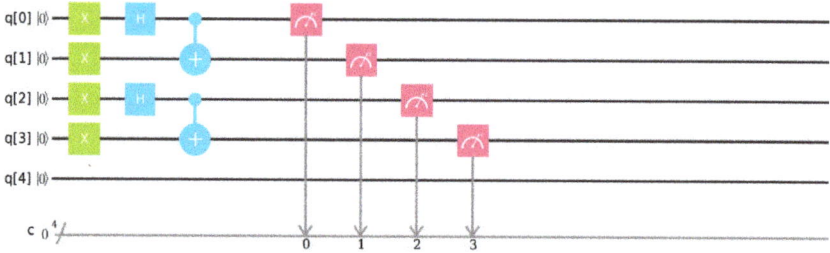

Figure 3. Quantum circuit used to generate $|0_s\rangle$ state from the initial state $|0000\rangle$. The (green) boxes with letter X represent the bit-flip gates, the (blue) boxes with letter H represents the Hadamard gates, while the next operations from the left are the CNOT 2-qubit gates. Finally, the (pink) boxes on the right represent measurements performed at every one of the four involved qubits.

The final state can be written as a superposition of 16 basis states in the product space of four qubit Hilbert spaces:

$$|\Psi\rangle = \sum_{ijkl\in\{0,1\}} a_{ijkl}|ijkl\rangle, \qquad (15)$$

where the normalization condition implies that $\sum_{ijkl\in\{0,1\}} |a_{ijkl}|^2 = 1$.

We have executed the quantum algorithm (14) with the use of both the IBM simulator of quantum computer and the real IBM Q 5-qubit quantum chip ibmqx4. In both cases the algorithm has been executed 1024 times. Moreover, the algorithm (14) has also been executed (1024 times) on the QX quantum computer simulator. Results of the measurements of probabilities $P(i) = |a_i|^2$ are summarized in Table 1.

Table 1. Results of measurements of $P(i) = |a_i|^2$ for the quantum circuit presented in Figure 3.

No.	Probability	Theory	IBM Simulator	IBM Q ibmqx4	QX Simulator
1	$|a_{0000}|^2$	0	0	0.014	0
2	$|a_{0001}|^2$	0	0	0.058	0
3	$|a_{0010}|^2$	0	0	0.050	0
4	$|a_{0011}|^2$	0	0	0.004	0
5	$|a_{0100}|^2$	0	0	0.023	0
6	$|a_{0101}|^2$	0.25	0.264	0.109	0.252
7	$|a_{0110}|^2$	0.25	0.232	0.091	0.241
8	$|a_{0111}|^2$	0	0	0.009	0
9	$|a_{1000}|^2$	0	0	0.034	0
10	$|a_{1001}|^2$	0.25	0.248	0.159	0.230
11	$|a_{1010}|^2$	0.25	0.256	0.158	0.276
12	$|a_{1011}|^2$	0	0	0.012	0
13	$|a_{1100}|^2$	0	0	0.034	0
14	$|a_{1101}|^2$	0	0	0.132	0
15	$|a_{1110}|^2$	0	0	0.110	0
16	$|a_{1111}|^2$	0	0	0.003	0

Clearly, the results obtained from the simulator match well with the values predicted in Equation (14). By increasing the number of shots, the accuracy of the results can be improved. In fact, because the number of shots in a single round was limited (either to 8192 in the case of the IBM simulator or to 1024 in the case of the QX simulator) the computational rounds had to be repeated in order to achieve better convergence to theoretical predictions. Furthermore, the results were also verified with the use of two other publicly available quantum simulators of quantum circuits, i.e., Quirk [35] and Q-Kit [36].

On the other hand, the errors of the ibmqx4 quantum processor are more significant, leading even to 10% contribution from the undesired states, such as $|1101\rangle$ and $|1110\rangle$. The errors have two main sources. The first are instrumental errors associated with uncertainty of gates, uncertainty of initial state preparation and uncertainty of readouts. For the IBM Q ibmqx4 quantum processor the single-qubit gate errors are at the level of 0.001 and the errors of readouts are reaching even 0.086 for some of the qubits. The two-qubit gates are less accurate than the single qubit gates, with the errors approximately equal to 0.035. The concrete values for every qubit and pairs of qubits are provided via the IBM website [22]. The second source of error is due to statistical nature of quantum mechanics and the limited number of measurements. For a single qubit, the problem of estimating corresponding error is equivalent to the 1D random walk, which leads to uncertainty of the estimation of probability equal $\frac{s}{n} = \sqrt{\frac{p(1-p)}{n}}$, where n is the number of measurements and p is a probability of one of the two basis states. As an example, for $n = 1024$ and $p = 1/2$ we obtain $\frac{s}{n} \approx 0.016$. In the considered case of 16 basis states the uncertainty is expected to be lower roughly by the factor 0.5^2, leading to an approximate error equal 0.008 (the value is smaller because the average number of counts per basis states decreased). Summing up both the instrumental error and the uncertainty of measurement, we may estimate the cumulative uncertainty to be at the level of ~15%, which is in agreement with the experimental data. With the current setup, the errors can be slightly reduced by increasing the number of measurements and by optimization of the quantum circuit, e.g., by placing (less noisy) single-qubit gates after (more noisy) two-qubit gates. In further studies, the circuits should also be equipped with quantum error correction algorithms. This will, however, require additional qubits to be involved. Moreover, reduction of the instrumental error will be a crucial challenge for the future utility of the quantum processors.

Let us end this section with quantitative comparison of the results from the Table 1 with the use of classical Fidelity (Bhattacharyya distance) $F(q,p) := \sum_i \sqrt{p_i q_i}$, where $\{p_i\}$ and $\{q_i\}$ are two sets of probabilities. Comparison of the theoretical values with the results obtained from the IBM Simulator gives us $F \approx 99.9\%$. However, comparing the theoretical values with the results of IBM Q ibmqx4 quantum computer we find much smaller Fidelity $F \approx 71.4\%$. Worth keeping in mind is that the employed Fidelity function concerns classical probabilities and further analysis of the quantum state obtained from the quantum computer should include also analysis of the quantum Fidelity $F(\hat{\rho}_1, \hat{\rho}_2) := \mathrm{tr}\sqrt{\sqrt{\hat{\rho}_1}\hat{\rho}_2\sqrt{\hat{\rho}_1}}$, where $\hat{\rho}_1$ and $\hat{\rho}_2$ are density matrices of the compared states [37]. For this purpose (i.e., reconstruction of the density matrix), full tomography of the obtained quantum state has to be performed.

3. Vertex Amplitude

Gravity is a theory of constraints. Specifically, in LQG three types of constraints are involved. The first is the mentioned Gauss constraint, which has already been imposed at the stage of constructing spin networks states. The second is the spatial diffeomorphism constraint which is satisfied by introducing equivalence relation between all spin-networks characterized by the same topology. The third is the so-called scalar or Hamiltonian constraint, which encodes temporal dynamics and is the most difficult to satisfy. In quantum theory, this constraint takes a form of an operator. Let us denote this operator as $\hat{C}$. Following the Dirac procedure for constrained quantum systems, the physical states are those belonging to the kernel of the constraints, i.e., $\hat{C}|\Psi\rangle = 0$. Due to the complicated form of the gravitational scalar constraint (see e.g., [6]), finding the physical states is in general a difficult task. However, for certain simplified scalar constraints, such as for the symmetry reduced cosmological models, the physical states are possible to extract. Furthermore, it has recently been proposed in Ref. [21] that the problem of solving simple constraints can be implemented on Adiabatic Quantum Computers.

Another approach to the problem of constraints is to consider a projection operator

$$\hat{P} := \lim_{T \to \infty} \frac{1}{2T} \int_{-T}^{T} d\tau e^{i\tau \hat{C}}, \tag{16}$$

which projects kinematical states onto physical subspace. In particular, the Formula (16) is valid for $\hat{C}$ characterized by discrete spectrum of eigenvalues. Specifically, the projection operator (16) can be used to evaluate transition amplitude between any two kinematical states $|x\rangle$ and $|x'\rangle$:

$$W(x, x') = \langle x'|\hat{P}|x\rangle. \tag{17}$$

---

[2] In order to prove it let us consider an asymmetric 1D random walk with probabilities $p = \frac{1}{16}$ (one of the basis states) and $q = 1 - p = \frac{15}{16}$ (rest of the 15 basis sates), for which $\frac{s}{n} = \sqrt{\frac{1}{16}\frac{15}{16}\frac{1}{n}} \approx \frac{1}{4\sqrt{n}}$.

The state $|x\rangle$ might correspond to the initial and $|x'\rangle$ to the final boundary spin network states (confront with Figure 1). While the notion of the boundary initial and final hypersurfaces is well defined in the case with preferred time foliation, the general relativistic case deserves generalization of the transition amplitude to the form being independent of the background time variable. This leads to the concept of *boundary formulation* [38] of transition amplitudes in which the transition amplitude is a function of boundary state only. Taking the particular boundary physical spin network state $|\Psi\rangle$ the transition amplitude can be, therefore, written as

$$W(\Psi) = \langle W|\Psi\rangle, \qquad (18)$$

where the state $|\Psi\rangle$ corresponds to representation in which the amplitude is evaluated. Worth stressing at this point is that physical interpretation of the transition amplitude (18) is still under debate. In particular, very little is known about relation of the amplitude with the physical states of the theory under consideration.

The object of our interest in this article, namely the vertex amplitude, is the amplitude (18) of boundary enclosing a single vertex. As we have already explained in the introduction, the spin network enclosing the single vertex has pentagram structure and can be written as:

$$\langle W|\Psi\rangle = A(\mathcal{I}_1, \mathcal{I}_2, \mathcal{I}_3, \mathcal{I}_4, \mathcal{I}_5). \qquad (19)$$

The associated spin network is shown in Figure 4.

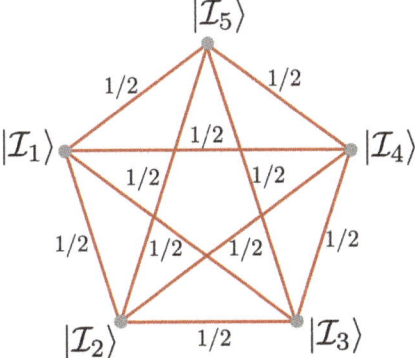

Figure 4. Pentagram spin network corresponding to boundary of a single vertex of spin foam. The boundary geometry has topology of three-sphere.

The pentagram spin network state is a tensor product of the five intertwiner qubits:

$$|\Psi\rangle = \bigotimes_{n=1}^{5} |\mathcal{I}_n\rangle. \qquad (20)$$

Since in the vertex amplitude (18) physical states have to be considered the intertwiner qubits $|\mathcal{I}_n\rangle$ have to be selected such that the state is annihilated by the scalar constraint: $\hat{C}|\Psi\rangle = 0$. Due to the difficulty of the issue for general form of the scalar constraint operator, we do not address the problem of selecting $|\Psi\rangle$ states here. As we have mentioned, for either symmetry reduced or simplified scalar constraints the physical states can be identified with the use of existing methods.

Another issue is the choice of the state $|W\rangle$. Usually the representation of holonomies associated with the links of the spin networks are considered. Here, following Ref. [19] we will evaluate the boundary spin network state in the state:

$$|W\rangle = \bigotimes_{l=1}^{10} |\mathcal{E}_l\rangle, \tag{21}$$

where

$$|\mathcal{E}_l\rangle = \frac{1}{\sqrt{2}} (|01\rangle - |10\rangle) \tag{22}$$

are Bell states associated with the links. This choice is interesting since the Bell states introduce entanglement between faces of the adjacent tetrahedra. Such a way of "gluing" tetrahedra by quantum entanglement has been recently studied in Ref. [39], where it was shown that the $|W\rangle$ state is a superposition of spin network states (with $\{15j\}$ symbols as coefficients of the decomposition). Since the spin network state $|\Psi\rangle$ is disentangled one can also interpret $\langle W|\Psi\rangle$ as an amplitude of transition between disentangled and maximally entangled piece of quantum geometry. Going further, possibly the quantum entanglement is the key ingredient which merges the chunks of space associated with the nodes of spin networks into a geometric structure. This reasoning is consistent with the recent advances in the domain of entanglement/gravity duality, an example of which is provided by the AdS/CFT correspondence [40], ER = EPR conjecture [41] and considerations of holographic entanglement entropy [42–44]. Interestingly, it has recently been argued that indeed the spin networks may represent structure of quantum entanglement [45], indicating relation between spin networks and tensor networks [46]. This is actually not very surprising since the holonomies associated with links of the spin-networks can be interpreted as "mediators" of entanglement.

Namely, the holonomies are maps between two vector (Hilbert) spaces at the ends of a curve $e(\lambda)$, where the affine parameter $\lambda \in [0,1]$. Let us denote the initial point as $a = e(0)$ and the final one as $b = e(1)$. Then, holonomy h_e is a map between two Hilbert spaces $\mathcal{H}^{(a)}$ and $\mathcal{H}^{(b)}$ associated with two (in general) different spatial locations:

$$h_e : \mathcal{H}^{(a)} \to \mathcal{H}^{(b)}. \tag{23}$$

In the case considered in this article, the Hilbert spaces are related with the elementary qubits $\mathcal{H}_{1/2}$ "living" at the ends of the links of the spin-networks (keep in mind that these are not the intertwiner qubits but the elementary qubits out of which the intertwiner qubits are built). As an example of the holonomy of the Ashtekar connection $\mathbf{A}$ considered in LQG (i.e., $h_e := \mathcal{P} \exp \int_e \mathbf{A}$, see e.g., Ref. [6]) let us consider

$$h_x(\alpha) := e^{i\sigma_x \alpha} = \mathbb{I}\cos(\alpha) + i\sigma_x \sin(\alpha), \tag{24}$$

where α is an angle variable and σ_x is the Pauli matrix. The holonomies as the one given by Equation (24) are associated with homogeneous models and are considered in Loop Quantum Cosmology [47] and Spinfoam Cosmology [48,49]. The special case is when $\alpha = \pi/2$ for which $h_x(\pi/2) = i\sigma_x$, which written as an operator

$$\hat{h}_x(\pi/2) = i\hat{X}, \tag{25}$$

where $\hat{X}$ is the bit-flip operator introduced earlier. Therefore, having, e.g., the elementary qubit $|0\rangle \in \mathcal{H}_{1/2}^{(a)}$ at point a, the operator (25) maps this state into $\hat{h}_x(\pi/2)|0\rangle = i\hat{X}|0\rangle = i|1\rangle \in \mathcal{H}_{1/2}^{(b)}$ at the point b[3]. This introduces a relation between the quantum states at distant points a and b, which possibly can be associated with entanglement. However, the issue of relation between holonomies and entanglement requires further more detailed studies, also in the spirit of the recent proposal of *Entanglement holonomies* [50].

---

[3] Performing an inverse mapping $\hat{h}_x^\dagger(\pi/2)$ we can map the state $i|1\rangle$ back to $\hat{h}_x^\dagger(\pi/2)i|1\rangle = -i\hat{X}i|1\rangle = |0\rangle$.

4. A Quantum Algorithm

Having the vertex amplitude (19) defined we may proceed to the task of determining $|\langle W|\Psi\rangle|^2$ with the use of quantum computers. Here, we will show how to obtain amplitude modulus square (the probability), while extraction of the phase factor will be a subject of our further investigations.

Let us begin with preparation of a suitable quantum register. Because each of the intertwiner qubits is a superposition of four elementary qubits, evaluation of the spin network with N nodes requires $4N$ qubits in the quantum register[4]. The corresponding Hilbert space is spanned by 2^{4N} basis states $|i\rangle$, where $i \in \{0, \ldots, 2^{4N} - 1\}$. The initial state for the quantum algorithm is:

$$|0\rangle = \underbrace{|0\rangle \otimes \cdots \otimes |0\rangle}_{4N}. \tag{26}$$

Now, we have to find unitary operators $\hat{U}_\Psi$ and $\hat{U}_W$ defined such that

$$|\Psi\rangle = \hat{U}_\Psi |0\rangle, \tag{27}$$
$$|W\rangle = \hat{U}_W |0\rangle, \tag{28}$$

where $|0\rangle$ is given by Equation (26). Utilizing the operators $\hat{U}_\Psi$ and $\hat{U}_W$ we introduce an operator $\hat{U} := \hat{U}_W^\dagger \hat{U}_\Psi$. Action of this operator on the initial state (26) can be expressed as a superposition of the basis states with some amplitudes $a_i \in \mathbb{C}$:

$$\hat{U}|0\rangle = \sum_{i=0}^{2^{4N}-1} a_i |i\rangle. \tag{29}$$

It is now easy to show that the a_0 coefficient in this superposition is the transition amplitude we are looking for. Namely:

$$a_0 = \langle 0|\hat{U}|0\rangle = \langle 0|\hat{U}_W^\dagger \hat{U}_\Psi|0\rangle = \langle W|\Psi\rangle. \tag{30}$$

By performing measurements on the final state we find the probabilities $P(i) = |a_i|^2$. The first of these probabilities is the modulus square of the vertex amplitude.

Before we will proceed to the discussion of the pentagram spin network associated with the vertex amplitude, let us first demonstrate the algorithm on two simpler examples of spin networks with one and two nodes.

4.1. Example 1—Single Tetrahedron

As a first example let us consider the case of a single-node spin network presented in Figure 5.

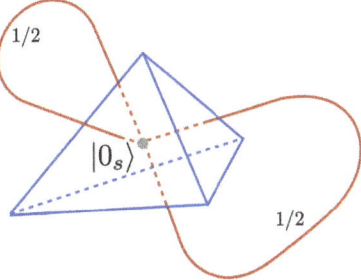

Figure 5. A single-node spin network associated with identification of the pairs of face of a tetrahedron.

---

[4] This statement is made under assumption that no ancilla qubits are required.

Here, the intertwiner qubit is in the $|\Psi\rangle = |0_s\rangle$ state composed out of the four elementary qubits according to Equation (14). The representation state $|W\rangle$ is a tensor product of two Bell states (22). There are basically two different choices of pairing faces of the tetrahedron. The first choice is according to the pairing of qubits entering to the two-qubit singlets $|S\rangle$ out of which the $|0_s\rangle$ state is built. The second choice is by linking qubits contributing to the two different singlets. The first choice is trivial since in that case $\hat{U}_W = \hat{U}_\Psi$ and in consequence the amplitude $\langle W|\Psi\rangle = \langle 0|\hat{U}_W^\dagger \hat{U}_\Psi|0\rangle = \langle 0|0\rangle = 1$. Therefore, we will consider the second case for which the quantum circuit associated with the $\hat{U} = \hat{U}_W^\dagger \hat{U}_\Psi$ operator is presented in Figure 6.

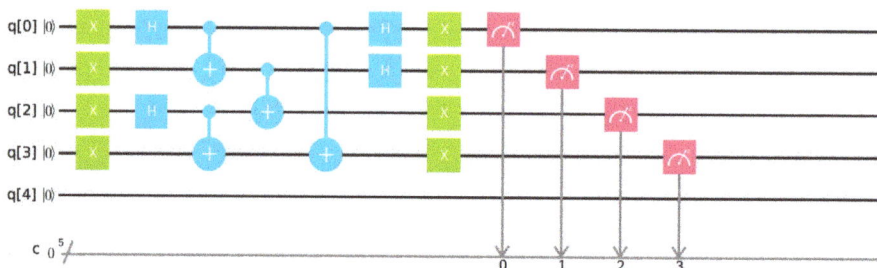

Figure 6. Quantum circuit used to evaluate $|\langle W|\Psi\rangle|^2$ the boundary transition amplitude of the spin network presented in Figure 5.

The simulations were performed on both the IBM simulator and the QX simulator, with 1024 shots in each computational round. The rounds have been repeated 10 times. The results obtained are collected in Table 2.

Table 2. Results of measurements of $P(0) = |a_0|^2$ for the quantum circuit presented in Figure 6, using both the IBM simulator and the QX simulator. Each measurement corresponds to the number of shots equal to 1024.

No.	P_0 (QX)	Hits of $|0\rangle$ (QX)	P_0 (IBM)	Hits of $|0\rangle$ (IBM)
1	0.255859375	262	0.263671875	270
2	0.248046875	254	0.2529296875	259
3	0.267578125	274	0.2578125	264
4	0.2568359375	263	0.27734375	284
5	0.2568359375	263	0.232421875	238
6	0.25	256	0.263671875	270
7	0.2578125	264	0.25	256
8	0.23828125	244	0.244140625	250
9	0.240234375	246	0.2607421875	267
10	0.25	256	0.2763671875	283

Averaging over the 10 rounds, the following values of modulus square of the amplitudes are obtained:

$$|\langle W|\Psi\rangle|^2 = |a_0|^2 = \begin{cases} 0.252 \pm 0.009 & \text{for QX simulator} \\ 0.256 \pm 0.014 & \text{for IBM simulator} \end{cases} \quad (31)$$

The results are consistent with the theoretically expected value $|a_0|^2 = 0.25$. Finally, worth mentioning is that the algorithm cannot directly be executed using the IBM Q 5-qubit quantum chip due to the topological constraints of the structure of coupling between qubits. Additional ancilla qubits have to be involved for this purpose.

4.2. Example 2—Two Tetrahedra

The second example concerns a bit more complex situation with two-node spin network presented in Figure 7.

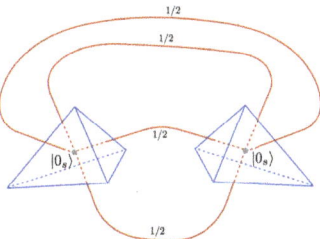

Figure 7. A two-node spin network associated with two tetrahedra.

Here, the representation state $|W\rangle$ similarly to the previous example is associated with the Bell sates (22) entangling faces of the two tetrahedra one into another. The corresponding choice of the quantum circuit used to evaluate the boundary amplitude is presented in Figure 8.

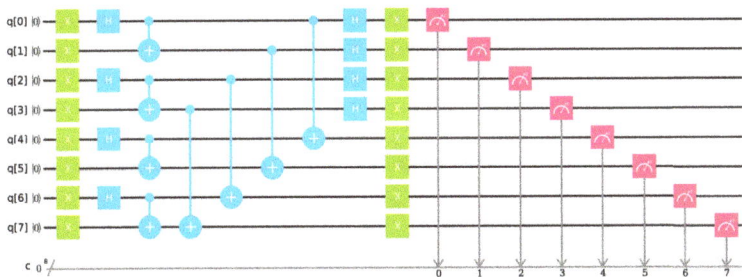

Figure 8. Quantum circuit used to evaluate $|\langle W|\Psi\rangle|^2$ the boundary transition amplitude of the spin network presented in Figure 7. The qubits $\{0,1,2,3\}$ belong to the one node while the qubits $\{4,5,6,7\}$ belong to the another. The links are between the pairs of qubits: $\{0,4\}, \{1,5\}, \{2,6\}$ and $\{3,7\}$.

The simulations were performed on both the IBM simulator and the QX simulator, with 1024 shots in each round. As in the previous example, the computational rounds have been repeated 10 times. The results obtained are collected in Table 3.

Table 3. Results of measurements of $P(0) = |a_0|^2$ for the quantum circuit presented in Figure 8 using both the IBM simulator and the QX simulator. Each measurement corresponds to the number of shots equal 1024.

No.	P_0 (QX)	Hits of $|0\rangle$ (QX)	P_0 (IBM)	Hits of $|0\rangle$ (IBM)
1	0.0595703125	61	0.0556640625	57
2	0.0595703125	61	0.06640625	68
3	0.06640625	68	0.060546875	62
4	0.0615234375	63	0.064453125	66
5	0.080078125	82	0.0634765625	65
6	0.0498046875	51	0.0595703125	61
7	0.052734375	54	0.0615234375	63
8	0.072265625	74	0.0537109375	55
9	0.0673828125	69	0.052734375	54
10	0.0634765625	65	0.0654296875	67

Performing averaging over the computational rounds the following values of $|\langle W|\Psi\rangle|^2$ are obtained:

$$|\langle W|\Psi\rangle|^2 = |a_0|^2 = \begin{cases} 0.063 \pm 0.009 & \text{for QX simulator} \\ 0.060 \pm 0.005 & \text{for IBM simulator} \end{cases}. \quad (32)$$

The results are in agreement with the theoretically expected value $|\langle W|\Psi\rangle|^2 = |a_0|^2 = 0.625$ (obtained using Quirk [35]).

5. Evaluation of Vertex Amplitude

We are now ready to address the task of determining the vertex amplitude (19) associated with the boundary spin network state:

$$|\Psi\rangle = |0_s\rangle \otimes |0_s\rangle \otimes |0_s\rangle \otimes |0_s\rangle \otimes |0_s\rangle. \tag{33}$$

The other possible choices of the spin network state will be discussed in our further work [34].

The $|W\rangle$ is given by Equation (21), representing entanglement between faces of tetrahedra being connected by the links of the spin network. Due to anti-symmetricity of the Bell states (22) for the 10 links under consideration we have in general $2^{10} = 1024$ ways to order the states between the nodes of the spin network. Here, in order to not distinguish any of the nodes, the configuration in which every node is entangled with two other nodes by the state $|\mathcal{E}_l\rangle = \frac{1}{\sqrt{2}}(|01\rangle - |10\rangle)$ and another two nodes by the state $e^{i\pi}|\mathcal{E}_l\rangle = \frac{1}{\sqrt{2}}(|10\rangle - |01\rangle)$ is considered. The resulting quantum circuit corresponding to the operator $\hat{U} = \hat{U}_W^\dagger \hat{U}_\Psi$, together with the measurements necessary to find $|a_0|^2 = |\langle W|\Psi\rangle|^2$ is shown in Figure 9.

The quantum circuit employs 20-qubit quantum register with the initial state:

$$|\mathbf{0}\rangle = \otimes_{n=0}^{19} |0\rangle. \tag{34}$$

The algorithm introduced in Section 4 requires finding amplitude of the initial state (34) in the final state. One has to keep in mind that the Hilbert space of the 20-qubit system is spanned by over 1,000,000 basis states: $2^{20} = 1048576$. Therefore, selecting amplitude of one of the basis states (i.e., $|\mathbf{0}\rangle$) is not an easy task.

The first attempt to determine the value of $|\langle W|\Psi\rangle|^2 = |A(0_s, 0_s, 0_s, 0_s, 0_s)|^2$ has been made by using the IBM simulator of quantum computer. Ten rounds of simulation, each of 1024 shots, have been performed. However, no single event with the $|\mathbf{0}\rangle$ state in the final state has been observed. Assuming that the probability is evenly distributed between the basis states, the probability $1/2^{20} \approx 10^{-6}$ per basis states can be expected. With the 10,240 measurements made, this gives roughly 1% chance to observe the state.

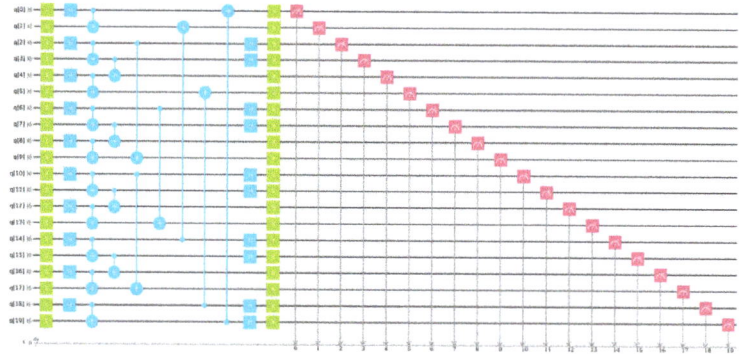

Figure 9. Quantum circuit used to determine $|A(0_s, 0_s, 0_s, 0_s, 0_s)|^2$. Nodes of the spin network correspond to the following sets of qubits: $\{0,1,2,3\}$, $\{4,5,6,7\}$, $\{8,9,10,11\}$, $\{12,13,14,15\}$, $\{16,17,18,19\}$. The links are between the pairs of qubits: $\{0,19\}$, $\{1,14\}$, $\{2,9\}$, $\{3,4\}$, $\{5,18\}$, $\{6,13\}$, $\{7,8\}$, $\{10,17\}$, $\{11,12\}$ and $\{15,16\}$.

The second attempt to determine value of the vertex amplitude has been made with the use of the QX quantum computer simulator. Similarly to the simulations performed on the IBM simulator, 10 computational rounds, each of 1024 shots, have been performed. In this case, the events with $|0\rangle$ have been observed and are collected in Table 4.

Table 4. Results of measurements of $P(0) = |a_0|^2$ for the quantum circuit presented in Figure 9 using the QX simulator. Each measurement corresponds to the number of shots equal to 1024.

No.	P_0	Hits of $|0\rangle$
1	0.0009765625	1
2	0.0029296875	3
3	0	0
4	0.0009765625	1
5	0.001953125	2
6	0.0009765625	1
7	0.001953125	2
8	0.0009765625	1
9	0.0029296875	3
10	0.0009765625	1

By averaging the results from Table 4, the following value of the modulus square of the vertex amplitude $\langle W|\Psi\rangle$ can be found:

$$|\langle W|\Psi\rangle|^2 = |A(0_s, 0_s, 0_s, 0_s, 0_s)|^2 = P_0 = 0.00147 \pm 0.00095. \tag{35}$$

The results obtained from the IBM and QX simulators are contradictory. However, the QX simulator result (35) is much closer to the theoretically expected value. Namely, the spin foam amplitude considered in this section can be determined using recoupling theory for $SU(2)$ group. Following the discussion in Ref. [39] on can find that the amplitude (19) is given by the $\{15j\}$ symbol. Employing definition of the symbol (see e.g., Equation (17) in Ref. [51]) for all the spin labels $j_{ab} = 1/2$ and the intertwiners $i_a = 0$ (which correspond to the $|0_s\rangle$ states), where $a, b \in \{1, 2, 3, 4, 5\}$, one can find that $\{15j\} = 0.0625$. In consequence, $|\langle W|\Psi\rangle|^2 = |\{15j\}|^2 = 0.0625^2 = 0.00390625$. The difference between this prediction and the result of simulations (35) goes beyond the statistical error and, therefore, one can expect that systematic error was involved. Resolution of this issue requires further investigation, especially analysis of the sampling methods used in the quantum simulator. The same concerns the IBM quantum simulator, where discrepancy between the theoretically predicted value and the results of measurements is even more serious. Certainly, the employed publicly available platforms exhibit significant limitations. Nevertheless, they can be overcome by using more advanced simulators run on up-to-date computational clusters.

A more serious challenge is that the algorithm, in its current form, requires making measurements on all qubits involved in the circuit. In consequence, the number of measurements is growing exponentially with the number of quibits. Therefore, proposals of new algorithms, which will allow to reduce the number of measurements, are welcome (one of the possibilities is to use Quantum Phase Estimation Algorithm).

Taking the above into account, a comment on utility of the applied methodology is desirable. First of all, we already used the fact that the vertex amplitude discussed in this section can easily be evaluated without the need of quantum circuits. The vertex amplitude, associated with the pentagram spin network is given by the $\{15j\}$ symbol. Recently, significant progress has been made in the development of numerical methods of evaluation of spin foam vertex amplitudes [52–54]. However, there are still some obstacles, e.g., oscillating nature of the spin foam amplitudes, which motivate search for alternative computational methods.

The aim of our study was to provide the proof of the concept of applicability of quantum circuits to determine quantities being of relevance in Loop Quantum Gravity and Spin Foam approaches. We have shown that indeed, despite the current hardware limitations (see Appendix A), interesting quantities can already be evaluated on simulators of quantum computers. As we demonstrated, 2 qubits per link of a spin network are needed for this purpose. Therefore, in case of the 5-node pentargam spin network (with 10 links) studied here, 20 qubit register was used. Utilizing the available commercial quantum simulators (e.g., 37 qubit QX multi-node simulator SurfSara [25]), amplitudes of spin networks with up to nine 4-valent nodes can potentially be computed.

Our plan is to perform such simulations in our further studies. Furthermore, our goal is to extend computational capabilities to 40 qubits using academic and commercial supercomputing resources. This will allow to simulate spin networks with 10 nodes. Worth mentioning is that while scaling the system size is straightforward for the method based on quantum circuits, application of the standard methods based on recoupling theory may turn out to be inefficient. This will become even more evident when advantageous quantum computers will become available (as expected) in the second half on the coming decade (see Appendix A), providing 100 and more fault tolerant qubits. Until that time, there is still a lot of potential for improving simulator-based computations and development of methods which will ultimately be applied on real quantum processors.

Furthermore, the introduced circuits for spin networks not only allow to study amplitudes but also other relevant quantities. In particular, analysis of quantum fluctuations and quantum entanglement between subsystems is possible to investigate. One of the open problems which is becoming possible to study by extending the methodology introduced here is the entanglement entropy between subsystems of spin networks and the issue of area law for entropy of entanglement.

In addition, while in the current setup we considered the spin labels $j = 1/2$, the case of higher spins can potentially also be implemented. This would be relevant especially from the point of view of the semi-classical, large j limit. Taking into account the restricted computational resources, this can be done only by the cost of keeping the number of nodes small. Therefore, basically we have two domains which seem to be plausible to investigate in the coming decade. The first is the one with the small spin labels (e.g., $j = 1/2$) and the increasing number of nodes of a spin network—the direction towards the thermodynamic limit of a quantum system. The second is the case with small number of nodes (e.g., dipole spin network) and increasing values of the spin labels. This corresponds to a semi-classical limit of a microscopic chunk of space. On the other hand, the case where both the number of nodes and the spin labels are large is the most interesting one. Analysis of this limit will allow to investigate emergence of classical space(time) in loop quantum gravity and spin foam approaches. This will provide an opportunity to (at least partially) verify physical relevance of the candidates for theory of quantum gravity. However, sufficient quantum computational resources are not expected to be available earlier than in the fourth decade of this century (see Appendix A).

6. Summary

The purpose of this article was to explore the possibility of computing vertex amplitudes in the spin foam models of quantum gravity with the use of quantum algorithms. The notion of *intertwiner qubit* being crucial to implement the vertex amplitudes on quantum computers has been pedagogically introduced. It has been shown how one of the two basis states of the intertwiner qubit can be implemented with the use available IBM 5-qubit quantum computer. To the best of our knowledge it was the first time ever a quantum gravitational quantity has been simulated on superconducting quantum chip.

Thereafter, a quantum algorithm allowing to determine modulus square of spin foam vertex amplitude ($|\langle W|\Psi\rangle|^2$) has been introduced. Utility of the algorithm has been demonstrated on examples of single-node and two-node spin networks. For the two cases, probabilities of the associated boundary states have been determined with the use of IBM and QX quantum computer simulators. Finally, the algorithm has been applied to the case of pentagram spin network, representing boundary of the spin foam vertex. Value of the modulus square of the amplitude in a certain quantum state has been measured with the use of 20-qubit register of the IBM and QX quantum simulators. While the QX results were close to the analytically predicted value, the outcomes of the IBM simulator failed to reproduce theoretical predictions.

The presented results are the first step towards simulating spin foam models (associated with Loop Quantum Gravity and Group Field Theories) with the use of universal quantum computers. In particular, the vertex amplitudes can be applied as elementary building blocks in construction of more complex transition amplitudes. The aim of the developed direction is to achieve the possibility of studying collective behavior of the Planck scale systems composed of huge number of elementary constituents ("atoms of space/spacetime"). Exploration of the many-body Planck scale quantum systems [55] may allow to extract continuous and semi-classical limits from the dynamics of the "fundamental" degrees of freedom. This is crucial from the perspective of making contact between Planck scale physics and empirical sciences.

Worth stressing is that the results presented in this article are rather preliminary and only set up the stage for further, more detailed studies. In particular, the following points have to be addressed:

- Introduction of a quantum circuit for the general intertwiner qubit $|\mathcal{I}\rangle$ (Equation (11)).
- Determination of the phase of vertex amplitude with the use of quantum algorithms (e.g., Quantum Phase Estimation Algorithm [56]).
- Investigation of different types of the state $|W\rangle$.
- Analysis of spin networks with up to 10 nodes on quantum computers simulators.
- A possibility of solving quantum constraints with the use of quantum circuits.
- Application to Spinfoam cosmology [48,49].
- Investigation of the architectures of forthcoming quantum processors (with the $N > 100$ number of qubits) in terms of application to spin foam transition amplitudes.

Some of the tasks will be subject of a sequel to this article [34].

Funding: This research was funded by the Sonata Bis Grant DEC-2017/26/E/ST2/00763 of the National Science Centre Poland and the Mobilność Plus Grant 1641/MON/V/2017/0 of the Polish Ministry of Science and Higher Education.

Conflicts of Interest: The author declare no conflict of interest.

Appendix A. Quantum Computing Technologies

The domain of quantum computing is currently experiencing an unprecedented speedup. The recent progress is mainly due to advances in development of the superconducting qubits [23]. In particular, utilizing the superconducting circuits, the IBM company has developed 5 and 20 qubit (noisy) quantum computers, which are accessible in cloud [22]. Furthermore, a prototype of 50 qubit quantum computer by this company has been built and is currently in the phase of tests. The company has recently also unveiled its first commercial 20-qubit quantum computer IBM Q System One. This is intended to be the first ever commercial universal quantum computer, and the second commercially available quantum computer after the adiabatic quantum computer (quantum annealer [57]) provided by D-Wave Systems [58]. The latests D-Wave 2000Q annealer uses a quantum chip with 2048 superconducting qubits, connected in the form of the so-called *chimera* graph. Another important player in the quantum race is Intel, which recently developed its 49-qubit superconducting quantum chip named Tangle Lake [59]. However, outside of the superconducting qubits, the company in collaboration with QuTech [60] advanced centre for Quantum Computing and Quantum Internet is also developing an approach to quantum computing based on electron's spin based qubits, stored in quantum dots. Further advances in the area of superconducting quantum circuits come from Google [61] and Rigetti Computing [62]. The first company has recently announced their 72-qubit quantum chip, while the second one is currently developing its 128-qubit universal quantum chip. On the other hand, the world's leading software company—Microsoft has focused its approach to quantum computing on topological qubits through Majorana fermions [63].

The alternative to superconducting qubits is also developed by IonQ Inc. startup, which is developing a trapped ion quantum computer based on ytterbium atoms [64]. The most recent quantum computer by this company allows to operate on 79 qubits, which is the current world record. The above are only the most sound examples of the advancement which has been made in the recent years in the area of hardware dedicated to quantum computing. There are still many challenges to be addressed, including reduction of the gate errors and increase of fidelity of the quantum states. However, even in pessimistic scenarios, the current momentum of the quantum computing technologies will undoubtedly lead to emergence of reliable and advantageous quantum machines (which cannot be emulated on classical supercomputers) in the coming decade. There are no fundamental physical reasons identified, which could stop the progress. However, the rate of the progress will depend on whether commercial applications of quantum computing technologies will emerge in the coming five years, stimulating further funding of research and development. See e.g., Ref. [65] for more detailed discussion of this issue.

Major players in the field, with large financial resources, such as IBM or governments may sustain the progress independently on short-term returns (which is not the case for start-ups). This may allow for a stable long term progress. In particular, IBM has recently announced a possibility of doubling a measure called *Quantum Volume* [66] every year [67]. The Quantum Volume V_Q is basically a maximal size of a certain random circuit, with equal width and depth, which can be successfully implemented on a given quantum computer. The current (2019) IBM's value of V_Q is 16 and corresponds to the IBM Q System One quantum computer mentioned above. This means that any quantum algorithm employing 4 qubits and four layers (time steps) of quantum circuit can be successfully implemented on the computer. If the trend will follow the hypothesized geometric trajectory (a sort of a new Moore's law [68] for integrated circuits), then one could expect the quantum volume V_Q to be of the order of 10^3 in 2025 and 10^5 in 2030. This means that in 2025, algorithms employing roughly $\log_2 10^3 \approx 10$ qubits and the same number of time steps will be possible to execute. This number will increase to approximately 16 qubits until the end of the coming decade. While this may not sound very optimistic, the prediction is very conservative and does not rule out that much bigger (non-random) circuits (especially well-fitted to the hardware) will be possible to execute at the same time.

Appendix B. Basics of Quantum Computing

The aim of the appendix is to provide a basic introduction of the concepts in quantum computing used in this article. This appendix will allow quantum gravity researchers who are not familiar with quantum computing, to grasp the relevant concepts.

The quantum computing is basically processing of quantum information. While the elementary portion of classical information is a *bit* $\{0,1\}$, its quantum counterpart is what we call a *qubit*. A single qubit is a state $|\Psi\rangle$ in two-dimensional Hilbert space, which we denote as $\mathcal{H} = \text{span}\{|0\rangle, |1\rangle\}$. The space is spanned by two orthonormal basis states $|0\rangle$ and $|1\rangle$, so that $\langle 1|0\rangle = 0$ and $\langle 0|0\rangle = 1 = \langle 1|1\rangle$. A general qubit is a superposition of the two basis states:

$$|\Psi\rangle = \alpha|0\rangle + \beta|1\rangle, \tag{A1}$$

where, $\alpha, \beta \in \mathbb{C}$ (complex numbers), and the normalization condition $\langle \Psi|\Psi\rangle = 1$ implies that $|\alpha|^2 + |\beta|^2 = 1$.

There are different unitary quantum operations which may be performed on the quantum state $|\Psi\rangle$. The elementary quantum operations are called *gates*, in analogy to electric circuits implementing Boolean logic. For instance, the so-called bit-flip operator $\hat{X}$ which transforms $|0\rangle$ into $|1\rangle$ and $|1\rangle$ into $|0\rangle$ ($\hat{X}|0\rangle = |1\rangle$ and $\hat{X}|1\rangle = |0\rangle$) can be introduced. The $\hat{X}$ operator introduces the NOT operation on a single qubit, and has representation in the form of the Pauli x matrix. Similarly, one can introduce $\hat{Y}$ and $\hat{Z}$ operators corresponding to the other two Pauli matrices. The *computational basis* $\{|0\rangle, |1\rangle\}$ is usually introduced such that the basis states are eigenvectors of the $\hat{Z}$ operator: $\hat{Z}|0\rangle = |0\rangle$ and $\hat{Z}|1\rangle = -|1\rangle$.

Another important operator (which does not have its classical counterpart) is the Hadamard operator $\hat{H}$ which is defined by the following action on the qubit basis states:

$$\hat{H}|0\rangle = \frac{1}{\sqrt{2}}\left(|0\rangle + |1\rangle\right) \text{ and } \hat{H}|1\rangle = \frac{1}{\sqrt{2}}\left(|0\rangle - |1\rangle\right). \tag{A2}$$

The above are examples of operators acting on a single qubit. However, quantum information processing usually concerns a multiple qubit system called *quantum register*. The quantum state of the register of N qubits belongs to a tensor product of N copies of single qubit Hilbert spaces: $\otimes_{i=1}^{N} \mathcal{H}_i$. The dimension of the product Hilbert space is $\dim\left(\otimes_{i=1}^{N} \mathcal{H}_i\right) = 2^N$. This exponential dependence of the dimensionality on N is the main obstacle behind simulating quantum systems on classical computers.

A *quantum algorithm* is a unitary operator $\hat{U}$ acting on the initial state of the quantum registes $|0\rangle := \otimes_{i=1}^{N}|0\rangle \in \otimes_{i=1}^{N} \mathcal{H}_i$, together with a sequence of measurements. The outcome of the quantum algorithm is obtained by performing measurements on the final sate: $\hat{U}|0\rangle$. Because of the probabilistic nature of quantum mechanics, the procedure has to be performed repeatedly in order to reconstruct the final state. In general, full reconstruction of the final state $\hat{U}|0\rangle$ requires the so-called *quantum tomography* to be applied. In the procedure, states of the qubits are measured in different bases (not only in the computational basis). The quantum state tomography, which reconstructs the density matrix $\hat{\rho} = \hat{U}|0\rangle\langle 0|\hat{U}^{\dagger}$, is however not always required. In most of the considered quantum algorithms, only probabilities (not complex amplitudes) of the basis states are necessary to measure, which is much simpler and faster than the quantum state tomography.

As already mentioned, the unitary operator $\hat{U}$ can be decomposed into elementary operations called quantum gates. The already introduced $\hat{X}$ and $\hat{H}$ operators are examples of single-qubit gates. However, the gates may also act on two or more qubits. An example of 2-qubit gate relevant for the purpose of this article is the so-called controlled-NOT (CNOT) gate, which we denote as $\hat{C}$. The operator is acting on 2-qubit state $|ab\rangle \equiv |a\rangle \otimes |b\rangle$, where $|a\rangle$ and $|b\rangle$ are single qubit states. Action of the CNOT operator on the basis states can be expressed as follows: $\hat{C}(|a\rangle \otimes |b\rangle) = |a\rangle \otimes |a \oplus b\rangle$, where $a, b \in \{0, 1\}$. The $\oplus$ is the XOR (exclusive or) logical operation (equivalent to addition modulo 2), defined as $0 \oplus b = b$, and $1 \oplus b = \neg b$, where by $\neg b$ we denote negation (NOT) of b. This explains why the gate is called the controlled-NOT (CNOT). The first qubit ($|a\rangle$) is a control qubit, while the second ($|b\rangle$) is a target qubit. The first qubits acts as a switch, which turns on negation of the second qubit if $a = 1$ and remain the second qubit unchanged if $a = 0$.

The diagrammatic representation of the of the unitary operator $\hat{U}$ composed of elementary quantum gates is called a *quantum circuit*, examples of which can be found through this article. Each computational qubit is associated with a horizontal line, which arranges the order at which the operations are performed (direction of time). The operations are executed from the left to the right. Then, the symbols representing gates can be place on either a single-qubit line (e.g., X,Y,Z,H gates) or by joining two or more lines (e.g., CNOT, Toffoli gates).

One of the advantages of quantum algorithms is the possibility of implementing the so-called *quantum parallelism*, which allows to reduce computational complexity of certain problems. The most known example is the Shor algorithm [69] which allows to reduce classical NP complexity of the factorization problem into the BQP complexity class (see e.g., Ref. [70] for definitions of complexity classes). Another seminal example is the Grover algorithm [71] which, statistically, reduces the number of steps needed to find an element in the random database containing N elements from classical $N/2$ to $\mathcal{O}(\sqrt{N})$. Even if we do not make use of quantum parallelism and the resulting reduction of computational complexity in this paper, the methods may also find application in the context of simulations of spin networks. This especially concerns the Quantum Phase Estimation Algorithm [56] which may possibly be applied to effectively measure phases of the spin foam amplitudes.

References

1. Ambjorn, J.; Goerlich, A.; Jurkiewicz, J.; Loll, R. Nonperturbative Quantum Gravity. *Phys. Rept.* **2012**, *519*, 127–210. [CrossRef]
2. Ambjorn, J.; Jordan, S.; Jurkiewicz, J.; Loll, R. A Second-order phase transition in CDT. *Phys. Rev. Lett.* **2011**, *107*, 211303. [CrossRef] [PubMed]
3. Bilke, S.; Burda, Z.; Jurkiewicz, J. Simplicial quantum gravity on a computer. *Comput. Phys. Commun.* **1995**, *85*, 278–292. [CrossRef]
4. Di Francesco, P.; Ginsparg, P.H.; Zinn-Justin, J. 2-D Gravity and random matrices. *Phys. Rept.* **1995**, *254*, 1–133. [CrossRef]
5. 't Hooft, G. A Planar Diagram Theory for Strong Interactions. *Nucl. Phys. B* **1974**, *72*, 461–473. [CrossRef]
6. Ashtekar, A.; Lewandowski, J. Background independent quantum gravity: A Status report. *Class. Quantum Gravity* **2004**, *21*, R53. [CrossRef]
7. Rovelli, C.; Vidotto, F. *Covariant Loop Quantum Gravity: An Elementary Introduction to Quantum Gravity and Spinfoam Theory*; Cambridge Monographs on Mathematical Physics; Cambridge University Press: Cambridge, UK, 2014.
8. Rovelli, C.; Smolin, L. Spin networks and quantum gravity. *Phys. Rev. D* **1995**, *52*, 5743–5759. [CrossRef] [PubMed]
9. Baez, J.C. Spin foam models. *Class. Quantum Gravity* **1998**, *15*, 1827–1858. [CrossRef]
10. Perez, A. The Spin Foam Approach to Quantum Gravity. *Living Rev. Relat.* **2013**, *16*, 3. [CrossRef] [PubMed]
11. Engle, J.; Pereira, R.; Rovelli, C. The Loop-quantum-gravity vertex-amplitude. *Phys. Rev. Lett.* **2007**, *99*, 161301. [CrossRef] [PubMed]
12. Bianchi, E.; Hellmann, F. The Construction of Spin Foam Vertex Amplitudes. *SIGMA* **2013**, *9*, 008 [CrossRef]
13. Oriti, D. The Group field theory approach to quantum gravity. In *Approaches to Quantum Gravity*; Oriti, D., Ed.; Cambridge University Press: Cambridge, UK, 2009; pp. 310–331.
14. Freidel, L. Group field theory: An Overview. *Int. J. Theor. Phys.* **2005**, *44*, 1769–1783. [CrossRef]
15. Krajewski, T. Group field theories. Presented at the 3rd Quantum Gravity and Quantum Geometry School (QGQGS 2011), Zakopane, Poland, 28 February–13 March 2011.
16. Ooguri, H. Topological lattice models in four-dimensions. *Mod. Phys. Lett. A* **1992**, *7*, 2799–2810. [CrossRef]
17. Gielen, S.; Oriti, D.; Sindoni, L. Cosmology from Group Field Theory Formalism for Quantum Gravity. *Phys. Rev. Lett.* **2013**, *111*, 031301. [CrossRef] [PubMed]
18. Benedetti, D.; Geloun, J.B.; Oriti, D. Functional Renormalisation Group Approach for Tensorial Group Field Theory: A Rank-3 Model. *J. High Energy Phys.* **2015**, *2015*, 084. [CrossRef]
19. Li, K.; Li, Y.; Han, M.; Lu, S.; Zhou, J.; Ruan, D.; Long, G.; Wan, Y.; Lu, D.; Zeng, B.; et al. Quantum Spacetime on a Quantum Simulator. *arXiv* **2017**, arXiv:1712.08711.
20. Feller, A.; Livine, E.R. Ising Spin Network States for Loop Quantum Gravity: A Toy Model for Phase Transitions. *Class. Quantum Gravity* **2016**, *33*, 065005. [CrossRef]
21. Mielczarek, J. Spin networks on adiabatic quantum computer. *arXiv* **2018**, arXiv:1801.06017.
22. Available online: https://www.research.ibm.com/ibm-q/ (accessed on 25 July 2019).
23. You, J.Q.; Nori, F. Superconducting Circuits and Quantum Information. *Phys. Today* **2005**, *58*, 42–47. [CrossRef]
24. Available online: http://quantum-studio.net/ (accessed on 25 July 2019).
25. Available online: https://www.quantum-inspire.com/ (accessed on 25 July 2019).
26. Feynman, R.P. Simulating physics with computers. *Int. J. Theor. Phys.* **1982**, *21*, 467–488. [CrossRef]
27. Mielczarek, J. Quantum Gravity on a Quantum Chip. *arXiv* **2018**, arXiv:1803.10592.
28. Albash, T.; Lidar, D.A. Adiabatic Quantum Computing. *Rev. Mod. Phys.* **2018**, *90*, 015002. [CrossRef]
29. Deutsch, D.; Barenco, A.; Ekert, A. Universality in quantum computation. *Proc. R. Soc. Lond. A* **1995**, *449*, 669–677. [CrossRef]
30. Ekert, A.; Hayden, P.; Inamori, H. Basic concepts in quantum computation. *arXiv* **2011**, arXiv:quant-ph/0011013.
31. Chen, Z.Y.; Zhou, Q.; Xue, C.; Yang, X.; Guo, G.C.; Guo, G.P. 64-Qubit Quantum Circuit Simulation. *Sci. Bull.* **2018**, *63*, 964–971. [CrossRef]
32. Biamonte, J.D.; Morales, M.E.S.; Koh, D.E. Quantum Supremacy Lower Bounds by Entanglement Scaling. *arXiv* **2018**, arXiv:1808.00460.
33. Long, G.L.; Sun, Y. Efficient Scheme for Initializing a Quantum Register with an Arbitrary Superposed State. *Phys. Rev. A* **2001**, *64*, 014303. [CrossRef]

34. Mielczarek, J. Quantum circuits for a qubit of space. 2019, In preparation.
35. Available online: https://algassert.com/quirk (accessed on 25 July 2019).
36. Available online: https://sites.google.com/view/quantum-kit/home (accessed on 25 July 2019).
37. Nielsen, M.A.; Chuang, I.L. *Quantum Computation and Quantum Information*; Cambridge University Press: Cambridge, UK, 2000.
38. Oeckl, R. A 'General boundary' formulation for quantum mechanics and quantum gravity. *Phys. Lett. B* **2003**, *575*, 318–324. [CrossRef]
39. Baytas, B.; Bianchi, E.; Yokomizo, N. Gluing polyhedra with entanglement in loop quantum gravity. *Phys. Rev. D* **2018**, *98*, 026001. [CrossRef]
40. Maldacena, J.M. The Large N limit of superconformal field theories and supergravity. *Int. J. Theor. Phys.* **1999**, *38*, 1113–1133. [CrossRef]
41. Susskind, L. Dear Qubitzers, GR=QM. *arXiv* **2017**, arXiv:1708.03040.
42. Swingle, B. Entanglement Renormalization and Holography. *Phys. Rev. D* **2012**, *86*, 065007. [CrossRef]
43. Ryu, S.; Takayanagi, T. Holographic derivation of entanglement entropy from AdS/CFT. *Phys. Rev. Lett.* **2006**, *96*, 181602. [CrossRef] [PubMed]
44. Rangamani, M.; Takayanagi, T. *Holographic Entanglement Entropy*; Lecture Notes in Physics; Springer: Cham, Switzerland, 2017; Volume 931, p. 1.
45. Han, M.; Hung, L.Y. Loop Quantum Gravity, Exact Holographic Mapping, and Holographic Entanglement Entropy. *Phys. Rev. D* **2017**, *95*, 024011. [CrossRef]
46. Biamonte, J.; Bergholm, V. Tensor Networks in a Nutshell. *arXiv* **2017**, arXiv:1708.00006.
47. Bojowald, M. Loop quantum cosmology. *Living Rev. Relat.* **2008**, *11*, 4. [CrossRef] [PubMed]
48. Bianchi, E.; Rovelli, C.; Vidotto, F. Towards Spinfoam Cosmology. *Phys. Rev. D* **2010**, *82*, 084035. [CrossRef]
49. Vidotto, F. Spinfoam Cosmology: quantum cosmology from the full theory. *J. Phys. Conf. Ser.* **2011**, *314*, 012049. [CrossRef]
50. Czech, B.; Lamprou, L.; Susskind, L. Entanglement Holonomies. *arXiv* **2018**, arXiv:1807.04276.
51. Donà, P.; Fanizza, M.; Sarno, G.; Speziale, S. SU(2) graph invariants, Regge actions and polytopes. *Class. Quantum Gravity* **2018**, *35*, 045011. [CrossRef]
52. Sarno, G.; Speziale, S.; Stagno, G.V. 2-vertex Lorentzian Spin Foam Amplitudes for Dipole Transitions. *Gen. Relat. Gravit.* **2018**, *50*, 43. [CrossRef]
53. Dona, P.; Sarno, G. Numerical methods for EPRL spin foam transition amplitudes and Lorentzian recoupling theory. *Gen. Relat. Gravit.* **2018**, *50*, 127. [CrossRef]
54. Dona, P.; Fanizza, M.; Sarno, G.; Speziale, S. Numerical study of the Lorentzian EPRL spin foam amplitude. *arXiv* **2019**, arXiv:1903.12624.
55. Oriti, D. Spacetime as a quantum many-body system. *arXiv* **2017**, arXiv:1710.02807.
56. Cleve, R.; Ekert, A.; Macchiavello, C.; Mosca, M. Quantum algorithms revisited. *Proc. R. Soc. Lond. A* **1998**, *454*, 339–354. [CrossRef]
57. Kadowaki, T.; Nishimori, H. Quantum annealing in the transverse Ising model. *Phys. Rev. E* **1998**, *58*, 5355. [CrossRef]
58. Available online: https://www.dwavesys.com/home (accessed on 25 July 2019).
59. Available online: https://newsroom.intel.com/press-kits/quantum-computing/#quantum-computing (accessed on 25 July 2019).
60. Available online: https://qutech.nl/ (accessed on 25 July 2019).
61. Available online: https://ai.google/research/teams/applied-science/quantum-ai/ (accessed on 25 July 2019).
62. Available online: https://www.rigetti.com/ (accessed on 25 July 2019).
63. Available online: https://www.microsoft.com/en-us/quantum/ (accessed on 25 July 2019).
64. Available online: https://ionq.co/ (accessed on 25 July 2019).
65. Grumbling, E.E.; Horowitz, M. *Quantum Computing—Progress and Prospects*; The National Academies Press: Washington, DC, USA, 2019.
66. Cross, A.W.; Bishop, L.S.; Sheldon, S.; Nation, P.D.; Gambetta, J.M. Validating quantum computers using randomized model circuits. *arXiv* **2018**, arXiv:1811.12926.
67. Available online: https://www.ibm.com/blogs/research/2019/03/power-quantum-device/ (accessed on 25 July 2019).
68. Moore, G.E. Cramming more components onto integrated circuits. *Electronics* **1965**, *38*, 114–117. [CrossRef]

69. Shor, P.W. Algorithms for quantum computation: Discrete logarithms and factoring. In Proceedings of the 35th Annual Symposium on Foundations of Computer Science, Santa Fe, NM, USA, 20–22 November 1994; pp. 124–134.
70. Available online: https://complexityzoo.uwaterloo.ca/Complexity_Zoo (accessed on 25 July 2019).
71. Grover, L.K. A Fast quantum mechanical algorithm for database search. In Proceedings of the 28th Annual ACM Symposium on the Theory of Computing, Philadelphia, PA, USA, 22–24 May 1996.

© 2019 by the author. Licensee MDPI, Basel, Switzerland. This article is an open access article distributed under the terms and conditions of the Creative Commons Attribution (CC BY) license (http://creativecommons.org/licenses/by/4.0/).

Review

Group Field Theory Condensate Cosmology: An Appetizer

Andreas G. A. Pithis * and Mairi Sakellariadou

Department of Physics, King's College London, University of London, Strand, London WC2R 2LS, UK; mairi.sakellariadou@kcl.ac.uk
* Correspondence: andreas.pithis@kcl.ac.uk

Received: 2 April 2019; Accepted: 8 June 2019; Published: 13 June 2019

Abstract: This contribution is an appetizer to the relatively young and fast-evolving approach to quantum cosmology based on group field theory condensate states. We summarize the main assumptions and pillars of this approach which has revealed new perspectives on the long-standing question of how to recover the continuum from discrete geometric building blocks. Among others, we give a snapshot of recent work on isotropic cosmological solutions exhibiting an accelerated expansion, a bounce where anisotropies are shown to be under control, and inhomogeneities with an approximately scale-invariant power spectrum. Finally, we point to open issues in the condensate cosmology approach.

Keywords: quantum geometry; quantum gravity; quantum cosmology; group field theory; loop quantum gravity

---

> Most important part of doing physics is the knowledge of *approximation*.
>
> Lev Davidovich Landau

1. Introduction

Current observational evidence strongly suggests that our universe is accurately described by the standard model of cosmology [1]. This model relies on Einstein's theory of general relativity (GR) and assumes its validity on all scales. However, this picture proves fully inadequate to describe the earliest stages of our universe, as our concepts of spacetime and its geometry as given by GR are then expected to break down due to the extreme physical conditions encountered in the vicinity of and shortly after the Big Bang. More explicitly, it appears that our universe emerged from a singularity, as implied by the famous theorems of Penrose and Hawking [2]. From a fundamental point of view, such a singularity is unphysical, and it is expected that quantum effects lead to its resolution [3]. This motivates the development of a quantum theory of gravity in which the quintessential features of GR and quantum field theory (QFT) are consistently unified. Such a theory will revolutionize our understanding of spacetime and gravity at a microscopic level and should be able to give a complete and consistent picture of cosmic evolution.

The difficulty in making progress in this field is ultimately rooted in the lack of experiments which have access to the physics at the smallest length scales and highest energies and so would provide a clear empirical guideline for the construction of such a theory. In turn, this severe underdetermination of theory by experiment is a reason for the current presence of a plethora of contesting approaches to quantum gravity [4]. Against this backdrop, the cosmology of the very early universe represents a unique window of opportunity out of this impasse [5]. For instance, it is naturally expected that traces of quantum gravity have left a fingerprint on the spectrum of the cosmic microwave background radiation, see e.g., Refs. [6–11]. Hence, cosmology provides an ideal testbed where the predictions of such competing theories can be compared and tested against forthcoming cosmological data.

A central conviction of *some* approaches to quantum gravity is that it should be a non-perturbative, background-independent, and diffeomorphism-invariant theory of quantum geometry. In this sense, the spacetime continuum is renunciated and is instead replaced by degrees of freedom of a discrete and combinatorial nature.[1] Particular representatives of this class of theories are the closely related canonical and covariant loop quantum gravity (LQG) [20–23], group field theory (GFT) [24,25], tensor models (TM) [26–32], and simplicial quantum gravity approaches such as quantum Regge calculus (QRC) [33] and Euclidean and causal dynamical triangulations (EDT, CDT) [34–36]. The perturbative expansion of their path integrals each yields a sum over discrete geometries and the most difficult problem for all of them then lies in the recovery of continuous spacetime geometry and GR describing its dynamics in an appropriate limit. This is challenging because it ideally requires formulating statements about the continuum by only calling upon notions rooted in the discontinuum. Taking the continuum limit in these approaches crucially depends on whether the discreteness of geometry is considered physical or unphysical therein, the proper weighting of configurations in the partition function, and the precise specification of the continuum limit itself. Consequently, strategies to reach this goal differ among them strongly, see e.g., Ref. [37] for an overview.

In the light of the above, it is vitally important to consider these approaches in a cosmological context, which has been accomplished to a varying degree of success by them. In this contribution, we give a brief and rather non-technical panorama of the GFT condensate cosmology program [38,39] which has been developed over the last few years and has so far borne promising fruits.[2] This program is motivated by the idea that the mechanism for regaining a continuum geometry from a physically discrete quantum gravity substratum in GFT is provided by a phase transition to a condensate phase [40,42]. Research on the phase structure of different GFT models in terms of functional renormalization group analyses finds support for such a conjecture in terms of IR fixed points [43–50]. Further backing is provided by saddle point studies [51] and Landau–Ginzburg mean field analyses [52] which probe non-perturbative aspects of GFT models, see also Ref. [37]. In this picture, a condensate would correspond to a non-perturbative vacuum which comprises of many bosonic GFT quanta and in the context of GFT models of four-dimensional quantum gravity is tentatively interpreted as a continuum geometry. Given this basic premise, the most striking successes and milestone results of the condensate cosmology approach are the recovery of Friedmann-like dynamics of an emergent homogeneous and isotropic geometry [53], an extended accelerated phase of expansion [54,55] right after a cosmological bounce [56], a simple yet effective mechanism for dynamical isotropization of microscopic anisotropies [57,58] and the finding of an approximately scale-invariant and small-amplitude power spectrum of quantum fluctuations of the local volume over a homogeneous background geometry perturbed by small inhomogeneities [59]. In the following, we will quickly review the GFT formalism, give the basic structures behind its condensate cosmology spin-off, highlight the main results, and give an outlook for future challenges of this program.[3]

---

[1] The introduction of discrete structures can be motivated to bypass the issue of perturbative non-renormalizability of GR within the continuum path integral formulation. Alternative points of view of dealing with this issue would be to assume the existence of a non-perturbative (i.e., interacting) fixed point for gravity in the UV as done by the asymptotic safety program [12–15] or to increase the amount of symmetries as compared to GR and QFT with the aim to regain perturbative renormalizability as proposed by string theory [16]. Yet another view, as presented by non-commutative geometry, is that above the Planck scale the concept of geometry collapses and spacetime is replaced by a non-commutative manifold [17–19].
[2] For previous articles giving a review account of the program, we refer to Refs. [40,41].
[3] Another way to relate GFT to cosmology was brought forward in Ref. [60]. This work is closer to canonical quantum cosmology (either Wheeler–DeWitt or loop quantum cosmology) in the sense that it is built on a minisuperspace model, i.e., symmetry reduction is applied before quantization and not afterwards as in the condensate program.

2. Group Field Theory

GFTs are quantum field theories which live on group configuration spaces, possess a gauge symmetry and are in particular characterized by combinatorially non-local interactions [24,25]. More precisely, the real- or complex-valued scalar field φ lives on d copies of a Lie group G. In models for quantum gravity, G corresponds to the local gauge group of GR and the gauge symmetry leads to an invariance of the GFT action under the (right) diagonal action of G which acts on the fields as

$$\varphi(g_1, ..., g_d) = \varphi(g_1 h, ..., g_d h), \quad \forall g_i, h \in G. \tag{1}$$

For 4d quantum gravity models G is typically $SO(3,1)$ (or $SL(2,\mathbb{C})$) in the Lorentzian case, $SO(4)$ (or $Spin(4)$) in the Riemannian case or their rotation subgroup $SU(2)$ which is the gauge group of Ashtekar–Barbero gravity. The group elements g_I with $I = 1, ..., d$ are parallel transports $\mathcal{P}e^{i\int_{e_I} A}$ which are associated with d links e_I and A denotes a gravitational connection 1-form. The gauge symmetry guarantees the closure of the faces dual to the links e_I to form a $d-1$-simplex. For the most discussed case where $d = 4$, one obtains tetrahedra in this way. The metric information encoded in the fields can be retrieved via a non-commutative Fourier transform [38,39,41,57,61–63].

For a complex-valued field the action has the structure

$$S[\varphi, \bar{\varphi}] = \int (\mathrm{d}g)^d \, \bar{\varphi}(g_I) \mathcal{K}(g_I) \varphi(g_I) + \mathcal{V}[\varphi, \bar{\varphi}], \tag{2}$$

where we used the shorthand notation $\varphi(g_I) \equiv \varphi(g_1, ..., g_d)$. Since further details about the action are specified below, here the following suffices to say. The local kinetic term typically incorporates a Laplacian and a "mass term" contribution. The former is motivated by renormalization studies on GFT [64] while the latter can be related to spin foam edge weights via the GFT/spin foam correspondence [65] (see below) and hence should not be confused with a physical mass. The so-called simplicial interaction term consists of products of fields paired via convolution according to a combinatorial non-local pattern which for $d = 4$ encodes the combinatorics of a 4-simplex. The precise details of the kinetic and interaction term are supposed to encode the Euclidean or Lorentzian embeddings of the theory [53,58,66].

With this, the perturbative expansion of the partition function

$$Z_{\text{GFT}} = \int [\mathcal{D}\varphi][\mathcal{D}\bar{\varphi}] e^{-S[\varphi, \bar{\varphi}]} \tag{3}$$

is indexed by Feynman diagrams which are dual to gluings of d-simplices.[4] In this way, it provides a generating function for the covariant quantization of LQG in terms of spin foam models [22,23]. In LQG, boundary spin network states of a spin foam correspond to 3-dimensional discrete quantum geometries while the spin foam transition amplitudes interpolate in between two such boundary configurations. There, a proper imposition of the so-called simplicity constraints guarantees that $SL(2,\mathbb{C})$- or $Spin(4)$-data in the bulk is reduced to $SU(2)$-valued data on the boundary [70–83]. The main aspects of the GFT formulation of the currently most studied spin foam model for Lorentzian 4d quantum gravity, the so-called EPRL model [22,23], are specified by the aforementioned simplicial interaction term, that the GFT fields are defined over $SU(2)^4$ (thus encoding the boundary geometry) and finally the proper embedding of these data into $SL(2,\mathbb{C})$ which is realized by the dynamics and thus is encoded by the details of the kinetic and interaction term in the action. We further specify this action in Sections 3.2 and 4.1 but note here that all the details of the interaction term have so far not

---

[4] Notice that by attributing an additional combinatorial degree of freedom named color to the fields, one can guarantee that the terms of the perturbative expansion are free of topological pathologies [67–69].

been put down in terms of its boundary data [58] which, for what matters in this review, does not pose a limitation.

In the second-quantized formulation of GFT, introduced in Ref. [84], motivated by the origins of GFT in LQG, spin network boundary states are viewed as elements of the GFT Fock space wherein spin network vertices, i.e., atoms of space, correspond to fundamental quanta which are created or annihilated by the field operators of GFT.[5] The GFT Fock space

$$\mathcal{F}(\mathcal{H}_v) = \bigoplus_{N=0}^{\infty} \operatorname{sym}(\otimes_{i=1}^{N} \mathcal{H}_v^{(i)}), \tag{4}$$

is built by means of the fundamental Hilbert space $\mathcal{H}_v = L^2(G^d)$ of a GFT quantum which is assumed to obey bosonic statistics.[6] Clearly, for $G = SU(2)$ and imposing gauge invariance as in Equation (1), a state in $\mathcal{H}_v$ represents an open LQG spin network vertex or its dual quantum polyhedron. In particular, for $d = 4$ a GFT quantum corresponds to a quantum tetrahedron which also is the most studied case within the condensate cosmology program [38,39,41]. In the remainder, we stick to this choice for G and d.

In this picture, many particle GFT states can be excited over the Fock vacuum $|\emptyset\rangle$ which is the state devoid of any topological and quantum geometric information. Standardly, it is defined via the action of an annihilation field operator, namely

$$\hat{\varphi}(g_I)|\emptyset\rangle = 0, \tag{5}$$

where the vacuum is normalized to 1. Given their bosonic statistics, the GFT field operators obey the canonical commutation relations

$$[\hat{\varphi}(g_I), \hat{\varphi}^\dagger(g_I')] = \int dh \prod_I \delta(g_I h g_I'^{-1}) \text{ and } [\hat{\varphi}^{(\dagger)}(g_I), \hat{\varphi}^{(\dagger)}(g_I')] = 0, \tag{6}$$

where the form of the delta distribution accounts for the imposition of gauge invariance, Equation (1).

In this framework, quantum geometric observable data can be retrieved from such states via second-quantized Hermitian operators [87], e.g., the number operator is given by

$$\hat{N} = \int (dg)^d \hat{\varphi}^\dagger(g_I)\hat{\varphi}(g_I) \tag{7}$$

while more general one-body operators read as

$$\hat{O} = \int (dg)^d \int (dg')^d \, \hat{\varphi}^\dagger(g_I) O(g_I, g_I') \hat{\varphi}(g_I'), \tag{8}$$

wherein $O(g_I, g_I')$ denote the matrix elements of a corresponding first-quantized operator. In this way, the area and volume operator of LQG can be imported into the GFT context, which is typically done by working in the spin representation introduced below.[7] Hence, in GFT the discreteness of geometry is considered to be being real, rooted in its strong connections to LQG where the spectra of geometric operators are discrete (as shown to hold at the kinematical level) [88–91].

---

[5] A detailed discussion on the subtle differences in between the Fock space of GFT and the kinematical Hilbert space of LQG, which are mostly related to the absence of the so-called cylindrical consistency and equivalence in the former, is found in Ref. [84].

[6] The assumption of bosonic statistics is crucial for the condensate cosmology program where spacetime is thought to arise from a GFT condensate. To justify this choice of statistics from a fundamental point of view is an open problem, see Ref. [84] for a discussion and Refs. [67–69,85,86] for explorations into other statistics.

[7] We refer e.g., to Appendix C of Ref. [58] for an extensive discussion of this matter for the case of the volume operator.

3. Group Field Theory Condensate Cosmology

The general aim of the GFT condensate cosmology program is to describe cosmologically relevant geometries by means of the formalism given above. Concretely, the goal is to *approximate* 3-dimensional homogeneous and extended geometries as well as their cosmological evolution in terms of GFT condensate states and their effective dynamics.

3.1. Motivation for Condensate States

As initially stated in the introduction, indications for the formation of a condensate phase have been found through the analyses of non-perturbative aspects of GFT models. In particular, functional renormalization group analyses of so-called tensorial GFTs [43–49] indicate a phase transition separating a symmetric from a broken/condensate phase as the "mass parameter" tends to negative values in the IR limit which is analogous to a Wilson-Fisher fixed point in the corresponding local QFT. This is illustrated in terms of the phase diagrams in Figure 1. Building on these, more work must be devoted to studying the phase structure of a (potentially colored) GFT model enriched with additional geometric data and an available simplicial quantum gravity interpretation. The hope would be that in the phase diagram of such a theory at least one phase can be found which can be interpreted as a physical continuum geometry of relevance to cosmology.[8]

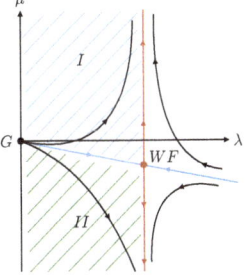
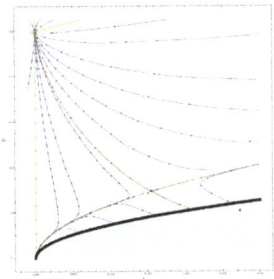

Figure 1. Left: Phase diagram of a local scalar field theory with quartic interaction on $\mathbb{R}^3$. The "mass parameter" is denoted by μ while the interaction couples with λ. G denotes the Gaussian fixed point and WF the Wilson-Fisher fixed point. In the region hatched in green $\langle \hat{\varphi} \rangle \neq 0$ holds. Right: Exemplary phase diagram of a quartic tensorial GFT on $\mathbb{R}^d$ (cf. Refs. [48,49]). In the analogue of region II on the left-hand side, a non-vanishing expectation value of the field operator is expected to be found.

Given this central hypothesis of the GFT condensate cosmology program, the goal is to directly derive the effective dynamics for GFT condensate states from the microscopic quantum dynamics using mean field techniques inspired by the theory on Bose-Einstein condensates [96–98] and to extract a cosmological interpretation thereafter. Generally, a condensate phase corresponds to a non-perturbative vacuum of a theory where the expectation value of the field operator is non-vanishing, i.e., $\langle \varphi(g_I) \rangle \neq 0$. Since such a vacuum is described by a large number N of quanta, in the GFT context this would make it suitable to model extended geometries. In addition, these quanta are occupying the same quantum geometric configuration which is desirable if a homogeneous background geometry is to emerge from the condensate. Simple trial states which capture these features are field coherent states of the form

$$|\sigma\rangle = A\, e^{\hat{\sigma}} |\emptyset\rangle, \quad \hat{\sigma} = \int (dg)^d\, \sigma(g_I) \hat{\varphi}^\dagger(g_I) \quad \text{and} \quad A = e^{-\frac{1}{2} \int (dg)^d\, |\sigma(g_I)|^2} \tag{9}$$

---

[8] Complementarily to the application of functional methods to study the notion of phases in this context, research on the algebraic foundations of GFT has shown the existence of representations which are unitarily inequivalent to the one of the GFT Fock space and that are potentially related to different phases of GFT models, in particular to condensate phases [92–95].

corresponding to an infinite superposition of states which for $d = 4$ describe disconnected quantum tetrahedra labeled by the same discrete geometric data. The latter is encoded by a single collective function, the condensate wave function σ. These states are field coherent since they are eigenstates of the field operator,

$$\hat{\varphi}(g_I)|\sigma\rangle = \sigma(g_I)|\sigma\rangle, \qquad (10)$$

for which $\langle \hat{\varphi}(g_I) \rangle = \sigma(g_I) \neq 0$ holds. Finally, in addition to the right invariance as in Equation (1), we require the invariance under the left diagonal action of G, i.e., $\sigma(kg_I) = \sigma(g_I)$ for all $k \in G$ to guarantee that the domain of the condensate wave function is isomorphic to the minisuperspace of homogeneous geometries $GL(3, \mathbb{R})/O(3)$ [63].[9,10]

3.2. Effective Condensate Dynamics

The effective dynamics of such states can be obtained by taking alternative but equivalent roads. One can either study the GFT path integral in saddle point approximation or use the lowest-order truncation of the Schwinger-Dyson equations of the GFT model under consideration [38,39,41,51,52,102]. These equations can be derived when using

$$0 = \delta_{\bar{\varphi}} \langle \mathcal{O}[\varphi, \bar{\varphi}] \rangle = \int [\mathcal{D}\varphi][\mathcal{D}\bar{\varphi}] \frac{\delta}{\delta\bar{\varphi}(g_I)} \left(\mathcal{O}[\varphi, \bar{\varphi}] e^{-S[\varphi,\bar{\varphi}]} \right) = \left\langle \frac{\delta \mathcal{O}[\varphi, \bar{\varphi}]}{\delta\bar{\varphi}(g_I)} - \mathcal{O}[\varphi, \bar{\varphi}] \frac{\delta S[\varphi, \bar{\varphi}]}{\delta\bar{\varphi}(g_I)} \right\rangle, \qquad (11)$$

where $\mathcal{O}$ is a functional of the fields. An expression encoding the effective dynamics is then extracted by setting $\mathcal{O}$ to the identity, giving

$$\left\langle \frac{\delta S[\varphi, \bar{\varphi}]}{\delta\bar{\varphi}(g_I)} \right\rangle = 0. \qquad (12)$$

If we evaluate the expectation value with respect to the condensate state, one yields

$$K(g_I)\sigma(g_I) + \frac{\delta V}{\delta\bar{\sigma}(g_I)} = 0 \qquad (13)$$

which is the classical equation of motion for the condensate wave function. Its solution would amount to solving the theory at tree-level. In general, this is a non-linear and non-local equation for the dynamics of the mean field σ and is given the interpretation of a quantum cosmology equation despite the fact that it has no direct probabilistic interpretation as compared to the equations of motion of Wheeler–DeWitt (WdW) quantum cosmology [103] and loop quantum cosmology (LQC) [104,105]. However, this does not pose a problem to extract cosmological predictions from the full theory, as we will review below.

In a next step, to extract information regarding the dynamics of such condensate systems, we extend the set of degrees of freedom of the formalism and couple a free, massless, minimally coupled real-valued scalar field to the GFT field,

$$\sigma : G^d \times \mathbb{R} \to \mathbb{C} \text{ (or } \mathbb{R}\text{)}. \qquad (14)$$

This scalar field serves as a relational clock, i.e., an internal time variable, with respect to which the latter evolves. Such a procedure is common practice in classical and quantum gravity [104–109].

---

[9] In principle, more complex composite states can be constructed so as to encode connectivity information in between GFT quanta and topological information to model e.g., spherical geometries [87,99,100].
[10] One may also take the view that the existence of a condensate phase transition is of less pronounced importance for such condensate states to be suitable non-perturbative states of physical relevance. We refer to Ref. [101] for a detailed discussion.

Notice that the expectation values of the above-introduced observables will then obviously depend on the relational clock ϕ. The precise introduction of this degree of freedom is based on the expression of the Feynman amplitudes of a given simplicial GFT model which take the form of simplicial gravity path integrals for gravity when coupled to such a scalar field, as explained in detail in Refs. [53,110]. In this discretized setting, the matter field sits on the vertices which are dual to the 4-simplices of the simplicial complex.

In this way, the action takes the general form

$$S[\sigma, \bar{\sigma}] = \int (dg)^d d\phi \, \bar{\sigma}(g_I, \phi) \mathcal{K}(g_I, \phi) \sigma(g_I, \phi) + \mathcal{V}[\sigma, \bar{\sigma}] \tag{15}$$

where $\mathcal{K}$ is local in g_I and ϕ and the interaction term is given by

$$\mathcal{V}[\sigma, \bar{\sigma}] = \frac{\lambda}{5} \int \left(\prod_{a=1}^{5} dg_{I_a} \sigma(g_{I_a}, \phi) \right) \mathcal{V}_5 + \text{c.c.,} \tag{16}$$

where each g_I corresponds to four group elements. The object $\mathcal{V}_5 = \mathcal{V}_5(g_{I_1}, g_{I_2}, g_{I_3}, g_{I_4}, g_{I_5})$ is a function of all group elements, encoding the combinatorics of a 4-simplex, which when appropriately specified together with $\mathcal{K}$ are supposed to yield the GFT formulation of the EPRL spin foam model for Lorentzian quantum gravity in $4d$ [53,58,80,81].

For the remainder of this review it is important to introduce the spin representation of GFT fields. For left- and right-invariant configurations as considered in the context of the condensate program, we may give the Peter-Weyl decomposition of the condensate field as

$$\sigma(g_1, g_2, g_3, g_4, \phi) = \sum_{\substack{j_1,...,j_4 \\ m_1,...,m_4 \\ n_1,...,n_4 \\ l_l, l_r}} \sigma^{j_1 j_2 j_3 j_4, l_l l_r}(\phi) \mathcal{I}^{j_1 j_2 j_3 j_4, l_l}_{m_1 m_2 m_3 m_4} \mathcal{I}^{j_1 j_2 j_3 j_4, l_r}_{n_1 n_2 n_3 n_4} \prod_{i=1}^{4} d_{j_i} D^{j_i}_{m_i n_i}(g_i), \tag{17}$$

where $D^j_{mn}(g)$ are the Wigner matrices and $d_j = 2j+1$ is the dimension of the corresponding irreducible representation. The representation label j is an element of the set $\{0, 1/2, 1, 3/2, ...\}$ while the indices m, n assume the values $-j \leq m, n \leq j$. The objects $\mathcal{I}^{j_1 j_2 j_3 j_4, l}$ are called intertwiners and are elements of the Hilbert space of states of a single tetrahedron, i.e.,

$$\mathcal{H} = L^2(G^4/G) = \bigoplus_{j_i \in \frac{\mathbb{N}}{2}} \text{Inv}\left(\otimes_{i=1}^{4} \mathcal{H}^{j_i}\right), \tag{18}$$

where $\mathcal{H}^{j_i}$ corresponds to the Hilbert space of an irreducible unitary representation of $G = \text{SU}(2)$. The index ι labels elements in a basis in $\mathcal{H}$. In this way, it is clear that the presence of the intertwiners with label ι_r is due to the imposition of the right invariance onto the field, while the left-invariance leads to the label ι_l, respectively. Hence, the quantum geometric content of the field is stored in the scalar functions $\sigma^{j_1...j_4, \iota_l \iota_r}(\phi)$ in the spin representation

When Equation (17) is injected into the action (15), one obtains an equation of motion for the condensate field which is a non-linear tensor equation and as such is notoriously difficult to solve, see Refs. [51,85,86,111,112]. We refrain from explicating the full details of this equation in the general case here and direct the reader to the original literature where these are given in depth [53,58]. Instead, we will focus on giving some details of a specific scenario relevant to cosmology the elaboration of which has led to most of the results of the condensate program.

4. Overview of Important Results

4.1. Recovery of Friedmann-Like Dynamics and Bouncing Solutions

What allows to make progress is the focusing on the case where the condensate wave function only depends on a single-spin variable, as discussed in the following. In fact, this corresponds to an isotropic restriction which leads to a highly symmetric configuration: In this way the condensate is made of equilateral tetrahedra which are the most "isotropic" configurations in a simplicial context. In effect, the domain of the left- and right-invariant field is reduced to a 1-dimensional manifold which is parametrized by a single variable, interpreted as the volume and the configuration space is that of a homogeneous and isotropic universe [53,58].[11] One thus requires the mean field to be of the form

$$\sigma^{j,\iota}(\phi) = \sigma^{j_1 j_2 j_3 j_4, \iota_l \iota_r}(\phi) \delta^{\iota_l \iota_r} \prod_{i=1}^{4} \delta^{j j_i}. \tag{19}$$

where the identification of intertwiner labels is due to the requirement that the volume be maximized in equilateral tetrahedra [53,58]. For such field configurations the action (15) (when dropping all repeated intertwiner labels for convenience) reads as

$$S = \int d\phi \sum_j \bar{\sigma}^{j,\iota} \mathcal{K}^{j,\iota} \sigma^{j,\iota} + \frac{\lambda}{5} \int d\phi \sum_j \left(\sigma^{j,\iota}\right)^5 \mathcal{V}_5(j;\iota) + \text{c.c.} \tag{20}$$

with

$$\mathcal{V}_5(j;\iota) = \mathcal{V}_5(\underbrace{j,...,j}_{10};\underbrace{\iota,...,\iota}_{5}) = f(j;\iota)w(j,\iota) \tag{21}$$

and

$$w(j,\iota) = \sum_{m_i} \prod_{i=1}^{10} (-1)^{j_i - m_i} \mathcal{I}^{jjjj,\iota}_{m_1 m_2 m_3 m_4} \mathcal{I}^{jjjj,\iota}_{-m_4 m_5 m_6 m_7} \mathcal{I}^{jjjj,\iota}_{-m_7 -m_3 m_8 m_9} \mathcal{I}^{jjjj,\iota}_{-m_9 -m_6 -m_2 m_{10}} \mathcal{I}^{jjjj,\iota}_{-m_{10} -m_8 -m_5 -m_1}. \tag{22}$$

The latter product of intertwiners can be cast into the form of a $\{15j\}$-symbol. The details of these calculations can be found in Ref. [53] and in greater detail in Ref. [58]. Again, the specific aspects of the EPRL GFT model would be encoded in the details of the objects $\mathcal{K}^j$ and $\mathcal{V}_5(j;\iota)$ (and thus $f(j;\iota)$). The interaction kernel is supposed to encode the Lorentzian embedding of the theory and thus what is known as the spin foam vertex amplitude with boundary SU(2)-states. Though its details are yet to be put down in the GFT context, its explicit form is not relevant to the results presented below.

With the above, one obtains the equation of motion of the condensate field, i.e.,

$$\mathcal{K}^j \sigma_j(\phi) + \mathcal{V}_5^j \bar{\sigma}_j(\phi)^4 = 0. \tag{23}$$

Most generally, the contribution of the kinetic term takes the form $\mathcal{K}^j = A_j \partial_\phi^2 - B_j$, where A_j and B_j parametrize ambiguities in the EPRL GFT model [53,58] and the partial derivatives with respect to the relational clock follow from a derivative expansion with respect to the same variable [53,110]. In what follows, we will see that the requirement that the Friedmann equations be recovered allows constraining of the form of A_j and B_j.

---

[11] We comment below on an alternative notion of isotropy which has been explored so far in the literature. However, notice that such a reduction is a common simplification also applied in the closely related contexts of tensor models for quantum gravity [26–32] and lattice gravity approaches [33–36].

To this aim, we follow Refs. [53,56] and consider the regime of the dynamics where the interaction term is sufficiently small as compared to the kinetic term. Since higher powers of the condensate field are directly proportional to the number of condensate constituents, we may refer to a regime where the interaction term is sub-dominant as being mesoscopic.[12] This is a crucial approximation to recover the Friedmann equations below. Then, the equation of motion reduces to

$$\partial_\phi^2 \sigma_j(\phi) - m_j^2 \sigma_j(\phi) = 0, \quad \text{with} \quad m_j^2 = \frac{B_j}{A_j} \tag{24}$$

and when using the polar decomposition of the field as $\sigma_j(\phi) = \rho_j(\phi) e^{i\theta_j(\phi)}$, yields

$$\rho_j'' - \frac{Q_j^2}{\rho_j^3} - m_j^2 \rho_j = 0 \tag{25}$$

together with the conserved quantities

$$Q_j = \rho_j^2 \theta_j' \quad \text{and} \quad E_j = (\rho_j')^2 + \rho_j^2 (\theta_j')^2 - m_j^2 \rho_j^2. \tag{26}$$

Notice that the central term in Equation (25) diverges towards $\rho_j \to 0$ to the effect that the system exhibits a quantum bounce (elaborated further below), as long as at least one Q_j is non-vanishing, see Figure 2.

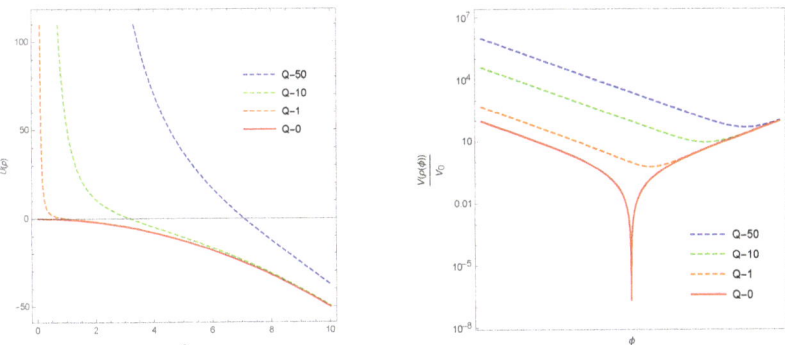

Figure 2. Equation (25) has the form of the equation of motion of a classical point particle with potential $U(\rho) = -\frac{1}{2} m^2 \rho^2 + \frac{Q^2}{2\rho^2}$. The potential is plotted for different values for Q while m and is kept fixed (**left**). Solutions to Equation (25) (initial conditions are arbitrarily chosen) lead to a plot for the volume (**right**). The point in relational time where the minimum of solutions is reached corresponds to the bounce. The solution for $Q = 0$ does not exhibit a bounce since the volume vanishes when $U(\rho)$ turns to zero. In the plots and in this caption the label j is suppressed.

This is the case when requiring the energy density of the clock field to be non-zero. This energy density is given in terms of its conserved momentum $\pi_\phi = \sum_j Q_j$ by

$$\rho_\phi = \frac{\pi_\phi^2}{2V^2} \tag{27}$$

---

[12] Notice that the term "mesoscopic" used here only refers to the number of quanta N so far. Detailed studies must determine the exact range of N for such a regime to hold true and relate it to a range of length scales in the future. Conversely, this would necessitate to study the regimes of very small and very large N where the simple field coherent state ansatz is expected to be inapplicable.

where V denotes the expectation value of the volume operator. (Please note that ρ_ϕ is not to be confused with ρ or ρ_j ascribed to the mean field.) It is explicitly given by

$$V(\phi) = \sum_j V_j \rho_j(\phi)^2, \quad \text{with} \quad V_j \sim j^{3/2} \ell_{\text{Pl}}^3. \tag{28}$$

We are now ready to give the dynamics of the volume of the emergent space, namely

$$\left(\frac{V'}{3V}\right)^2 = \left(\frac{2\sum_j V_j \rho_j \text{sgn}(\rho_j') \sqrt{E_j - Q_j^2/\rho_j^2 + m_j^2 \rho_j^2}}{3\sum_j V_j \rho_j^2}\right)^2 \quad \text{and} \quad \frac{V''}{V} = \frac{2\sum_j V_j \left(E_j + 2m_j^2 \rho_j^2\right)}{\sum_j V_j \rho_j^2}, \tag{29}$$

as obtained in Ref. [53] and call these the generalized Friedmann equations.

The classical limit of these equations is obtained when considering sufficiently large volumes for which the terms with E_j and Q_j in Equations (29) are suppressed. If one identifies then also $m_j^2 \equiv 3\pi G_N$ (where G_N denotes Newton's constant), one recovers the classical Friedmann equations of GR for a flat universe in terms of the relational clock ϕ[13],

$$\left(\frac{V'}{3V}\right)^2 = \frac{4\pi G_N}{3} \quad \text{and} \quad \frac{V''}{V} = 12\pi G_N. \tag{31}$$

Notice that the definition of G_N is understood as a definition in terms of the microscopic parameters m_j (or A_j and B_j) and not as an interpretation of the latter.

Another relevant situation where the dynamics of the volume can be solved exactly, is when the condensate is dominated by a single spin j_o.[14] In this case, the Equations (29) yield

$$\left(\frac{V'}{3V}\right)^2 = \frac{4\pi G_N}{3}\left(1 - \frac{\rho_\phi}{\rho_c}\right) + \frac{4V_{j_o} E_{j_o}}{9V} \quad \text{and} \quad \frac{V''}{V} = 12\pi G_N + \frac{2V_{j_o} E_{j_o}}{V} \tag{32}$$

where $\rho_c \sim \frac{3\pi}{2j_o^3}\rho_{\text{Pl}}$ is a critical density [56]. The terms involving ρ_c and E_{j_o} correspond to quantum corrections where the one involving ρ_c is responsible for the quantum bounce. To the past of this event, the emergent space contracts while it expands to the future. It should be remarked that up to the terms depending on E_{j_o} these equations are exactly the modified Friedmann equations derived in LQC.[15,16] For $E_{j_o} > 0$ the bounce takes place at an energy density larger than ρ_c, while for $E_{j_o} < 0$ the bounce is realized for an energy density smaller than ρ_c. Independently of the exact value of E_{j_o}, a bounce will occur. It should nevertheless be clear that the physical meaning of the conserved quantity E_j, from a fundamental point of view, is yet to be clarified. In future research it would also be important

---

[13] We exemplify the link to the standard Friedmann equations of GR for a flat universe in proper time t as compared to those in relational time ϕ via the first Friedmann equation, i.e.,

$$H^2 = \left(\frac{V'}{3V}\right)^2 \left(\frac{d\phi}{dt}\right)^2 \quad \text{with} \quad \pi_\phi = V\dot{\phi}. \tag{30}$$

The second Friedmann equation can be rewritten in a similar way, see e.g., Ref. [53]. This makes transparent that the dynamical equations for the volume as derived by GR and the condensate program take the same form. Notice that the concept of proper time does not exist in the GFT context.

[14] This is akin to what is done in LQC, where one assumes that the links of the underlying spin network are all identically labeled, with $j = \frac{1}{2}$ being the most studied case [104,105]. We will present a possible dynamical mechanism leading to a single-spin condensate further below.

[15] In fact, these background dynamics are understood to generalize the effective dynamics of LQC which can be retained as a special case. We refer to Refs. [113,114] where this point was further explored.

[16] In the given picture, the cosmological dynamics expressed by the expansion of the volume is vastly driven by a growing occupation number [115,116]. It should be remarked that this is a GFT realization of the lattice refinement of LQC [104,105].

to consider the impact of different j-modes onto the dynamics. This is in principle straightforward but would then require solving Equations (29) numerically.

Finally, to contextualize, notice that a quantum gravity induced bounce falls into the more general class of bouncing cosmologies which present tentative alternatives to the standard inflationary scenario to resolve the problems of the standard model of cosmology, see Ref. [117] for an overview.[17]

We may list important side-results which support the findings presented above:

- In Ref. [121], it is shown that for growing relational time, the condensate dynamically settles into a low-spin configuration, i.e., it will be dominated by the lowest non-trivial representations labeled by j. This goes in hand with a classicalization of the emergent geometry [57]. Following Ref. [121], this can be seen from the general solutions to Equation (24), i.e.,

$$\sigma_j(\phi) = \alpha_j^+ \exp\left(\sqrt{\frac{B_j}{A_j}}\phi\right) + \alpha_j^- \exp\left(-\sqrt{\frac{B_j}{A_j}}\phi\right) \tag{33}$$

which either lead to exponentially expanding and contracting or oscillating solutions depending on the sign of the argument of the root function. All models for which B_j/A_j has a positive maximum for some $j = j_o$ (as long as $j = 0$ is excluded[18]) lead to

$$\lim_{\phi \to \pm\infty} V(\phi) = V_{j_o} |\alpha_{j_o}^{\pm}|^2 \exp\left(\pm 2\sqrt{\frac{B_{j_o}}{A_{j_o}}}\phi\right). \tag{34}$$

A low-spin configuration is dynamically reached if the maximum of B_j/A_j occurs at a low j_o. This was demonstrated for reasonable choices of A_j and B_j in Refs. [121] and [57] to lead to $j_o = \frac{1}{2}$. One may then argue that the type of configuration which is usually assumed in the LQC literature can be derived from the quantum dynamics of GFT.

- A careful analysis shows that the identification $m_j^2 \equiv 3\pi G_N$ only holds asymptotically for large ϕ, rendering G_N a state-dependent function [54]. This is illustrated in Figure 3.

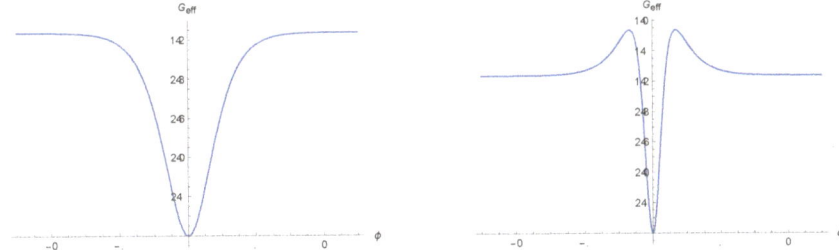

Figure 3. The effective gravitational constant as a function of relational time ϕ for $E < 0$ to the left and $E > 0$ to the right given in arbitrary units, taken from Ref. [54]. A bounce occurs towards $\phi = \Phi$ situated at the origin of both plots. For large ϕ Newton's constant is asymptotically attained. In the plots and in this caption the label j is suppressed.

---

[17] In Ref. [118] the relations between the condensate program and mimetic gravity were explored. Mimetic gravity is a Weyl-symmetric extension of GR [119] proposed to mimic the effects of cold dark matter within the context of modifications of GR. In the context of limiting curvature mimetic gravity, it is possible to realize non-singular bouncing cosmologies in the sense that it is possible to reproduce their background dynamics. This has been shown for the case of LQC [120] and very recently for the case of the effective dynamics of GFT condensates [118].

[18] We refer to Refs. [84,121] for a discussion touching on the subtle differences in between the Hilbert spaces of GFT and LQG especially relevant to the point of the zero-mode $j = 0$.

- Another related notion of isotropic restriction has been studied in the literature so far where the condensate is built from tri-rectangular tetrahedra [57]. This produces physically equivalent results in terms of the dynamics of the volume, as one would expect when invoking naive universality arguments. Notice that both isotropic restrictions correspond to symmetry reductions applied to the quantum state and thus should by no means be equated with those performed in WdW quantum cosmology or LQC. In the latter cases, symmetry reductions are imposed before quantization and this procedure is expected to violate the uncertainty principle [103].[19] In light of the above, it would be important to give a precise notion of isotropy in terms of a properly defined GFT curvature operator.
- In a related model which does not make use of the relational clock, the field content has been explicitly studied for free and effectively interacting scenarios [92]. For such static configurations one finds that the condensate consists of many GFT quanta residing in the lowest spin configurations. This is indicated by the analysis of the discrete spectra of the geometric operators, as illustrated by Figure 4. This also supports the idea that under the given isotropic restrictions, such GFT condensate states are suitable candidates to describe effectively continuous homogeneous and isotropic 3-spaces built from many smallest building blocks of the quantum geometry.

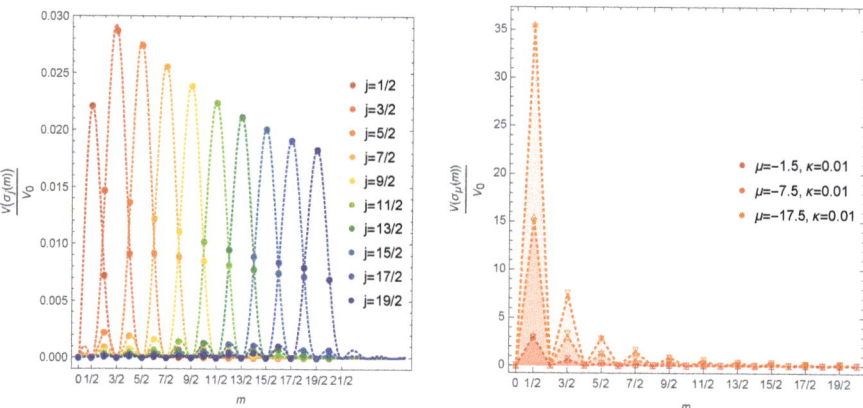

Figure 4. The spectrum to the volume operator in the free case to the left and effectively interacting case to the right. These illustrate that the geometric operators are dominated by the lowest non-trivial modes characterizing the condensate field. These plots were taken from Ref. [92] where a detailed discussion of their underlying computations can be found.

4.2. Cyclic Cosmologies and Accelerated Expansion

In a series of works the effect of simplified GFT interactions onto the cosmological dynamics has been investigated under the assumption that the spin representation j is fixed [55,57] (which may be motivated by the above-described process of reaching a low-spin configuration). Given these phenomenologically motivated interactions, the equations of motion take a simple non-linear form. Despite the fact that from a GFT point of view such interactions seem to be somewhat artificial due to their lack of a discrete geometric interpretation, they bring us nearer to the physics which we want to

---

[19] For a recent attempt at imposing a quantum counterpart of the classical symmetry reduction in the LQG context, we refer to Refs. [122,123].

probe as they capture the basic non-linearity of the original GFT interactions. The form of the effective potential is given by

$$V = B|\sigma(\phi)|^2 + \frac{2}{n}w|\sigma(\phi)|^n + \frac{2}{n'}w'|\sigma(\phi)|^{n'}, \quad \text{with} \quad n > n' \tag{35}$$

and we require $w' > 0$ so that the potential is bounded from below. Using the polar form of the field, we obtain the equation of motion

$$\rho'' - \frac{Q^2}{\rho^3} - m^2\rho + \lambda\rho^{n-1} + \mu\rho^{n'-1} = 0 \tag{36}$$

where we have set

$$\lambda \equiv -\frac{w}{A} \quad \text{and} \quad \mu \equiv -\frac{w'}{A}. \tag{37}$$

To guarantee that this equation does not lead to an open cosmology expanding at a faster than exponential rate, we have $\mu > 0$. In consistency with the free case discussed above, $m^2 > 0$ while the sign of λ can be left unconstrained. A first observation from this equation of motion is its resemblance with that of a classical point particle in the potential

$$U(\rho) = -\frac{m^2}{2}\rho^2 + \frac{Q^2}{2\rho^2} + \frac{\lambda}{n}\rho^n + \frac{\mu}{n'}\rho^{n'} \tag{38}$$

so that with the given signs and the bouncing contribution of strength Q^2 the solutions to Equation (36) yield cyclic motions. Via Equation (28) these correspond to cyclic solutions for the dynamics of the emergent universe. Hence, we observe that bounded interactions induce a recollapse. Given that in the classical theory a recollapsing solution follows from a closed topology of 3-space, this might give an indication of how to obtain such topologies from these simple GFT condensates.

Regarding the expansion behavior of the emergent geometry, using the above-given interactions it is possible to obtain a long-lasting accelerated phase after the bounce. In fact, the free parameters may be fine-tuned to achieve any desirable value of e-folds so that this behavior can be understood as an inflationary expansion of quantum geometric origin. This becomes transparent when writing for the number of e-folds

$$N = \frac{1}{3}\log\left(\frac{V_{end}}{V_{bounce}}\right) = \frac{2}{3}\log\left(\frac{\rho_{end}}{\rho_{bounce}}\right) \tag{39}$$

and incorporating it in an expression for the acceleration. Since there is no notion of proper time in GFT, a sensible definition of acceleration can only be given in relational terms. In particular, we seek a definition that agrees with the standard one given in ordinary cosmology via the Raychaudhuri equation which allows us to define the acceleration as

$$\mathfrak{a}(\rho) = \frac{V''}{V} - \frac{5}{3}\left(\frac{V'}{V}\right)^2 \tag{40}$$

the derivation of which is discussed in detail in Refs. [54,55]. Using this, one finds that the free case does not lead to a value of N large enough to supplant the standard inflationary mechanism in cosmology. However, the careful analysis of Ref. [55] demonstrates that with a hierarchy $\mu \ll |\lambda|$ together with $\lambda > 0$ leads to $n' > n \geq 5$ which allows room for an era of accelerated expansion analogous to that of models of inflationary cosmology. If phantom energy is ruled out, only $n = 5$ and $n' = 6$ are admissible selecting an interaction term which is in principle compatible with simplicial interactions, introduced above. These results are illustrated in Figure 5. Notice that these works emphasize the role that phenomenology can take for model building in quantum gravity.

It should be emphasized that these findings have a purely quantum geometric origin and in particular are not based in any way on the assumption of a specific potential for the minimally coupled massless scalar field ϕ, the relational clock. This is in stark contrast to inflation which depends on the choice of potential and initial conditions of the inflaton field to yield the desired expansion behavior [124–127].

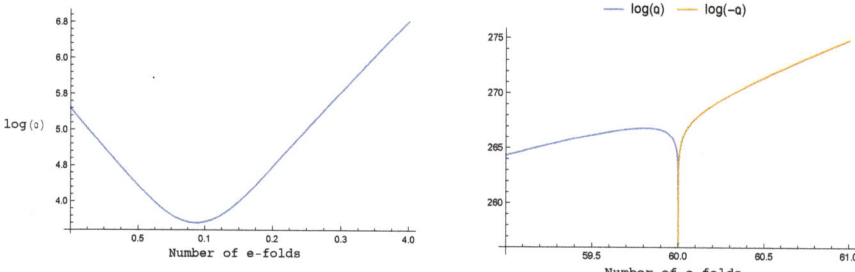

Figure 5. The plot to the left shows the behavior of the acceleration close to the bounce while the one to the right illustrates it towards the end of inflation. These plots were taken from Ref. [55] to which we direct for details.

4.3. Anisotropies and Inhomogeneities

If quantum gravity is to offer the picture of the earliest moments of our universe, it must include an approximately homogeneous and isotropic background with superimposed perturbations. Given the above results an important step for the condensate cosmology program is to go beyond the considered isotropic restriction and homogeneous configurations and to study more general configurations and their dynamics. In the following, we want to briefly discuss recent advances in which the exploration of anisotropies [57,58] and inhomogeneities [59] has been commenced.

The study of anisotropic GFT configurations and their dynamics is of general importance since it would be desirable to see if at least a subset of these can agree with the observed isotropy of our universe at late times. For these one must show that anisotropies do not grow in the expanding phase. Apart from that, it is well known that bouncing cosmologies are haunted by the notorious problem of uncontrolled growth of anisotropies when the universe contracts [117]. This is the problem of the Belinsky–Khalatnikov–Lifshitz instability [128,129]. In light of this, it is interesting to understand the fate of anisotropies when approaching the quantum bounce as predicted by GFT.

Leaving the technical details aside, in Ref. [57] it has indeed been shown for rather general configurations that they dynamically isotropize in relational time by means of a simple mechanism (which is akin to the one responsible for settling the system into a low-spin configuration, as described in Section 4.1). Conversely, it is demonstrated that anisotropic contributions to the condensate become increasingly pronounced towards small volumes. This paved the way to a systematic investigation of anisotropic perturbations over an isotropic background in the vicinity of the bounce in Ref. [58]. In particular, a region in the parameter space is identified such that these anisotropies can be large at the bounce but are fully under control. From this it also follows that towards the bounce the quantum geometry of the emergent universe is rather degenerate. Furthermore, this analysis shows that the anisotropic perturbations become negligible the further away the system is from the bouncing phase and can be completely irrelevant to the dynamics before interactions kick in. Hence, after the bounce a cosmological background emerges the dynamics of which can again be cast into the form of the above-given effective Friedmann equations, thus corroborating the results of Refs. [53,56,57]. These results are illustrated in Figure 6. On more general grounds, these studies form a crucial starting point towards identifying anisotropic cosmologies, i.e., Bianchi models, within this approach and allow establishment of contact with corresponding studies in WdW [103], spin foam [130–142] and loop quantum cosmology [104,105].

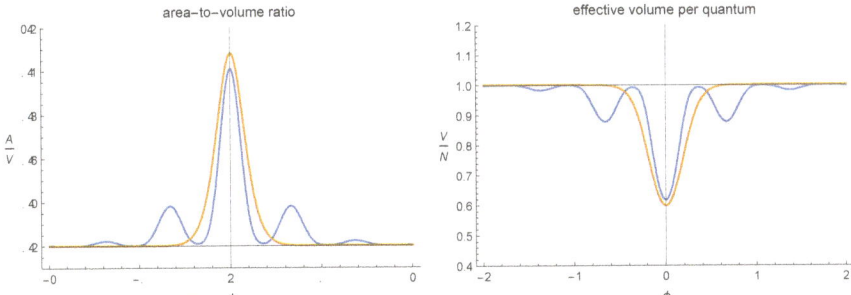

Figure 6. The surface-area-to-volume ratio (**left**) and the effective volume per quantum (**right**) as a function of relational time. These figures demonstrate the degenerate character of the quantum geometry towards the bounce and the damping of the anisotropies in the outgoing phase. The plots were taken from Ref. [58] where detailed explanations are given.

Modern Cosmology teaches us that the seeds for structure formation are represented by inhomogeneities in the very early universe [143]. Hence, the identification and study of cosmological inhomogeneities in the condensate approach is mandatory to promote it to a realistic contestant theory of quantum cosmology. In particular, the goal would be to find a mechanism rooted in quantum geometry which explains the origins of inhomogeneities without referring to the inflationary paradigm where the inhomogeneities correspond to quantum fluctuations of the inflaton field [124].

Recent progress building on Refs. [144,145], allows extension of the formalism beyond homogeneity. In Refs. [59,146] the formalism of including a free massless scalar field [53,110] is extended to incorporate four reference scalar fields which are used as relational clocks and rods, i.e., as a physical coordinate system. Again, this procedure is common in classical and quantum gravity approaches [108,109]. In this setting quantum fluctuations (i.e., small inhomogeneities) of the local 3-volume around a nearly homogeneous background geometry are studied. Their power spectrum can be calculated, and this is shown to be approximately scale-invariant (where the scale is defined by the reference matter), the amplitude is small, and decreases as the emergent universe expands. However, it was also demonstrated that analogous statements do not hold for perturbations in the total density of the scalar fields when the gradient energy is non-negligible. Notice that the details of these arguments relied on the specific choice of condensate state which solves the condensate dynamics and thus depends on the approximation scheme summarized in Sections 3.1 and 3.2. More recent work [147] has shown how the transition from the initial quantum fluctuations present in the deep quantum gravity regime to classical observable inhomogeneities can be accomplished. By and large, it is striking that features of the spectrum of cosmologically relevant observables can be recognized using the condensate formalism. Future research has to bridge the gap between observations of the early universe and the condensate formalism and the hope is that the incorporation of more complicated matter dynamics can reproduce observationally viable results.[20]

5. Discussion and Outlook

In this brief review we wanted to draw attention to key results of the GFT condensate cosmology program which illustrate its potential to provide a quantum gravitational foundation for early universe cosmology. Finally, we would like to point to open directions (if not already stated in the main body of this appetizer) and address some relations with other non-perturbative discrete quantum gravity approaches trying to extract cosmological solutions from their path integral formulations.

---

[20] Notice that this proposal of incorporating inhomogeneities has recently been further developed using the separate universe approach to describe long-wavelength scalar perturbations [148].

Beyond its application to cosmology showing a rich phenomenology, this program has revealed an interesting and powerful perspective on the extraction of continuum information from a discrete geometric setting: The field theoretic setting of GFT and specifically the use of field coherent states prove extremely useful and elegant to this aim.

To contextualize this, we may compare with the other non-perturbative and discrete path integral approaches to quantum gravity. In EDT, CDT, tensor models, the discreteness of geometry is regarded as a mathematical tool allowing us to rewrite the continuum path integral in a discrete form. The philosophy behind taking the continuum limit is rather similar among them and the goal is to study the phase structure of the respective theories via analyticity properties of the partition function. The CDT approach has been able to produce physically relevant, i.e., extended macroscopic geometries which obey effective minisuperspace dynamics [34–36]. This potentially highlights the role of causality in facilitating the escape from the sector of unphysical continuum geometries EDT [34–36] and TMs [26–32] are so far stuck with.

A point of criticism often invoked regarding these approaches, concerns the lack of a clear interpretation of expectation values of observables rigorously defined on a physical Hilbert space, which is in principle available in covariant LQG and GFT. Given this, GFT condensate cosmology is not the only approach which tries to extract the cosmological sector of LQG from a covariant formulation of its dynamics. In the spin foam cosmology approach, one uses the spin foam expansion which is an expansion in terms of the number of degrees of freedom [130–137]. It has mostly been studied for so-called dipole graphs (corresponding to the simplest cellular decomposition of the 3-sphere) [138–140] and can be extended to more general regular graphs [141,142]. A central assumption of this approach is that a *fixed* number of quanta of geometry captures all relevant physics. In contrast, in the condensate program one looks for continuum physics away from the Fock vacuum and does not restrict the number of quanta which can be *rather large and dynamical*.

Although GFT allows studying of an infinite class of simplicial complexes by construction, it provides the field theoretic approximation tools to study the physics of many LQG degrees of freedom while bypassing the treatment of highly complicated spin networks. In addition, in the spin foam context one typically studies the semi-classical limit by requiring the configurations to peak on some triangulated classical geometry [20–23].[21] This point of view is also not assumed in GFT condensate cosmology, motivated by the condensate hypothesis which fixes the states to be described by the simple condensate wave function.

Being aware that the choice of simple trial states and the disregard of proper simplicial interactions so far neglects all the connectivity information of those spin networks that would be considered important for a realistic definition of a non-perturbative continuum vacuum state, it is pressing to go beyond the given simplifications. The exploration of the phenomenologically motivated interactions presented here as well as the study of dipole condensates [38,39] goes into this direction. Notice that recent work [51] in the context of the dynamical Boulatov model which is a model for Euclidean quantum gravity in 3d has shown that non-trivial condensate solutions can be produced where simplicial interactions are fully considered. The background quantum geometries one yields in this way must be better understood but this procedure could in principle be carried over to the case in 4d. Also, it is clear from this example that the choice of simple states does not pose a major problem since more complicated ones associated with connected graphs are then easily generated by the simplicial interaction term. In light of this, it would be important to study the relational evolution of such properly interacting condensates to see if the intriguing results regarding the accelerated expansion of the emergent geometry found via exploring the simplified interactions can be reproduced. Furthermore, studying the effect of the simplicial interaction of the Lorentzian EPRL GFT model (or

---

[21] A different and interesting take on regaining the continuum in spin foam models is presented by the spin foam coarse graining and renormalization program for which we refer to Refs. [149–157].

any related model) onto the condensate will require putting it explicitly down in terms of its boundary data [58]. It could also be interesting to consider a colored version of such a model, given the insight that in perturbative expansion of the partition function the color degree of freedom guarantees that all terms are free of topological pathologies [67–69].

Apart from their impact on the dynamics, a better understanding of interactions will also allow construction of more sophisticated observables capturing curvature and cosmological anisotropies which in turn will prove indispensable to classify different emergent geometries from one another. It is nevertheless remarkable that in the regime where interactions are sub-dominant and which has been explored most so far, rich Friedmann-like dynamics can be obtained.

A related point to be focused on, touches on higher order corrections to the so-far considered condensate equation of motion and to understand if they can be neglected or if they would have a drastic impact on the cosmological interpretation of this approach. Understanding the full quantum dynamics will then shed light onto the phase structure of interesting GFT models of $4d$ quantum gravity. In other words, it must be checked by means of non-perturbative techniques if a simplicial GFT model for $4d$ quantum gravity can truly exhibit a phase or phases which are related to $(3+1)$-dimensional Lorentzian continuum geometries. In the context of exploring the notion of phases in GFT, it would also be important to understand the relation in between a potential phase transition into a geometric phase for which the order parameter should vanish and the occurrence of a bounce which (by definition) forbids a zero-volume state.

At the very end, this approach will be judged by the ability to extract phenomenological signatures to see if it can be a realistic contestant theory of quantum cosmology. The development of a scenario to explain the origin of cosmological perturbations neither by the mechanism provided by inflation nor as in ordinary bounce models but rather via the quantum fluctuations of the geometry itself as given by GFT condensates is an important step into this direction.

Author Contributions: All authors contributed equally to all aspects of this manuscript.

Funding: M. S. is supported in part by the Science and Technology Facility Council (STFC), United Kingdom, under the research grant ST/P000258/1.

Acknowledgments: The authors thank the referees for their remarks which led to an improvement of the manuscript.

Conflicts of Interest: The authors declare no conflict of interest.

References

1. Akrami, Y.; Arroja, F.; Ashdown, M.; Aumont, J.; Baccigalupi, C.; Ballardini, M.; Banday, A.J.; Barreiro, R.B.; Bartolo, N.; Basak, S.; Battye, R. Planck 2018 results. I. Overview and the cosmological legacy of Planck. *arXiv* **2018**, arXiv:1807.06205.
2. Hawking, S.W.; Ellis, G.F.R. *The Large Scale Structure of Space-Time*; Cambridge Monographs on Mathematical Physics; CUP: Cambridge, UK, 2011.
3. De Witt, B.S. Quantum theory of gravity. i. the canonical theory. *Phys. Rev.* **1967**, *160*, 1113–1148. [CrossRef]
4. Oriti, D. (Ed.) *Approaches to Quantum Gravity*; Cambridge University Press: Cambridge, UK, 2009.
5. Sakellariadou, M. Quantum Gravity and Cosmology: An intimate interplay. *IOP Conf. Ser. J. Phys. Conf. Ser.* **2017**, *880*, 012003. [CrossRef]
6. Kiefer, C.; Krämer, M. Quantum Gravitational Contributions to the CMB Anisotropy Spectrum. *Phys. Rev. Lett.* **2012**, *108*, 021301. [CrossRef] [PubMed]
7. Kiefer, C.; Krämer, M. Can effects of quantum gravity be observed in the cosmic microwave background? *Int. J. Mod. Phys.* **2012**, *D21*, 1241001. [CrossRef]
8. Kiefer, C.; Krämer, M. On the Observability of Quantum-Gravitational effects in the Cosmic Microwave Background. *Springer Proc. Phys.* **2014**, *157*, 531–538.
9. Agullo, I.; Ashtekar, A.; Nelson, W. A Quantum Gravity Extension of the Inflationary Scenario. *Phys. Rev. Lett.* **2012**, *109*, 251301. [CrossRef]
10. Agullo, I.; Ashtekar, A.; Nelson, W. Extension of the quantum theory of cosmological perturbations to the Planck era. *Phys. Rev. D* **2013**, *87*, 043507. [CrossRef]

11. Agullo, I.; Ashtekar, A.; Nelson, W. The pre-inflationary dynamics of loop quantum cosmology: Confronting quantum gravity with observations. *Class. Quantum Gravity* **2013**, *30*, 085014. [CrossRef]
12. Weinberg, S. Effective Field Theory, Past and Future. *arXiv* **2009**, arXiv:0908.1964.
13. Eichhorn, A. Status of the asymptotic safety paradigm for quantum gravity and matter. *Found. Phys.* **2018**, *48*, 1407–1429.
14. Percacci, R. *Introduction To Covariant Quantum Gravity And Asymptotic Safety*; World Scientific: Singapore, 2017.
15. Reuter, M.; Saueressig, F. *Quantum Gravity and the Functional Renormalization Group*; Cambridge University Press: Cambridge, UK, 2019.
16. Blumenhagen, R.; Lüst, D.; Theisen, S. *Basic Concepts of String Theory*; Springer: Berlin/Heidelberg, Germany, 2012.
17. Connes, A. *Noncommutative Geometry*; Academic Press: Amsterdam, The Netherlands, 1994
18. Chamseddine, A.H.; Fröhlich, J.; Grandjean, O. The Gravitational Sector in the Connes-Lott Formulation of the Standard Model. *J. Math. Phys.* **1995**, *36*, 6255–6275. [CrossRef]
19. Chamseddine, A.H.; Connes, A. The Spectral Action Principle. *Commun. Math. Phys.* **1997**, *186*, 731. [CrossRef]
20. Ashtekar, A.; Lewandowski, J. Background Independent Quantum Gravity: A Status Report. *Class. Quantum Gravity* **2004**, *21*, R53. [CrossRef]
21. Thiemann, T. *Modern Canonical Quantum General Relativity*; Cambridge University Press: Cambridge, UK, 2007.
22. Rovelli, C. Zakopane lectures on loop gravity. *arXiv* **2011**, arXiv:1102.3660.
23. Perez, A. The Spin Foam Approach to Quantum Gravity. *Living Rev. Relativ.* **2013**, *16*, 3. [CrossRef]
24. Freidel, L. Group field theory: An Overview. *Int. J. Theor. Phys.* **2005**, *44*, 1769–1783. [CrossRef]
25. Oriti, D. (Ed.) The Group field theory approach to quantum gravity. In *Approaches to Quantum Gravity*; Cambridge University Press: Cambridge, UK, 2009; pp. 310–331.
26. Ambjørn, J.; Durhuus, B.; Jonsson, T. Three-dimensional simplicial quantum gravity and generalized matrix models. *Mod. Phys. Lett.* **1991**, *A6*, 1133–1146. [CrossRef]
27. Gross, M. Tensor models and simplicial quantum gravity in > 2-D. *Nucl. Phys. Proc. Suppl.* **1992**, *25A*, 144–149. [CrossRef]
28. Gurau, R. Invitation to Random Tensors. *SIGMA* **2016**, *12*, 094. [CrossRef]
29. Gurau, R. *Random Tensors*; Oxford University Press: Oxford, UK, 2016.
30. Rivasseau, V. The Tensor Track, IV. *arXiv* **2016**, arXiv:1604.07860.
31. Rivasseau, V. Random Tensors and Quantum Gravity. *SIGMA* **2016**, *12*, 069. [CrossRef]
32. Delporte, N.; Rivasseau, V. The Tensor Track V: Holographic Tensors. *arXiv* **2018**, arXiv:1804.11101.
33. Williams, R.M. Quantum Regge Calculus. In *Approaches to Quantum Gravity*; Oriti, D., Ed.; Cambridge University Press: Cambridge, UK, 2009; pp. 360–377.
34. Loll, R. Discrete approaches to quantum gravity in four dimensions. *Living Rev. Relativ.* **1998**, *1*, 13. [CrossRef] [PubMed]
35. Ambjørn, J.; Görlich, A.; Jurkiewicz, J.; Loll, R. Nonperturbative quantum gravity. *Phys. Rep.* **2012**, *519*, 127–210. [CrossRef]
36. Ambjørn, J.; Görlich, A.; Jurkiewicz, J.; Loll, R. Quantum gravity via causal dynamical triangulations. In *Springer Handbook of Spacetime*; Springer: Berlin, Germany, 2014; pp. 723–741.
37. Pithis, A.G.A. Aspects of Quantum Gravity. Ph.D. Thesis, University of London, King's College. Available online: https://kclpure.kcl.ac.uk/portal/en/theses/aspects-of-quantum-gravity(904af0f7-5dcc-4387-b905-32b5869db8c9).html (accessed on 31 March 2019).
38. Gielen, S.; Oriti, D.; Sindoni, L. Cosmology from Group Field Theory Formalism for Quantum Gravity. *Phys. Rev. Lett.* **2013**, *111*, 031301. [CrossRef] [PubMed]
39. Gielen, S.; Oriti, D.; Sindoni, L. Homogeneous cosmologies as group field theory condensates. *J. High Energy Phys.* **2014**, *1406*, 013. [CrossRef]
40. Oriti, D. The Universe as a Quantum Gravity Condensate. In *"Testing Quantum Gravity with Cosmology" of Comptes Rendus Physique*; Barrau, A., Ed.; Académie des Sciences: Paris, France, 2017.
41. Gielen, S.; Sindoni, L. Quantum Cosmology from Group Field Theory Condensates: A Review. *SIGMA* **2016**, *12*, 082. [CrossRef]

42. Oriti, D. Disappearance and emergence of space and time in quantum gravity. *Stud. Hist. Philos. Sci. Part B Stud. Hist. Philos. Mod. Phys.* **2014**, *46*, 186–199. [CrossRef]
43. Benedetti, D.; Geloun, J.B.; Oriti, D. Functional Renormalisation Group Approach for Tensorial Group Field Theory: A Rank-3 Model. *J. High Energy Phys.* **2015**, *3*, 084. [CrossRef]
44. Benedetti, D.; Lahoche, V. Functional Renormalization Group Approach for Tensorial Group Field Theory: A Rank-6 Model with Closure Constraint. *Class. Quantum Gravity* **2016**, *33*, 095003. [CrossRef]
45. Geloun, J.B.; Koslowski, T.A. Nontrivial UV behavior of rank-4 tensor field models for quantum gravity. *arXiv* **2016**, arXiv:1606.04044.
46. Carrozza, S.; Lahoche, V. Asymptotic safety in three-dimensional SU(2)-Group Field Theory: Evidence in the local potential approximation. *Class. Quantum Gravity* **2017**, *34*, 115004. [CrossRef]
47. Geloun, J.B.; Koslowski, T.A.; Oriti, D.; Pereira, A.D. Functional Renormalization Group analysis of rank 3 tensorial group field theory: The full quartic invariant truncation. *Phys. Rev. D* **2018**, *97*, 126018. [CrossRef]
48. Geloun, J.B.; Martini, R.; Oriti, D. Functional Renormalisation Group analysis of a Tensorial Group Field Theory on R3. *EPL (Europhys. Lett.)* **2015**, *112*, 31001. [CrossRef]
49. Geloun, J.B.; Martini, R.; Oriti, D. Functional Renormalisation Group analysis of Tensorial Group Field Theories on R^d. *Phys. Rev. D* **2016**, *94*, 024017. [CrossRef]
50. Carrozza, S. Flowing in Group Field Theory Space: A Review. *SIGMA* **2016**, *12*, 070. [CrossRef]
51. Geloun, J.B.; Kegeles, A.; Pithis, A.G.A. Minimizers of the equilateral dynamical Boulatov model. *Eur. Phys. J. C* **2018**, *78*, 996. [CrossRef]
52. Pithis, A.G.A.; Thürigen, J. Phase transitions in group field theory: The Landau perspective. *Phys. Rev. D* **2018**, *98*, 126006. [CrossRef]
53. Oriti, D.; Sindoni, L.; Wilson-Ewing, E. Emergent Friedmann dynamics with a quantum bounce from quantum gravity condensates. *Class. Quantum Gravity* **2016**, *33*, 224001. [CrossRef]
54. de Cesare, M.; Sakellariadou, M. Accelerated expansion of the Universe without an inflaton and resolution of the initial singularity from Group Field Theory condensates. *Phys. Lett. B* **2017**, *764*, 49–53. [CrossRef]
55. de Cesare, M.; Pithis, A.G.A.; Sakellariadou, M. Cosmological implications of interacting Group Field Theory models: Cyclic Universe and accelerated expansion. *Phys. Rev. D* **2016**, *94*, 064051. [CrossRef]
56. Oriti, D.; Sindoni, L.; Wilson-Ewing, E. Bouncing cosmologies from quantum gravity condensates. *Class. Quantum Gravity* **2017**, *34*, 04LT01. [CrossRef]
57. Pithis, A.G.A.; Sakellariadou, M. Relational evolution of effectively interacting GFT quantum gravity condensates. *Phys. Rev. D* **2017**, *95*, 064004. [CrossRef]
58. de Cesare, M.; Sakellariadou, M.; Pithis, A.G.A.; Oriti, D. Dynamics of anisotropies close to a cosmological bounce in quantum gravity. *Class. Quantum Gravity* **2018**, *35*, 015014. [CrossRef]
59. Gielen, S.; Oriti, D. Cosmological perturbations from full quantum gravity. *Phys. Rev. D* **2018**, *98*, 106019. [CrossRef]
60. Calcagni, G.; Gielen, S.; Oriti, D. Group field cosmology: A cosmological field theory of quantum geometry. *Class. Quantum Gravity* **2012**, *29*, 105005. [CrossRef]
61. Freidel, L.; Livine, E.R. 3d Quantum Gravity and Effective Non-Commutative Quantum Field Theory. *Phys. Rev. Lett.* **2006**, *96*, 221301. [CrossRef]
62. Baratin, A.; Dittrich, B.; Oriti, D.; Tambornino, J. Non-commutative flux representation for loop quantum gravity. *Class. Quantum Gravity* **2011**, *28*, 175011. [CrossRef]
63. Gielen, S. Quantum cosmology of (loop) quantum gravity condensates: An example. *Class. Quantum Gravity* **2014**, *31*, 155009. [CrossRef]
64. Ben Geloun, J. On the finite amplitudes for open graphs in Abelian dynamical colored Boulatov-Ooguri models. *J. Phys. A* **2013**, *46*, 402002. [CrossRef]
65. Reisenberger, M.P.; Rovelli, C. Spacetime as a Feynman diagram: The connection formulation. *Class. Quantum Gravity* **2001**, *18*, 121. [CrossRef]
66. Oriti, D.; Rosati, G. Non-commutative Fourier transform for the Lorentz group via the Duflo map. *Phys. Rev. D* **2019**, *99*, 106005. [CrossRef]
67. Gurau, R. Colored Group Field Theory. *Commun. Math. Phys.* **2011**, *304*, 69–93. [CrossRef]
68. Bonzom, V.; Gurau, R.; Rivasseau, V. Random tensor models in the large N limit: Uncoloring the colored tensor models. *Phys. Rev. D* **2012**, *85*, 084037. [CrossRef]

69. Gurau, R. Lost in translation: Topological singularities in group field theory. *Class. Quantum Gravity* **2010**, *27*, 235023. [CrossRef]
70. De Pietri, R.; Freidel, L. so(4) Plebanski Action and Relativistic Spin Foam Model. *Class. Quantum Gravity* **1999**, *16*, 2187–2196. [CrossRef]
71. Barrett, J.W.; Crane, L. A Lorentzian Signature Model for Quantum General Relativity. *Class. Quantum Gravity* **2000**, *17*, 3101–3118. [CrossRef]
72. Perez, A.; Rovelli, C. Spin foam model for Lorentzian General Relativity. *Phys. Rev. D* **2001**, *63*, 041501. [CrossRef]
73. Freidel, L.; Krasnov, K. A New Spin Foam Model for 4d Gravity. *Class. Quantum Gravity* **2008**, *25*, 125018. [CrossRef]
74. Engle, J.; Livine, E.; Pereira, R.; Rovelli, C. LQG vertex with finite Immirzi parameter. *Nucl. Phys.* **2008**, *B799*, 136. [CrossRef]
75. Dupuis, M.; Livine, E.R. Lifting SU(2) Spin Networks to Projected Spin Networks. *Phys. Rev. D* **2010**, *82*, 064044. [CrossRef]
76. Ding, Y.; Rovelli, C. Physical boundary Hilbert space and volume operator in the Lorentzian new spin-foam theory. *Class. Quantum Gravity* **2010**, *27*, 205003. [CrossRef]
77. Freidel, L.; Speziale, S. On the Relations between Gravity and BF Theories. *SIGMA* **2012**, *8*, 032. [CrossRef]
78. Kaminksi, W.; Kisielowski, M.; Lewandowski, J. Spin-Foams for All Loop Quantum Gravity. *Class. Quantum Gravity* **2010**, *27*, 095006, Erratum in **2012**, *29*, 049502.
79. Speziale, S.; Wieland, W.M. The twistorial structure of loop-gravity transition amplitudes. *Phys. Rev. D* **2012**, *86*, 124023. [CrossRef]
80. Geloun, J.B.; Gurau, R.; Rivasseau, V. EPRL/FK Group Field Theory. *Europhys. Lett.* **2010**, *92*, 60008. [CrossRef]
81. Baratin, A.; Oriti, D. Quantum simplicial geometry in the group field theory formalism: Reconsidering the Barrett-Crane model. *New J. Phys.* **2011**, *13*, 125011. [CrossRef]
82. Han, M.; Zhang, M. Asymptotics of Spinfoam Amplitude on Simplicial Manifold: Euclidean Theory. *Class. Quantum Gravity* **2012**, *29*, 165004. [CrossRef]
83. Baratin, A.; Oriti, D. Group field theory with non-commutative metric variables. *Phys. Rev. Lett.* **2010**, *105*, 221302. [CrossRef]
84. Oriti, D. Group field theory as the 2nd quantization of Loop Quantum Gravity. *Class. Quantum Gravity* **2016**, *33*, 085005. [CrossRef]
85. Girelli, F.; Livine, E.R.; Oriti, D. 4d Deformed Special Relativity from Group Field Theories. *Phys. Rev. D* **2010**, *81*, 024015. [CrossRef]
86. Girelli, F.; Livine, E.R. A Deformed Poincare Invariance for Group Field Theories. *Class. Quantum Gravity* **2010**, *27*, 245018. [CrossRef]
87. Oriti, D.; Pranzetti, D.; Ryan, J.P.; Sindoni, L. Generalized quantum gravity condensates for homogeneous geometries and cosmology. *Class. Quantum Gravity* **2015**, *32*, 235016. [CrossRef]
88. Rovelli, C.; Smolin, L. Knot Theory and Quantum Gravity. *Phys. Rev. Lett.* **1988**, *61*, 1155. [CrossRef] [PubMed]
89. Rovelli, C.; Smolin, L. Loop space representation of quantum general relativity. *Nucl. Phys. B* **1990**, *331*, 80152. [CrossRef]
90. Rovelli, C.; Smolin, L. Discreteness of area and volume in quantum gravity. *Nucl. Phys. B* **1995**, *442*, 593–622; Erratum in **1995**, *456*, 753. [CrossRef]
91. Rovelli, C.; Smolin, L. Spin networks and quantum gravity. *Phys. Rev. D* **1995**, *52*, 57435759. [CrossRef]
92. Pithis, A.G.A.; Sakellariadou, M.; Tomov, P. Impact of nonlinear effective interactions on GFT quantum gravity condensates. *Phys. Rev. D* **2016**, *94*, 064056. [CrossRef]
93. Kegeles, A.; Oriti, D.; Tomlin, C. Inequivalent coherent state representations in group field theory. *Class. Quantum Gravity* **2018**, *35*, 125011. [CrossRef]
94. Kotecha, I.; Oriti, D. Statistical Equilibrium in Quantum Gravity: Gibbs states in Group Field Theory. *New J. Phys.* **2018**, *20*, 073009. [CrossRef]
95. Chirco, G.; Kotecha, I.; Oriti, D. Statistical equilibrium of tetrahedra from maximum entropy principle. *Phys. Rev. D* **2019**, *99*, 086011. [CrossRef]
96. Pitaevskii, L.; Stringari, S. *Bose-Einstein Condensation*, 1st ed.; Oxford University Press: Oxford, UK, 2003.

97. Kevrekidis, P.G., Frantzeskakis, D.J., Carretero-Gonzalez, R., Eds. *Emergent Nonlinear Phenomena in Bose-Einstein Condensates Theory and Experiment*; Springer: Berlin, Germany, 2008.
98. Yukalov, V.I. Theory of cold atmos: Bose-Einstein statistics. *Laser Phys.* **2016**, *26*, 062001. [CrossRef]
99. Oriti, D.; Pranzetti, D.; Sindoni, L. Horizon entropy from quantum gravity condensates. *Phys. Rev. Lett.* **2016**, *116*, 211301. [CrossRef] [PubMed]
100. Oriti, D.; Pranzetti, D.; Sindoni, L. Black Holes as Quantum Gravity Condensates. *Phys. Rev. D* **2018**, *97*, 066017. [CrossRef]
101. Oriti, D. *The Universe as a Quantum Gravity Condensate, Extended Version of the Invited Contribution to the Special Issue "Testing Quantum Gravity with Cosmology" of Comptes Rendus Physique*; Académie des Sciences: Paris, France, 2017.
102. Sindoni, L. Effective equations for GFT condensates from fidelity. *arXiv* **2014**, arXiv:1408.3095.
103. Kiefer, C. *Quantum Gravity*, 3rd ed.; Oxford University Press: Oxford, UK, 2012.
104. Ashtekar, A.; Singh, P. Loop Quantum Cosmology: A Status Report. *Class. Quantum Gravity* **2011**, *28*, 213001. [CrossRef]
105. Banerjee, K.; Calcagni, G.; Martin-Benito, M. Introduction to Loop Quantum Cosmology. *SIGMA* **2012**, *8*, 016. [CrossRef]
106. Dittrich, B. Partial and complete observables for canonical General Relativity. *Class. Quantum Gravity* **2006**, *23*, 6155–6184. [CrossRef]
107. Rovelli, C. Partial observables. *Phys. Rev. D* **2002**, *65*, 124013. [CrossRef]
108. Brown, D.; Kuchar, K. Dust as a standard of space and time in canonical quantumgravity. *Phys. Rev. D* **1995**, *51*, 5600–5629. [CrossRef]
109. Giesel, K.; Thiemann, T. Scalar material referencesystems and loop quantum gravity. *Class. Quantum Gravity* **2015**, *32*, 135015. [CrossRef]
110. Li, Y.; Oriti, D.; Zhang, M. Group field theory for quantum gravity minimally coupled to a scalar field. *Class. Quantum Gravity* **2017**, *34*, 195001. [CrossRef]
111. Fairbairn, W.J.; Livine, E.R. 3d spinfoam quantum gravity: Matter as a phase of the group field theory. *Class. Quantum Gravity* **2007**, *24*, 5277. [CrossRef]
112. Livine, E.R.; Oriti, D.; Ryan, J.P. Effective Hamiltonian Constraint from Group Field Theory. *Class. Quantum Gravity* **2011**, *28*, 245010. [CrossRef]
113. Adjei, E.; Gielen, S.; Wieland, W. Cosmological evolution as squeezing: A toy model for groupfield cosmology. *Class. Quantum Gravity* **2018**, *35*, 105016. [CrossRef]
114. Wilson-Ewing, E. A relational Hamiltonian for group field theory. *Phys. Rev. D* **2019**, *99*, 086017. [CrossRef]
115. Gielen, S.; Oriti, D. Quantum cosmology from quantum gravity condensates: Cosmological variables and lattice-refined dynamics. *New J. Phys.* **2014**, *16*, 123004. [CrossRef]
116. Calcagni, G. Loop quantum cosmology from group field theory. *Phys. Rev. D* **2014**, *90*, 064047. [CrossRef]
117. Brandenberger, R.; Peter, P. Bouncing Cosmologies: Progress and Problems. *Found. Phys.* **2017**, *47*, 797–850. [CrossRef]
118. de Cesare, M. Limiting curvature mimetic gravity for group field theory condensates. *Phys. Rev. D* **2019**, *99*, 063505. [CrossRef]
119. Sebastiani, L.; Vagnozzi, S.; Myrzakulov, R. Mimetic gravity: A review of recent developments and applications to cosmology and astrophysics. *Adv. High Energy Phys.* **2017**, *2017*, 3156915. [CrossRef]
120. Chamseddine, A.H.; Mukhanov, V. Resolving Cosmological Singularities. *J. Cosmol. Astropart. Phys.* **2017**, *1703*, 009. [CrossRef]
121. Gielen, S. Emergence of a low spin phase in group field theory condensates. *Class. Quantum Gravity* **2016**, *33*, 224002. [CrossRef]
122. Alesci, E.; Cianfrani, F. Quantum-Reduced Loop Gravity: Cosmology. *Phys. Rev. D* **2013**, *87*, 083521. [CrossRef]
123. Alesci, E.; Cianfrani, F. Quantum reduced loop gravity: Universe on a lattice. *Phys. Rev. D* **2015**, *92*, 084065. [CrossRef]
124. Mukhanov, V. *Physical Foundations of Cosmology*; Cambridge University Press: Cambrige, UK, 2005.
125. Linde, A. Inflationary Cosmology after Planck 2013. *arXiv* **2014**, arXiv:1402.0526.
126. Ijjas, A.; Steinhardt, P.J.; Loeb, A. Inflationary paradigm in trouble after Planck 2013. *Phys. Lett. B* **2013**, *723*, 261–266. [CrossRef]

127. Ijjas, A.; Steinhardt, P.J. Implications of Planck 2015 for inflationary, ekpyrotic and anamorphic bouncing cosmologies. *Class. Quantum Gravity* **2016**, *33*, 044001. [CrossRef]
128. Belinsky, V.; Khalatnikov, I.; Lifshitz, E. Oscillatory approach to a singular point in the relativistic cosmology. *Adv. Phys.* **1970**, *19*, 525. [CrossRef]
129. Lifshitz, E.; Khalatnikov, I. Investigations in relativistic cosmology. *Ad. Phys.* **1963**, *12*, 185. [CrossRef]
130. Bianchi, E.; Rovelli, C.; Vidotto, F. Towards Spin-foam Cosmology. *Phys. Rev. D* **2010**, *82*, 084035. [CrossRef]
131. Rovelli, C.; Vidotto, F. On the spinfoam expansion in cosmology. *Class. Quantum Gravity* **2010**, *27*, 145005. [CrossRef]
132. Bianchi, E.; Krajewski, T.; Rovelli, C.; Vidotto, F. Cosmological constant in spinfoam cosmology. *Phys. Rev. D* **2011**, *83*, 104015. [CrossRef]
133. Hellmann, F. On the Expansions in Spin Foam Cosmology. *Phys. Rev. D* **2011**, *84*, 103516. [CrossRef]
134. Livine, E.R.; Martin-Benito, M. Classical Setting and Effective Dynamics for Spinfoam Cosmology. *Class. Quantum Gravity* **2013**, *30*, 035006. [CrossRef]
135. Schroeren, D. Decoherent Histories of Spin Networks. *Found. Phys.* **2013**, *43*, 310–328. [CrossRef]
136. Rennert, J.; Sloan, D. A Homogeneous Model of Spinfoam Cosmology. *Class. Quantum Gravity* **2013**, *30*, 235019. [CrossRef]
137. Rennert, J.; Sloan, D. Anisotropic Spinfoam Cosmology. *Class. Quantum Gravity* **2014**, *31*, 015017. [CrossRef]
138. Rovelli, C.; Vidotto, F. Stepping out of Homogeneity in Loop Quantum Cosmology. *Class. Quantum Gravity* **2008**, *25*, 225024. [CrossRef]
139. Battisti, M.V.; Marciano, A.; Rovelli, C. Triangulated Loop Quantum Cosmology: Bianchi IX and inhomogenous perturbations. *Phys. Rev. D* **2010**, *81*, 064019. [CrossRef]
140. Borja, E.F.; Diaz-Polo, J.; Garay, I.; Livine, E.R. Dynamics for a 2-vertex Quantum Gravity Model. *Class. Quantum Gravity* **2010**, *27*, 235010. [CrossRef]
141. Vidotto, F. Many-nodes/many-links spinfoam: The homogeneous and isotropic case. *Class. Quantum Gravity* **2011**, *28*, 245005. [CrossRef]
142. Borja, E.F.; Garay, I.; Vidotto, F. Learning about Quantum Gravity with a Couple of Nodes. *SIGMA* **2012**, *8*, 015. [CrossRef]
143. Mukhanov, V.; Feldman, H.A.; Brandenberger, R. Theory of cosmological perturbations. *Phys. Rep.* **1992**, *215*, 203–333. [CrossRef]
144. Gielen, S. Perturbing a quantum gravity condensate. *Phys. Rev. D* **2015**, *91*, 043526. [CrossRef]
145. Gielen, S. Identifying cosmological perturbations in group field theory condensates. *J. High Energy Phys.* **2015**, *1508*, 010. [CrossRef]
146. Gielen, S. Group field theory and its cosmology in a matter reference frame. *Universe* **2018**, *4* 103. [CrossRef]
147. Gielen, S. Inhomogeneous universe from group field theory condensate. *J. Cosmol. Astropart. Phys.* **2019**, *1902*, 013. [CrossRef]
148. Gerhardt, F.; Oriti, D.; Wilson-Ewing, E. The separate universe framework in group field theory condensate cosmology. *Phys. Rev. D* **2018**, *98*, 066011. [CrossRef]
149. Dittrich, B. The continuum limit of loop quantum gravity—A framework forsolving the theory. *arXiv* **2014**, arXiv:1409.1450.
150. Dittrich, B. From the discrete to the continuous: Towards a cylindrically con-sistent dynamics. *New J. Phys.* **2012**, *14*, 123004. [CrossRef]
151. Dittrich, B.; Eckert, F.C.; Martin-Benito, M. Coarse graining methods for spin net and spin foam models. *New J. Phys.* **2012**, *14*, 035008. [CrossRef]
152. Dittrich, B.; Schnetter, E.; Seth, C.J.; Steinhaus, S. Coarse graining flow of spin foam intertwiners. *Phys. Rev. D* **2016**, *94*, 124050. [CrossRef]
153. Bahr, B. On background-independent renormalization of spin foam models. *Class. Quantum Gravity* **2017**, *34*, 075001. [CrossRef]
154. Bahr, B.; Steinhaus, S. Numerical evidence for a phase transition in 4d spin foam quantum gravity. *Phys. Rev. Lett.* **2016**, *117*, 141302. [CrossRef]
155. Bahr, B.; Steinhaus, S. Hypercuboidal renormalization in spinfoam quantum gravity. *Phys. Rev. D* **2017**, *95*, 126006. [CrossRef]

156. Steinhaus, S.; Thürigen, J. Emergence of Spacetime in a restricted Spin-foam model. *Phys. Rev. D* **2018**, *98*, 026013. [CrossRef]
157. Bahr, B.; Rabuffo, G.; Steinhaus, S. Renormalization of symmetry restricted spin foam models with curvature in the asymptotic regime. *Phys. Rev. D* **2018**, *98*, 106026. [CrossRef]

© 2019 by the authors. Licensee MDPI, Basel, Switzerland. This article is an open access article distributed under the terms and conditions of the Creative Commons Attribution (CC BY) license (http://creativecommons.org/licenses/by/4.0/).

Article

Equivalence of Models in Loop Quantum Cosmology and Group Field Theory

Bekir Baytaş, Martin Bojowald * and Sean Crowe

Institute for Gravitation and the Cosmos, The Pennsylvania State University, 104 Davey Lab, University Park, PA 16802, USA; bub188@psu.edu (B.B.); stc151@psu.edu (S.C.)
* Correspondence: bojowald@gravity.psu.edu

Received: 29 November 2018; Accepted: 18 January 2019; Published: 23 January 2019

Abstract: The paradigmatic models often used to highlight cosmological features of loop quantum gravity and group field theory are shown to be equivalent, in the sense that they are different realizations of the same model given by harmonic cosmology. The loop version of harmonic cosmology is a canonical realization, while the group-field version is a bosonic realization. The existence of a large number of bosonic realizations suggests generalizations of models in group field cosmology.

Keywords: loop quantum cosmology; group field theory; bosonic realizations

1. Introduction

Consider a dynamical system given by a real variable, V, and a complex variable, J, with Poisson brackets:

$$\{V, J\} = i\delta J \quad , \quad \{V, \bar{J}\} = -i\delta \bar{J} \quad , \quad \{J, \bar{J}\} = 2i\delta V \tag{1}$$

for a fixed real δ. We identify $H_\varphi^\delta = \delta^{-1} \mathrm{Im} J = -i(2\delta)^{-1}(J - \bar{J})$ as the Hamiltonian of the system and interpret V as the volume of a cosmological model. The third (real) variable, $\mathrm{Re} J$, is not independent provided we fix the value of the Casimir $R = V^2 - |J|^2$ of the Lie algebra $su(1,1)$ given by the brackets (1). To be specific, we will choose $R = 0$.

Writing evolution with respect to some parameter φ, the equations of motion are solved by:

$$V(\varphi) = A \cosh(\delta\varphi) - B \sinh(\delta\varphi) \tag{2}$$
$$\mathrm{Re} J(\varphi) = A \sinh(\delta\varphi) - B \cosh(\delta\varphi). \tag{3}$$

Since R is required to be zero, we obtain $A^2 - B^2 - (\delta H_\varphi^\delta)^2 = 0$, and therefore, there is some φ_0 such that $A/(\delta H_\varphi^\delta) = \cosh(\delta\varphi_0)$ and $B/(\delta H_\varphi^\delta) = -\sinh(\delta\varphi_0)$. The solution (2) then reads:

$$V(\varphi) = \delta H_\varphi^\delta \cosh(\delta(\varphi - \varphi_0)) \tag{4}$$

and displays the paradigmatic behavior of the volume of a bouncing universe model. This construction defines harmonic cosmology [1,2]. See also [3] for further properties related to $su(1,1)$, in particular group coherent states, and [4] for an application to coarse graining.

The bouncing behavior can also be inferred from an effective Friedmann equation that describes modified evolution of the scale factor giving rise to the volume V. To do so, we should provide a physical interpretation to the time parameter φ used so far. A temporal description, shared by some models of loop quantum cosmology [5,6] and group field cosmology [7–11], is a so-called internal time [12]: The parameter φ is proportional to the value of a scalar field ϕ as a specific matter contribution devised such that ϕ is in one-to-one correspondence with some time coordinate such as proper time τ. The scalar ϕ itself can then be used as a global time. Its dynamics must be such that its momentum p_ϕ never becomes zero; "time" ϕ then never stops. With a standard isotropic scalar Hamiltonian:

$$h_\phi = \frac{1}{2}\frac{p_\phi^2}{V} + VW(\phi), \qquad (5)$$

this condition is fulfilled only for vanishing potential $W(\phi)$, such that p_ϕ is conserved. The scalar should therefore be massless and without self-interactions. With these conditions, the conserved momentum p_ϕ generates "time" translations in ϕ and can therefore be identified with the evolution generator H_ϕ^δ introduced above. In order to match with coefficients in the Friedmann equation derived below, we set:

$$p_\phi = \sqrt{12\pi G} H_\phi^\delta. \qquad (6)$$

The Hamiltonian (5) also allows us to derive a relationship between ϕ and proper time τ, measured by co-moving observers in an isotropic cosmological model. Proper-time equations of motion are determined by Poisson brackets with the Hamiltonian constraint, to which (5) provides the matter contribution. Therefore,

$$\frac{d\phi}{d\tau} = \{\phi, h_\phi\} = \frac{p_\phi}{V}. \qquad (7)$$

Writing proper-time derivatives with a dot and using $V = a^3$ to introduce the scale factor a, the chain rule then implies:

$$\left(\frac{\dot a}{a}\right)^2 = \left(\frac{\phi}{3V}\frac{dV}{d\phi}\right)^2 = \frac{p_\phi^2}{9V^4}\left(\frac{dV}{d\phi}\right)^2 \qquad (8)$$

in which:

$$\frac{1}{V^2}\left(\frac{dV}{d\phi}\right)^2 = \frac{1}{V^2}\{V,p_\phi\}^2 = 12\pi G\frac{(\mathrm{Re}J)^2}{V^2} = 12\pi G\left(1 - \frac{\delta^2 p_\phi^2}{12\pi GV^2}\right) \qquad (9)$$

follows from the ϕ-equations of motion, the zero Casimir $R = 0$, and the identification (6) with $H_\phi^\delta = \delta^{-1}\mathrm{Im}J$. Putting everything together,

$$\left(\frac{\dot a}{a}\right)^2 = \frac{4\pi G}{3}\frac{p_\phi^2}{V^2}\left(1 - \frac{\delta^2 p_\phi^2}{12\pi GV^2}\right) = \frac{8\pi G}{3}\rho_\phi\left(1 - \frac{\delta^2 \rho_\phi}{6\pi G}\right) \qquad (10)$$

with the energy density $\rho_\phi = \frac{1}{2}p_\phi^2/a^6$ of the free, massless scalar. Upon rescaling $\tilde\delta = 4\pi G\delta$, this effective Friedmann equation agrees with what has been derived in loop quantum cosmology, following [13].

Harmonic cosmology can be obtained as a deformation of a certain model of classical cosmology. In the limit of vanishing δ, $H_\phi^0 = \lim_{\delta \to 0} H_\phi^\delta$ has Poisson bracket:

$$\{V, H_\phi^0\} = \lim_{\delta \to 0}\mathrm{Re}J. \qquad (11)$$

For finite H_ϕ^0, we must have $\lim_{\delta \to 0}\mathrm{Im}J = 0$, such that the vanishing Casimir implies $\lim_{\delta \to 0}\mathrm{Re}J = V$. Therefore,

$$\{V, H_\phi^0\} = V \qquad (12)$$

with an exponential solution $V(\phi) = \exp(\sqrt{12\pi G}\phi)$ that no longer exhibits a bounce. Moreover, noticing that:

$$\{V, V^{-1}H_\phi^0\} = 1, \qquad (13)$$

we can identify $H_\phi^0/V = P$ with the momentum canonically conjugate to V in the limit of $\delta \to 0$. Therefore,

$$H_\phi^0 = VP \qquad (14)$$

is quadratic. Squaring this equation, we find:

$$P^2 = \frac{(H_\phi^0)^2}{V^2} = \frac{p_\phi^2}{12\pi G V^2} \tag{15}$$

which, upon relating $P = \dot{a}/(4\pi G a)$ to the Hubble parameter and V to the scale factor cubed, is equivalent to the Friedmann equation of an isotropic, spatially-flat model sourced by a free, massless scalar field with momentum p_ϕ:

$$\left(\frac{\dot{a}}{a}\right)^2 = \frac{8\pi G}{3} \rho_\phi . \tag{16}$$

2. Loop Quantum Cosmology as a Canonical Realization of Harmonic Cosmology

It is of interest to construct a canonical momentum P of V also in the case of non-zero δ. The pair (V, P) will then be Darboux coordinates on symplectic leaves of the Poisson manifold defined by (1), and the full (real) three-dimensional manifold will have Casimir–Darboux coordinates (V, P, R). Following the methods of [14], we can construct such a momentum directly from the brackets (1).

Suppose we already know the momentum P. The Poisson bracket of any function on our manifold with V then equals the negative derivative by P. In particular,

$$\frac{\partial \mathrm{Im} J}{\partial P} = -\{\mathrm{Im} J, V\} = \delta \mathrm{Re} J \tag{17}$$

$$\frac{\partial \mathrm{Re} J}{\partial P} = -\{\mathrm{Re} J, V\} = -\delta \mathrm{Im} J \tag{18}$$

while $\partial V/\partial P = 0$. Up to a crucial sign, these equations are very similar to our equations of motion in the preceding section, and the same is true for their solutions:

$$\mathrm{Im} J(V, P) = A(V) \cos(\delta P) - B(V) \sin(\delta P) \tag{19}$$

$$\mathrm{Re} J(V, P) = -A(V) \sin(\delta P) - B(V) \cos(\delta P) . \tag{20}$$

Since we are now dealing with partial differential equations, the previous constants A and B are allowed to depend on V.

Given these solutions, we can evaluate the Casimir:

$$R = V^2 - |J|^2 = V^2 - A(V)^2 - B(V)^2 . \tag{21}$$

If it equals zero, we have $A(V)^2 + B(V)^2 = V^2$, and there is a P_0 such that $A(V)/V = -\sin(\delta P_0)$ and $B(V)/V = -\cos(\delta P_0)$. Thus,

$$\mathrm{Im} J(V, P) = V \sin(\delta(P - P_0)) \tag{22}$$

$$\mathrm{Re} J(V, P) = V \cos(\delta(P - P_0)) \tag{23}$$

or:

$$J(V, P) = V \exp(i\delta(P - P_0)) . \tag{24}$$

The canonical realization of (1), given by Casimir–Darboux coordinates (V, P, R), identifies J as a "holonomy modification" of the classical Hamiltonian (14), in which the Hubble parameter represented by the momentum P is replaced by a periodic function of P.[1] The vanishing Casimir, $R = 0$, then appears as a reality condition for P in (24).

We conclude that the paradigmatic bounce model of loop quantum cosmology, analyzed numerically in [19], is a canonical realization of harmonic cosmology.

3. Group Field Theory as a Bosonic Realization of Harmonic Cosmology

The canonical realization constructed in the preceding section is faithful: the number of Darboux coordinates agrees with the rank of the Poisson tensor given by (1), and the number of Casimir coordinates agrees with the co-rank. If one drops the condition of faithfulness, inequivalent realizations can be constructed which even locally are not related to the original system by canonical transformations. We will call "realization equivalent" any two systems that are realizations of the same model. This notion of equivalence therefore generalizes canonical equivalence. As we will show now, this generalization is crucial in relating loop quantum cosmology to group field theory.

3.1. Bosonic Realizations

Instead of canonical realizations, one may consider bosonic realizations, replacing canonical variables, (q, p) such that $\{q, p\} = 1$, with classical versions of creation and annihilation operators, $(z, \bar{z})$ such that $\{\bar{z}, z\} = i$. The map $z = 2^{-1/2}(q + ip)$ defines a bijection between canonical and bosonic realizations.

The brackets (1) correspond to the Lie algebra su(1,1). A different real form of this algebra, sp(2, $\mathbb{R}$), has a large number of (non-faithful) bosonic realizations given by the special case of $N = 1$ in the family of realizations:

$$A^{(n)}_{ab} = \sum_{\alpha=1}^{n} \bar{z}_{a\alpha} \bar{z}_{b\alpha} \quad , \quad B^{(n)}_{ab} = \sum_{\alpha=1}^{n} z_{a\alpha} z_{b\alpha} \quad , \quad C^{(n)}_{ab} = \frac{1}{2} \sum_{\alpha=1}^{n} (\bar{z}_{a\alpha} z_{b\alpha} + z_{b\alpha} \bar{z}_{a\alpha}) \tag{25}$$

of sp(2N, $\mathbb{R}$) [20–23] with relations:

$$[A_{ab}, A_{a'b'}] = 0 = [B_{ab}, B_{a'b'}] \tag{26}$$

$$[B_{ab}, A_{a'b'}] = C_{b'b}\delta_{aa'} + C_{a'v}\delta_{ab'} + C_{b'a}\delta_{ba'} + C_{aa'}\delta_{bb'} \tag{27}$$

$$[C_{ab}, A_{a'b'}] = A_{ab'}\delta_{ba'} + A_{aa'}\delta_{bb'} \tag{28}$$

$$[C_{ab}, B_{a'b'}] = -B_{bb'}\delta_{aa'} - B_{ba'}\delta_{ab'} \tag{29}$$

$$[C_{ab}, C_{a'b'}] = C_{ab'}\delta_{a'b} - C_{a'b}\delta_{ab'} . \tag{30}$$

The indices take values in the ranges $\alpha = 1, \ldots, n$ and $a, b = 1, \ldots, N$, where $a \leq b$ in A_{ab} and B_{ab}. There are $2nN$ real degrees of freedom in the bosonic coordinates $z_{a\alpha}$, while sp(2N, $\mathbb{R}$) has dimension $N(2N + 1)$.

For $N = 1$, we have three generators:

$$A^{(n)} = \sum_{\alpha=1}^{n} \bar{z}_\alpha \bar{z}_\alpha \quad , \quad B^{(n)} = \sum_{\alpha=1}^{n} z_\alpha z_\alpha \quad , \quad C^{(n)} = \frac{1}{2} \sum_{\alpha=1}^{n} (\bar{z}_\alpha z_\alpha + z_\alpha \bar{z}_\alpha) \tag{31}$$

---

[1] In loop quantum cosmology [15], the Hubble parameter is obtained from a component of the Ashtekar connection, which, as in the full theory of loop quantum gravity, is not represented directly as an operator, but only indirectly through holonomies [16,17]. In isotropic models, matrix elements of holonomies along straight lines are of the form $\exp(i\delta P)$, as it appears in (24). In loop quantum cosmology, going back to [18], the parameter δ has sometimes been related to the area spectrum in loop quantum gravity. However, this relationship is ad-hoc, and therefore, it is not surprising that no such role of δ can be seen in the present realization.

with relations:

$$[A^{(n)}, B^{(n)}] = C^{(n)} \quad , \quad [A^{(n)}, C^{(n)}] = -2A^{(n)} \quad , \quad [B^{(n)}, C^{(n)}] = 2B^{(n)}. \tag{32}$$

For any n, the identification:

$$A^{(n)} = i\bar{J}/\delta \quad , \quad B^{(n)} = iJ/\delta \quad , \quad C^{(n)} = 2iV/\delta \tag{33}$$

relates these brackets to (1).

3.2. Model of Group Field Theory

In [24], a toy model of group field theory has been derived that produces bouncing cosmological dynamics for the number observable of certain microscopic degrees of freedom. Starting with a tetrahedron, the model assigns annihilation and creation operators to the sides, which change the area in discrete increments. For an isotropic model, the four areas should be identical, and their minimal non-zero value is determined by a quantum number $j = 1/2$, modeling the discrete nature through a spin system following the loop paradigm [25]. Each isotropic excitation has the "single-particle" Hilbert space $(1/2)^{\otimes 4}$, which contains a unique spin-two subspace. Since this subspace consists of totally-symmetric products of the individual states, it is preferred by the condition of isotropy. Restriction to the spin-two subspace then implies a five-dimensional single-particle Hilbert space with complex-valued bosonic variables A_i.

A simple non-trivial dynamics is then proposed [24] by the action:

$$S = \int d\phi \left(\frac{1}{2} i \left(A_i^* \frac{dA^i}{d\phi} - \frac{dA_i^*}{d\phi} A^i \right) - \mathcal{H}(A^i, A_j^*) \right) \tag{34}$$

in internal time ϕ. The first term indeed implies bosonic Poisson brackets $\{A_i^*, A^j\} = i\delta_i^j$. The second term is fixed by proposing a squeezing Hamiltonian:

$$\mathcal{H}(A^i, A_j^*) = \frac{1}{2} i \lambda \left(A_i^* A_j^* g^{ij} - A^i A^j g_{ij} \right) \tag{35}$$

with a coupling constant λ and a constant metric g_{ij} with inverse g^{ij}. The metric is defined through an identification of the spin-two index i with all totally-symmetric combinations of four indices $B_I \in \{1,2\}$ taking two values, such that:

$$g_{(B_1 B_2 B_3 B_4)(C_1 C_2 C_3 C_4)} = \epsilon_{(B_1(C_1} \epsilon_{B_2 C_2} \epsilon_{B_3 C_3} \epsilon_{B_4)C_4)} \tag{36}$$

with separate total symmetrizations of $\{B_1, B_2, B_3, B_4\}$ and $\{C_1, C_2, C_3, C_4\}$, respectively, and the usual totally antisymmetric ϵ_{BC}. Ordering index combinations as:

$$i \in (1,2,3,4,5) = (1111, (1112), (1122), (1222), (2222)), \tag{37}$$

the metric can be determined explicitly as the matrix:

$$g = \begin{pmatrix} 0 & 0 & 0 & 0 & 1 \\ 0 & 0 & 0 & -1 & 0 \\ 0 & 0 & 1 & 0 & 0 \\ 0 & -1 & 0 & 0 & 0 \\ 1 & 0 & 0 & 0 & 0 \end{pmatrix}. \tag{38}$$

A second crucial observable, in addition to the Hamiltonian, is the excitation number,

$$V = \frac{1}{2}\left(A_i^* A^i + A^i A_i^*\right), \tag{39}$$

identified with the cosmological volume following group field cosmology. This volume evolves in internal time ϕ according to the Hamiltonian $\mathcal{H}$. Solutions for $V(\phi)$, derived in [24], show bouncing behavior (4) that can be modeled by the effective Friedmann equation (10).

We can now readily show that this behavior is not a coincidence: The metric (38) has eigenvalues $+1$ with three-fold degeneracy and -1 with two-fold degeneracy. Diagonalizing it by an orthogonal matrix gives linear combinations z_α of the A^i and A_i^* that preserve the bosonic bracket $\{A_i^*, A^j\} = i\delta_i^j$, defining a bosonic transformation:

$$z_1 = \frac{1}{\sqrt{2}}(A^1 + A^5), \quad z_2 = \frac{1}{\sqrt{2}}(A^2 - A^4), \quad z_3 = A^3 \tag{40}$$

for eigenvalue $+1$, and:

$$z_4 = \frac{1}{\sqrt{2}}(A^1 - A^5), \quad z_5 = \frac{1}{\sqrt{2}}(A^2 + A^4) \tag{41}$$

for eigenvalue -1.

We can deal with the negative eigenvalues in two ways. First, multiplication of z_4 and z_5 with i preserves the bosonic bracket and leads to a metric $g'_{ij} = \delta_{ij}$. We then have $\mathcal{H} = \frac{1}{2}i\lambda(A^{(5)} - B^{(5)})$ for (35) and $V = C^{(5)}$ for (39). Alternatively, using only diagonalization by an orthogonal matrix, we have:

$$\mathcal{H} = \frac{1}{2}i\lambda\left(A^{(3)} - B^{(3)} - (A^{(2)} - B^{(2)})\right) \tag{42}$$

and:

$$V = C^{(3)} + C^{(2)} \tag{43}$$

where z_1, z_2, and z_3 contribute to the $n = 3$ realization and z_4 and z_5 to $n = 2$. Observing (33) and the fact that the relations (1) are invariant under changing the sign of J, the volumes and Hamiltonians in both loop quantum cosmology and group field theory are identified with the same generators in harmonic cosmology. The models of loop quantum cosmology and group field theory are therefore realization equivalent.

4. Implications and Further Directions

There is an immediate application of our result to the appearance of singularities in the model [24] of group field cosmology. As argued in this paper, because the volume is derived from the positive number operator of microscopic excitations A^i, it can be zero only at a local minimum, which requires $V(\phi_{min}) = 0$ and $dV/d\phi = 0$ at some internal time ϕ_{min}. The combination of these two conditions is quite restrictive, and [24] concludes that a singularity (zero volume) can be reached only for a small number of initial conditions.

However, our identification of the model of [24] as a bosonic realization of harmonic cosmology suggests a more cautious approach to the singularity problem. In su(1,1), there is no positivity condition on the generator that corresponds to the volume V. The bosonic realization in terms of microscopic excitations A^i is therefore local, in the sense that the A^i are local coordinates on the Poisson manifold that realizes harmonic cosmology, and $V = 0$ is at the boundary of a local chart. Accompanying $V(\phi_{min}) = 0$ by $dV/d\phi = 0$ is therefore unjustified unless one can show that evolution never leaves a local chart. The condition $V(\phi_{min}) = 0$ is not as restrictive as the combination, and it leaves more room for solutions that reach zero volume (These solutions may still be considered non-singular if there is a unique Hamiltonian that evolves solutions through zero volume.

In loop quantum cosmology, evolving through $V = 0$ is interpreted as changing the orientation of space [15,26].).

In harmonic cosmology, further generalizations of the model used here have already been explored in some detail. The new relationship with group field theory suggests similar generalizations also on the group-field side of the equivalence. For instance, harmonic cosmology can be defined for any power-law $Q = a^p$ replacing $V = a^3$, describing a quantization ambiguity that corresponds to lattice refinement of an underlying discrete geometry [27,28]. The same algebra, with arbitrary exponent p, can then be realized bosonically, suggesting related group-field models (while the power-law $V = a^3$ is preferred at large volume because it avoids an expansion of the discrete scale to macroscopic size, a different power-law may well be relevant near a spacelike singularity).

Another parameter related to the relation $V = a^3$ is the averaging volume V_0 used to define the isotropic model. We have implicitly assumed $V_0 = 1$ in order to focus on algebraic properties; in general, we have $V = V_0 a^3$ where V_0 is computed as the coordinate volume of the averaging region. Classical equations do not depend on V_0, but quantum corrections do, as can be seen here from the fact that in the action (34), the Hamiltonian $\mathcal{H}$ is proportional to V_0, but the symplectic term is not. The microscopic action is then not invariant under changing V_0. The implications of a relation between V_0 and the infrared scale of an underlying field theory [29] are of importance for the interpretation of quantum cosmology [30], and similar conclusions should hold true in group-field cosmology.

In classical harmonic cosmology, the Casimir $R = 0$ is exactly zero, but this value usually changes in the presence of quantum corrections [1,2,31]. The bouncing behavior (2) is no longer guaranteed if $R < 0$ and $|R| > (\delta H_\phi^\delta)^2$, because $V(\phi)$ behaves like a sinh under these conditions. These conditions require large quantum corrections, greater than the matter density related to p_ϕ^2. They are therefore unlikely to be fulfilled in a macroscopic universe. However, as pointed out in [30], an appeal to the BKLscenario [32] near a spacelike singularity shows that a homogeneous model is a good approximation only if it has small co-moving volume, given by the averaging volume V_0 mentioned above. Such a tiny region does not contain much matter energy, which can then easily be surpassed by quantum corrections in a high-curvature regime: $p_\phi \propto V_0$ is suppressed for small V_0, while volume fluctuations ΔV are not proportional to V_0 because they are bounded from below by the V_0-independent $\hbar$ in uncertainty relations. The genericness of bouncing solutions in loop quantum cosmology or group-field cosmology is then not guaranteed.

Finally, a large class of microscopic models can be constructed from the bosonic realizations of harmonic cosmology with arbitrary n in (31) (the fact that group field cosmology leads to non-faithful realizations of harmonic cosmology with a potentially large number of microscopic degrees of freedom is a consequence of the "second quantization" made use of in group field theory). The question of whether these are related to group field cosmology in some way appears to be of interest.

Author Contributions: Conceptualization, M.B.; writing—original draft preparation, M.B.; formal analysis, B.B., M.B. and S.C.; writing—review and editing, B.B., M.B. and S.C.

Funding: This research was funded by NSF Grant Number PHY-1607414.

Conflicts of Interest: The authors declare no conflict of interest.

References

1. Bojowald, M. Large scale effective theory for cosmological bounces. *Phys. Rev. D* **2007**, *75*, 081301. [CrossRef]
2. Bojowald, M. Harmonic cosmology: How much can we know about a universe before the big bang? *Proc. R. Soc. A* **2008**, *464*, 2135–2150. [CrossRef]
3. Livine, E.R.; Martín-Benito, M. Group theoretical Quantization of Isotropic Loop Cosmology. *Phys. Rev. D* **2012**, *85*, 124052. [CrossRef]
4. Bodendorfer, N.; Haneder, F. Coarse graining as a representation change. *arXiv* **2018**, arXiv:1811.02792.
5. Bojowald, M. Loop Quantum Cosmology. *Living Rev. Relat.* **2008**, *11*, 4. [CrossRef]
6. Bojowald, M. Quantum cosmology: A review. *Rep. Prog. Phys.* **2015**, *78*, 023901. [CrossRef]

7. Calcagni, G.; Gielen, S.; Oriti, D. Group field cosmology: A cosmological field theory of quantum geometry. *Class. Quantum Grav.* **2012**, *29*, 105005. [CrossRef]
8. Gielen, S.; Oriti, D.; Sindoni, L. Cosmology from Group Field Theory Formalism for Quantum Gravity. *Phys. Rev. Lett.* **2013**, *111*, 031301. [CrossRef]
9. Gielen, S.; Oriti, D.; Sindoni, L. Homogeneous cosmologies as group field theory condensates. *JHEP* **2014**, *1406*, 013. [CrossRef]
10. Gielen, S.; Oriti, D. Quantum cosmology from quantum gravity condensates: Cosmological variables and lattice-refined dynamics. *New J. Phys.* **2014**, *16*, 123004. [CrossRef]
11. Gielen, S. Perturbing a quantum gravity condensate. *Phys. Rev. D* **2015**, *91*, 043526. [CrossRef]
12. Dirac, P.A.M. Generalized Hamiltonian dynamics. *Can. J. Math.* **1950**, *2*, 129–148. [CrossRef]
13. Vandersloot, K. On the Hamiltonian Constraint of Loop Quantum Cosmology. *Phys. Rev. D* **2005**, *71*, 103506. [CrossRef]
14. Baytaş, B.; Bojowald, M.; Crowe, S. Faithful realizations of semiclassical truncations. *arXiv* **2018**, arXiv:1810.12127.
15. Bojowald, M. Isotropic Loop Quantum Cosmology. *Class. Quantum Grav.* **2002**, *19*, 2717–2741. [CrossRef]
16. Rovelli, C.; Smolin, L. Loop Space Representation of Quantum General Relativity. *Nucl. Phys. B* **1990**, *331*, 80–152. [CrossRef]
17. Ashtekar, A.; Lewandowski, J.; Marolf, D.; Mourão, J.; Thiemann, T. Quantization of Diffeomorphism Invariant Theories of Connections with Local Degrees of Freedom. *J. Math. Phys.* **1995**, *36*, 6456–6493. [CrossRef]
18. Ashtekar, A.; Bojowald, M.; Lewandowski, J. Mathematical structure of loop quantum cosmology. *Adv. Theor. Math. Phys.* **2003**, *7*, 233–268. [CrossRef]
19. Ashtekar, A.; Pawlowski, T.; Singh, P. Quantum Nature of the Big Bang: An Analytical and Numerical Investigation. *Phys. Rev. D* **2006**, *73*, 124038. [CrossRef]
20. Goshen, S.; Lipkin, H.J. A simple independent-particle system having collective properties. *Ann. Phys.* **1959**, *6*, 301–309. [CrossRef]
21. Deenen, J.; Quesne, C. Dynamical group of microscopic collective states. I. One-dimensional case. *J. Math. Phys.* **1982**, *23*, 878–889. [CrossRef]
22. Castaños, O.; Chacón, E.; Moshinsky, M.; Quesne, C. Boson realization of sp(4). I. The matrix formulation. *J. Math. Phys.* **1985**, *28*, 2107–2123. [CrossRef]
23. Moshinsky, M. Boson realization of symplectic algebras. *J. Phys. A* **1985**, *18*, L1–L6. [CrossRef]
24. Adjei, E.; Gielen, S.; Wieland, W. Cosmological evolution as squeezing: A toy model for group field cosmology. *Class. Quantum Grav.* **2018**, *35*, 105016. [CrossRef]
25. Rovelli, C.; Smolin, L. Discreteness of Area and Volume in Quantum Gravity. *Nucl. Phys. B* **1995**, *442*, 593–619, [CrossRef]
26. Bojowald, M. Absence of a Singularity in Loop Quantum Cosmology. *Phys. Rev. Lett.* **2001**, *86*, 5227–5230. [CrossRef] [PubMed]
27. Bojowald, M. Loop quantum cosmology and inhomogeneities. *Gen. Rel. Grav.* **2006**, *38*, 1771–1795. [CrossRef]
28. Bojowald, M. The dark side of a patchwork universe. *Gen. Rel. Grav.* **2008**, *40*, 639–660. [CrossRef]
29. Bojowald, M.; Brahma, S. Minisuperspace models as infrared contributions. *Phys. Rev. D* **2015**, *92*, 065002. [CrossRef]
30. Bojowald, M. The BKL scenario, infrared renormalization, and quantum cosmology. *JCAP* **2019**, *01*, 026. [CrossRef]
31. Bojowald, M. How quantum is the big bang? *Phys. Rev. Lett.* **2008**, *100*, 221301. [CrossRef]
32. Belinskii, V.A.; Khalatnikov, I.M.; Lifschitz, E.M. A general solution of the Einstein equations with a time singularity. *Adv. Phys.* **1982**, *13*, 639–667. [CrossRef]

 © 2019 by the authors. Licensee MDPI, Basel, Switzerland. This article is an open access article distributed under the terms and conditions of the Creative Commons Attribution (CC BY) license (http://creativecommons.org/licenses/by/4.0/).

Article

Primordial Power Spectra from an Emergent Universe: Basic Results and Clarifications

Killian Martineau and Aurélien Barrau *

Laboratoire de Physique Subatomique et de Cosmologie, Université Grenoble-Alpes, CNRS/IN2P3 53, avenue des Martyrs, 38026 Grenoble CEDEX, France; killian.martineau@gmail.com
* Correspondence: barrau@in2p3.fr

Received: 30 November 2018; Accepted: 15 December 2018; Published: 18 December 2018

Abstract: Emergent cosmological models, together with the Big Bang and bouncing scenarios, are among the possible descriptions of the early Universe. This work aims at clarifying some general features of the primordial tensor power spectrum in this specific framework. In particular, some naive beliefs are corrected. Using a toy model, we investigate the conditions required to produce a scale-invariant spectrum and show to what extent this spectrum can exhibit local features sensitive to the details of the scale factor evolution near the transition time.

Keywords: Emergent universe; quantum cosmology; primordial tensor spectrum

1. Introduction

The term "Big Bang" is somewhat ambiguous. In a sense, it just refers to the expansion of space and to the fact that the entire observable universe was, in the past, much smaller, denser, and hotter. This is obviously non-controversial. In another sense, it refers to the initial singularity in and of itself. In this stronger meaning, the very idea of the Big Bang is far from obvious. It is a generic prediction of general relativity (GR)—remaining usually true in the inflationary paradigm [1,2]—which can, however, be violated in some circumstances.

The first important class of models without a Big Bang (in the strong sense) are bouncing models. Among the very numerous ways to get a bounce (an excellent review can be found in [3]), it is worth mentioning the violation of the null energy condition [4], the violation of the strong energy condition [5], the existence of ghost condensates [6], galileons [7], S-branes [8], quintom fields [9], higher derivatives [10,11], non-standard couplings in the Lagrangian [12], supergravity [13], and loop quantum cosmology [14,15]. These are only some examples, and an exhaustive list should also include the ekpyrotic and cyclic scenarios [16,17] and, in a way, string gas cosmology [18]. Those ideas are also being investigated in the black hole sector; see [19] and the references therein.

The second important class of models beyond the Big Bang are those based on an emergent scenario. Instead of decreasing and then increasing, the scale factor is, in this case, constant until, at some point, a transition occurs and leads to the current expansion of the Universe. As examples, one can think of (some versions of) nonlinear sigma models [20], Horava–Lifshitz gravity [21,22], Einstein–Gauss–Bonnet theory [23], exotic matter [24], branes [25], Kaluza–Klein cosmology [26], the particle creation mechanism [27], microscopic effects [28], quantum reduced loop gravity [29], and quintom matter (see [30] for the background dynamics and [31] for the associated perturbations). This leads to interesting consequences reviewed for example in [32–36].

In this article, we focus on emergent models. We do not choose a specific theory, but instead, we try to highlight generic features from a purely phenomenological approach. The aim is not to demonstrate new outstanding results. It simply consists of clarifying the situation, correcting some common misunderstandings and explaining the expected observational features, which, to the best of our knowledge, have not been presented so far in a systematic way in the literature. We basically use an "ad hoc" evolution of the scale factor from a static phase ($a = cte$) to an inflationary phase ($a \propto e^{H_0 t}$) (the subscript 0 does not refer in this context to the value of the Hubble parameter now, but to its nearly constant value during inflation). As this transition is expected to be triggered by some event occurring in the evolution of the Universe, we add a small distortion of the scale factor evolution around the transition time. This distortion can be a bounce (i.e., a phase of contraction followed by a phase of expansion) or an anti-bounce (the opposite), which we usually also call a bounce. Both are expected to capture some basic features of emergent models, but are also motivated by explicit results obtained, e.g., in loop quantum cosmology or in quantum reduced loop gravity (see the detailed behavior of the scale factor in [29,37]). The existence of an inflationary stage is natural as soon as a massive scalar field is assumed to be the dominant content of the universe. This will be our implicit hypothesis. In this case, inflation is a strong attractor [38] and occurs nearly inevitably.

We investigate how the primordial tensor power spectrum is affected by variations in the physical characteristics of the features present in the evolution of the scale factor so as to draw a wide picture of the observational characteristics of emergent models. We deliberately decide to focus on tensor perturbations, as the scalar spectrum does not depend only on the scale factor evolution.

Throughout this work, we use Planck units.

2. Primordial Tensor Power Spectra

2.1. The Mukhanov–Sasaki Equation for Tensor Perturbations

The first order perturbed Einstein equations are equivalent, for a flat FLRW universe and a single matter content modeled by a scalar field, to the gauge-invariant Mukhanov–Sasaki equation:

$$v''(\eta, \vec{x}) - \Delta v(\eta, \vec{x}) - \frac{z''_{T/S}(\eta)}{z_{T/S}(\eta)} v(\eta, \vec{x}) = 0 . \tag{1}$$

The $'$ symbol refers to a derivative with respect to conformal time η such that $a d\eta = dt$. This equation depends on two variables v and $z_{T/S}$, called the Mukhanov variables. The canonical variable, v, is obtained from a gauge-invariant combination of both the metric coordinate perturbations and the perturbations of the scalar field. The nature of the considered perturbations is encoded in the background variable $z_{T/S}$, in which the T/S indices refer either to tensor or scalar modes.

Since the background variable writes $z_S(t) = a(t)\Phi(t)/H(t)$ for scalar modes, Φ being the scalar field background, the associated evolution highly depends on the matter evolution. We will therefore not consider scalar perturbations anymore in this study, even if they are currently the most relevant ones for observations. Instead, we will focus on tensor modes, for which the background variable is simply given by $z_T(t) = a(t)$. The results and conclusions will therefore be fully generic and usable for any model in which the scale factor behaves, at least partially, in the way described below, independent of the cause.

The Mukhanov–Sasaki equation, that is Equation (1), reduces the cosmological evolution of perturbations to the propagation equation of a free scalar field, v, with a time-dependent mass $m^2 = -z''_T/z_T$ in the Minkowski space-time. The time-dependence of the mass represents the perturbations' sensitivity to the dynamical background.

During the quantization procedure, the variable v is promoted to be the operator. Its associated Fourier modes satisfy:

$$v''_k(\eta) + \left(k_c^2 - \frac{z''_T(\eta)}{z_T(\eta)}\right) v_k(\eta) = 0 , \tag{2}$$

where k_c refers to comoving wavenumbers. This equation can be re-written in cosmic time:

$$\ddot{v}_k(t) + H(t)\dot{v}_k(t)$$
$$+ \left(\frac{k_c^2}{a(t)^2} - \frac{\dot{z}_T(t)}{z_T(t)}H(t) - \frac{\ddot{z}_T(t)}{z_T(t)}\right)v_k = 0\,. \tag{3}$$
$$\Leftrightarrow \ddot{v}_k(t) + H(t)\dot{v}_k(t) + \left(\frac{k_c^2}{a(t)^2} - H(t)^2 - \frac{\ddot{a}(t)}{a(t)}\right)v_k = 0\,.$$

We introduce a new parameter $h_k(t) = v_k(t)/a(t)$ such that Equation (3) becomes:

$$\ddot{h}_k(t) + 3H(t)\dot{h}_k(t) + \frac{k_c^2}{a(t)^2}h_k(t) = 0\,. \tag{4}$$

It is convenient to introduce a second parameter $g_k(t) = a(t)\dot{h}_k(t)$ in order to rewrite Equation (4) as a set of two first order ordinary differential equations (ODEs):

$$\begin{cases} \dot{h}_k(t) = \frac{1}{a(t)}g_k(t)\,, \\ \dot{g}_k(t) = -2H(t)g_k(t) - \frac{k_c^2}{a(t)}h_k(t)\,. \end{cases} \tag{5}$$

2.2. Initial Conditions

By definition, in the static phase, the scale factor is constant. The propagation equation is then the one of a standard harmonic oscillator,

$$v_k''(\eta) + k_c^2 v_k(\eta) = 0\,, \tag{6}$$

which can be used to set the usual Bunch–Davies vacuum. The initial conditions chosen in this work are therefore of the usual type, comparable to what is done in the remote past of a de Sitter state (inflationary model) or in the remote past of a bouncing scenario. Whatever the considered wavenumber, even in the bouncing case, it is always possible to find a time such that the curvature radius can be neglected: the mode effectively "feels" a Minkowski-like spacetime. As far as initial conditions for the perturbations are concerned, the emergent universe is not different from other usual models. This is true only for tensor modes, as the situation is much trickier for scalar ones [39].

3. Purely Emergent Universe

We model the evolution of an emergent universe by a static phase followed by an inflationary stage:

$$a(t) = A + Ae^{H_0(t - t_{\text{transition}})}\,, \tag{7}$$

in which A and H_0 are two constants and $t_{\text{transition}}$ characterizes the time at which the transition between the static and the inflationary phase occurs. If we arbitrarily set $t_{\text{transition}} = 0$, without any loss of generality, then the scale factor is simply given by $a(t) = A + Ae^{H_0 t}$. The corresponding evolution, with the constants set to $A = 1$ and $H_0 = 0.01$, is plotted in arbitrary units in Figure 1.

Obviously, the constant A in itself has no meaning and can be absorbed in any rescaling of the scale factor. In addition, a modification of the constant in front of the exponential term, such that $a(t) = A + AC_1 e^{H_0 t}$, $C_1 \in \mathbb{R}^{*+}$, is simply equivalent to the definition of a new $t_{\text{transition}} = -\ln(C_1)/H_0$.

The primordial tensor power spectrum, defined by:

$$\mathcal{P}_T(k_c) = \frac{32k_c^3}{\pi}\left|\frac{v_k(t_e)}{z_T(t_e)}\right|^2, \tag{8}$$

where t_e refers to a post-inflationary time (chosen so that the considered modes have exited the horizon), can be explicitly calculated for the evolution of the scale factor given by Figure 1. The result, obtained for different values of H_0, is shown in Figure 2.

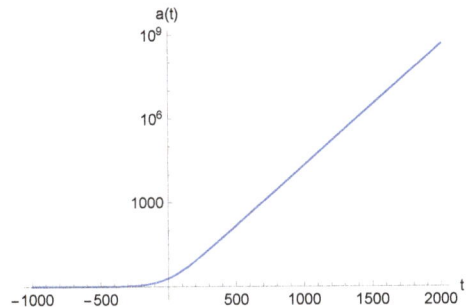

Figure 1. Scale factor evolution with $A = 1$ and $H_0 = 0.01$.

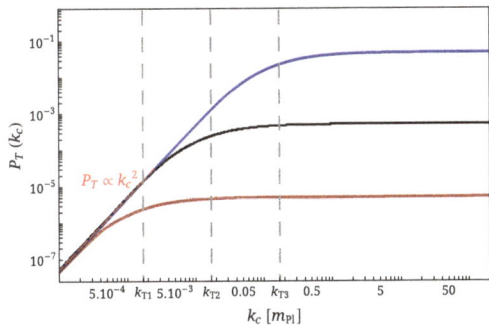

Figure 2. Primordial tensor power spectra from the emergent evolution for different values of H_0. Lower curve (red): $H_0 = 0.001$; middle curve (black): $H_0 = 0.01$; and upper curve (blue): $H_0 = 0.1$.

Clearly, two regimes do appear in those spectra. First, one can notice a scale-invariant behavior in the ultraviolet (UV), that is for large values of k_c. Then, a power law behavior appears in the infrared (IR), corresponding to low k_c values. The transition scale k_T between those two regimes corresponds to the square root of the tensor potential at the transition time, i.e, at $t = 0$ in our setting. The tensor potential is given by:

$$\frac{z_T''}{z_T} = \ddot{a}a + \dot{a}^2 = A^2 H_0^2 e^{H_0 t} + 2A^2 H_0^2 e^{2H_0 t} \tag{9}$$

and its value at the transition is then:

$$\left.\frac{z_T''}{z_T}\right|_{t=0} = 3A^2 H_0^2 . \tag{10}$$

For example, in the case $H_0 = 0.01$ displayed in the middle curve of Figure 2, the transition scale is $k_{T2} = \sqrt{\left.\frac{z_T''}{z_T}\right|_{t=0}} = \sqrt{3.10^{-4}} \simeq 1.7 \times 10^{-2}$. The dependence of the transition scale on H_0 also appears clearly since $k_{T1} \simeq 1.7 \times 10^{-3}$ and $k_{T3} \simeq 1.7 \times 10^{-1}$.

This already raises two basic points. First, the naive view according to which the causal contact made possible by the static phase, where $H = 0$ ($R_H \to \infty$), would be sufficient to ensure a spectrum compatible with observation is obviously wrong. Inflation (or other processes leading to scale invariance) is still needed. Second, the way inflation begins does matter and sets the scale above which the spectrum becomes (nearly) flat.

4. Emergent Universe with a Bounce

The previously-considered situation is clearly over-simplified. We now make the model slightly more complicated by adding a "feature" in the evolution of the scale factor before the transition to the inflationary period. As mentioned in the previous section, the interesting–and somehow usually under-estimated—fact about emergent models is that the spectrum does depend on the details of the transition period. Some information on this specific period might therefore be observationally attainable. In addition, some concrete models of quantum gravity lead to a "mini-bounce" before the transition. This is, for example, the case in quantum reduced loop gravity [29,37]. This model was designed to study symmetry reduced systems consistently within the loop quantum gravity framework (see, e.g., [40]). In particular, it bridges the gap between effective cosmological models of loop quantum cosmology [41] and the full theory, addressing the dynamics before any minisuperspace reduction [42]. This basically preserves the graph structure and SU(2) quantum numbers. It was explicitly shown that this model leads to a little bounce (or even to several mini-bounces) preceding the inflationary stage. Beyond this specific case, one can generically expect a footprint in the evolution of the scale factor of whatever physical phenomenon has triggered the transition. In the following, we therefore perturb the scale factor evolution just before the inflationary stage to study how the primordial tensor power spectrum is sensitive to the details of this distortion.

The scale factor evolution is now modeled by the following function:

$$a(t) = A + Ae^{H_0 t} + \frac{A \times C}{\arctan(B_1 \sigma_1) - \arctan(B_2 \sigma_2)} \times \quad (11)$$
$$\{\arctan[B_1(t - (\mu - \sigma_1))] - \arctan[B_2(t - (\mu - \sigma_2))]\}.$$

The constant C characterizes the bounce amplitude, and μ is its mean value; σ_1 and σ_2 allow setting the width, and B_1 and B_2 correspond to the steepness. The term $[\arctan(B\sigma_1) - \arctan(B\sigma_2)]^{-1}$ is just a normalization to ensure that the bounce amplitude remains constant under variations of B, σ_1, and σ_2. In the following, we set $B_1 = B_2 = B$, to focus on symmetrical bounces. The influence of an asymmetry is a higher order effect, which is beyond the scope of this study. We also choose $\sigma_1 = -\sigma_2 = \sigma$. The scale factor is finally expressed as:

$$a(t) = A + Ae^{H_0 t} + \frac{A \times C}{2\arctan(B\sigma)} \times \quad (12)$$
$$\{\arctan[B(t - (\mu - \sigma))] - \arctan[B(t - (\mu + \sigma))]\}.$$

Arbitrarily choosing $A = 1$ and $H_0 = 10^{-2}$, as in the case without any bounce, and fixing $C = 1$, $\mu = -400$, $\sigma = 2$, and $B = 0.4$, the scale factor evolution is displayed in the first panel of Figure 3. The second panel shows the associated tensor potential around the bounce. It is worth noticing that the "sign" of the bounce has no influence on the spectrum. It is displayed in Figure 3 as a local increase of the scale factor, but should we choose the other sign, leading to a decrease of the scale factor, the spectrum would remain the same, as will be shown later.

The primordial tensor power spectrum computed with this background evolution is given in Figure 4.

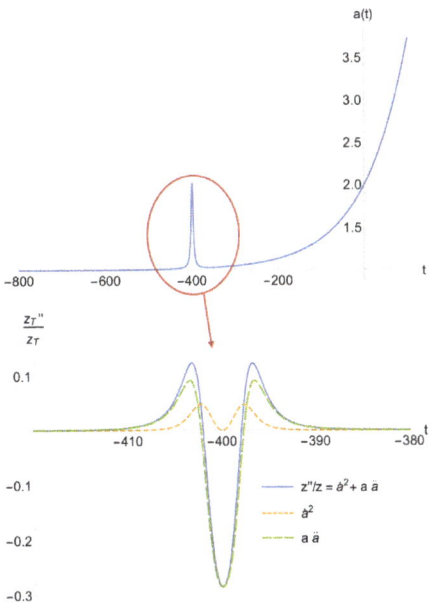

Figure 3. (**First panel**) Scale factor evolution with one bounce characterized by $C_1 = 1$, $\mu = -400$, $\sigma = 2$, and $B = 0.4$. (**Second panel**) Tensor potential around the bounce.

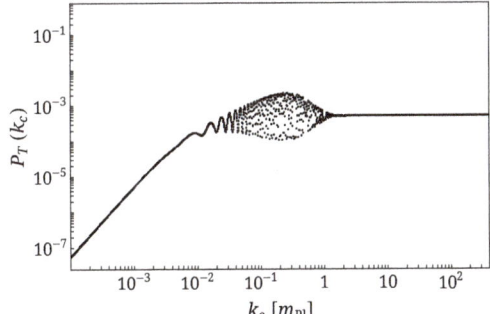

Figure 4. Primordial tensor power spectrum obtained from the scale factor evolution with one bounce characterized by $C = 1$, $\mu = -400$, $\sigma = 2$, and $B = 0.4$.

First, one can notice that the general trend is the same as in the case without bounce, which is not surprising, as the main tendencies are driven by the choice of the initial state and the existence of an inflationary stage. The spectrum is still scale-invariant in the UV and grows proportionally to k_c^2 in the IR. However, the bounce does have an impact on the spectrum: it induces oscillations in the k_c space. The envelope of the oscillations forms a kind of "bullet" in the spectrum. Those oscillations can be traced back to the time evolution of the mode functions, which becomes highly k_c-dependent in the presence of a bounce. This clearly establishes an observational window on the detailed behavior of the Universe close to the emergent time (and even before). This also contradicts a second naive belief according to which whatever happens before inflation is washed out by inflation. The details of the transition regime might be observationally probed.

This spectrum will be the reference one for the rest of this study. We now investigate how it depends (amplitude, IR and UV limits, shape of the "bullet", etc.) on the different parameters of the model.

As previously mentioned, all the results derived in this work remain valid if the bounce is of negative sign, that is it corresponds to a transition between a (locally) contracting and an expanding phase. A negative bounce of this kind is shown in Figure 5, together with the corresponding potential. The resulting spectrum is displayed in Figure 6 and can hardly be distinguished from the reference one.

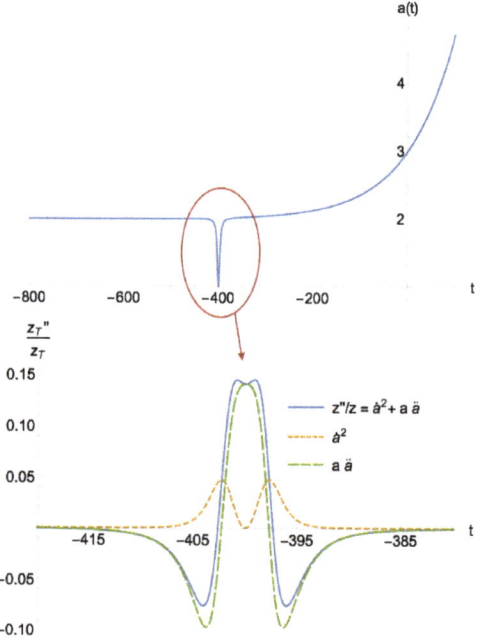

Figure 5. (**First panel**) Scale factor evolution with one bounce of negative sign characterized by $C = 1$, $\mu = -400$, $\sigma = 2$, and $B = 0.4$. (**Second panel**) Tensor potential around the bounce.

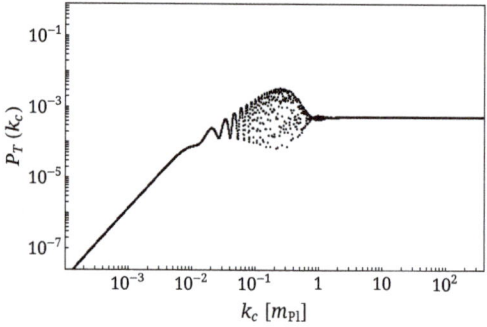

Figure 6. Primordial tensor power spectrum associated with the scale factor evolution with one bounce of negative sign characterized by $C = 1$, $\mu = -400$, $\sigma = 2$, and $B = 0.4$.

4.1. Impact of the Bounce Parameters on the Primordial Tensor Power Spectra

The aim of this section is to study how variations of the bounce parameters B, C, μ, and σ modify the shape of the primordial tensor power spectrum. Even if those oscillations cannot be currently observed, it is still interesting to see if general trends appear. Many experiments are being operated or considered to measure B-modes in the cosmological microwave background (CMB). Since the toy-model presented in this article is basically independent of the details of the quantum cosmology or modified gravity theory considered (as long as the Mukhanov–Sasaki equation remains valid), the results presented can easily be applied or adapted to forthcoming emergent, bouncing, or emergent-bouncing cosmological models.

4.1.1. The Position of the Bounce μ

First, let us study whether the position of the bounce in the static phase plays an important role in the characteristics of the spectrum. To this aim, it is enough to change the value of μ. It appears that the bounce position in the static phase has almost no consequence on the primordial tensor power spectrum. For example, Figure 7 shows the spectrum obtained for a bounce similar to the reference one, but shifted to $\mu = -800$. It can easily be seen that this spectrum is very close to the reference one. The numerical investigations show, beyond this particular example, that the position of the bounce has no significant influence on the spectrum whatever its position in the static phase. This, in principle, opens an observational window on arbitrarily-remote times in the history of the Universe. Once the "instability" is triggered, the time at which it takes place if basically of no relevance.

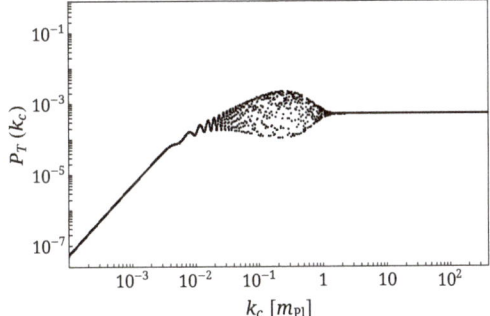

Figure 7. Primordial tensor power spectrum associated with an evolution with a bounce identical to the reference case, i.e., $C = 1$, $\sigma = 2$, and $B = 0.4$, but shifted to $\mu = -800$.

4.1.2. The Steepness of the Bounce B

To study the impact of the bounce steepness, i.e., its "slope", we vary the parameter B. The bounce dependence on this parameter is presented in Figure 8, together with the associated tensor potentials. Since the tensor potential is highly sensitive to variations of the bounce steepness (as it includes derivatives of the scale factor), only small variations of B are represented.

The larger the value of B, the steeper the bounce. The (local) maximum value of the potential at the bounce thus increases with B. We therefore expect that the range of k_c corresponding to modes impacted by (or sensitive to) the bounce is shifted to higher k_c values compared to the reference case. Let us consider a cosmic evolution where the bounce has been highly steepened when compared to the reference case. We choose $B = 40$ (one hundred times higher than the value of the reference case), the values of the other parameters being the same as in the reference case. The resulting power spectrum is shown in the first panel of Figure 9. The two frequencies appearing in the plot are associated with the two scales of the problem (width of the bounce and rise time of the edge).

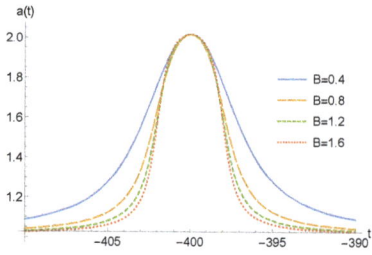
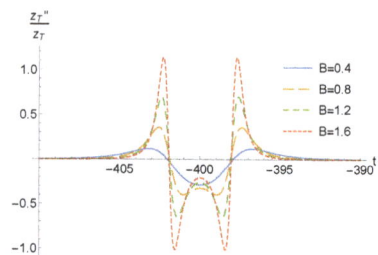

Figure 8. (**First panel**) Evolution of the scale factor with $\mathcal{C} = 1$, $\mu = -400$, $\sigma = 2$. but different values of B. (**Second panel**) Associated tensor potentials around the bounce.

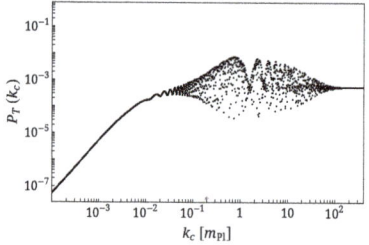
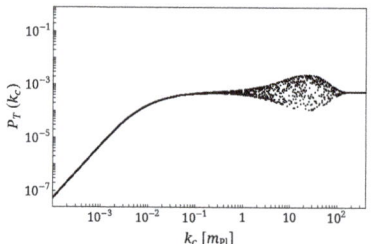

Figure 9. (**First panel**) Primordial tensor power spectrum obtained with a steep bounce characterized by $\mathcal{C} = 1$, $\mu = -400$, $\sigma = 2$, and $B = 40$. (**Second panel**) Spectrum with a narrower bounce (than in the first panel): $\sigma = 0.02$.

The size of the "bullet" (or range of oscillations) in the k_c space is extended up to higher values. This is an interesting point as the "low"-k_c features are often considered to be hard to probe experimentally. For example, in loop quantum cosmology, deviations from scale invariance, in the form of oscillations, happen in the IR (see [43,44]). They are often considered as extremely difficult to probe as this would require a very high level of fine-tuning. The comobile values of the wavenumbers that can be seen in the CMB are actually set by the duration (number of e-folds) of inflation. In loop quantum cosmology, the interesting IR features can only be seen if this number is arbitrarily set to its lowest experimentally-allowed value [45]. This makes the model difficult to probe unless new specific features appear in the UV, e.g., through trans-Planckian effects [46] or because of a change of signature [47]. The effect underlined here, that is the displacement or widening of the "bullet" to larger values of k_c because of the steepness of the mini-bounce, is therefore of potential observational significance. The specific features might be probed without fine-tuning the number of inflationary e-folds to its lowest allowed value (around $N \sim 60$). It is worth reminding that, in principle, if the evolution starts at the Planck density and if the Universe is filled with a massive scalar field, the number of e-folds can be anything between zero and a few 10^{14} (and remains compatible with observations). In some bouncing cases, this number of e-folds can be predicted [48,49] by the model, but this remains an open issue for emergent scenarios (finding a known probability distribution function for initial conditions is tricky unless the existence of an oscillating phase for the field is demonstrated).

The second panel of Figure 9 corresponds to a reduction of the width of the bounce (by a factor of one hundred) with respect to the previous case. The shape of the distortion gets closer to the reference one but, as expected, the "bullet" is translated toward the higher k_c regime.

Obviously, if the bounce is smoothed (by a decrease of B), the maximum of the tensor potential decreases, and the opposite effect occurs: the oscillations are shifted to the IR regime, which is far less interesting for phenomenology.

A modification of the steepness of the bounce—presumably associated with the triggering of the transition from the static to the inflationary phase—has a strong impact on the shape of the tensor potential at the bounce. The range of comoving modes sensitive to the bounce thus highly depends on the steepness of the evolution of the scale factor. This establishes that, as far as phenomenology is concerned, a very steep bounce is more likely to be observable, even if it occurs in the most remote past of the Universe.

4.1.3. The Amplitude of the Bounce C

We now study the consequences of variations of the bounce amplitude on the primordial tensor spectrum. Figure 10 displays, in the first panel, the effect of a variation of the factor C entering the scale factor evolution. The second panel shows the associated potentials.

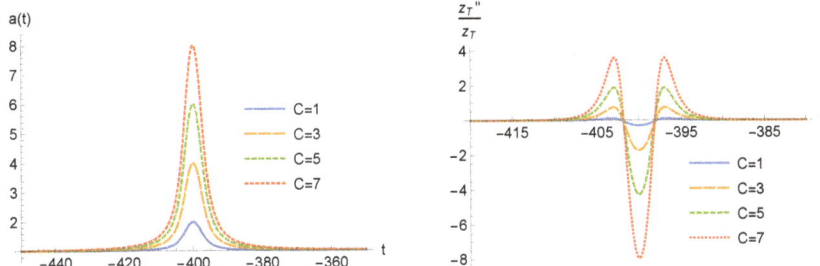

Figure 10. (**First panel**) Scale factor evolution around the bounce. From bottom to top: increasing values of C. (**Second panel**) Associated tensor potentials.

The amplitude of the bounce in itself has no meaning. The important parameter is the ratio between the extremal value of the scale factor at the bounce and its value in the static phase. This is the relevant parameter, which is varied.

The primordial tensor power spectra, for A, H_0, B, and σ taken as in the reference case but for different values of the bounce amplitude, given by $C = 0.1$, $C = 1$ (reference case), and $C = 10$, are shown in Figure 11. It can easily be seen that an increase in the amplitude of the bounce amplifies the oscillations in the k_c space. This both opens a possible observational window and allows, in principle, putting constraints on the amplitude of the bounce using upper limits on the tensor-scalar ratio.

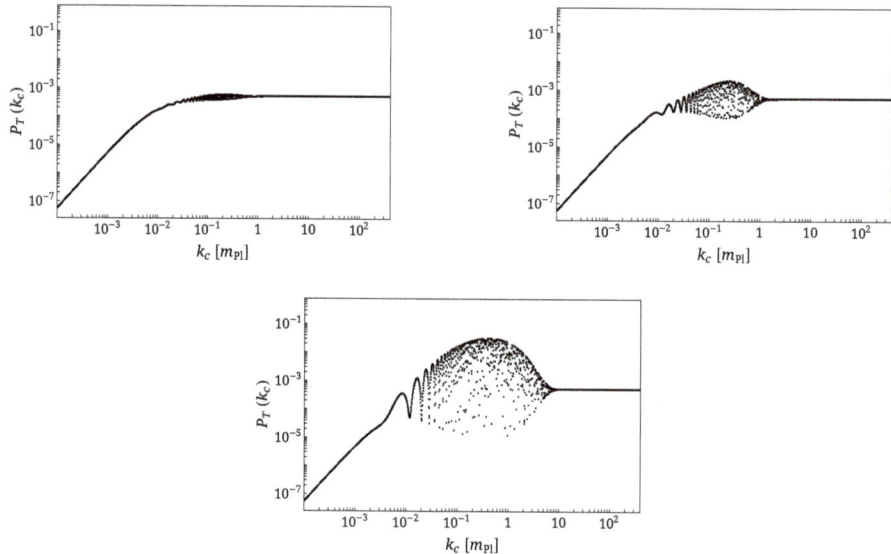

Figure 11. Primordial tensor spectra for different values of the bounce amplitude $\mathcal{C}$, the other parameters being unchanged with respect to the reference case. (**First panel**) $\mathcal{C} = 0.1$, (**Second panel**) $\mathcal{C} = 1$ (reference case), and (**Third panel**) $\mathcal{C} = 10$.

4.1.4. The Width of the Bounce σ

We turn to the study of the bounce width. The considered variations and their consequences on the tensor potential are shown in Figure 12.

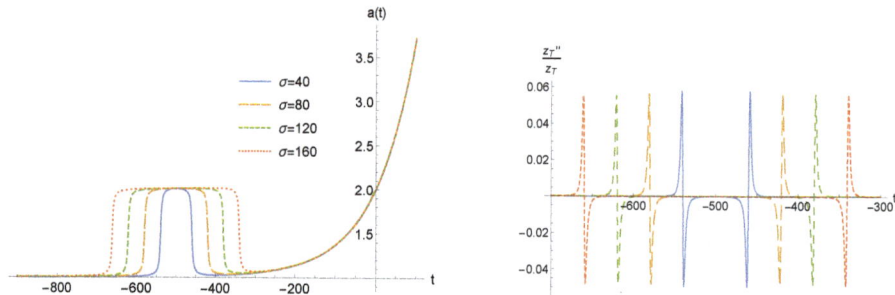

Figure 12. (**First panel**) Evolution of the scale factor with a bounce centered on $\mu = -500$ and different values of σ. The other parameters of the model are unchanged with respect to the reference case. (**Second panel**) Associated tensor potentials.

The impact of a modification of the width of bounce on the primordial tensor spectrum is shown in Figure 13. For clarity and without any explicit consequence, the bounce position has been shifted to $\mu = -500$. The values of σ are varied, the amplitude and the steepness remaining, as usual, unchanged. The main characteristics of the power spectrum are not significantly affected. Unless compensating for the steepness variation, as explained previously, the width of the bounce is unlikely to produce any spectacular observational consequence.

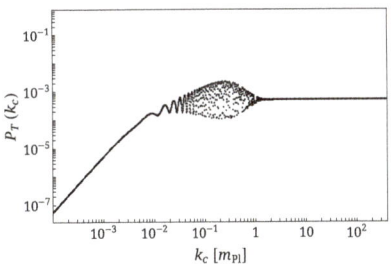 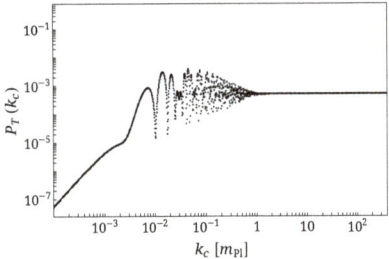

Figure 13. (**First panel**) Spectrum of reference. (**Second panel**) Spectrum with a wider bounce described by $\sigma = 100$ and shifted to $\mu = -500$, the other bounce parameters remaining unchanged with respect to the reference case.

4.2. Impact of the Parameters Unrelated to the Bounce

4.2.1. The Hubble Parameter during Inflation H_0

In this section, we focus on the consequences of the inflationary stage on the tensor spectrum. As is well known, a long enough inflationary phase leads, when combined with an appropriate choice of initial vacuum, to a scale-invariant spectrum. We have varied H_0 and studied the impact on the spectrum. The results for $H_0 = 0.01$, $H_0 = 0.1$, and $H_0 = 1$ are given in Figure 14.

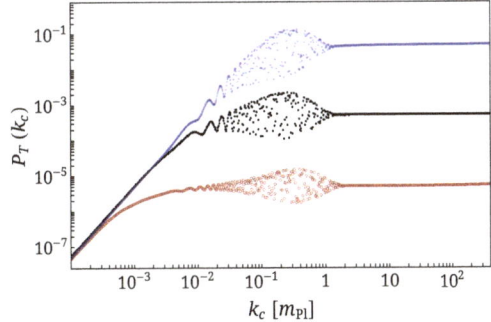

Figure 14. Primordial tensor power spectra obtained from an evolution with a bounce characterized by $C = 1$, $\mu = -400$, $\sigma = 2$, and $B = 0.4$, and three different values of the Hubble parameter during inflation. From top to bottom: $H_0 = 0.1$ (blue stars), $H_0 = 0.01$ (reference case, black disks), and $H_0 = 0.001$ (red circles).

Changing the value of H_0 modifies the amplitude of the power spectrum in the ultraviolet regime. More precisely, exactly as expected, the power is proportional to H_0^2, as in standard cosmology. Varying the value of the Hubble parameter does not change the picture beyond any standard and expected effect.

4.2.2. The Normalization of the Scale Factor

In this section, H_0 is kept fixed to $H_0 = 0.01$, and we investigate the impact of the variations of the constant A. In principle, this is just an unphysical rescaling of the scale factor. However, the case of effective quantum cosmology is slightly more subtle, as an extra fundamental scale (presumably of the order of the Planck length) might enter the game. This is not the case in the toy model we consider here, but this is clearly the case in quantum reduced loop gravity [29,37]. In this situation, the conjugate variables (a and H) should not be understood as describing the Universe as a whole, but instead as referring to a fundamental (or elementary) cell [50,51]. To make the use of our results simple in another context, we therefore present in Figure 15 the effects of a variation of the constant A. As expected, a rescaling of the scale factor just shifts the spectrum such that the physical wavenumber values remain unchanged.

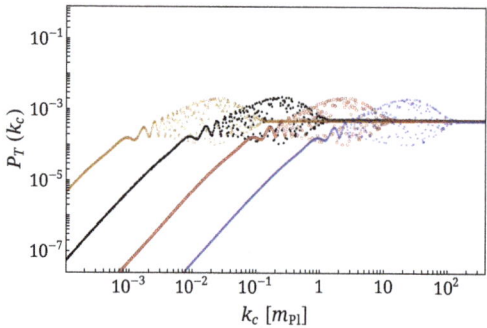

Figure 15. Primordial tensor power spectra obtained from an evolution with a bounce characterized by $C = 1$, $\mu = -600$, $\sigma = 2$, and $B = 0.4$, and different values of A. From left to right: $A = 0.1$ (orange triangles), $A = 1$ (reference case, black disks), $A = 10$ (red circles) and $A = 100$ (blue stars).

5. Multiple Bounces

It is worth considering, in addition to the first reference bounce, another perturbation of the scale factor in the static phase, that is a scale factor given by:

$$a(t) = A + Ae^{H_0 t} + A \frac{C}{2\arctan(B\sigma)} \times$$
$$\{\arctan[B(t - (\mu - \sigma))] - \arctan[B(t - (\mu + \sigma))]\}$$
$$+ A \frac{C^*}{2\arctan(B^*\sigma^*)} \{\arctan[B^*(t - (\mu^* - \sigma^*))]$$
$$- \arctan[B^*(t - (\mu^* + \sigma^*))]\},$$
(13)

in which the parameters labeled with the "*" symbol are the analogues of C, B, μ, and σ for the new additional bounce. Once again, although this possibility is in principle generic and fully phenomenological, it is also motivated by some quantum gravity results.

The spectrum corresponding to an evolution with two bounces of different steepnesses is shown in Figure 16. One can notice the presence of two "bullet" features, one for each bounce of the scale factor. It is possible to vary the characteristics of each bounce, thus the characteristics of each "bullet", independently by adjusting the parameters appropriately. If the two bounces have the same width, even if their positions are far from the one another, then the two bumps are perfectly superposed in the spectrum. If, however, the shapes of the bounces differ, they might be distinguishable in the tensor spectrum, and observational footprints of the details of the transition phase might be expected.

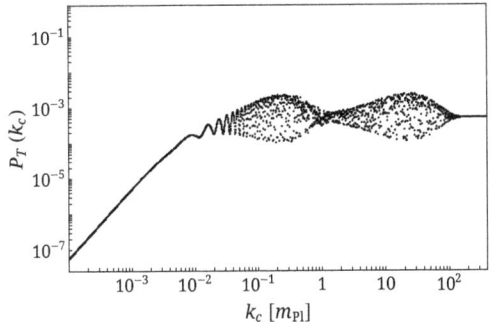

Figure 16. Primordial tensor power spectrum obtained from an evolution with two bounces of different steepness, the first one being described by $\mathcal{C} = 1$, $\mu = -400$, $\sigma = 2$, and $B = 0.4$ and the second by $\mathcal{C}^\star = 1$, $\mu^\star = -400$, $\sigma^\star = 2$, and $B^\star = 40$. The Hubble parameter during inflation is $H_0 = 0.01$.

6. Conclusions

In this article, we have clarified some general properties of the primordial cosmological tensor power spectrum in emergent models. Following a purely phenomenological approach, we have studied how different features in the behavior of the scale factor around the transition time (or before) can affect the spectrum. The main results are the following:

- in and of itself, the existence of a static phase in the remote past of the Universe does not lead to a scale-invariant power spectrum.
- if the static phase is followed by a long enough stage of inflation, the spectrum might become flat in the observable range of wavenumbers.
- the consequences of the details of the evolution of the scale factor around the transition time, modeled as a mini-bounce (or anti-bounce), are not erased by inflation and appear as a "bullet" feature in the spectrum.
- the position of the mini-bounce has only a small influence on the shape of the "bullet", but its steepness and amplitude control, respectively, the comobile position and the size of the bullet.
- multiple bounces can leave complex features in the spectrum. Bounces with different characteristics might leave distinguishable imprints in the tensor spectrum.

This work establishes that non-trivial features occurring at the transition time in an emergent universe might be detectable in the primordial tensor spectrum. The detection of the CMB B-modes is a very active field involving big collaborations. On the ground, progress is expected from the BICEPor POLARBEAR(now grouped into Stage 4) experiments and, in space, potentially from LiteBIRD. At this stage, trying to detect those modes is probably the best path toward finding traces of quantum gravity effects in the CMB. The features studied in this work may therefore be observable in the not so distant future, if the duration and energy scale of inflation are favorable.

It would clearly be interesting to go beyond the tensor spectrum and to investigate scalar perturbations that are currently observed. This, however, requires an explicit specific model as the evolution of the scale factor is no longer enough to compute the evolution of perturbations.

Author Contributions: conceptualization, K.M. and A.B.; methodology, K.M. and A.B.; software, K.M.; validation, K.M.; formal analysis, K.M. and A.B.; investigation, K.M. and A.B.; resources, A.B.; writing—original draft preparation, K.M. and A.B.; writing—review and editing, K.M. and A.B.; project administration, A.B.

Funding: This research received no external funding.

Acknowledgments: K.M. is supported by a grant from the CFM foundation.

Conflicts of Interest: The authors declare no conflict of interest.

References

1. Borde, A.; Vilenkin, A. Eternal inflation and the initial singularity. *Phys. Rev. Lett.* **1994**, *72*, 3305–3309. [CrossRef] [PubMed]
2. Borde, A.; Vilenkin, A. Singularities in Inflationary Cosmology: A Review. *Int. J. Mod. Phys. D* **1996**, *5*, 813–824. [CrossRef]
3. Brandenberger, R.; Peter. P. Bouncing cosmologies: Progress and problems. *Found. Phys.* **2017**, *47*, 797–850. [CrossRef]
4. Peter, P.; Pinto-Neto, N. Has the Universe always expanded? *Phys. Rev. D* **2002**, *65*, 023513. [CrossRef]
5. Falciano, F.T.; Lilley, M.; Peter, P. Primordial Black Holes from Inflaton Fragmentation into Oscillons. *Phys. Rev. D* **2008**, *77*, 083513. [CrossRef]
6. Lin, C.; Brandenberger, R.H.; Perreault Levasseur, L. A Matter Bounce By Means of Ghost Condensation. *J. Cosmol. Astropart. Phys.* **2011**, *2011*, 19. [CrossRef]
7. Qiu, T.; Evslin, J.; Cai, Y.-F.; Li, M.; Zhang, X. Bouncing Galileon Cosmologies. *J. Cosmol. Astropart. Phys.* **2011**, *2011*, 36. [CrossRef]
8. Kounnas, C.; Partouche, H.; Toumbas, N. Thermal duality and non-singular cosmology in d-dimensional superstrings. *Nucl. Phys. B* **2012**, *855*, 280–307. [CrossRef]
9. Cai, Y.-F.; Qiu, T.; Piao, Y.-S.; Li, M.; Zhang, X. Bouncing Universe with Quintom Matter. *J. High Energy Phys.* **2007**, *2007*, 71. [CrossRef]
10. Biswas, T.; Mazumdar, A.; Siegel, W. Bouncing Universes in String-inspired Gravity. *J. Cosmol. Astropart. Phys.* **2006**, *2006*, 9. [CrossRef]
11. Biswas, T.; Brandenberger, R.; Mazumdar, A.; Siegel, W. Reconstruction of the Scalar-Tensor Lagrangian from a LCDM Background and Noether Symmetry. *J. Cosmol. Astropart. Phys.* **2007**, *2007*, 11. [CrossRef]
12. Langlois, D.; Naruko, A. Bouncing cosmologies in massive gravity on de Sitter. *Class. Quant. Grav.* **2013**, *30*, 205012. [CrossRef]
13. Koehn, M.; Lehners, J.-L.; Ovrut, B.A. A Cosmological Super-Bounce. *Phys. Rev. D* **2014**, *90*, 025005. [CrossRef]
14. Bojowald, M. Absence of Singularity in Loop Quantum Cosmology. *Phys. Rev. Lett.* **2001**, *86*, 5227–5230. [CrossRef] [PubMed]
15. Ashtekar, A. Barrau, A. Loop quantum cosmology: From pre-inflationary dynamics to observations. *Class. Quant. Grav.* **2015**, *32*, 234001. [CrossRef]
16. Khoury, J.; Ovrut, B.A.; Steinhardt, P.J.; Turok, N. The Ekpyrotic Universe: Colliding Branes and the Origin of the Hot Big Bang. *Phys. Rev. D* **2001**, *64*, 123522. [CrossRef]
17. Steinhardt, P.J.; Turok, N. Cosmic Evolution in a Cyclic Universe. *Phys. Rev. D* **2002**, *65*, 126003. [CrossRef]
18. Battefeld, T.; Watson, S. String Gas Cosmology. *Rev. Mod. Phys.* **2006**, *78*, 435. [CrossRef]
19. Barceló, C.; Carballo-Rubio, R.; Garay, L.J. Gravitational echoes from macroscopic quantum gravity effects. *J. High Energy Phys.* **2017**, *2017*, 54. [CrossRef]
20. Beesham, A.; Chervon, S.V.; Maharaj, S.D. An emergent universe supported by a nonlinear sigma model. *Class. Quant. Grav.* **2009**, *26*, 075017. [CrossRef]
21. Wu, P.; Yu, H.W. Emergent universe from the Horava-Lifshitz gravity. *Phys. Rev. D* **2010**, *81*, 103522. [CrossRef]
22. Mukerji, S.; Chakraborty, S. Emergent universe in Horava gravity. *Astrophys. Space Sci.* **2011**, *331*, 665–671. [CrossRef]
23. Mukerji, S.; Chakraborty, S. Emergent Universe in Einstein-Gauss-Bonnet Theory. *Int. J. Theor. Phys.* **2010**, *49*, 2446–2455 . [CrossRef]
24. Paul, B.C.; Ghose, S.; Thakur, P. Emergent Universe from A Composition of Matter, Exotic Matter and Dark Energy. *Mon. Not. R. Astron. Soc.* **2011**, *413*, 686–690. [CrossRef]
25. Debnath, U.; Chakraborty, S. Emergent Universe with Exotic Matter in Brane World Scenario. *Int. J. Theor. Phys.* **2011**, *50*, 2892–2898. [CrossRef]
26. Rudra, P. Emergent Universe with Exotic Matter in Loop Quantum Cosmology, DGP Brane World and Kaluza-Klein Cosmology. *Mod. Phys. Lett. A* **2012**, *27*, 1250189. [CrossRef]
27. Chakraborty, S. Is Emergent Universe a Consequence of Particle Creation Process? *Phys. Lett. B* **2014**, *732*, 81–84. [CrossRef]

28. Perez, A.; Sudarsky, D.; Bjorken, J.D. A microscopic model for an emergent cosmological constant. *arXiv* **2018**, arXiv:1804.07162.
29. Alesci, E.; Botta, G.; Cianfrani, F.; Liberati, S. Cosmological singularity resolution from quantum gravity: The emergent-bouncing universe *Phys. Rev. D* **2017**, *96*, 046008. [CrossRef]
30. Cai, Y.-F.; Li, M.; Zhang, X. Emergent Universe Scenario via Quintom Matter. *Phys. Lett. B* **2012**, *718*, 248–254. [CrossRef]
31. Cai, Y.-F.; Wan, Y.; Zhang, X. Cosmology of the Spinor Emergent Universe and Scale-invariant Perturbations. *Phys. Lett. B* **2014**, *731*, 217–226. [CrossRef]
32. Paul, B.C.; Thakur, P. Observational Constraints on EoS parameters of Emergent Universe. *Astrophys. Space Sci.* **2017**, *362*, 73. [CrossRef]
33. Labraña, P. Emergent universe scenario and the low CMB multipoles. *Phys. Rev. D* **2015**, *91*, 083534. [CrossRef]
34. Zhang, K.; Wu, P.; Yu, H. Emergent universe in spatially flat cosmological model. *J. Cosmol. Astropart. Phys.* **2014**, *2014*, 48. [CrossRef]
35. Ghose, S.; Thakur, P.; Paul, B.C. Observational Constraints on the Model Parameters of a Class of Emergent Universe. *Mon. Not. R. Astron. Soc.* **2012**, *421*, 20–24. [CrossRef]
36. del Campo, S.; Guendelman, E.I.; Herrera, R.; Labrana, P. Emerging Universe from Scale Invariance. *J. Cosmol. Astropart. Phys.* **2010**, *2010*, 26. [CrossRef]
37. Alesci, E.; Barrau, A.; Botta, G.; Martineau, K.; Stagno, G. Phenomenology of Quantum Reduced Loop Gravity in the isotropic cosmological sector. *arXiv* **2018**, arXiv:1808.10225.
38. Bolliet, B.; Barrau, A.; Martineau, K.; Moulin, F. Some Clarifications on the Duration of Inflation in Loop Quantum Cosmology. *Class. Quant. Grav.* **2017**, *34*, 145003. [CrossRef]
39. Barrau, A.; Jamet, P.; Martineau, K.; Moulin, F. Scalar spectra of primordial perturbations in loop quantum cosmology. *Phys. Rev. D* **2018**, *98*, 086003. [CrossRef]
40. Rovelli, C. Zakopane lectures on loop gravity. *arXiv* **2011**, arXiv:1102.3660.
41. Ashtekar, A. ; Singh, P. Loop Quantum Cosmology: A Status Report. *Class. Quant. Grav.* **2011**, *28*, 213001. [CrossRef]
42. Alesci, E.; Cianfrani, F. Loop Quantum Cosmology from Loop Quantum Gravity. *Europhys. Lett.* **2015**, *111*, 40002. [CrossRef]
43. Ashtekar, A.; Wilson-Ewing, E. Loop quantum cosmology of Bianchi I models. *Phys. Rev. D* **2009**, *79*, 083535. [CrossRef]
44. Bolliet, B.; Grain, J.; Stahl, C.; Linsefors, L.; Barrau, A. Comparison of primordial tensor power spectra from the deformed algebra and dressed metric approaches in loop quantum cosmology. *Phys. Rev. D* **2015**, *91*, 084035. [CrossRef]
45. Barrau, A.; Bolliet, B. Some conceptual issues in loop quantum cosmology. *arXiv* **2016**, arXiv:1602.04452.
46. Martineau, K.; Barrau, A.; Grain, J. A first step towards the inflationary trans-Planckian problem treatment in Loop Quantum Cosmology. *Int. J. Mod. Phys. D* **2018**, *27*, 1850067. [CrossRef]
47. Schander, S.; Barrau, A.; Bolliet, B.; Linsefors, L.; Mielczarek, J.; Grain, J. Primordial scalar power spectrum from the Euclidean Big Bounce. *Phys. Rev. D* **2016**, *93*, 023531. [CrossRef]
48. Linsefors, L.; Barrau, A. Duration of inflation and conditions at the bounce as a prediction of effective isotropic loop quantum cosmolog. *Phys. Rev. D* **2013**, *87*, 123509. [CrossRef]
49. Martineau, K.; Barrau, A.; Schander, S. Detailed investigation of the duration of inflation in loop quantum cosmology for a Bianchi-I universe with different inflaton potentials and initial conditions. *Phys. Rev. D* **2017**, *95*, 083507. [CrossRef]
50. Barrau, A.; Bojowald, M.; Calcagni, G.; Grain, J.; Kagan, M. Anomaly-free cosmological perturbations in effective canonical quantum gravity. *J. Cosmol. Astropart. Phys.* **2015**, *2015*, 51. [CrossRef]
51. Bojowald, M. Quantum cosmology: A review. *Rept. Prog. Phys.* **2015**, *78*, 023901. [CrossRef] [PubMed]

© 2018 by the authors. Licensee MDPI, Basel, Switzerland. This article is an open access article distributed under the terms and conditions of the Creative Commons Attribution (CC BY) license (http://creativecommons.org/licenses/by/4.0/).

Article

Reconstruction of Mimetic Gravity in a Non-Singular Bouncing Universe from Quantum Gravity

Marco de Cesare

Department of Mathematics and Statistics, University of New Brunswick, Fredericton, NB E3B 5A3, Canada; marco.de_cesare@unb.ca

Received: 1 April 2019; Accepted: 4 May 2019; Published: 7 May 2019

Abstract: We illustrate a general reconstruction procedure for mimetic gravity. Focusing on a bouncing cosmological background, we derive general properties that must be satisfied by the function $f(\Box \phi)$ implementing the limiting curvature hypothesis. We show how relevant physical information can be extracted from power-law expansions of f in different regimes, corresponding e.g., to the very early universe or to late times. Our results are then applied to two specific models reproducing the cosmological background dynamics obtained in group field theory and in loop quantum cosmology, and we discuss the possibility of using this framework as providing an effective field theory description of quantum gravity. We study the evolution of anisotropies near the bounce, and discuss instabilities of scalar perturbations. Furthermore, we provide two equivalent formulations of mimetic gravity: one in terms of an effective fluid with exotic properties, the other featuring *two* distinct time-varying gravitational "constants" in the cosmological equations.

Keywords: mimetic gravity; limiting curvature; bouncing cosmology; quantum gravity; effective field theory

1. Introduction

The resolution of spacetime singularities is one of the main expected consequences of quantum gravity. In cosmology, the realization of such a possibility would lead to the replacement of the Big Bang singularity by a smooth spacetime region, e.g., a bounce, with profound implications for our understanding of the earliest stages of cosmic expansion and of the initial conditions for our universe. Non-singular bouncing cosmologies have been extensively studied and may represent an alternative to the inflationary scenario [1] with specific observational signatures (see also [2]). Resolution of the initial singularity in cosmology has been achieved in various approaches based on a loop quantization of the gravitational field, such as loop quantum cosmology (LQC) [3,4], group field theory (GFT) condensate cosmology [5,6], and quantum reduced loop gravity [7]; more specifically, both in LQC and in GFT the initial singularity is replaced by a regular bounce, marking the transition from a contracting phase to an expanding one.

One of the main open problems that is common to all background-independent approaches to quantum gravity is the derivation of an effective field theory taking into account effects due to the underlying discreteness of spacetime at the Planck scale. In fact, at present very little is known about quantum gravity beyond perfect homogeneity, although efforts to include inhomogeneities in the description of an emergent universe from full quantum gravity are underway [8–10]. One possible alternative approach then consists of considering modifications of general relativity that can reproduce known features of a given quantum gravity theory. The hope is that by doing so, we can gain insight (at least qualitatively) into the consequences of quantum gravitational effects in different regimes. In this work, we adopt the framework of limiting curvature mimetic gravity and examine in detail the problem of reconstructing the theory from the evolution of the cosmological background, with particular attention to the case of a bouncing background. Such a theory should then be regarded as a toy model for an effective description of quantum gravity [11,12] and can be used to study its phenomenological consequences. Possible applications include e.g., the dynamics of inhomogeneous and anisotropic degrees of freedom in cosmology, and black holes.

The idea of limiting curvature as a possible solution to the singularities of general relativity was first envisaged in Ref. [13–16], and subsequently implemented in modifications of the Einstein-Hilbert action including higher-order curvature invariants in Refs. [17–20]. An alternative proposal for constructing a gravitational theory with a built-in limiting curvature scale was put forward in Ref. [21] as an extension of mimetic gravity. This is achieved by including in the action functional a (multivalued) potential term f depending on the d'Alembertian of a scalar field ϕ. Upon closer inspection, such a potential turns out to depend on the expansion scalar χ of a privileged irrotational congruence of time-like geodesics, singled out by the so-called mimetic constraint [22]. On a cosmological spacetime, $f(\chi)$ reduces to a function of the Hubble rate [23]. Multivaluedness of the potential is necessary for a consistent realization of bouncing cosmologies in this framework [22,24–26]. Non-singular black hole solutions have been studied in Refs. [27,28].

The particular model proposed in Ref. [21] exactly reproduces the effective dynamics obtained in (flat, isotropic) homogeneous LQC. Thus, all curvature invariants are bounded throughout spacetime by a limiting curvature scale, which is in turn related to the existence of a critical value for the energy density of matter at the bounce. From the point of view of quantum gravity, it is natural to require that the limiting curvature scale be Planckian. In Ref. [12] a broader class of theories was identified in the degenerate higher-order scalar tensor theories (DHOST) family, all reproducing the effective dynamics of LQC; these models can be further extended by the inclusion of a term corresponding to the spatial curvature. The relation between the model of Ref. [21] and effective LQC was further investigated in Refs. [11,29] from a Hamiltonian perspective, showing that the equivalence holds in the spatially flat, homogeneous, and isotropic sector; however, the correspondence is lost in the anisotropic case. Nevertheless, even for anisotropic cosmologies the solutions of the two models are qualitatively similar [11,29]. The mimetic model of Ref. [21] has been recently generalized in Ref. [22], where a limiting curvature mimetic gravity theory was reconstructed so as to exactly reproduce the background evolution obtained from GFT condensates in Ref. [5,6]; the effective dynamics of homogeneous LQC is then recovered as a particular case for some specific choice of the parameters of the model.

This paper has two main goals. The first one is to give a general account of theory reconstruction in mimetic gravity, showing how essential information about background evolution (e.g., the critical energy density, the bounce duration, and the equation of state of effective fluids) is encoded in the function $f(\chi)$, particularly in its asymptotic behavior in regimes of physical interest. The case of a generic bouncing background is examined in detail, although our methods have a much broader applicability. We provide general prescriptions for the matching of the different branches of the multivalued function $f(\chi)$, which are necessary in order to obtain a smooth evolution of the universe, thus generalizing the analysis of matching conditions in Ref. [22]. Our second goal is to study in detail the properties of mimetic gravity theories with the same background evolution as obtained in non-perturbative approaches to quantum gravity. Specifically, we analyze the model of Ref. [22] reproducing the background evolution obtained from GFT condensates, and compare it to the special case corresponding to the LQC effective dynamics. We study the evolution of anisotropies near the bounce in a Bianchi I spacetime, including the effects of hydrodynamic matter with generic equation of state, thus extending the results of Ref. [21]. As in the model of Ref. [21], our more general results also show that the smooth bounce is not spoiled by anisotropies, which stay bounded during the bounce era. Instabilities in the inhomogeneous sector are also discussed. Moreover, given its relevance and simplicity, the particular case corresponding to the effective dynamics of LQC is analyzed separately.

The plan of the paper is as follows. The formulation of mimetic gravity is briefly reviewed in Section 2. In Section 3 we discuss the reconstruction procedure. In Section 4 we focus on the model of Ref. [22]:

we discuss the background evolution, exhibit the form of the function $f(\chi)$ and derive its expansion in its two branches, corresponding to the region around the bounce and to a large universe. The model of Ref. [22], which can be obtained as a particular case from our more general model, is discussed separately due to its relevance and simplicity. Section 5 is devoted to the study of anisotropies in a bouncing background. In Section 6 we provide an alternative description of the cosmological dynamics of mimetic gravity in terms of two effective gravitational "constants", both depending on the expansion rate of the universe. In Section 7 we discuss instabilities of scalar perturbations. We conclude with a discussion of our results in Section 8.

We choose units such that $8\pi G = 1$. Landau-Lifshitz conventions for the metric signature $(+---)$ are adopted.

2. Mimetic Gravity and Its Cosmology

The version of mimetic gravity considered in Ref. [21] is based on the action

$$S[g_{\mu\nu}, \phi, \lambda, \psi] = \int d^4x \sqrt{-g} \left(-\frac{1}{2}R + \lambda(g^{\mu\nu}\partial_\mu\phi\partial_\nu\phi - 1) + f(\chi) + L_m(\psi, g_{\mu\nu}) \right), \quad (1)$$

with $\chi = \Box\phi$. The gravitational sector consists of the metric $g_{\mu\nu}$ and the scalar field ϕ. The Lagrange multiplier λ enforces the mimetic constraint

$$g^{\mu\nu}\partial_\mu\phi\partial_\nu\phi = 1. \quad (2)$$

We have included a matter Lagrangian L_m, where ψ represents a generic matter field, coupled to $g_{\mu\nu}$ only and not to ϕ. Due to the term $f(\chi)$, the action (1) represents a higher-derivative extension of the original mimetic gravity theory of Ref. [30].[1]

Due to the mimetic constraint, the vector field $u^\mu = g^{\mu\nu}\partial_\nu\phi$ has unit norm and generates an irrotational congruence of time-like geodesics (see Ref. [22] for more details). Thus, the theory admits a preferred foliation[2] with time function $t = \phi$ and time-flow vector field $u^\mu \frac{\partial}{\partial x^\mu} = \frac{\partial}{\partial t}$. The quantity χ, defined above, can be expressed as $\chi = \nabla^\mu u_\mu$ and represents the expansion of the geodesic congruence generated by u^μ. In FLRW spacetime, one has $\chi = 3H$, where H denotes the Hubble rate. It is for this reason that the term $f(\chi)$ in the action (1) plays an important role in the cosmological applications of the model, since for a homogenous and isotropic background $f(\chi)$ reduces to a function of the Hubble rate only. This is a crucial property of the model, which allows for a straightforward theory reconstruction procedure, starting from a given cosmological background evolution. This aspect will be analyzed in detail in Section 3.

It is worth stressing that although the action for mimetic gravity includes higher-derivative terms through $f(\chi)$, the equations of motion are second order. In fact, mimetic gravity is a particular case of so-called DHOST, which are characterized by the absence of Ostrogradski ghost [23,35]. Nevertheless, compared to general relativity, the mimetic gravity theory described by (1) has an extra propagating scalar degree of freedom if $f_{\chi\chi} \neq 0$ [33,36]. Importantly, this is always a source of instabilities in the theory, as discussed in Section 7.

The field equations read as [21]

$$G_{\mu\nu} = T_{\mu\nu}^\psi + \tilde{T}_{\mu\nu}, \quad (3)$$

where the matter stress-energy tensor is defined as usual

---

[1] The original formulation of mimetic gravity of Ref. [30] relied on a singular disformal transformation [31] (see also Ref. [23]). An equivalent formulation with a Lagrange multiplier implementing the constraint (2) was given in Ref. [32]. The latter represents the starting point for further generalizations of the model considered in Refs. [21,33]. See also the review [34].
[2] Such a gauge choice corresponds to unit lapse and vanishing shift, i.e., $N = 1$ and $N^i = 0$.

$$T^\psi_{\mu\nu} = \frac{2}{\sqrt{-g}} \frac{\delta S_m}{\delta g^{\mu\nu}}, \tag{4}$$

and the extra term in Equation (3) is an effective stress-energy tensor arising from the ϕ-sector of the action (1)

$$\tilde{T}_{\mu\nu} = 2\lambda \partial_\mu \phi \partial_\nu \phi + g_{\mu\nu}(\chi f_\chi - f + g^{\rho\sigma}\partial_\rho f_\chi \partial_\sigma \phi) - (\partial_\mu f_\chi \partial_\nu \phi + \partial_\nu f_\chi \partial_\mu \phi). \tag{5}$$

The Lagrange multiplier λ can be eliminated by solving the following equation

$$\Box f_\chi - 2\nabla^\mu(\lambda \partial_\mu \phi) = 0, \tag{6}$$

which can be obtained by varying the action with respect to ϕ. Equation (6) can be interpreted as a conservation law for the Noether current associated with the global shift-symmetry of the action (1), see Refs. [22,37]

Considering a flat FLRW model $ds^2 = dt^2 - a^2(t)\delta_{ij}dx^i dx^j$, the field Equation (3) lead to a modification of the Friedmann and Raychaudhuri equations

$$\frac{1}{3}\chi^2 = \rho + \tilde{\rho} + M, \tag{7}$$

$$\dot{\chi} = -\frac{3}{2}\left[(\rho + P) + (\tilde{\rho} + \tilde{P}) + M\right]. \tag{8}$$

The quantities introduced in Equations (7) and (8) are defined as follows: ρ and P denote the energy density and pressure of ordinary matter, whereas $\tilde{\rho}$ and $\tilde{P}$ represent the corresponding quantities for the effective fluid, given by

$$\tilde{\rho} = \chi f_\chi - f, \tag{9}$$

$$\tilde{P} = -(\tilde{\rho} + f_{\chi\chi}\dot{\chi}). \tag{10}$$

The properties of the effective fluid for a quadratic $f(\chi)$ were studied in Ref. [37]. Lastly, we have $M = \frac{C}{a^3}$, where C is an integration constant for Equation (6). The quantity M represents the energy density of so-called mimetic dark matter [30]. We note that for vanishing f the action (1) describes irrotational dust minimally coupled to gravity, corresponding to a particular case of the Brown-Kuchař action [38].[3] Finally, we observe that the effective fluid satisfies the continuity equation

$$\dot{\tilde{\rho}} + \chi(\tilde{\rho} + \tilde{P}) = 0. \tag{11}$$

3. Theory Reconstruction

We henceforth consider a spatially flat, homogeneous, and isotropic universe, as described by the FLRW line element $ds^2 = dt^2 - a^2(t)\delta_{ij}dx^i dx^j$. The proper time gauge $N = 1$ will be used throughout. The spacetime geometry is then fully characterized by the evolution of a single degree of freedom: the scale factor $a(t)$. Given a theory of gravity with second order field equations, cosmological solutions can be represented as trajectories in the plane (a, χ). In general relativity, the trajectories are determined by the Friedmann equation

$$\frac{1}{3}\chi^2 = \sum_i \rho_i. \tag{12}$$

Here the quantities ρ_i denote the energy density of different matter species. For the sake of simplicity, we can assume that all matter species are non-interacting and have constant equation of state parameters w_i. Thus, we have $\rho_i = c_i V^{-(w_i+1)}$, where c_i are constants depending on the initial

---

[3] See also Refs. [39,40].

conditions and $V = a^3$ is the proper volume of a unit comoving cell. It is convenient to introduce a new variable $\eta = V^{-1}$, so that the Friedmann equation can be re-expressed as

$$\frac{1}{3}\dot{\chi}^2 = \sum_i c_i \eta^{w_i+1}. \tag{13}$$

Such a parametrization is particularly useful in bouncing cosmologies, where η has a bounded range. In the following, we will denote by Γ the trajectory in the (η, χ) plane given by Equation (13).

Despite the derivation given above, based on the standard Friedmann equation, Equation (13) has a broader applicability. In fact, it also holds in a more general class of modified gravity theories and quantum cosmological models, provided that the corrections to the standard Friedmann equation can be described—at an effective level—as perfect fluids. Such effective fluids may have exotic properties and, depending on the model, can violate the energy conditions. This is the case, for instance, in the effective dynamics of both LQC and GFT condensate cosmology. In fact, having a bounce requires that both the weak and the null energy conditions must be violated due to the effective fluids. The former violation is necessary to accommodate for a vanishing expansion, see Equation (13). The latter violation follows instead from the requirement that $\dot{\chi} > 0$ at the bounce, and from the Raychaudhuri equation including effective fluids contributions

$$\dot{\chi} = -\frac{3}{2}\sum_i(\rho_i + P_i). \tag{14}$$

It is important to observe that in general Equation (13) allows us to define χ as a function of η only locally. In fact, in bouncing models, the function $\chi(\eta)$ has (at least) two branches. More branches are possible if one allows, e.g., for intermediate recollapse eras; we shall disregard this possibility in the following for simplicity. For a universe undergoing a single bounce, the trajectory Γ has the profile depicted in Figure 1. The bounce is represented by the point $B = (\eta_{\max}, 0)$, where Γ and the η axis intersect orthogonally. Since we are assuming a flat spatial geometry, both endpoints of Γ will have $\eta = 0$ if the weak energy condition is satisfied for a large universe. The value of χ at the endpoints is determined by the equation of state of the dominant matter species in such a regime: for $w > -1$ one has that χ vanishes as η tends to zero, for $w = -1$ (cosmological constant) χ approaches a constant value. We note that for $w + 1 > 0$ the two endpoints coincide with the origin; moreover, for $-1 < w < 1$ the trajectory Γ intersects the η axis orthogonally at the origin, whereas for $w \geq 1$ it has a cusp.

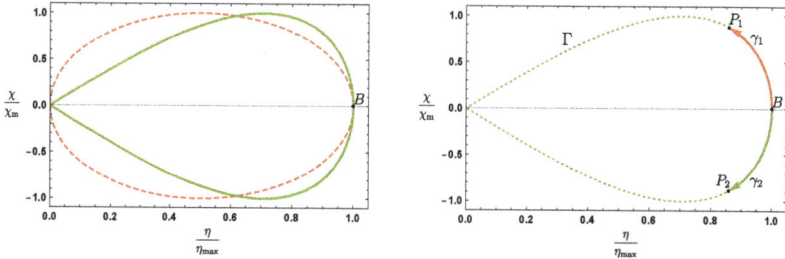

Figure 1. Trajectories Γ in the (η, χ) plane for a (symmetric) bouncing universe. The upper half-plane corresponds to the expanding phase, whereas the lower half-plane describes the contracting phase. The bounce is represented by the point B, where the expansion rate vanishes, and the scale factor attains its minimum (correspondingly η is maximized). The left figure shows the trajectory Γ for a universe filled with a scalar field (thick green line), or dust (dashed orange line); parameters are chosen so that the two trajectories are characterized by the same critical density ρ_c and limiting expansion rate χ_m. The right figure shows the two integration contours γ_1, γ_2 used in Equation (17).

3.1. Reconstruction Procedure

Given a background evolution as specified by the trajectory Γ, it is possible to apply a reconstruction procedure that allows us to uniquely determine the function $f(\chi)$ in the mimetic gravity action (1). The method illustrated in this section extends to a generic background evolution the procedure applied in Refs. [22,25,26] and ensures that appropriate matching conditions are implemented at the branching points.[4] We start by rewriting Equation (7), using Equation (9), as

$$\frac{\chi^2}{3}\left[1 - 3\frac{\mathrm{d}}{\mathrm{d}\chi}\left(\frac{f}{\chi}\right)\right] = \rho. \tag{15}$$

The solution to this equation can be obtained by quadrature, and is given by

$$f(\chi) = \frac{\chi}{3}\int_A^P \mathrm{d}\chi\left(1 - \frac{3\rho(\eta(\chi))}{\chi^2}\right) + \bar{c}\chi, \tag{16}$$

where $\bar{c}$ is an integration constant. The integral is computed along an arc of curve $\gamma \subseteq \Gamma$ with endpoints A and P, representing a fixed reference point and a generic point on Γ, respectively.

In bouncing cosmologies, the background dynamics is characterized by the existence of a limiting curvature scale, which is attained at the bounce. In turn, this scale is related to the existence of a maximum expansion rate, which will be denoted by $\chi_m \equiv \max_\Gamma \chi$, see Figure 1. In this class of models, it is convenient to take the bounce as a reference point, i.e., $A \equiv B$ in Equation (16). Since the energy density of matter is given as a function of the inverse volume, i.e., $\rho = \rho(\eta)$, the explicit computation of the integral (16) requires the determination of the inverse function $\eta(\chi)$. In general, such an inverse function exists only locally. This implies that in bouncing models the function $f(\chi)$ given by Equation (16) must be multivalued as a function of χ.[5] More precisely, in models with a single bounce $f(\chi)$ has two branching points where χ attains its extrema, one in the expanding phase, the other in the contracting phase. For a generic bouncing background $f(\chi)$ would have three branches, each corresponding to one of the three branches of the inverse functions $\eta(\chi)$. Thus, one branch f_B corresponds to the bounce phase, and two (a priori distinct) branches f_L^c, f_L^e correspond to the regions away from the bounce in the contracting and expanding phase, respectively. We will refer to the latter as the large universe branches, characterized by $\dot{\chi} < 0$. As shown in Section 3.2, for symmetric bounces the two branches f_L^c, f_L^e can be identified, provided that an appropriate choice is made for the integration constant in Equation (16).

We remark that our solution for f is continuous on Γ by construction. The derivative f_χ is also continuous, except at the origin $\chi = \eta = 0$.[6] This ensures that the energy density of the effective fluid, Equation (9), is continuous throughout cosmic history. Thus, the matching conditions prescribed in Ref. [22] are automatically implemented in Equation (16). As a general property of this class of models $f_{\chi\chi}$ diverges at the branching points, see discussion in Section 4.

After computing the integral in Equation (16), the reconstructed action for mimetic gravity can then be obtained by replacing $\chi \to \Box\phi$ in the result. Clearly, the value of the integration constant $\bar{c}$ has no influence on the equations of motion, since the linear term contributes a total divergence to the action (1).

---

[4] We note that a different version of mimetic gravity is considered in Ref. [26] that agrees at the background level with the one presently considered. However, the two theories will differ in general at the level of perturbations.

[5] We observe that for models entailing a single bounce, the solution (16) is single-valued if regarded as a function of the pair $(\chi, \dot{\chi})$.

[6] The fact that the origin is a singular point in the parametrization adopted here should not be too surprising: in fact, it corresponds to the infinite volume limit of both contracting and expanding branches. In a flat universe these are clearly two disconnected regimes.

3.2. Bounce Asymptotics

For a symmetric bounce model, the function $f(\chi)$ is even, provided that an appropriate choice of the integration constant is made in Equation (16). In fact, defining $P_1 = (\eta, \chi)$ and $P_2 = (\eta, -\chi)$, with η and χ satisfying the background equation, one has

$$\int_B^{P_1} d\chi \left(1 - \frac{3\rho(\eta(\chi))}{\chi^2}\right) = -\int_B^{P_2} d\chi \left(1 - \frac{3\rho(\eta(\chi))}{\chi^2}\right). \tag{17}$$

Thus, the integral is odd. The curves γ_1 and γ_2 are depicted in Figure 1. Using Equations (17) and (16), it is then straightforward to show that setting $\tilde{c} = 0$ leads to $f(\chi) = f(-\chi)$. In the following, we shall restrict our attention to symmetric bounce models and assume that $f(\chi)$ be even, unless otherwise stated.

The value of the function f at the bounce is independent from all other details of cosmic history. It can be computed as a limit of Equation (16). Denoting by f_B the bounce branch of the multivalued function f, we have

$$f_B(0) = \lim_{P \to B} \frac{\chi}{3} \int_B^P d\chi \left(1 - \frac{3\rho(\eta(\chi))}{\chi^2}\right) = \lim_{\chi \to 0} \frac{\chi}{3} \int_0^\chi d\chi \left(1 - \frac{3\rho(\eta(\chi))}{\chi^2}\right) = \rho_c, \tag{18}$$

where ρ_c is the critical density, i.e., the maximum of the energy density of matter, which is attained at the bounce. Since f_B is even by hypothesis, we have for $\chi \simeq 0$

$$f_B(\chi) = \rho_c + \frac{1}{2!} \vartheta \, \chi^2 + \mathcal{O}(\chi^4), \tag{19}$$

where we introduced the notation $\vartheta = (f_B)_{\chi\chi}|_0$. Hence, it follows that the energy density of the effective fluid at the bounce is given by $\tilde{\rho} = -\rho_c$. The sign of the second derivative can be determined by the requirement that the effective fluid must also violate the null energy condition (NEC) at the bounce. In fact, using Equation (10) we have

$$\tilde{\rho} + \tilde{P} = -(f_B)_{\chi\chi} \, \dot{\chi} < 0 \tag{20}$$

Since $\dot{\chi} > 0$ at the bounce, we conclude $\vartheta > 0$.

NEC violation also allows us to derive an upper bound for the duration of the bounce in limiting curvature mimetic gravity. To prove such a statement, let us assume that at the bounce the most relevant contributions to the energy density are due to the effective fluid and to a perfect fluid with equation of state parameter w. The condition $\dot{\chi} > 0$, which must be valid in a neighborhood of the bounce, implies

$$\rho + p + \tilde{\rho} + \tilde{P} < 0. \tag{21}$$

In turn, Equation (21) implies

$$(1+w)\rho_c - (f_B)_{\chi\chi} \dot{\chi} < 0. \tag{22}$$

For first-order bounce models[7] during the bounce phase the expansion χ is well approximated by a linear function of time. We can estimate the time derivative of χ at the bounce as $\dot{\chi} \sim \frac{\chi_m}{T}$, where T is the bounce duration. Therefore, in this case we obtain from Equation (22)

$$T \lesssim \frac{\vartheta \, \chi_m}{\rho_c (1+w)}. \tag{23}$$

Typically, $\rho_c \sim \chi_m^2$ and $\vartheta \sim \mathcal{O}(1)$, so that $T \lesssim \chi_m^{-1}$. When such an approximation applies, the number of e-folds of expansion during the bounce phase is $N = \log\left(\frac{a(T)}{a_B}\right) \lesssim \mathcal{O}(1)$. These considerations also apply to the models studied in Section 4 (see Equation (33) for the corresponding expansion of f near the bounce). In fact, the estimate (23) agrees with the upper bound for the number of e-folds obtained in Ref. [42] for the so-called non-interacting model. We mention that the so-called fast-bounce models, considered e.g., in Ref. [43], are first-order bounces whose duration is much shorter than the time-scale linked to the maximum expansion rate, i.e., such that $T \ll \chi_m^{-1}$; such a scenario can be realized in mimetic gravity by requiring $\left.\frac{(f_B)_{\chi\chi}}{f_B}\right|_{\chi=0} \ll \chi_m^{-2}$.

3.3. Late Time Asymptotics

Considerations on the evolution of the universe at late times allow us to put restrictions on the leading order terms of the branch f_L around $\chi \simeq 0$. In fact, we observe that the effective fluid is characterized by a time-dependent equation of state parameter $\tilde{w}$, given by

$$\tilde{w} = \frac{\tilde{p}}{\tilde{\rho}} = -\left(1 + \frac{f_{\chi\chi}}{\tilde{\rho}}\dot{\chi}\right), \tag{24}$$

where we used Equations (9) and (10). It is interesting to examine the case where the universe at late times is dominated by matter with equation of state w and the effective fluid is sub-dominant, with $\tilde{w}$ approaching a constant value as $\chi \to 0$. Clearly, consistency of such assumptions requires $w < \tilde{w}$. The leading order term in the expansion of $f_L(\chi)$ around $\chi \simeq 0$ is then given by

$$f_L(\chi) \simeq \lambda \, \chi^{2\left(\frac{\tilde{w}+1}{w+1}\right)}, \tag{25}$$

where λ is a constant. In fact, since by hypothesis we must leading order $\chi \sim \eta^{\frac{1+w}{2}}$, Equation (25) implies $\tilde{\rho} \sim \eta^{1+\tilde{w}}$, consistently with our assumptions.

4. Effective Approach to Quantum Gravitational Bouncing Cosmologies

In Ref. [22] the reconstruction procedure outlined in Section 3 was successfully applied to the cosmological dynamics obtained from group field theory condensates in [5,6]. The evolution equation for such a model can be expressed in relational form by introducing a minimally coupled massless scalar field ψ [44]. In fact, provided that its momentum be non-vanishing $p_\psi \neq 0$, ψ is a monotonic function of t and thus represents a perfect clock. For definiteness, we will assume $p_\psi > 0$. Using the relational clock ψ as time, the FLRW line element can be expressed as

$$ds^2 = N^2(\psi) \, d\psi^2 - a^2(\psi) \delta_{ij} dx^i dx^j, \tag{26}$$

where the lapse function reads as

$$N(\psi) = (\dot\psi)^{-1} = p_\psi^{-1} a^3(\psi). \tag{27}$$

---

[7] The order of the bounce is defined as the positive integer n such that $a^{(2n)}(t_B) > 0$ is the lowest-order non-vanishing derivative of the scale factor at the bounce [41].

We can define a relational Hubble rate as $\mathcal{H} = \frac{a'}{a}$, where a prime denotes differentiation with respect to ψ. The expansion χ is related to $\mathcal{H}$ as follows

$$\chi = 3 p_\psi \frac{\mathcal{H}}{a^3} . \tag{28}$$

The relational Friedmann equation governing the dynamics of GFT condensates reads as (recall $V = a^3$)

$$\mathcal{H}^2 = \frac{1}{6} + \frac{\alpha}{V} - \frac{\beta}{V^2} , \tag{29}$$

where α and $\beta > 0$ are parameters depending on the details of the microscopic model, see Ref. [5,6].[8] An effective Friedmann equation with the same form as Equation (29) was obtained in the GFT models of Refs. [45,46]. The first term in Equation (29) is the contribution of the massless scalar field ψ, whereas the remaining two terms represent quantum gravitational corrections; in particular, the α term represents a correction to the effective dynamics of LQC. It must be stressed that for simplicity, we are neglecting interactions between GFT quanta, which would contribute additional terms to Equation (29). The cosmological consequences of interactions were considered in Ref. [42].

Changing time parametrization back to proper time and recalling $\eta = V^{-1}$, we have

$$\frac{1}{3}\chi^2 = p_\psi^2 \left(\frac{1}{2}\eta^2 + 3\alpha\,\eta^3 - 3\beta\,\eta^4 \right) . \tag{30}$$

The first term to the r.h.s. of Equation (30) gives the energy density ρ_ψ of the scalar ψ; the quantum gravitational corrections (second and third terms) correspond instead to effective fluids with equation of state parameter $w = 2, 3$. The third term becomes important for large values of η (i.e., small values of the scale factor); moreover, since $\beta > 0$ such a term violates both the weak and the NECs, and is therefore responsible for the bounce. It must be noted that the bounce is symmetric for any choice of parameters in this model. The equation for $\dot\chi$ is

$$\dot\chi = -\frac{3}{2}p_\psi^2 \left(\eta^2 + 9\alpha\,\eta^3 - 12\beta\,\eta^4 \right) . \tag{31}$$

For further details on the effective fluid description of quantum gravity corrections in the effective Friedmann equation arising in the GFT approach, including interactions between quanta, the reader is referred to Refs. [42,47]. For a large universe (i.e., small η) the first term in Equation (30) becomes the dominant one: the standard Friedmann evolution is thus recovered, and the quantum gravity corrections are sub-leading.

The background evolution (30) can be exactly reproduced in mimetic gravity if the function $f(\chi)$ is given by [22]

$$f(\chi) = \rho_\psi(\chi) + \frac{1}{3}\chi^2 + \frac{p_\psi}{3\sqrt{\beta}}|\chi|\left[\arctan\left(\frac{1}{\sqrt{\beta}}\frac{d|\mathcal{H}|}{d\eta}\right) + \frac{\pi}{2}\right] . \tag{32}$$

By construction, the different branches of the multivalued function in Equation (32) satisfy matching conditions at the branching points, to ensure the regularity of cosmological evolution. Around the bounce the following expansion holds

$$f_{\rm B}(\chi) = \rho_c + \frac{1}{3}\left(\frac{2V_{\rm B} + 3\alpha}{V_{\rm B} + 3\alpha}\right)\chi^2 + \mathcal{O}(\chi^4) , \tag{33}$$

---

[8] It is worth remarking that α and β are defined only up to arbitrary constant rescalings of the comoving volume V_0, which was set equal to one above. We have in general $V = V_0\, a^3$. Under the transformation $V_0 \to k V_0$ with constant k, α and β transform according to $\alpha \to k\alpha$, $\beta \to k^2\beta$. Thus, the scale invariance property of the standard Friedmann equation is preserved by the quantum corrections. In the GFT formalism such rescaling properties correspond to the invariance of the dynamics under constant rescalings of the number of quanta, cf. Ref. [5,6].

where $V_B = -3\alpha + \sqrt{9\alpha^2 + 6\beta}$ is the volume at the bounce and $\rho_c = \frac{p_\psi^2}{2V_B^2}$. For the asymptotic expansion of $f(\chi)$ around the branching points at maximum expansion rate $|\dot{\chi}| = \chi_m$, see Ref. [22]. Both f and f_χ are continuously matched at the branching points. However, the second derivative $f_{\chi\chi}$ has a discontinuity there: this is a general property of mimetic gravity theories with a limiting curvature scale. Nevertheless, the effective pressure $\tilde{P}(\chi)$ is guaranteed to be finite even when $f_{\chi\chi}$ diverges, since Equation (8) implies

$$\tilde{P}(\pm\chi_m) = -(\rho + P) = -(w+1)\rho. \tag{34}$$

When the universe is large (i.e., in the regime χ, $\eta \sim 0$) one has the expansion (disregarding the linear term, which does not affect the equations of motion)

$$f_L(\chi) = \sqrt{\frac{2}{3}\frac{\alpha}{p_\psi}}|\chi|^3 - \frac{4}{p_\psi^2}\left(\alpha^2 + \frac{1}{9}\beta\right)\chi^4 + \mathcal{O}(|\chi|^5), \tag{35}$$

which can be rewritten as

$$f_L(\chi) = \frac{\alpha}{2V_*}\sqrt{2 + \frac{6\alpha}{V_*}}\frac{|\chi|^3}{\chi_m} - \frac{(V_* + 3\alpha)(V_*^2 + 9\alpha V_* + 108\alpha^2)}{36V_*^3}\frac{\chi^4}{\chi_m^2} + \mathcal{O}(|\chi|^5), \tag{36}$$

where $\chi_m = \frac{p_\psi}{2V_*}\sqrt{3 + \frac{9\alpha}{V_*}}$, and $V_* = \frac{1}{2}\left(\sqrt{81\alpha^2 + 48\beta} - 9\alpha\right)$ is the volume at $\chi = \chi_m$. Please note that if $\alpha = 0$, the next non-vanishing term in the expansion is $\mathcal{O}(\chi^6)$.

Once the function $f(\chi)$ has been reconstructed from a given background evolution, one can also consider different matter species coupled to gravity. It must be pointed out that when matter species other than a minimally coupled massless scalar field are considered, parameters such as p_ψ and V_B in Equation (32) lose their usual interpretation. This is to be expected, since the relation between χ and η will be different from Equation (30) in the general case. Nevertheless, the values of the critical energy density ρ_c and the maximum expansion rate χ_m are not affected by the different matter species, and represent universal features of the model.

Let us now assume hydrodynamic matter with constant equation of state parameter w. Comparing Equations (36) and (25), at late times we obtain a simple description of the effective fluid corresponding to the mimetic gravity corrections as a sum of perfect fluid contributions, each with a constant equation of state. Specifically, we find for the third order term in Equation (36) $\tilde{w}_3 = \frac{1}{2}(3w + 1)$, whereas for the fourth order term we have $\tilde{w}_4 = 2w + 1$. Clearly, for a massless scalar field $w = 1$ one recovers the effective fluid corrections given in Equation (30).

A Special Case: Reproducing the LQC Effective Dynamics

The case $\alpha = 0$ is special and deserves being discussed separately. In fact, in this case one recovers the model of Ref. [21], which reproduces the effective dynamics of LQC for a spatially flat, isotropic universe. After locally inverting $\chi = \chi(\eta)$, one finds the two branches of the function $f(\chi)$

$$f_B = \frac{2}{3}\chi_m^2\left\{1 + \frac{1}{2}q^2 + \sqrt{1 - q^2} + q\arcsin(q)\right\}, \tag{37}$$

$$f_L = \frac{2}{3}\chi_m^2\left\{1 + \frac{1}{2}q^2 - \sqrt{1 - q^2} - |q|(\arcsin|q| - \pi)\right\}, \tag{38}$$

where $\chi_m = p_\psi\sqrt{\frac{3}{48\beta}}$ and we defined $q = \frac{\chi}{\chi_m}$ to make the notation lighter. It must be noted that Equations (37) and (38) do not make any reference to the scalar field ψ, which was assumed as the only matter species coupled to gravity in the derivation of Equation (29) in Ref. [5,6]. Thus, for $\alpha = 0$ the effective Friedmann equation will take the same universal form regardless of the matter species considered. Using Equation (18), the critical energy density is determined as $\rho_c = f_B(0) = \frac{4}{3}\chi_m^2$. The energy density of the effective fluid can be computed using Equation (9); the result is $\tilde{\rho} = -\frac{\rho_c}{2}\left(1 - \frac{q^2}{2} \pm \sqrt{1-q^2}\right)$, where the upper sign corresponds to the bounce branch and the lower one corresponds to a large universe. After some straightforward algebraic manipulations, the Friedmann Equation (7) can then be recast in the following form

$$\frac{1}{3}\chi^2 = \rho\left(1 - \frac{\rho}{\rho_c}\right),\qquad(39)$$

where ρ denotes the total energy density of all matter species that are present. Similarly, using Equations (8) and (10) we can obtain the equation for $\dot{\chi}$. We have, for a general $f(\chi)$

$$\left(1 - \frac{3}{2}f_{\chi\chi}\right)\dot{\chi} = -\frac{3}{2}(\rho + P).\qquad(40)$$

The bracket to the r.h.s. of Equation (40) can be evaluated using Equations (37) and (38)

$$1 - \frac{3}{2}f_{\chi\chi} = \mp\frac{1}{\sqrt{1-q^2}} = \left(1 - \frac{2\rho}{\rho_c}\right)^{-1},\qquad(41)$$

where we used Equation (39) in the last equality. Finally, we have

$$\dot{\chi} = -\frac{3}{2}(\rho + P)\left(1 - \frac{2\rho}{\rho_c}\right).\qquad(42)$$

Thus, the time derivative of the expansion is positive for $\frac{\rho_c}{2} < \rho \leq \rho_c$ (super-inflation). This is to be contrasted with general relativity, where one always has $\dot{\chi} < 0$ for matter satisfying the NEC. Equations (39) and (42) coincide with the effective dynamics of (flat, isotropic) LQC, see e.g., Ref. [4].

It is important to observe that one must change branch of $f(\chi)$ when $\dot{\chi} = 0$ [24]. This happens when the density reaches the value $\frac{\rho_c}{2}$, see Equation (42), whereby the expansion attains its extremum $\chi^2 = \chi_m^2$. It must be noted that in both branches, as given by Equations (37) and (38), $f_{\chi\chi}$ diverges as $|\chi| \to \chi_m$; however, the effective pressure $\tilde{P}$ is continuous in the limit since $\tilde{P} = -\frac{\rho}{\rho_c}(\rho + 2P)$.

Exact solutions of the effective Friedmann Equation (39) can be derived for hydrodynamic matter (see Ref. [21])

$$a(t) = a_B\left(1 + \frac{3}{4}\rho_c(w+1)^2(t-t_B)^2\right)^{\frac{1}{3(1+w)}},\qquad(43)$$

where the origin of time has been set to have the bounce at $t = 0$. Provided that matter satisfies the NEC, one finds for the bounce duration (defined to have $\chi(T) = \chi_m$)

$$T = \frac{1}{\chi_m(1+w)},\qquad(44)$$

which is in good agreement with the estimate given by Equation (23).

Finally, the expansions (33) and (36) for $\alpha = 0$ become, respectively

$$f_B(\chi) = \rho_c + \frac{2}{3}\chi^2 + \mathcal{O}(\chi^4),\qquad(45)$$

and

$$f_L(\chi) = -\frac{1}{36}\frac{\chi^4}{\chi_m^2} + \mathcal{O}(\chi^6).\qquad(46)$$

5. Anisotropies near the Bounce

In this Section we generalize the analysis of Ref. [21], studying the evolution of anisotropies near the bounce in a non-singular Bianchi I spacetime, for the model of Section 4 and in the presence of hydrodynamic matter with generic equation of state.

The line element of Bianchi I in proper time gauge is

$$ds^2 = dt^2 - a^2(t) \sum_i e^{2\beta_{(i)}(t)} (dx^i)^2 , \tag{47}$$

where $a(t)$ is the mean scale factor, and the variables $\beta_{(i)}$ representing the anisotropies satisfy $\sum_i \beta_{(i)} = 0$. We will assume hydrodynamical matter with barotropic equation of state. Using the field Equations (3), it can be shown that the $\beta_{(i)}$ evolve according to

$$\ddot{\beta}_{(i)} + \chi \dot{\beta}_{(i)} = 0 . \tag{48}$$

The solution of Equation (48) gives

$$\dot{\beta}_{(i)} = \frac{\lambda_{(i)}}{a^3(t)}, \tag{49}$$

with $\lambda_{(i)}$ integration constants satisfying $\sum_i \lambda_{(i)} = 0$. The field equations lead to an effective Friedmann equation for the mean scale factor, which includes the contribution of anisotropies

$$\frac{1}{3}\chi^2 = \rho + \tilde{\rho} + \frac{1}{2} \sum_i \dot{\beta}_{(i)}^2 . \tag{50}$$

The last term of Equation (50) represents the effective energy density of anisotropies (cf. e.g., Ref. [48]), which will be denoted by ρ_Σ. Using Equation (49), we have

$$\rho_\Sigma = \frac{\Sigma^2}{2a^6}, \tag{51}$$

having defined the shear scalar as $\Sigma^2 = \sum_i \lambda_{(i)}^2$. Thus, the contribution of anisotropies to the modified Friedmann equation is described as a perfect fluid with stiff equation of state $w = 1$, as in general relativity.

The evolution of anisotropies, as represented by the $\beta_{(i)}$, is obtained by integrating Equation (49)

$$\beta_{(i)}(t) = \lambda_{(i)} \int \frac{dt}{a^3(t)}, \tag{52}$$

where $a(t)$ in the integrand is a solution of Equation (50). In the remainder of this Section, we will determine the evolution of anisotropies during the bounce phase for the function $f(\chi)$ given by Equation (32). Since we are only interested in the region around the bounce, it is convenient to use the expansion (33). The energy density of the effective fluid then reads as

$$\tilde{\rho} \simeq -\rho_c + \frac{1}{3}\left(\frac{2V_B + 3\alpha}{V_B + 3\alpha}\right)\chi^2 . \tag{53}$$

The effective Friedmann equations in this regime can then be recast as

$$\frac{\chi^2}{3} \simeq \left(\frac{V_B + 3\alpha}{V_B}\right)(\rho_c - \rho - \rho_\Sigma) , \tag{54}$$

$$\dot{\chi} \simeq \frac{3}{2}\left(\frac{V_B + 3\alpha}{V_B}\right)(\rho + p + 2\rho_\Sigma) . \tag{55}$$

At the bounce, the scale factor attains its minimum a_B, and the r.h.s. of Equation (54) must vanish. We can use this condition to determine the energy density of matter at the bounce ρ_B (not to be confused with the critical energy density ρ_c, which includes the contribution of anisotropies). We have

$$\rho_B = \rho_c - \rho_{\Sigma,B}, \tag{56}$$

with $\rho_{\Sigma,B} = \frac{\Sigma^2}{2a_B^6}$ being the energy density of anisotropies at the bounce. The r.h.s. of Equation (54) can be expanded around a_B; taking into account that $\rho = \rho_B \left(\frac{a_B}{a}\right)^{3(w+1)}$, this gives

$$\frac{\chi^2}{3} \simeq 3\left(\frac{V_B + 3\alpha}{V_B}\right) (\rho_B(w+1) + 2\rho_{\Sigma,B}) \left(\frac{a}{a_B} - 1\right). \tag{57}$$

Taking into account Equation (56), we can rewrite Equation (57) as

$$\frac{\chi^2}{3} \simeq 3(w+1)\left(\frac{V_B + 3\alpha}{V_B}\right)\left(\rho_c - \frac{w-1}{w+1}\rho_{\Sigma,B}\right)\left(\frac{a}{a_B} - 1\right). \tag{58}$$

The solution is

$$a(t) \simeq a_B \left(1 + \frac{1}{4}\Omega^2 t^2\right), \tag{59}$$

where we defined

$$\Omega^2 = (w+1)\left(\frac{V_B + 3\alpha}{V_B}\right)\left(\rho_c - \frac{w-1}{w+1}\rho_{\Sigma,B}\right). \tag{60}$$

The solution (59) for the scale factor shows that regardless of the presence of anisotropies, the model features a first-order bounce, according to the definition given in Ref. [41]. From Equation (59), we find that the mean expansion rate evolves as

$$\chi(t) \simeq \frac{3}{2}\Omega^2 t. \tag{61}$$

Finally, using Equations (59) and (52) we find that the $\beta_{(i)}$ evolve linearly during the bounce

$$\beta_{(i)}(t) \simeq \beta_{(i)}^0 + \frac{\lambda_{(i)}}{a_B^3} t, \tag{62}$$

where $\beta_{(i)}^0$ are integration constants. Our solution (62) shows that anisotropies stay bounded during the bounce, and can be kept under control by means of a suitable choice of parameters for the model. It is interesting to compare this result with a similar one obtained in Ref. [48] for a non-singular bouncing model based on kinetic gravity braiding theories [49].

6. Effective Gravitational Constant(s)

The cosmological background equations of mimetic gravity, Equations (7) and (8), can be recast in an alternative form which makes no reference to perfect fluids. The effects introduced by the function $f(\chi)$ in the action (1) are then included in two effective gravitational "constants" G_F^{eff} and G_R^{eff}, representing respectively the effective coupling of matter to gravity in the Friedmann and the Raychaudhuri equations

$$\frac{1}{3}\chi^2 = 8\pi G_F^{eff}(\chi)\rho, \tag{63}$$

$$\dot{\chi} = -12\pi G_R^{eff}(\chi)(\rho + P). \tag{64}$$

The effective couplings are functions of the expansion rate, and are defined as

$$8\pi G_F^{eff}(\chi) = \left(1 - 3\frac{d}{d\chi}\left(\frac{f}{\chi}\right)\right)^{-1}, \qquad (65)$$

$$8\pi G_R^{eff}(\chi) = \left(1 - \frac{3}{2}f_{\chi\chi}\right)^{-1}. \qquad (66)$$

It is worth remarking that variable gravitational constants arise in this framework despite of the fact that the action (1) contains no dilaton couplings. In fact, the reformulation provided here hinges on the presence of a function of the expansion rate $f(\chi)$.

From Equations (63) and (64), and the continuity equation for matter, we find the following equation relating the change of G_F^{eff} over time to the difference between the two gravitational constants

$$\dot{G}_F^{eff} \rho = \chi(G_F^{eff} - G_R^{eff})(\rho + p). \qquad (67)$$

We observe that $G_F^{eff} = G_R^{eff}$ if and only if $f(\chi) = k_1 \chi + \frac{k_2}{2}\chi^2$. In this case, the linear term in χ has no effect, while the quadratic one leads to a finite redefinition of the Newton constant $8\pi G_F^{eff} = (1 - \frac{3}{2}k_2)^{-1}$ (see Ref. [37]); thus, in a large universe we must require $k_2 < \frac{2}{3}$ to ensure that the gravitational interaction remains attractive.[9] In the general case, both G_F^{eff} and G_R^{eff} will evolve with χ. For instance, assuming that in the large universe branch one has $f(\chi) \simeq k\chi^p$ with $p > 2$ to leading order in χ, leads to

$$8\pi G_F^{eff}(\chi) \simeq 1 + 3k(p-1)\chi^{p-2}, \qquad (68)$$

$$8\pi G_R^{eff}(\chi) \simeq 1 + \frac{3}{2}kp(p-1)\chi^{p-2}. \qquad (69)$$

If we assume that the universe (away from the bounce) is dominated by hydrodynamic matter with equation of state parameter w, we have

$$8\pi G_F^{eff}(t) \simeq 1 + 3k(p-1)\left(\frac{2}{(w+1)t}\right)^{p-2}, \qquad (70)$$

$$8\pi G_R^{eff}(t) \simeq 1 + \frac{3}{2}kp(p-1)\left(\frac{2}{(w+1)t}\right)^{p-2}. \qquad (71)$$

The reformulation of the cosmological equations of mimetic gravity offered by Equation (63) and (64) suggests that the coefficients of the leading order terms in the expansion of the branch f_L can be constrained using observational bounds on the time variation of the gravitational constant. We have from Equation (70), for a small k and retaining only the main contribution (corresponding to the radiation dominated era, $w = \frac{1}{3}$)

$$\frac{\Delta G_F^{eff}}{G_F^{eff}} = 1 - \frac{G_F^{eff}(t_{BBN})}{G_F^{eff}(t_0)} \simeq 3k(p-1)\left(\frac{3}{2}\right)^{p-2}(t_{BBN})^{2-p}. \qquad (72)$$

where t_0 is the age of the universe and t_{BBN} is the time of nucleosynthesis. Bounds on the time variation of the gravitational constant G_F^{eff} can be derived from primordial nucleosynthesis: $-0.10 < \frac{\Delta G_F^{eff}}{G_F^{eff}} < 0.13$ [50,51]. For a given $p > 2$, such a bound can be translated into a constraint on k. However, such a constraint is very weak for bouncing models. In fact, if the limiting curvature

---

[9] This must be contrasted with the case of bouncing models examined in Sections 3.2 and 4, where the coefficient of the quadratic term must satisfy an opposite inequality in order to guarantee that gravity becomes repulsive at the bounce.

hypothesis is made, dimensional arguments suggest that $k \sim \chi_m^{2-p}$. This is in fact the case for the models considered in Section 4, see Equations (46) and (36). Moreover, typically one has for the limiting value of the expansion rate $\chi_m \sim t_{Pl}^{-1}$, where t_{Pl} is Planck time. Therefore, the time variation of the gravitational constant is extremely small in such models $\frac{\Delta G_F^{eff}}{G_F^{eff}} \sim \left(\frac{t_{Pl}}{t_{BBN}}\right)^{p-2}$.

A more detailed investigation of the phenomenological consequences of the time variation of G_F^{eff} and G_R^{eff} is beyond the scope of the present article and will be left for future work.

7. Instabilities

Our presentation of mimetic gravity would not be complete without a discussion of perturbative instabilities. Instabilities of cosmological perturbations for the mimetic gravity theory with action (1) have been studied in Refs. [36,52] for a generic $f(\chi)$; for earlier studies focused on the case of a quadratic f see Ref. [53,54].[10] Compared to general relativity, the theory has one extra propagating scalar degree of freedom, whose speed of sound is given by

$$c_s^2 = \frac{1}{2}\frac{f_{\chi\chi}}{1 - \frac{3}{2}f_{\chi\chi}}. \tag{73}$$

Depending on the sign of the speed of sound, the theory has a ghost instability (for $c_s^2 > 0$) or a gradient instability (for $c_s^2 < 0$), see references above. The propagation speed of tensor perturbations is not affected by the term $f(\chi)$ in the action (1).[11]

In the following we will assume that the analytic properties of the function $f(\chi)$ are such as to accommodate for a bouncing background. Some general conclusions can then be drawn on the profile of the speed of sound as a function of the expansion, based on the results derived in Section 3.2. In fact, around the bounce $f(\chi)$ must admit the expansion (19). Moreover, since $\dot{\chi} > 0$ in a neighborhood of the bounce, Equation (40) implies that we must have $\vartheta > \frac{2}{3}$, provided that ordinary matter fields satisfy the NEC. Thus, at the bounce we have

$$c_s^2 = \frac{\vartheta}{2 - 3\vartheta} < 0, \tag{74}$$

which corresponds to a gradient instability. The expansion rate attains its extremum at $|\chi| = \chi_m$, where two different branches of the multivalued function $f(\chi)$ are joined together; at that point the second derivative $f_{\chi\chi}$ is divergent, whereby the speed of sound squared takes the universal value $c_s^2 = -\frac{1}{3}$. We conclude that a generic feature of bouncing models in mimetic gravity is that the bounce is always accompanied by a gradient instability of scalar perturbations, which extends beyond the onset of the standard decelerated expansion. The possibility that c_s^2 may turn to positive values at a later stage is not excluded, but depends on the details of the model, and specifically on the functional form of the branch $f_L(\chi)$ corresponding to a large universe.

It is interesting to study the behavior of c_s^2 in the models examined in Section 4, where a bouncing background is explicitly realized. To begin with, let us start from the special case $\alpha = 0$, which reproduces the LQC effective dynamics for the cosmological background. The two branches f_B, f_L in this case are given by Equations (37) and (38), respectively. We find, using Equations (73) and (41)

$$c_s^2 = -\frac{1}{3}\left(1 \pm \sqrt{1 - q^2}\right) = -\frac{2}{3}\frac{\rho}{\rho_c}. \tag{75}$$

---

[10] It must be noted that the quadratic case is equivalent with the IR limit of projectable Hořava-Lifshitz gravity [54], see also Ref. [55].
[11] The situation is different in other versions of mimetic gravity, see e.g., [23] for a general analysis based on the DHOST formulation of (extended) mimetic gravity theories.

In the second step of (75), the upper sign corresponds to f_B, whereas the lower one corresponds to f_L. We note that the speed of sound squared is always negative, has a minimum at the bounce $(c_s^2)_{\min} = -\frac{2}{3}$ when $\rho = \rho_c$, and approaches zero from below as $\rho \to 0$. Given Equation (75), and recalling that maximal expansion rate in this model is reached at $\rho = \frac{\rho_c}{2}$, it is straightforward to check the general feature $c_s^2(\pm \chi_m) = -\frac{1}{3}$. We observe that c_s^2 is negative throughout cosmic history for the model with $\alpha = 0$, and approaches zero from below in the large universe branch as χ tends to zero (cf. Ref. [25]). It is interesting to compare these results with those obtained in Ref. [56] for a model based on generalized Galileons [57], where the speed of sound squared becomes negative—although only for a short period—around the bounce; see also Ref. [58,59] for a comparison between such effective models and the dynamics of perturbations in LQC. In the models cited above gradient instabilities arise due to the violation of the NEC at the bounce (see also [60] and references therein). Recently, the possibility of establishing a theoretical no-go theorem regarding the realization of a healthy non-singular bounce (i.e., free of pathologies such as gradient instabilities) has been discussed in the context of generalized Galileons, see Refs. [61–63].

The example examined above is just a particular case of the model reproducing the background dynamics of GFT condensates, studied in Section 4. In the general case, i.e., for $\alpha \neq 0$, we have at the bounce

$$\left(c_s^2\right)_{\min} = -\frac{2}{3}\left(1 + \frac{\alpha}{V_B}\right). \tag{76}$$

In the large universe branch instead and for $\chi \simeq 0$ we have, to leading order in χ

$$c_s^2 \simeq \frac{3\alpha}{V_*}\sqrt{2 + \frac{6\alpha}{V_*}\frac{|\chi|}{\chi_m}}. \tag{77}$$

Thus, c_s^2 and α have the same sign in this regime. Therefore, for $\alpha < 0$ the situation is qualitatively similar to the $\alpha = 0$ case examined above, with a gradient instability extending also to the large universe branch. For $\alpha > 0$ the situation is different: there is a cross-over from $c_s^2 < 0$ near the bounce to $c_s^2 > 0$ when the universe is large. Such a cross-over must necessarily take place after the universe enters the phase of decelerated expansion, since $c_s^2 = -\frac{1}{3}$ when $\dot\chi = 0$ (see above). Thus, while the bounce is always accompanied by a gradient instability, the late universe branch would be characterized by a ghost instability for $\alpha > 0$. We remark that the cross-over point where $c_s^2 = 0$ corresponds to a regime of strong coupling [54].

8. Discussion

We conclude by reviewing the main results obtained in this work and indicating directions for future studies.

In Section 3 we illustrated in complete generality the theory reconstruction procedure for the function $f(\chi)$ in mimetic gravity. In the case of bouncing backgrounds, the implementation of the limiting curvature hypothesis requires that $f(\chi)$ be multivalued. This case was carefully examined and we gave general prescriptions to ensure continuity of $f(\chi)$ and its first derivative along the cosmic trajectory; in particular, by imposing suitable matching conditions at the branching points, both the energy density $\tilde\rho$ and pressure $\tilde P$ of the effective fluid are continuous throughout cosmic history. We showed that local properties of the function $f(\chi)$ are directly related to physically relevant quantities characterizing the evolution of the cosmic background, such as the critical energy density and the bounce duration, as well as the equation of state of the effective fluid. In particular, the latter was shown to approach a constant value at late times, which is determined by the dominant matter species and the leading order term in the asymptotic expansion of $f(\chi)$ in that regime.

In Section 4 we focused on a specific model obtained in Ref. [22], where the function $f(\chi)$ was suitably reconstructed in order to reproduce the background evolution obtained from GFT condensates in Ref. [5,6]. Quantities of physical interest were derived from local analysis of the two branches f_B, f_L, using the results of Section 3. The special case corresponding to the effective dynamics of LQC for a flat, isotropic universe was studied in detail. As an application, we studied the evolution of anisotropies near the bounce in a Bianchi I universe for the model of Ref. [22]: our results generalize those obtained in Ref. [21] and show that anisotropies do not grow significantly during the bounce, and therefore do not spoil the smoothness of the bounce. It would be interesting to compare the results obtained in the effective approach considered here, with those of Ref. [64], where the dynamics of GFT condensates of anisotropic quanta was studied (see also Ref. [65]). As discussed in Ref. [11,29], the evolution of anisotropies is qualitatively similar in LQC and the corresponding mimetic gravity theory. It is, therefore, natural to ask whether an analogous statement can be made for GFT cosmology and the related model in mimetic gravity. We leave this question for future work. Spherically symmetric geometries are also of interest and can be studied in the present framework by extending the analysis of Refs. [27,28].

In Section 6 we showed that there is an interesting reformulation of mimetic gravity involving two distinct time-varying effective gravitational constants G_F^{eff} and G_R^{eff}, featuring respectively in the Friedmann and the Raychaudhuri equations. Consistency of such a description with the Bianchi identities is ensured by Equation (67), which is identically satisfied in mimetic gravity by all choices of the function $f(\chi)$. We derived the time evolution of the effective gravitational constants during the phase of decelerated expansion for $f(\chi) \sim \chi^p$, with $p > 2$. We showed that the predicted time variation is too small to be observed if the limiting curvature hypothesis is realized. It would be of interest to further explore the consequences of the time variation of G_F^{eff} and G_R^{eff} in a more general and model independent setting.

Our discussion of perturbative instabilities in Section 7 highlights some serious limitations of bouncing models in mimetic gravity, which may hinder the possibility of using the simplest framework with the covariant action (1) for an effective description of quantum gravity in inhomogeneous spacetimes. The presence of gradient or ghost instabilities, which is a distinctive feature of mimetic gravity, seems to be even more serious in bouncing cosmologies; in fact, in such models the infinite age of the universe would offer no chance to keep instabilities under control. Remarkably, this issue has not been much appreciated in the literature on bouncing cosmologies in mimetic gravity. Based on the analogy with LQC (see Ref. [58]), we expect the bounce to be accompanied by a short-lived gradient instability around the bounce affecting short-wavelength modes; however, there should be no instabilities away from the bounce. Some proposals to cure the instabilities by means of further modification of the mimetic gravity action have been made in Refs. [66,67]; however, their correspondence with the effective dynamics of quantum gravity models is yet to be established and shall be investigated in future work.

Funding: This work was partially supported by the Atlantic Association for Research in the Mathematical Sciences (AARMS) and by the Natural Sciences and Engineering Research Council of Canada (NSERC).

Acknowledgments: It is a pleasure to thank Sabir Ramazanov and Edward Wilson-Ewing for helpful discussions on instabilities in mimetic gravity.

Conflicts of Interest: The authors declare no conflict of interest.

References

1. Brandenberger, R.; Peter, P. Bouncing Cosmologies: Progress and Problems. *Found. Phys.* **2017**, *47*, 797–850. [CrossRef]
2. Cai, Y.F. Exploring Bouncing Cosmologies with Cosmological Surveys. *Sci. China Phys. Mech. Astron.* **2014**, *57*, 1414–1430. [CrossRef]
3. Bojowald, M. Loop quantum cosmology. *Living Rev. Rel.* **2008**, *11*, 4. [CrossRef]
4. Ashtekar, A.; Singh, P. Loop Quantum Cosmology: A Status Report. *Class. Quant. Grav.* **2011**, *28*, 213001. [CrossRef]
5. Oriti, D.; Sindoni, L.; Wilson-Ewing, E. Emergent Friedmann dynamics with a quantum bounce from quantum gravity condensates. *Class. Quant. Grav.* **2016**, *33*, 224001. [CrossRef]

6. Oriti, D.; Sindoni, L.; Wilson-Ewing, E. Bouncing cosmologies from quantum gravity condensates. *Class. Quant. Grav.* **2017**, *34*, 04LT01. [CrossRef]
7. Alesci, E.; Botta, G.; Cianfrani, F.; Liberati, S. Cosmological singularity resolution from quantum gravity: The emergent-bouncing universe. *Phys. Rev. D* **2017**, *96*, 046008. [CrossRef]
8. Gielen, S.; Oriti, D. Cosmological perturbations from full quantum gravity. *Phys. Rev. D* **2018**, *98*, 106019. [CrossRef]
9. Gerhardt, F.; Oriti, D.; Wilson-Ewing, E. Separate universe framework in group field theory condensate cosmology. *Phys. Rev. D* **2018**, *98*, 066011. [CrossRef]
10. Gielen, S. Inhomogeneous universe from group field theory condensate. *J. Cosmol. Astropart. Phys.* **2019**, *2019*, 013. [CrossRef]
11. Bodendorfer, N.; Schäfer, A.; Schliemann, J. Canonical structure of general relativity with a limiting curvature and its relation to loop quantum gravity. *Phys. Rev. D* **2018**, *97*, 084057. [CrossRef]
12. Langlois, D.; Liu, H.; Noui, K.; Wilson-Ewing, E. Effective loop quantum cosmology as a higher-derivative scalar-tensor theory. *Class. Quant. Grav.* **2017**, *34*, 225004. [CrossRef]
13. Markov, M. Limiting density of matter as a universal law of nature. *J. Exp. Theor. Phys. Lett.* **1982**, *36*, 265.
14. Markov, M. Possible state of matter just before the collapse stage. *J. Exp. Theor. Phys. Lett.* **1987**, *46*, 431.
15. Frolov, V.P.; Markov, M.A.; Mukhanov, V.F. Black Holes as Possible Sources of Closed and Semiclosed Worlds. *Phys. Rev. D* **1990**, *41*, 383. [CrossRef]
16. Frolov, V.P.; Markov, M.A.; Mukhanov, V.F. Through a black hole into a new universe? *Phys. Lett. B* **1989**, *216*, 272–276. [CrossRef]
17. Mukhanov, V.F.; Brandenberger, R.H. A Nonsingular universe. *Phys. Rev. Lett.* **1992**, *68*, 1969–1972. [CrossRef] [PubMed]
18. Brandenberger, R.H.; Mukhanov, V.F.; Sornborger, A. A Cosmological theory without singularities. *Phys. Rev. D* **1993**, *48*, 1629–1642. [CrossRef]
19. Easson, D.A.; Brandenberger, R.H. Nonsingular dilaton cosmology in the string frame. *J. High Energy Phys.* **1999**, *1999*, 003. [CrossRef]
20. Yoshida, D.; Quintin, J.; Yamaguchi, M.; Brandenberger, R.H. Cosmological perturbations and stability of nonsingular cosmologies with limiting curvature. *Phys. Rev. D* **2017**, *96*, 043502. [CrossRef]
21. Chamseddine, A.H.; Mukhanov, V. Resolving Cosmological Singularities. *J. Cosmol. Astropart. Phys.* **2017**, *2017*, 009. [CrossRef]
22. de Cesare, M. Limiting curvature mimetic gravity for group field theory condensates. *Phys. Rev. D* **2019**, *99*, 063505. [CrossRef]
23. Langlois, D.; Mancarella, M.; Noui, K.; Vernizzi, F. Mimetic gravity as DHOST theories. *J. Cosmol. Astropart. Phys.* **2019**, *2019*, 036. [CrossRef]
24. Brahma, S.; Golovnev, A.; Yeom, D.H. On singularity-resolution in mimetic gravity. *Phys. Lett. B* **2018**, *782*, 280–284. [CrossRef]
25. De Haro, J.; Aresté Saló, L.; Pan, S. Limiting curvature mimetic gravity and its relation to Loop Quantum Cosmology. *arXiv* **2018**, arXiv:gr-qc/1803.09653.
26. De Haro, J.; Pan, S. Note on bouncing backgrounds. *Phys. Rev. D* **2018**, *97*, 103518. [CrossRef]
27. Chamseddine, A.H.; Mukhanov, V. Nonsingular Black Hole. *Eur. Phys. J. C* **2017**, *77*, 183. [CrossRef]
28. Ben Achour, J.; Lamy, F.; Liu, H.; Noui, K. Non-singular black holes and the Limiting Curvature Mechanism: A Hamiltonian perspective. *J. Cosmol. Astropart. Phys.* **2018**, *2018*, 072. [CrossRef]
29. Bodendorfer, N.; Mele, F.M.; Münch, J. Is limiting curvature mimetic gravity an effective polymer quantum gravity? *Class. Quant. Grav.* **2018**, *35*, 225001. [CrossRef]
30. Chamseddine, A.H.; Mukhanov, V. Mimetic Dark Matter. *J. High Energy Phys.* **2013**, *2013*, 135. [CrossRef]
31. Deruelle, N.; Rua, J. Disformal Transformations, Veiled General Relativity and Mimetic Gravity. *J. Cosmol. Astropart. Phys.* **2014**, *2014*, 002. [CrossRef]
32. Golovnev, A. On the recently proposed Mimetic Dark Matter. *Phys. Lett. B* **2014**, *728*, 39–40. [CrossRef]
33. Chamseddine, A.H.; Mukhanov, V.; Vikman, A. Cosmology with Mimetic Matter. *J. Cosmol. Astropart. Phys.* **2014**, *2014*, 017. [CrossRef]
34. Sebastiani, L.; Vagnozzi, S.; Myrzakulov, R. Mimetic gravity: A review of recent developments and applications to cosmology and astrophysics. *Adv. High Energy Phys.* **2017**, *2017*, 3156915. [CrossRef]

35. Langlois, D. Dark Energy and Modified Gravity in Degenerate Higher-Order Scalar-Tensor (DHOST) theories: A review. *arXiv* **2018**, arXiv:gr-qc/1811.06271.
36. Firouzjahi, H.; Gorji, M.A.; Hosseini Mansoori, S.A. Instabilities in Mimetic Matter Perturbations. *J. Cosmol. Astropart. Phys.* **2017**, *2017*, 031. [CrossRef]
37. Mirzagholi, L.; Vikman, A. Imperfect Dark Matter. *J. Cosmol. Astropart. Phys.* **2015**, *2015*, 028. [CrossRef]
38. Brown, J.D.; Kuchar, K.V. Dust as a standard of space and time in canonical quantum gravity. *Phys. Rev. D* **1995**, *51*, 5600–5629. [CrossRef]
39. Husain, V.; Pawlowski, T. Time and a physical Hamiltonian for quantum gravity. *Phys. Rev. Lett.* **2012**, *108*, 141301. [CrossRef]
40. Husain, V.; Pawlowski, T. Dust reference frame in quantum cosmology. *Class. Quant. Grav.* **2011**, *28*, 225014. [CrossRef]
41. Cattoen, C.; Visser, M. Necessary and sufficient conditions for big bangs, bounces, crunches, rips, sudden singularities, and extremality events. *Class. Quant. Grav.* **2005**, *22*, 4913–4930. [CrossRef]
42. de Cesare, M.; Pithis, A.G.A.; Sakellariadou, M. Cosmological implications of interacting Group Field Theory models: cyclic Universe and accelerated expansion. *Phys. Rev. D* **2016**, *94*, 064051. [CrossRef]
43. Lin, C.; Brandenberger, R.H.; Perreault Levasseur, L. A Matter Bounce By Means of Ghost Condensation. *J. Cosmol. Astropart. Phys.* **2011**, *2011*, 019. [CrossRef]
44. Gielen, S. Group field theory and its cosmology in a matter reference frame. *Universe* **2018**, *4*, 103. [CrossRef]
45. Adjei, E.; Gielen, S.; Wieland, W. Cosmological evolution as squeezing: A toy model for group field cosmology. *Class. Quant. Grav.* **2018**, *35*, 105016. [CrossRef]
46. Wilson-Ewing, E. A relational Hamiltonian for group field theory. *arXiv* **2018**, arXiv:gr-qc/1810.01259.
47. de Cesare, M.; Sakellariadou, M. Accelerated expansion of the Universe without an inflaton and resolution of the initial singularity from Group Field Theory condensates. *Phys. Lett. B* **2017**, *764*, 49–53. [CrossRef]
48. Cai, Y.F.; Brandenberger, R.; Peter, P. Anisotropy in a Nonsingular Bounce. *Class. Quant. Grav.* **2013**, *30*, 075019. [CrossRef]
49. Deffayet, C.; Pujolas, O.; Sawicki, I.; Vikman, A. Imperfect Dark Energy from Kinetic Gravity Braiding. *J. Cosmol. Astropart. Phys.* **2010**, *2010*, 026. [CrossRef]
50. Uzan, J.P. Varying Constants, Gravitation and Cosmology. *Living Rev. Rel.* **2011**, *14*, 2. [CrossRef]
51. Cyburt, R.H.; Fields, B.D.; Olive, K.A.; Skillman, E. New BBN limits on physics beyond the standard model from ^{4}He. *Astropart. Phys.* **2005**, *23*, 313–323. [CrossRef]
52. Takahashi, K.; Kobayashi, T. Extended mimetic gravity: Hamiltonian analysis and gradient instabilities. *J. Cosmol. Astropart. Phys.* **2017**, *2017*, 038. [CrossRef]
53. Ijjas, A.; Ripley, J.; Steinhardt, P.J. NEC violation in mimetic cosmology revisited. *Phys. Lett. B* **2016**, *760*, 132–138. [CrossRef]
54. Ramazanov, S.; Arroja, F.; Celoria, M.; Matarrese, S.; Pilo, L. Living with ghosts in Hoava-Lifshitz gravity. *J. High Energy Phys.* **2016**, *2016*, 020. [CrossRef]
55. Capela, F.; Ramazanov, S. Modified Dust and the Small Scale Crisis in CDM. *J. Cosmol. Astropart. Phys.* **2015**, *2015*, 051. [CrossRef]
56. Cai, Y.F.; Easson, D.A.; Brandenberger, R. Towards a Nonsingular Bouncing Cosmology. *J. Cosmol. Astropart. Phys.* **2012**, *2012*, 020. [CrossRef]
57. Deffayet, C.; Gao, X.; Steer, D.A.; Zahariade, G. From k-essence to generalised Galileons. *Phys. Rev. D* **2011**, *84*, 064039. [CrossRef]
58. Cai, Y.F.; Wilson-Ewing, E. Non-singular bounce scenarios in loop quantum cosmology and the effective field description. *J. Cosmol. Astropart. Phys.* **2014**, *2014*, 026. [CrossRef]
59. Cai, Y.F.; Marciano, A.; Wang, D.G.; Wilson-Ewing, E. Bouncing cosmologies with dark matter and dark energy. *Universe* **2016**, *3*, 1. [CrossRef]
60. Libanov, M.; Mironov, S.; Rubakov, V. Generalized Galileons: Instabilities of bouncing and Genesis cosmologies and modified Genesis. *J. Cosmol. Astropart. Phys.* **2016**, *2016*, 037. [CrossRef]
61. Kobayashi, T. Generic instabilities of nonsingular cosmologies in Horndeski theory: A no-go theorem. *Phys. Rev. D* **2016**, *94*, 043511. [CrossRef]
62. Akama, S.; Kobayashi, T. Generalized multi-Galileons, covariantized new terms, and the no-go theorem for nonsingular cosmologies. *Phys. Rev. D* **2017**, *95*, 064011. [CrossRef]

63. Banerjee, S.; Cai, Y.F.; Saridakis, E.N. Evading the theoretical no-go theorem for nonsingular bounces in Horndeski/Galileon cosmology. *arXiv* **2018**, arXiv:gr-qc/1808.01170.
64. de Cesare, M.; Oriti, D.; Pithis, A.G.A.; Sakellariadou, M. Dynamics of anisotropies close to a cosmological bounce in quantum gravity. *Class. Quant. Grav.* **2018**, *35*, 015014. [CrossRef]
65. Pithis, A.G.A.; Sakellariadou, M. Relational evolution of effectively interacting group field theory quantum gravity condensates. *Phys. Rev. D* **2017**, *95*, 064004. [CrossRef]
66. Gorji, M.A.; Hosseini Mansoori, S.A.; Firouzjahi, H. Higher Derivative Mimetic Gravity. *J. Cosmol. Astropart. Phys.* **2018**, *2018*, 020. [CrossRef]
67. Hirano, S.; Nishi, S.; Kobayashi, T. Healthy imperfect dark matter from effective theory of mimetic cosmological perturbations. *J. Cosmol. Astropart. Phys.* **2017**, *2017*, 009. [CrossRef]

© 2019 by the author. Licensee MDPI, Basel, Switzerland. This article is an open access article distributed under the terms and conditions of the Creative Commons Attribution (CC BY) license (http://creativecommons.org/licenses/by/4.0/).

Article

On the Geometry of No-Boundary Instantons in Loop Quantum Cosmology

Suddhasattwa Brahma [1],* and Dong-han Yeom [1,2]

1. Asia Pacific Center for Theoretical Physics, Pohang 37673, Korea
2. Department of Physics, Pohang University of Science and Technology (POSTECH), Pohang 37673, Korea; innocent.yeom@gmail.com
* Correspondence: suddhasattwa.brahma@gmail.com

Received: 5 November 2018; Accepted: 7 January 2019; Published: 10 January 2019

Abstract: We study the geometry of Euclidean instantons in loop quantum cosmology (LQC) such as those relevant for the no-boundary proposal. Confining ourselves to the simplest case of a cosmological constant in minisuperspace cosmologies, we analyze solutions of the semiclassical (Euclidean) path integral in LQC. We find that the geometry of LQC instantons have the peculiar feature of an infinite tail which distinguishes them from Einstein gravity. Moreover, due to quantum-geometry corrections, the small-a behaviour of these instantons seem to naturally favor a closing-off of the geometry in a regular fashion, as was originally proposed for the no-boundary wavefunction.

Keywords: no-boundary proposal; loop quantum cosmology; LQC instanton

1. Introduction

Recently, the introduction of the 'no-boundary' proposal in loop quantum cosmology (LQC), for minisuperspace models, has unveiled a lot of interesting physical possibilities [1]. It has been shown that the original Hartle-Hawking formulation [2], improved by an effective action which includes corrections due to LQC, can lead to an expanded solution space due to singularity-resolution [3,4] coming from the latter. In particular, it has been shown that not only is the probability for a de-Sitter (dS) universe nucleating from nothing increased in such a scenario, there can now be compact, non-singular instantonic solutions in cases where there were none in Einstein gravity. As an example, the model of a Friedmann-Robertson-Lemaitre-Walker (FLRW) closed universe, coupled to a massless scalar field, was considered in [1] and shown to have a nontrivial compact instanonic solution with a finite probability for nucleation. This study has opened the doors for revisiting the original no-boundary proposal augmented by quantum-geometry effects governing the dynamics of the early-universe which are, in any case, expected for a meaningful UV-completion. A detailed study of such effects for physically relevant questions such as the probability of inflation and number of e-folds predicted by the (improved) no-boundary measure can now be answered within the purview of LQC. However, this is not the intention of present work and shall be pursued later elsewhere.

In this work, we focus on the geometry of these Euclidean instanton solutions in LQC. This is a necessary first step before using such solutions to consider nucleation of universes from nothing and employing the measure provided by the associated wavefunction for predicting the probabilities of physically interesting phenomena for the Lorentzian histories. Our starting point shall be the (Euclidean) path integral for quantum gravity, along with the prescription that the initial conditions are provided by the no-boundary proposal. The main new ingredient, in comparison to the original Hartle-Hawking proposal, shall be the 'effective' action appearing in the path integral derived from LQC, as opposed to the usual Einstein-Hilbert one. (It was established in [5,6] that by replacing the standard FLRW action by the 'polymerized' version of it, the path integral formulation of LQC, in its phase space realization, retains all the crucial aspects of the quantum geometry which appear in the canonical LQC.) Other than this, our formalism shall be exactly the same as in the original no-boundary proposal: We shall look only at the saddle-point approximation of the Euclidean path integral and consider the wavefunction to be a functional of the value of the scale factor only at the final (spatial) boundary. Moreover, as shall be obvious throughout our paper, we make a minisuperspace approximation for all our calculations and consider the matter content to be only that due to a cosmological constant. The latter approximation ensures that we always have a compact instantonic solution and do not require dealing with subtleties which can give rise to Euclidean wormholes [7,8]. Since we want to show new features of the LQC instantons with respect to its geometry, as compared to the original Hartle-Hawking ones, these approximations shall help us emphasize our main result without unnecessarily complicating the system.

For our purposes, the effective action consists of two main types of quantum corrections specific to LQC—the holonomy and inverse-triad modifications [9]. The first appears due to the fact that there are no quantum operators corresponding to the connection or extrinsic curvature (in other words, the momenta conjugate to the spatial metric) on the kinematic[1] Hilbert space of the theory. On the other hand, there are well-defined operators corresponding to the holonomy (or parallel transport) of the connection [10]. Therefore, one expresses the curvature operator in terms of these holonomies instead of the connection itself. Classically, one can take the limit such that one recovers the expression of the curvature written in terms of the connection from the expression given for the holonomies. However, the geometrical operators in the full loop quantun gravity have discrete spectra for quantities such as area and volume [11] on the kinematical Hilbert space spanned by the spin-network states, rendering taking such a limit unviable. Therefore, one inherits an 'area-gap', in analogy with the minimum energy-gap of the harmonic oscillator, from the full theory in LQC [12,13]. The main effect of this regularization of the curvature in terms of holonomies, for symmetry-reduced models, lie in replacing the extrinsic curvature by matrix elements of $SU(2)$-holonomies which are periodic functions of the connection. Specifically, for minisuperspace cosmologies, we have $H \to \sin(\delta H)/\delta$, where H is the Hubble parameter and δ is related to the area-gap.

The discrete spectra of area and volume operators also lead to other type of corrections in LQC. The most significant of them are the inverse-triad corrections which arise from the requirement of having a well-defined operator corresponding to the inverse of some power of the scale factor whose spectra contains the zero eigenvalue. Naively, it is impossible to have a densely-defined operator in such a case. However, using the aforementioned holonomy operators and what is commonly known as the 'Thiemann trick' in the literature, one can express the relation [14,15]

$$\hat{h}^{-1}[\hat{h}, \sqrt{\hat{a}}] = -\frac{1}{2}\hbar\delta\widehat{a^{-1/2}}. \tag{1}$$

In this definition, $\hat{h} := \widehat{\exp(i\delta p_a)}$, for the momentum, $p_a \propto \dot{a}$, conjugate to the scale factor a, is precisely a $SU(2)$-valued holonomy operator mentioned previously. Using the usual properties of a commutator, it is clear from this relation that one can have an operator, whose classical limit is some inverse power of the scale factor on the RHS although we do not require any inverse operator on the LHS [16,17]. Using this, one gets rid of the singular behaviour of any function which contains some inverse power of a due to the replacement by these aforementioned inverse-triad corrections. Once again, their form for minisuperspace cosmologies is rather simple, as shall be explicitly demonstrated later.

Let us briefly summarize our main result. Conceptually, at least in the cosmological constant case, the main effect of the LQC quantum-geometry corrections lies in the small-a behaviour of the Euclidean

---

[1] 'Kinematic' here refers to the fact that the Gauss and the (spatial) diffeomorphism constraints have been solved whereas 'physical' would imply the solution of the Hamiltonian constraint as well. For a minisuperspace model, these distinctions are not very important since the only leftover symmetry of the system is time-reparameterization invariance.

LQC instantons. As shall be demonstrated, due to the inverse-triad corrections, the LQC-modified Friedmann equation is such that the solution tails off to zero at the symmetry point of the theory. The geometry of the LQC instanton will emerge to be quite different from the original Hartle-Hawking proposal with an infinitely stretched tail in Euclidean time; however, their *topology* remains the same. Moreover, such an infinitely long tail of the instanton (in imaginary time) is not an inherent problem since the only meaningful physical quantity is the probability of nucleation which remains finite for this system. The interesting fact is the quantum-geometry regularization is such that this tail closes the geometry in a regular way without requiring any additional fine-tuning even though the field equations are heavily modified in LQC. This is suggestive of the fact that the no-boundary proposal is robust and, if anything, such a necessary tail-off of LQC instantons to zero points towards it being more natural in the presence of quantum-geometry corrections.

2. The Hartle-Hawking Proposal Revisited

In this section, we first briefly review the geometry of the no-boundary instantons in Einstein gravity. In the process, we also fix our notation for the rest of the paper.

2.1. The Wheeler-de Witt Equation and Boundary Conditions

Let us consider Einstein gravity with a scalar field

$$S = \int \sqrt{-g}\,dx^4 \left[\frac{R}{16\pi} - \frac{1}{2}(\nabla\phi)^2 - V(\phi) \right]. \tag{2}$$

In this paper, we shall exclusively focus on a minisuperspace cosmological model [18]

$$ds^2 = \sigma^2 \left[-N^2(t)dt^2 + a^2(t)d\Omega_3^2 \right], \tag{3}$$

where $\sigma^2 = 2/3\pi$ is some normalization constant[2]. Throughout this paper, we shall work with a closed $k=1$ FLRW cosmology, as is typically required for the Hartle-Hawking proposal[3]. By assuming the slow-roll limit $\dot\phi \approx 0$, the Lagrangian can be simplified as

$$\mathcal{L} = \frac{1}{2}N\left[a\left(1 - \frac{\dot a^2}{N^2}\right) - \tilde V a^3\right], \tag{4}$$

where $\tilde V = 16V/9$. From this, one can get the conjugate momentum $p_a = -a\dot a/N$. The Hamiltonian $\mathcal{H}$ is obtained by the usual Legendre transform

$$\mathcal{L} = p_a \dot a - N\mathcal{H}, \tag{5}$$

where

$$\mathcal{H} = -\frac{1}{2}\left[\frac{p_a^2}{a} + a - \tilde V a^3\right]. \tag{6}$$

---

[2] Note that we choose this normalization at this point for historical reasons and to keep the resulting equations simple. However, we shall change this normalization later on to facilitate comparison with LQC.

[3] The reason why one traditionally has to consider a $k=1$ universe is because it is the only case which provides a potential barrier for the tunneling of the universe from nothing. In other words, from the Friedmann equation, only for the $k=1$ case can the right hand side go to zero for certain choices of matter (a flat potential or a pure cosmological constant). However, this can be generalized for LQC since one gets a "bounce" in all types of tolopologies for a FLRW universe. We intend to establish this generalization in future work.

On quantization, by replacing $p_a = -i\,(d/da)$, one gets the Wheeler-de Witt equation for the wave function of the scale factor

$$\left[\frac{d^2}{da^2} + \frac{\gamma}{a}\frac{d}{da} - U(a)\right]\psi(a) = 0, \quad (7)$$

where γ is a constant due to the ambiguity in operator-ordering and

$$U(a) = a^2\left(1 - \bar{V}a^2\right). \quad (8)$$

In the semi-classical regime, the ambiguity due to operator ordering is not that important and can be ignored in a first approximation [18]. It is straightforward to see that the behavior of the system in the classically allowed region $U < 0$ (hence, $a > 1/\sqrt{\bar{V}}$) and in the classically forbidden region $U > 0$ (hence, $a < 1/\sqrt{\bar{V}}$) are different. For the classically allowed region, the solution is essentially oscillatory and can be a superposition of in-going and out-going modes; for the classically forbidden region, the solution is a superposition of exponentially growing and decaying modes.

In order to extract a specific solution from these general solutions, one needs to impose boundary conditions. However, quantum cosmology is a *closed* system in which a set of boundary conditions cannot be determined by the environment external to the setup as is the normal practice. In general, there is no fundamental principle to assign all the boundary conditions necessary to specify the wave function of the universe Ψ. At best, the general consensus is that these boundary conditions need to be supplied as additional fundamental laws of nature. There are two famous, mathematically consistent wavefunctions corresponding to the following boundary conditions:

1. *The Hartle-Hawking proposal* [2]—If we choose the *exponentially growing mode* for $a < 1/\sqrt{\bar{V}}$, then the wave function becomes a superposition of in-going and out-going modes for $a > 1/\sqrt{\bar{V}}$.
2. *The tunneling proposal* [19]—If we choose the *out-going mode* for $a > 1/\sqrt{\bar{V}}$, then the wave function becomes a superposition of growing and decaying modes for $a < 1/\sqrt{\bar{V}}$.

The probability of the universe nucleating from 'nothing' mainly depends on the contribution from the classically disallowed (hence, quantum) regime. Therefore, the (leading-order contribution to the) probability distribution is approximately

$$P(a, \phi) \simeq \exp\left(\pm\frac{3}{8V(\phi)}\right), \quad (9)$$

where $+$ corresponds to the Hartle-Hawking wavefunction, and $-$ to the tunneling wavefunction, respectively.

2.2. Euclidean Path Integral

Since there is no fundamental principle to choose a boundary condition for the universe in general, we can look towards the path integral quantization for some guidance. In this paper, we shall take this route of quantizing gravity in the path integral formulation instead of the canonical one outlined above. We believe this is the most elegant presentation of the original Hartle-Hawking proposal [2] and shall, therefore, stick to it for the LQC-modified case [1]. However, as an aside, we note that the Wheeler-de Witt equation mentioned above transforms into a "difference" equation (of finite step-size) in LQC resulting from a modified Hamiltonian constraint on a non-separable Hilbert space arising in the theory. It is completely natural to ask if one might impose boundary conditions analogous to the ones mentioned above for this difference equation in a canonical quantization [20]. We are currently investigating if it is possible to impose the no-boundary conditions on the (minisuperspace of the) wave function in order to solve the difference equation in LQC.

Between two hypersurfaces (which take given values of the 3-metric and scalar field specified on it) at the initial boundary (h_i, ϕ_i) and the final one (h_f, ϕ_f), the (Feynmann) propagator is given by

$$\Psi[h_f, \phi_f; h_i, \phi_i] = \int \mathcal{D}[g]\mathcal{D}[\phi] \; e^{iS[g,\phi]}, \tag{10}$$

where we sum over all geometries, allowing for topology-changes, which have the specified initial and final boundaries. This integration is highly-oscillatory and ill-defined, and it was hoped that its convergence properties can perhaps be improved on introducing a Wick-rotation to Euclidean time $dt = -id\tau$, as is often done in standard quantum field theories,

$$\Psi_0[h_f, \phi_f; h_i, \phi_i] = \int \mathcal{D}[g]\mathcal{D}[\phi] \; e^{-S_E[g,\phi]}, \tag{11}$$

where now we sum over all Euclidean geometries and corresponding field combinations with the given boundaries (left of Figure 1). For usual quantum field theories, this corresponds to the ground state wavefunction. Although there is no straightforward way to define the ground state in quantum gravity (since the action is unbounded from below), but Hartle and Hawking proposed that the above form of the Euclidean path integral may correspond to the ground state wavefunction of the universe.

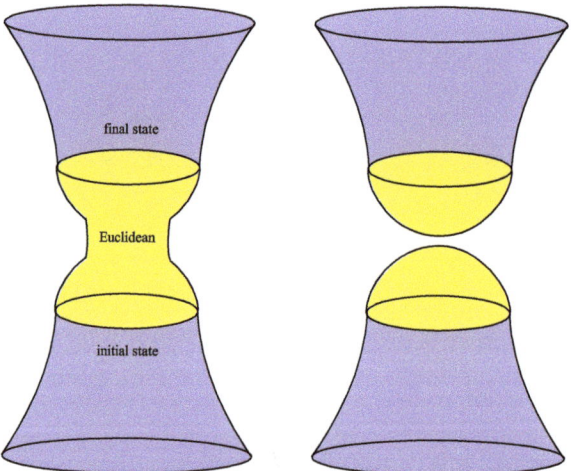

Figure 1. *Left*: Euclidean path integral that connects from the initial state to the final state. *Right*: If two states are disconnected at the Euclidean manifold, one can consider a wave function for the only final state.

Let us further assume that the intermediating Euclidean geometries between the initial and final boundaries are disconnected (right of Figure 1). Especially, if we consider a closed universe (i.e., $k = +1$ FLRW cosmology), then the wavefunction remains well-defined if we remove the initial boundary

$$\Psi_0[h_f, \phi_f] = \int \mathcal{D}[g]\mathcal{D}[\phi] \; e^{-S_E[g,\phi]}. \tag{12}$$

This unique boundary condition has been given the so-called nickname *no-boundary proposal*, because it has no initial boundary.

There is a lot of justifiable controversy whether this Euclidean path integral is a good approximation of the original Lorentzian path integral or not [21–27]. However, this no-boundary proposal, defined as an Euclidean path integral, is attractive because of several nice properties it possesses.

- The Euclidean path integral can be interpreted as the partition function of a thermal system [28]

$$Z = \text{Tr} \exp\left(-\beta\hat{\mathcal{H}}\right) = \int \mathcal{D}[g]\mathcal{D}[\phi]\ e^{-S_E[g,\phi]}, \tag{13}$$

where $\hat{\mathcal{H}}$ is the quantum Hamiltonian of the system and β is the inverse of the Hawking temperature. The right-hand side can be evaluated in the steepest-descent approximation as

$$Z = \exp\left(-\beta\mathcal{F}\right) \simeq \exp\left(+\frac{3}{8V_0}\right) = e^{\mathcal{A}/4}, \tag{14}$$

where $\mathcal{F} = E - T\mathcal{S}$ is the Helmholtz free energy, E and $\mathcal{S}$ being energy and entropy of the system, respectively. Here, $\mathcal{A}$ is the area of the cosmological horizon. Since the ADM energy E is zero for dS space, we can consistently recover the Bekenstein-Hawking entropy formula $\mathcal{S} = \mathcal{A}/4$. This reveals that the Euclidean path integral is consistent with the expected thermodynamic properties of gravity.

- Building on this result, one may consider other semi-classical effects in dS space as well. The classical (Lorentzian) equation of motion for a scalar field on a fixed dS background is given by

$$\ddot{\phi} = -3H\dot{\phi} - V', \tag{15}$$

where $H = \dot{a}/a$. On inserting random 'white noise' in the slow-roll limit as a thermal effect due to Hawking temperature, one obtains the Langevin equation [29,30]

$$\frac{d}{dt}\phi = -\frac{V'}{3H} + \frac{H^{3/2}}{2\pi}\xi(t), \tag{16}$$

where one imposes the white noise conditions:

$$\langle \xi(t) \rangle = 0, \tag{17}$$
$$\langle \xi(t)\xi(t') \rangle = \delta(t-t'). \tag{18}$$

Then, the probability to have the field value ϕ at t will follow the Fokker-Planck equation [31]

$$\frac{\partial P(\phi,t)}{\partial t} = \frac{2\sqrt{2}}{3\sqrt{3\pi}}\frac{\partial}{\partial \phi}\left[V^{3/4}(\phi)\frac{\partial}{\partial \phi}\left(V^{3/4}(\phi)P(\phi,t)\right) + \frac{3V'(\phi)}{8V^{1/2}(\phi)}P(\phi,t)\right], \tag{19}$$

while the probability to have the field value initially χ at $t = 0$ will follow the equation

$$\frac{\partial P(\phi,t|\chi)}{\partial t} = \frac{2\sqrt{2}}{3\sqrt{3\pi}}\left[V^{3/4}(\chi)\frac{\partial}{\partial \chi}\left(V^{3/4}(\chi)\frac{\partial P(\phi,t|\chi)}{\partial \chi}\right) - \frac{3V'(\chi)}{8V^{1/2}(\chi)}\frac{\partial P(\phi,t|\chi)}{\partial \chi}\right]. \tag{20}$$

In the static limit, a solution that satisfies both of these equations is given by

$$P(\phi,t|\chi) \sim V^{-3/4}(\phi) \exp\left[\frac{3}{8V(\phi)} - \frac{3}{8V(\chi)}\right]. \tag{21}$$

We can interpret this as the tunneling probability of a homogeneous part of a universe that tunnels from the field value χ to ϕ via stochastic quantum fluctuations when the wavelength is of the order of the Hubble radius.

This wave function is consistent with the Euclidean path integral approximated by the Hawking-Moss instantons [32,33]. On further normalizing the initial boundary, one can obtain the no-boundary wave function. Therefore, we can conclude that the Euclidean path integral describes the stationary limit, or thermal equilibrium, of quantum fluctuations of the Hubble-scale wavelength modes consistently, whereas in many situations, such a thermal equilibrium is coincident with the ground state of the system [30].

The above physical motivations are reason enough to investigate and apply the Euclidean path integral with the no-boundary condition as the wavefuntion of the universe, not only due to its mathematical simplicity but also due to its self-consistency in the low-energy limit with quantum field theory in curved spacetime.

2.3. Semiclassical Approximation: Instanton Solutions

One can calculate the Euclidean path integral in the saddle-point approximation by using Euclidean on-shell solutions, or so-called instantons, as

$$\int \mathcal{D}[g]\mathcal{D}[\phi]\ e^{-S_E[g,\phi]} \simeq \sum_{\text{instanton}} e^{-S_E^{\text{instanton}}} . \qquad (22)$$

The on-shell solution in pure dS space, on imposing the no-boundary condition $a(0) = 0$, becomes regular to give

$$a(\tau) = \frac{1}{H_0} \sin(H_0 \tau) , \qquad (23)$$

where $H_0^2 = 8\pi \tilde{V}/3$. This solution reveals that at the $a(0) = 0$ ("South Pole") point, we need to impose the condition $\dot{a} = 1$ from the Hamiltonian constraint.

Inserting this solution, one can evaluate the Euclidean action. In the phase space formulation, the action takes the form

$$S_E = \int d\tau\ \mathcal{L} = \int d\tau\ (p_a \dot{a} + p_\phi \dot{\phi} - N\mathcal{H}) . \qquad (24)$$

On plugging the on-shell condition, $\mathcal{H} = 0$ is automatically satisfied. Hence,

$$S_E^{\text{instanton}} = \int d\tau\ (p_a \dot{a} + p_\phi \dot{\phi}) . \qquad (25)$$

This solution can be analytically continued to Lorentzian time for any constant-τ hypersurface, but after the Wick-rotation, the metric is in general complex-valued except at the $\dot{a} = 0$ hypersurface (i.e., $\tau = \pi/2H_0$, left of Figure 2). If we Wick-rotate on this surface, then the metric becomes (right of Figure 2)

$$a(t) = \frac{1}{H_0} \cosh(H_0 t) . \qquad (26)$$

We give more mathematical details corresponding to this solution in the following section.

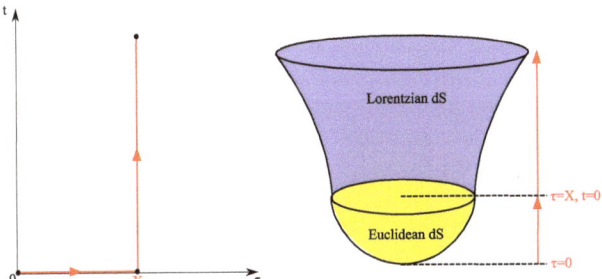

Figure 2. *Left:* A typical time contour over the complex time, where $X = \pi/2H_0$. *Right:* Euclidean and Lorentzian manifold along the given time contour.

3. Geometry of the Hartle-Hawking Instantons

We give a more detailed derivation of the schematics described for the minisuperspace model in the previous section. The Friedmann equation (or the Hamiltonian constraint) for the $k = 1$ FLRW universe, in Euclidean time, is given by

$$\frac{1}{N^2}\left(\frac{da}{d\eta}\right)^2 = \frac{8\pi}{3}\frac{a^2}{N^2}\left(\frac{d\phi}{d\eta}\right)^2 - 1 + \left(\frac{\Lambda}{3} + V(\phi)\right)a^2, \tag{27}$$

where we use the metric $ds^2 = N^2(\eta)d\eta^2 + a^2(\eta)\,d\Omega_3^2$ and set $G = 1$ throughout[4]. η denotes the Euclidean time parameter while reserving t for Lorentzian time, as before. We rewrite the other relevant equation which is the scalar field equation (also, in Euclidean time)

$$\frac{1}{N^2}\left(\frac{d^2\phi}{d\eta^2}\right) + \frac{3}{aN}\left(\frac{da}{d\eta}\right)\left(\frac{d\phi}{d\eta}\right) - V'(\phi) = 0. \tag{28}$$

First of all, let us make a gauge choice and fix the lapse function $N = 1$. As pointed out in [34], this can be rigorously achieved by introducing the complex variable $\tau(\eta) = \int_0^\eta d\eta'\, N(\eta')$. Given any lapse function, the variable τ defines a complex contour on the τ-plane. Once we rewrite the above equations in terms of the variable τ, the task of finding the no-boundary instantons is to solve these equations for the pair of complex analytic functions $a(\tau)$ and $\phi(\tau)$, given the appropriate boundary conditions.

The set of equations, in terms of this new variable, can be expressed as

$$\dot{a}^2 = 1 - \left(\frac{\Lambda}{3} + V(\phi) + \frac{8\pi}{3}\dot{\phi}^2\right)a^2 =: \mathcal{V}(a), \tag{29}$$

$$\ddot{\phi} + 3H\dot{\phi} - V'(\phi) = 0, \tag{30}$$

where a dot refers to a derivative with respect to τ and the Hubble parameter $H := \dot{a}/a$. The fact that we can write the RHS of (29) as a function of the scale factor alone is only possible for the simplest case of a massless scalar or a cosmological constant. There is always the Raychaudhuri equation involving the second derivative of a but only two of these three equations are linearly-independent. For our purposes of examining the on-shell Euclidean instantons, required for estimating the path integral by its saddle-points, considering these two equations is sufficient. The usual procedure is to solve the above equations for the 'no-boundary' boundary conditions [35]: $a(0) = 0$ and $\dot{\phi}(0) = 0$. The first condition is a requirement that the geometry must close in a regular fashion while the second is a necessary condition for keeping the solution for ϕ regular as $a \to 0$ [7,8]. It is often customary

---

[4] Note the new normalization chosen here to facilitate comparison with LQC later on.

to quote another condition $\dot{a}(0) = 1$; however, this is the consequence of the Friedmann equation. In general, requiring that the scale factor and the scalar field take some fixed value on the final surface, $(a(\tau_f) = b, \phi(\tau_f) = \chi)$, and some fixed value initially, $(a(0) = 0, \phi(0) = 0)$, exhausts all the conditions necessary to give a unique solution. The value of the derivative of the scalar field must be fixed from the scalar field Equation (30) to be zero while the value of the scalar field at the 'South Pole'—$\phi(0) = \phi_0$—gives the one-parameter family of instantonic solutions which satisfy the no-boundary proposal [34,35]. Additional tunings are necessary to ensure the classicality of our universe at late-times, the details of which are unimportant for our purposes (see [34,36–38]).

If we take the pure gravity model, in the absence of any scalar field, one can analytically solve for the Euclidean instanton to find that there is a $O(5)$–symmetric solution given by $a(\tau) = \sqrt{3/\Lambda}\sin\left(\sqrt{\Lambda/3}\,\tau\right)$ (this is Equation (23) above written in the new normalization). In the presence of a scalar field, one requires that the potential is sufficiently flat, i.e., of the inflationary type, for the solution to be regular. In the case of a slowly varying potential, the solution for $a(\tau)$ is a deformed version of the sine function. However, for the massless scalar field (i.e., in the absence of any potential term at all), there are no compact instantons which would give rise to a nontrivial universe. This is obvious from the fact that in this case, the scalar field is in the "no roll" condition and the energy density of the universe is trivial. However, this scenario, which is beyond the scope of this paper, leads to new solutions for the no-boundary proposal in the presence of LQC corrections [1,39].

Before going on to the LQC instantons, let us revisit the geometry of these Hartle-Hawking instantons in Einstein gravity. (This is schematically shown in the right panel of Figure 1). Restricting to the case of pure gravity is already sufficient to illustrate its salient features. For Lorentzian signatures, one has the usual dS solution of Einstein's equations with a positive cosmological constant as $ds_L^2 = -dt^2 + a_L^2(t)d\Omega^2$, with $a_L(t) = \sqrt{3/\Lambda}\cosh\left(\sqrt{\Lambda/3}\,t\right)$ (equivalent to (26) above in the new normalization). To get the $O(5)$ invariant Euclidean instanton from this result, one analytically continues t such that $\tau = \tau_f + it$, where τ_f is the point where $a(\tau)$ reaches its maximum value. In other words, one gets a Lorentzian dS spacetime from a Euclidean hemisphere (right of Figure 2), by matching the two hypersurfaces across the zero-extrinsic curvature ($\dot{a} = 0$) 'bounce' surface. The effective potential, $\mathcal{V}(a)$, goes to zero on this surface. Such a sharp transition between a real Euclidean half-sphere and a real Lorentzian part is only possible for the simplest example of a pure cosmological constant considered in this paper. In general, in the presence of an inflationary-type potential, the transition would be in terms of 'fuzzy' Euclidean instantons [34,35], whereby the solutions would be complex in some parts [40–42]. These details, however, are not important for us while focusing near the 'South Pole' to exhibit the general 'shuttle-cock' type shape of the no-boundary instanton in Euclidean gravity.

For the pure dS model, the Friedmann equation, in Euclidean time, takes the form $\dot{a}^2 = 1 - \Lambda a^2/3$ which clearly shows that as $a \to 0$, one gets $\dot{a} \to 1$. The geometric interpretation of this result goes as follows. The Euclidean 4-metric, possessing the $O(5)$ symmetry, $ds^2 = d\tau^2 + a^2\,d\Omega_3^2$ has to smoothly close-off in a regular manner into flat space (written in spherical coordinates) $ds^2 = dr^2 + r^2\,d\Omega_3^2$. For this to happen, one has to identify $a(\tau) \sim \tau$ as $a \to 0$. This suggests that $\dot{a} \to 1$ in this limit, as is required from the Hamiltonian constraint. However, as mentioned before, this requirement for the derivative of the scale factor automatically follows from the constraint and is not part of the no-boundary condition.

Let us make one last comment before presenting our new results for the LQC instantons. The no-boundary initial condition is simply that the geometry closes off smoothly, as denoted by $a(0) = 0$. This shall be important later on for the LQC instantons. If we try to solve the Friedmann equation, we get the (famous) unique solution (23) only on *imposing the no-boundary condition*. Of course, choosing the 'initial' point at $\tau = 0$ is only for convenience. A priori, there is no need for such a condition to be satisfied by the modified field equations in LQC. However, as we shall show in the LQC case, the initial condition that compact (Euclidean) instantons in LQC go off to $a \to 0$ is favored *naturally* even in the presence of quantum-geometry corrections, at least in the pure gravity case. This is the main result of our work which we shall elaborate on in the following sections.

4. No-Boundary Instantons in LQC

In [1], it was shown how the effective LQC action modifies the instantons in the theory even for a simple cosmological constant. Moreover, this leads to slight enhancement of the probability of nucleation of the dS universe from nothing due to the LQC corrections. The general shape of the instanton is reproduced in Figure 2. However, we shall only be interested in the small-a behaviour of this LQC instanton. In particular, what emerges to be intriguing is the infinite tail of the instanton. Although this infinite tail is in Euclidean time, and therefore not physically relevant directly, it does have certain distinguishing features which we shall demonstrate below.

However, before proceeding with the calculations, let us clarify a conceptual issue regarding the path integral formulation of LQC. In LQC, one often introduces 'holonomy' and 'inverse-triad' corrections in the equations of motion in a heuristic manner, adopting a semiclassical approximation. Naively, it might seem that we are also following such a semiclassical 'effective' Hamiltonian as the starting point for our path integral quantization. However, this is not correct. Following [5,6], we first note that the rigorously defined path integral for LQC, in its phase space version, deviates from the usual gravitational path integral in that the paths are weighted by a 'polymerized' action instead of the Einstein-Hilbert one. This is the crucial point for us—the relevant action for the LQC path integral is different than the one in Einstein gravity, and it remembers the effects of quantum-geometry such as holonomy and inverse-triad corrections. So why does it look like we start from the heuristic quantum-corrected equations in LQC? This is due to the subtlety of taking the saddle-point approximation of the LQC path integral. As also discussed in [5,6], the saddle-point approximation of the LQC path integral leads to the so-called 'semiclassical' limit, in which one keeps the 'area-gap' ($\propto \hbar \gamma^3$) fixed while taking $\hbar \to 0$ (whereas shrinking the area-gap to zero would lead to the Einstein-Hilbert action starting from the LQC one). Since we shall only be working in the saddle-point approximation in this paper, the resulting equations from the LQC path integral shall indeed be the semiclassical ones. However, this is not an *ad hoc* choice of including some LQC corrections but rather the result of working in the saddle-point approximation even when starting from the rigorous LQC path integral. We also emphasize that this is the same approximation one usually employs for the Einstein-Hilbert path integral, whereby ignoring the higher loop corrections. We refer the inquisitive reader to [5,6] for details on deriving the path integral for LQC along with more analysis of these subtleties.

In this work, our main novelty would be to impose the no-boundary condition on the LQC path integral thereby necessarily having to consider Euclidean histories, going beyond what exists in the literature [1]. We begin with the modified Friedmann in LQC [43] due to the quantum geometry corrections mentioned in the Introduction.

$$\dot{a}^2 = -a^2 \left(\frac{8\pi}{3}\right)\left(\frac{f^2(a)}{v^2(a)}\right)[\tilde{\rho}-\rho_1]\left[\frac{1}{\rho_c}(\rho_2-\tilde{\rho})\right] =: \mathcal{V}_{\text{LQC}}(a), \qquad (31)$$

where $\tilde{\rho}$ is the contribution from a positive cosmological constant.

$$\tilde{\rho} := \left(\frac{v(a)}{f(a)}\right)\frac{\Lambda}{8\pi}, \qquad (32)$$

$$\rho_1 := -\rho_c \left[\sin^2(\sqrt{\Delta}/a) - (1+\gamma^2)\frac{\Delta}{a^2}\right], \qquad (33)$$

$$\rho_2 := \rho_c \left[\cos^2(\sqrt{\Delta}/a) + (1+\gamma^2)\frac{\Delta}{a^2}\right], \qquad (34)$$

where

$$v(a) := K\left(\frac{3}{4\pi\gamma l_{Pl}^2}\right)^{3/2} a^3 V_0, \qquad (35)$$

$$f(a) := \left(\frac{1}{2}\right) v(a) ||v(a) - 1| - |v(a) + 1||, \qquad (36)$$

$$K = \frac{2\sqrt{2}}{3\sqrt{3\sqrt{3}}}, \qquad (37)$$

$$\rho_c = \frac{3}{8\pi\gamma^2 \Delta}, \qquad (38)$$

$$\Delta = 2\sqrt{3}\pi\gamma l_{Pl}^2, \qquad (39)$$

with $V_0 = 16\pi^2$. Although we have adhered to the conventions of [43], we have generalized their results by adding in non-perturbative expressions for the inverse-triad corrections.

On first look, the above set of equations look rather complicated due to the different terms involved. Here, $f(a)$ represents the inverse-triad corrections whereas holonomy modifications show up in ρ_1 and ρ_2. Importantly, note that due to the presence of the inverse-triad corrections, it is always possible to impose the no-boundary condition $a \to 0$. However, to gain some intuition into the modified equation, let us begin by setting the holonomy corrections to zero for simplicity. This would be like taking the area gap Δ to zero. In this limit, $\Delta \to 0$, we get

$$\dot{a}^2 = -a^2 \left(\frac{\Delta}{3}\right)\left(\frac{f(a)}{v(a)}\right) + \left(\frac{f(a)}{v(a)}\right)^2. \qquad (40)$$

This is the modified Friedmann equation only in the presence of inverse-triad corrections. It is easy to check that in the large $a \gg 1$ limit, one gets $f(a) \approx v(a)$, and therefore we get back the usual Friedmann equation for a closed universe. However, in the $a \ll 1$ limit, one cannot make such an approximation. Instead, in this limit, we get $f(a) \approx v(a)^2$. Our aim in this work is not to solve for those instantons which extremizes the Euclidean path integral but rather to examine its small-a behaviour. Therefore, considering $v(a) \propto a^3$ and reinstating the holonomy modifications, we get the leading order term for $a \approx 0$ as

$$\dot{a}^2 \sim Ca^2, \qquad (41)$$

for some constant $C > 0$. To obtain this result, we notice that the leading order term comes from the $\tilde{\rho}^2$ term in (31) whereas the remaining terms are subdominant. This is a term which arises only in the quantum-corrected Friedmann equation (there is no term quadratic in the energy density in the classical Friedmann equation). The $\tilde{\rho}^2$ term comes with an additional minus sign which leads to a $C > 0$. Moreover, note that the dominant contribution in the classical case comes from the curvature term $(1/a^2)$ whereas that term, contained in ρ_1, is now sub-dominant. As already argued in the previous section, the essential requirement of the no-boundary condition is the geometry should be closed off in a regular manner and the condition on $\dot{a}$ should follow from the Hamiltonian constraint. In the LQC case, the modified Hamiltonian constraint implies $\dot{a} = 0$ instead of 1. Nevertheless, $\mathcal{V}(a)$ remains regular even in this case.

The above findings for no-boundary instantons in LQC in quite remarkable. In order to appreciate this properly, let us make a few comments. Firstly, note that there was no reason that the modified Friedmann equation have to allow for the $a \to 0$ limit to be imposed consistently. It could easily have been that this limit is singular in LQC. To illustrate this, let us consider only the holonomy modifications while ignoring the inverse-triad ones. Typically, for the Lorentzian effective trajectories such an approximation is completely justified and valid even near the 'bounce' surface. In this case, the RHS of (31) has a singular term coming from '$\rho_1\rho_2$' (proportional to Δ^2), which is absent in the classical case. However, luckily for us, when one considers Euclidean histories as is required for our case, one cannot ignore the inverse-triad corrections any longer. Secondly, the structure of the modified Friedmann equation is such that the resulting instantons remain regular for all values of τ even on imposing the no-boundary condition. For a counterexample, imagine if the form of the equation was such that $\dot{a}^2 \propto a^n$ with $n > 2$, in that case, the limit $\tau \to 0$ would have been singular and there would not have been consistent no-boundary instantons in the theory. Moreover, these inverse-a modifications not only play a crucial role in ensuring that the no-boundary condition can be imposed but also modify the geometry of these instantons to distinguish them from the Einstein gravity. It could also have been the case that the inverse-a modifications are such that one still gets the same condition for $\dot{a}$ at the South Pole. In that case, although the explicit solutions of the instantons would have been different, there would have been no difference in the geometry of LQC and Einstein gravity instantons. The quantum-geometry corrections in LQC *conspire* to ensure that we have no-boundary instantons in the theory with such a geometry that is tapers off to the symmetry point in a novel fashion.

Concretely, the small-a solution for a is given by $a_0 e^{c\tau}$. Obviously, this implies that the point $\tau = 0$ is not a good point to impose the no-boundary condition. Rather both $(a(\tau), \dot{a}(\tau))$ goes to zero as $\tau \to -\infty$. However, this does not represent any difficulty since this infinite stretching is in Euclidean time and is, thus, not physically relevant directly. This novel feature of LQC instantons can be seen from Figure 3, where the tail of the compact instanton is stretched infinitely, asymptotically tapering off to zero. The tail *does* contribute to the probability of nucleation of the Lorentzian dS universe, although the path integral and consequently the probability remains finite and well-defined in spite of this infinitely stretched geometry.

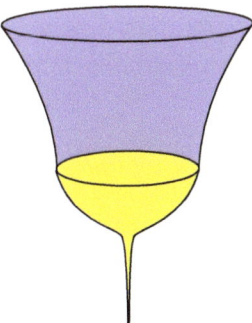

Figure 3. Euclidean and Lorentzian manifold with loop quantum cosmology (LQC) corrections.

4.1. Numerical Results

We give some sample numerical solutions for the LQC instantons, going beyond the small-a limit, to illustrate our claims regarding their infinite tail. Figures 4 and 5 show a typical shape of the effective potential $\mathcal{V}_{LQC}(a)$ and its solution $a(\tau)$ for the HR, respectively. This solution demonstrates that the instanton is indeed infinitely stretched (Figure 3). The approximate behavior of the nucleation probability is of the form (Figure 6)

$$S = -2S_E \simeq \frac{\mathcal{A}}{4} + c + d \log \mathcal{A} + \ldots \tag{42}$$

with a model dependent positive constant c, where $\mathcal{A} = 4\pi a_{\max}^2$. As mentioned before, $a_{\max}$ denotes the max value the instanton takes in the Euclidean regime, on which surface we analytically continue to the Lorentzian regime. The above constant c can easily be absorbed away in the normalization of the probability measure and the only relevant correction due to LQC, over the Einstein-Hilbert value, comes from the parameter d. This parameter d can be approximately expanded by $d \simeq 8.7 \times l_{Pl}^2/\gamma$

(Figure 7) in terms of the fundamental parameters—Planck length and Immirzi parameter—of the theory, as shown via numerical reconstruction.

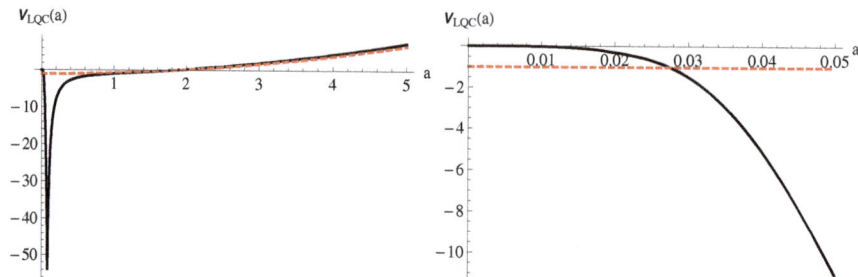

Figure 4. The black curve is an example of $\dot{a}^2$ for $\Lambda = 1$, $G = 1$, and $l_{Pl} = 0.1$, where the red dashed curve is the limit of the Einstein gravity with the same a_{max} that satisfies $\dot{a}_{max} = 0$. Right is the behavior near the $a = 0$ limit.

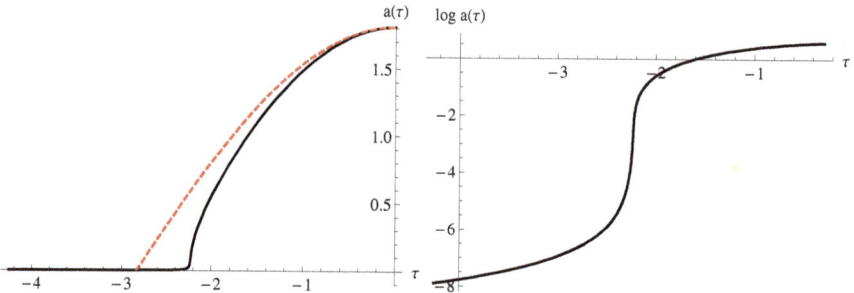

Figure 5. Left: $a(\tau)$ (red dashed curve is the limit of the Einstein gravity). **Right:** $\log a$ for small a limit. It has an infinitely long throat.

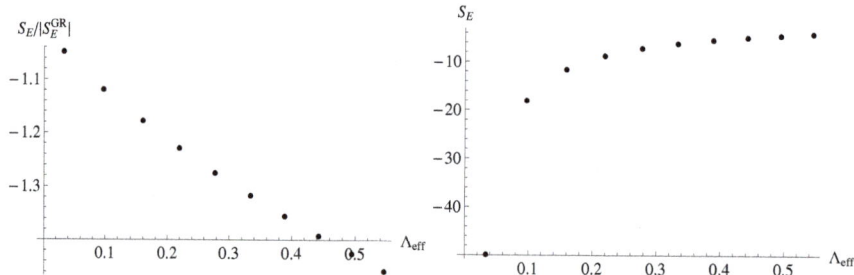

Figure 6. Left: $S_E/|S_E^{GR}|$, where for $G = 1$, and $l_{Pl} = 0.1$ by varying Λ (equivalently, varying $\Lambda_{eff} \equiv 1/a_{max}^2$), where S_E^{GR} is the Euclidean action for the corresponding Einstein limit. **Right:** S_E for the same parameters.

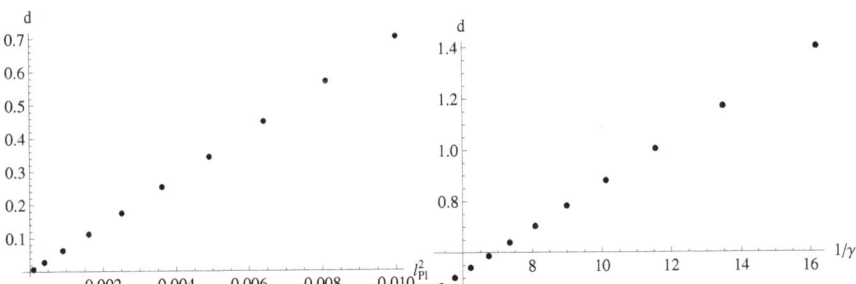

Figure 7. **Left**: By varying l_{Pl}, one can calculate $d \propto l_{Pl}^2$ numerically. **Right**: By varying γ (with $l_{Pl} = 0.1$), one can see a linear dependence $d \propto 1/\gamma$. We can numerically conclude that $d \simeq 8.7 \times l_{Pl}^2/\gamma$.

5. Robustness of the No-Boundary Condition

In Einstein gravity, when considering a positive cosmological constant as the only matter source, the only solution which is regular at the South Pole is given by (23). The crucial point here is that one gets this solution for the instanton in Einstein gravity from the (Euclidean) Friedmann equation on imposing the no-boundary condition. Interestingly, as was shown in previous section, the small-a behaviour for the LQC instantons following the modified Friedmann equation is always of the form

$$a(\tau) \propto e^{C\tau}, \qquad (43)$$

for some constant C. This shows that as $\tau \to -\infty$, we get $a \to 0$ implying a natural implementation of the no-boundary condition in LQC instantons, as a result of quantum-geometry regularizations, at least for the simplest case of a cosmological constant. In this sense, this shows that the no-boundary condition is robust and more natural when the Einstein-Hilbert action is augmented by LQG corrections.

The Euclidean instantons, in this pure gravity scenario, are always going to be compact and there is no risk of an Euclidean wormhole forming for some values of the parameter space. This conclusion is true both for Einstein gravity as well as the LQC case. Therefore, one always gets compact Euclidean instantons for the pure dS case in both cases. The overall conceptual picture of the nucleation of the universe from nothing is also the same in both cases. One quantitative difference is due to the modified equations in LQC: The 'bounce' surface ($\dot{a} = 0$), as predicted by the effective trajectories of sharply-peaked semiclassical states, in LQC is (slightly) different from the hypersurface joining the Euclidean and Lorentzian parts in Einstein gravity. However, the qualitative behaviour remains the same and this difference is reflected in enhancing the probability of nucleation of the universe manifesting as a next-to-leading-order correction in LQC. As mentioned in [1], these type of terms can also appear in Einstein gravity from going to higher order corrections and there should be a competition between the two terms in LQC—one appearing from radiative corrections and the other from the inherent quantum geometry. It is thus difficult to unambiguously state that LQC enhances the tunneling amplitude for such no-boundary universes.

6. Conclusions

It is an old expectation that *mathematical consistency* alone shall be sufficient to derive the boundary conditions in quantum cosmology [44]. The form of the LQC instantons suggest that it might indeed be possible to identify *typical* smooth initial conditions due to quantum-geometry corrections. This demonstrates that one of the most fundamental proposal for the initial condition in quantum cosmology can appear naturally in another—LQC—due to putative corrections coming from quantum geometry. The introduction of the no-boundary proposal in LQC has also opened new physical possibilities for the latter. Instead of replacing the big bang singularity with a deterministic bounce, as is predicted by semiclassical states in some restricted models of LQC, this opens up the opportunity to allow for Euclidean trajectories leading to a bubble nucleation of our universe. The semiclassical saddle-point approximation of the no-boundary proposal is distinct from the semiclassical sharply-peaked states in LQC and therefore the former acts as an example of how a new state can unveil novel features in a well-established theory. At this point, it is difficult to compare the probability of a bounce versus that of tunneling of the universe from nothing. However, the no-boundary proposal also provides a (set of) *natural* initial conditions for considering inhomogeneous perturbations in LQC leading to effects observable from early-universe cosmology without having to resort to *ad hoc* choices for the initial state.

Regarding the geometry of no-boundary instantons in LQC, we have demonstrated that the feature of having an infinite tail distinguishes these instantons from the original Hartle-Hawking ones in Einstein gravity. We end our discussion with a few caveats. Firstly, it has been pointed out recently that the Euclidean path integral in gravity is not a good approximation for the original Lorentzian path integral due to several conceptual reasons [21]. However, even if one works with the Lorentzian path integral and applies a different mathematical trick (Pecard-Lefshetz theory) to improve its convergence, the resulting theory typically has runaway perturbations due to the old conformal factor problem in gravity [22–24]. Interestingly, LQC can come to the rescue of the no-boundary proposal [45], written as a Lorentzian path integral, even in this case. However, the main physical effect from LQC responsible for this is 'dynamical signature-change' [46–48], something we have ignored in this work as a first pass. The second caveat is regarding the fact that our discussions were limited to the case of a pure cosmological constant in this paper. Indeed, the more interesting physical scenario is that of having a scalar field in some potential. However, preliminary investigations have already revealed that the solution space of the no-boundary wavefunction is greatly enhanced for such a system in the presence of LQG corrections. Finally, one can ask how physical is the fact that the tail of these instantons are stretched to infinity? As already mentioned, this is only true in Euclidean time and therefore not directly meaningful. However, it might even be possible that for some different gauge choice (i.e., $N \neq 1$), one can even avoid such an infinite stretching altogether. Nevertheless, all the interesting effects of having such a geometry as explained in this paper would still be valid in this case. Most importantly, no matter what the gauge choice, the remarkable conclusion that the modified Friedmann equation in LQC not only allows for the no-boundary condition to be imposed, but also somehow makes it more natural, seems to be robust and points towards a new paradigm in quantum cosmology merging these two mainstream approaches.

Author Contributions: All of the work was done equally by both the authors.

Funding: This research was supported in part by the Ministry of Science, ICT & Future Planning, Gyeongsangbuk-do and Pohang City and the National Research Foundation of Korea Grant No. 2018R1D1A1B07049126.

Conflicts of Interest: The authors declare no conflict of interest.

References

1. Brahma, S.; Yeom, D. The no-boundary wave function for loop quantum cosmology. *arXiv* **2018**, arXiv:1808.01744.
2. Hartle, J.B.; Hawking, S.W. Wave Function of the Universe. *Phys. Rev. D* **1983**, *28*, 2960–2975. [CrossRef]
3. Bojowald, M. Loop quantum cosmology. *Living Rev. Relativ.* **2005**, *8*, 11. [CrossRef] [PubMed]
4. Bojowald, M. Quantum cosmology: A review. *Rep. Prog. Phys.* **2015**, *78*, 023901. [CrossRef] [PubMed]
5. Ashtekar, A.; Campiglia, M.; Henderson, A. Path Integrals and the WKB approximation in Loop Quantum Cosmology. *Phys. Rev. D* **2010**, *82*, 124043. [CrossRef]
6. Huang, H.; Ma, Y.; Qin, L. Path Integral and Effective Hamiltonian in Loop Quantum Cosmology. *Gen. Relativ. Gravit.* **2013**, *45*, 1191–1210.
7. Chen, P.; Hu, Y.C.; Yeom, D. Fuzzy Euclidean wormholes in de Sitter space. *J. Cosmol. Astropart. Phys.* **2017**, *2017*, 001. [CrossRef]

8. Chen, P.; Yeom, D. Why concave rather than convex? *arXiv* **2018**, arXiv:1706.07784.
9. Bojowald, M. Isotropic loop quantum cosmology. *Class. Quantum Gravit.* **2002**, *19*, 2717. [CrossRef]
10. Ashtekar, A.; Lewandowski, J.; Marolf, D.; Mourao, J.; Thiemann, T. Quantization of diffeomorphism invariant theories of connections with local degrees of freedom. *J. Math. Phys.* **1995**, *36*, 6456–6493. [CrossRef]
11. Ashtekar, A.; Lewandowski, J. Quantum theory of geometry. 1: Area operators. *Class. Quantum Gravit.* **1997**, *14*, A55–A81. [CrossRef]
12. Ashtekar, A.; Bojowald, M.; Lewandowski, J. Mathematical structure of loop quantum cosmology. *Adv. Theor. Math. Phys.* **2003**, *7*, 233–268. [CrossRef]
13. Banerjee, K.; Calcagni, G.; Martin-Benito, M. Introduction to loop quantum cosmology. *SIGMA* **2012**, *8*, 016. [CrossRef]
14. Thiemann, T. QSD 5: Quantum gravity as the natural regulator of matter quantum field theories. *Class. Quantum Gravit.* **1998**, *15*, 1281–1314. [CrossRef]
15. Bojowald, M. The Inverse scale factor in isotropic quantum geometry. *Phys. Rev. D* **2001**, *64*, 084018. [CrossRef]
16. Bojowald, M. Quantization ambiguities in isotropic quantum geometry. *Class. Quantum Gravit.* **2002**, *19*, 5113–5230. [CrossRef]
17. Bojowald, M. Loop quantum cosmology: Recent progress. *Pramana* **2004**, *63*, 765–776. [CrossRef]
18. Vilenkin, A. Quantum Cosmology and the Initial State of the Universe. *Phys. Rev. D* **1988**, *37*, 888–897. [CrossRef]
19. Vilenkin, A. Quantum Creation of Universes. *Phys. Rev. D* **1984**, *30*, 509. [CrossRef]
20. Bojowald, M.; Vandersloot, K. Loop quantum cosmology, boundary proposals, and inflation. *Phys. Rev. D* **2003**, *67*, 124023. [CrossRef]
21. Feldbrugge, J.; Lehners, J.L.; Turok, N. Lorentzian Quantum Cosmology. *Phys. Rev. D* **2017**, *95*, 103508. [CrossRef]
22. Feldbrugge, J.; Lehners, J.L.; Turok, N. No smooth beginning for spacetime. *Phys. Rev. Lett.* **2017**, *119*, 171301. [CrossRef]
23. Feldbrugge, J.; Lehners, J.L.; Turok, N. No rescue for the no boundary proposal: Pointers to the future of quantum cosmology. *Phys. Rev. D* **2018**, *97*, 023509. [CrossRef]
24. Feldbrugge, J.; Lehners, J.L.; Turok, N. Inconsistencies of the New No-Boundary Proposal. *Universe* **2018**, *4*, 100. [CrossRef]
25. Diaz Dorronsoro, J.; Halliwell, J.J.; Hartle, J.B.; Hertog, T.; Janssen, O. Real no-boundary wave function in Lorentzian quantum cosmology. *Phys. Rev. D* **2017**, *96*, 043505. [CrossRef]
26. Diaz Dorronsoro, J.; Halliwell, J.J.; Hartle, J.B.; Hertog, T.; Janssen, O.; Vreys, Y. Damped perturbations in the no-boundary state. *Phys. Rev. Lett.* **2018**, *121*, 081302. [CrossRef]
27. Vilenkin, A.; Yamada, M. Tunneling wave function of the universe. *Phys. Rev. D* **2018**, *98*, 066003. [CrossRef]
28. Gibbons, G.W.; Hawking, S.W. Action Integrals and Partition Functions in Quantum Gravity. *Phys. Rev. D* **1977**, *15*, 2752–2756. [CrossRef]
29. Le Bellac, M.; Mortessagne, F.; Batrouni, G.G. *Equilibrium and Non-Equilibrium Statistical Thermodynamics*; Cambridge University Press: Cambridge, UK, 2004.
30. Hwang, D.; Lee, B.H.; Stewart, E.D.; Yeom, D.; Zoe, H. Euclidean quantum gravity and stochastic inflation. *Phys. Rev. D* **2013**, *87*, 063502. [CrossRef]
31. Linde, A.D.; Linde, D.A.; Mezhlumian, A. From the Big Bang theory to the theory of a stationary universe. *Phys. Rev. D* **1994**, *49*, 1783. [CrossRef]
32. Hawking, S.W.; Moss, I.G. Supercooled Phase Transitions in the Very Early Universe. *Phys. Lett. B* **1982**, *1100*, 35–38. [CrossRef]
33. Lee, B.H.; Lee, W.; Yeom, D. Oscillating instantons as homogeneous tunneling channels. *Int. J. Mod. Phys. A* **2013**, *28*, 1350082. [CrossRef]
34. Hartle, J.B.; Hawking, S.W.; Hertog, T. The Classical Universes of the No-Boundary Quantum State. *Phys. Rev. D* **2008**, *77*, 123537. [CrossRef]
35. Hartle, J.B.; Hawking, S.W.; Hertog, T. No-Boundary Measure of the Universe. *Phys. Rev. Lett.* **2008**, *100*, 201301. [CrossRef] [PubMed]

36. Hwang, D.; Kim, S.A.; Lee, B.H.; Sahlmann, H.; Yeom, D. No-boundary measure and preference for large *e*-foldings in multi-field inflation. *Class. Quantum Gravit.* **2013**, *30*, 165016. [CrossRef]
37. Hwang, D.; Kim, S.A.; Yeom, D. No-boundary wave function for two-field inflation. *Class. Quantum Gravit.* **2015**, *32*, 115006. [CrossRef]
38. Chen, P.; Qiu, T.; Yeom, D. Phantom of the Hartle–Hawking instanton: Connecting inflation with dark energy. *Eur. Phys. J. C* **2016**, *76*, 91. [CrossRef]
39. Brahma, S.; Yeom, D. New solutions for the no-boundary solution in loop quantum cosmology. Unpublished work, 2019.
40. Hwang, D.; Sahlmann, H.; Yeom, D. The No-boundary measure in scalar-tensor gravity. *Class. Quantum Gravit.* **2012**, *29*, 095005. [CrossRef]
41. Hwang, D.; Lee, B.H.; Sahlmann, H.; Yeom, D. The no-boundary measure in string theory: Applications to moduli stabilization, flux compactification, and cosmic landscape. *Class. Quantum Gravit.* **2012**, *29*, 175001. [CrossRef]
42. Hwang, D.; Yeom, D. Toward inflation models compatible with the no-boundary proposal. *J. Cosmol. Astropart. Phys.* **2014**, *2014*, 007. [CrossRef]
43. Ashtekar, A.; Pawlowski, T.; Singh, P.; Vandersloot, K. Loop quantum cosmology of $k = 1$ FRW models. *Phys. Rev. D* **2007**, *75*, 024035. [CrossRef]
44. Bojowald, M. Dynamical initial conditions in quantum cosmology. *Phys. Rev. Lett.* **2001**, *87*, 121301. [CrossRef]
45. Bojowald, M.; Brahma, S. Loops rescue the no-boundary proposal. *arXiv* **2018**, arXiv:1810.09871.
46. Bojowald, M.; Brahma, S.; Yeom, D.H. Effective line elements and black-hole models in canonical loop quantum gravity. *Phys. Rev. D* **2018**, *98*, 046015. [CrossRef]
47. Bojowald, M.; Mielczarek, J. Some implications of signature-change in cosmological models of loop quantum gravity. *J. Cosmol. Astropart. Phys.* **2015**, *2015*, 052. [CrossRef]
48. Brahma, S. Spherically symmetric canonical quantum gravity. *Phys. Rev. D* **2015**, *91*, 124003. [CrossRef]

© 2019 by the authors. Licensee MDPI, Basel, Switzerland. This article is an open access article distributed under the terms and conditions of the Creative Commons Attribution (CC BY) license (http://creativecommons.org/licenses/by/4.0/).

Article

Rainbow-Like Black-Hole Metric from Loop Quantum Gravity

Iarley P. Lobo [1,*] and Michele Ronco [2]

[1] Departamento de Física, Universidade Federal da Paraíba, Caixa Postal 5008, João Pessoa 58059-900, PB, Brazil
[2] Dipartimento di Fisica, Università di Roma "La Sapienza", P.le A. Moro 2, 00185 Roma, Italy; michele.ronco@roma1.infn.it
* Correspondence: iarley_lobo@fisica.ufpb.br

Received: 18 October 2018; Accepted: 28 November 2018; Published: 1 December 2018

Abstract: Hypersurface deformation algebra consists of a fruitful approach to derive deformed solutions of general relativity based on symmetry considerations with quantum-gravity effects, of which the linearization has been recently demonstrated to be connected to the DSR program by κ-Poincaré symmetry. Based on this approach, we analyzed the solution derived for the interior of a black hole and we found similarities with the so-called rainbow metrics, like a momentum-dependence of the metric functions. Moreover, we derived an effective, time-dependent Planck length and compared different regularization schemes.

Keywords: quantum-gravity phenomenology; hypersurface deformation algebra; loop quantum gravity; black holes

1. Introduction

Despite tenacious and enduring efforts over many years of research, the dream of quantizing gravity is still far from being accomplished. Various attempts, which seemed particularly promising at their birth, were faced with insurmountable obstacles in the form of several formal complexities [1–7]. In light of this, a more pragmatic approach to the problem of quantum gravity (QG) consisted in looking for simplified (or, better to say, effective) models, able to encode a few characteristics of what we expect to be the theory of QG [3,8–11]. Of course, these models could not provide us with the "final theory", but may capture some key ingredients of QG, optimistically those that may allow us to perform experimental tests needed to guide our intuition as well as the construction of more reliable formal approaches to the problem. Typically, fully fledged QG approaches and more phenomenological models moved along parallel tracks. However, in the last few years, some steps to shorten the gap between these two complementary views have been taken.

Given the complexity and variety of the QG panorama, it is useful and common to divide different approaches in two broad categories: covariant and canonical approaches. The former class is based on the assumption of diffeomorphism invariance, and seems to leave no room for quantum deformations of it. On the other hand, the canonical procedure makes the covariance of general relativity (GR) less evident by construction [12–16] and, indeed, symmetries need to be checked directly by means of the calculation of the Poisson brackets between gravitational constraints. Interestingly, such a procedure has been recently proven to allow for modifications of GR covariance that preserve a certain symmetry structure in a deformed sense [17,18]. We feel this could be insightful for the construction of QG models, as well as the relations between different models and, hopefully, also for its phenomenological signatures.

In particular, the approach of canonical loop QG (LQG) [14–16], that counts remarkable accomplishments, such as singularity resolution in various cosmological and black-hole scenarios and a meaningful space discretization, faces major difficulties in finding quantum realization of the Hamiltonian, difficulties that so far remained unsolved [19–21]. Given that a number of recent analyses [22–25] have tried to circumvent such a problem by constructing canonical effective theories of QG, analogously to what is being done for modified theories of gravity in the study of dark matter or dark energy with several toy models. In this way, one can write modified gravitational constraints (i.e., Hamiltonian density and momenta) which take into account quantum corrections in the form of nonlinear modifications of phase-space variables inspired by the LQG quantization techniques. Remarkably, at least in symmetry-reduced LQG models, it has been shown that modified constraints still form a closed set of Poisson brackets, which is alternatively deformed with respect to the case of Arnowitt–Deser–Misner (ADM) GR [24,26–31]. These first studies, performed within the framework of effective LQG, have attracted renewed interest in the possibility of QG-induced symmetry deformations, and inspired further analyses in other approaches beyond general relativity, namely, the gravity sector of multifractional models [32], a certain class of (minimally) modified theories of gravity in the canonical formulation [9], and finally canonical noncommutative gravity with $\star$-product deformations of algebra [33]. Thus, this gave additional support to claim that QG may require a deformation of GR covariance.

Intriguingly, the possibility of symmetry deformations induced by quantum effects is not something new in the QG research, but rather a recurring idea that, from time to time, has taken different concrete forms in the literature. Most significantly, it is at the core of the class of models that goes under the name of deformed (or doubly) special relativity (DSR), where the Planck length, the characteristic scale of QG physics, is supposed to play the role of a relativistic invariant scale, analogous to the speed of light [34–38]. Concretely, such a proposal has been realized in the studies of noncommutative spacetime geometries where, as a consequence of spacetime noncommutativity, the special relativistic symmetries are modified by Planckian corrections and in some cases, most notably in the so-called κ-Minkowski geometry, which is the noncommutative spacetime dual to the κ-Poincaré algebra, M_P actually represents a relativistic invariant quantity [39,40]. For the purposes of this work, it is of particular importance that, in the Minkowski limit, LQG-deformed symmetries are consistent with the κ-Minkowski noncommutative spacetime as shown by one of us in Reference [41].

The main importance of the results on covariance derived in modified canonical models is that they may serve to bridge the gap between LQG and observable low-energy physics. In fact, some studies [42,43] have outlined how modified dispersion relations (MDR), i.e., Planck-scale corrections to the on-shell relation, and a reduction of dimensions at the Planck scale [44,45] can be derived from the modified brackets of gravitational constraints. Moreover, the analysis of Reference [46] suggested that the type of modifications introduced in the gravitational constraints affects directly the form of the MDR in such a way that future tests of Planck-scale departures from special relativistic symmetries could hopefully distinguish different theoretical scenarios in the not-too-distant future. Within the context of deformed covariance, another strategy to extract phenomenology could be the computation of effective metrics from the LQG-deformed constraint equations. Such an approach has already proved its richness in the case of loop quantum cosmology where effective Friedmann–Robertson–Walker (FRW) spacetimes allowed researchers to find robust solutions to the singularity problem as in bouncing cosmological models. Very recently, effective line elements for black-hole models have been derived by solving deformed Einstein-like equations implied by the deformed algebra of constraints [47,48]. This opens the way to the investigation of semiclassical black-hole solutions with LQG corrections.

In both cases, one has to address the issue of coarse-graining at larger scales (i.e., lower energies) the microscopic texture of the geometry, which at the Planck scale is described nonperturbatively by quantum operators and the associated states on a Hilbert space. It is worth stressing that a satisfactory definition of the (semi-)classical continuum limit has not been accomplished yet by working within the full complexity of LQG formalism. However, several encouraging results have been obtained in the context of symmetry-reduced models. In those cases, the problem of dynamics is greatly simplified and an analytic expression for the scalar constraint can be found. Then, semiclassical states are defined by peaking around classical trajectories, and it has been shown that these states exponentially dominate the partition function that sums over geometries [49]. Thus, effective models can be eventually considered in analogy with gauge theories defined on a discrete lattice, whereby constraint operators are regularized by some lattice parameter identified with (or close to) Planck length. As a consequence, the continuum limit is automatically obtained once such a regulator is removed.

Here, building on the results of Reference [48], we show that LQG modifications of the black-hole metric can be written as functions of the total radial momentum, thereby introducing an explicit dependence of the metric on Poincaré charges as proposed in the approach of rainbow gravity (RG) [35,36,50–53]. Such an observation we here put forward to strengthen the lacking synergy between fundamental approaches and phenomenological toy models that can be important in order to both improve our intuition about the formal structures required by QG theory and, at the same time, conceive experimental tests of potential Planck-scale effects.

The paper is organized as follows: in Section 2 we review basics properties of the algebra of gravitational constraints (or HDA) and its deformations from LQG corrections, as well as the modified black-hole solutions derived in Reference [48]. In Section 3 we further analyze this map in order to map the effective metric found in Reference [48] and rainbow metrics. In Section 4 we compare our results with the ones previously found in the literature. Then, we conclude in Section 5.

2. Hypersurface Deformation Algebra

In the last decade, one of the most interesting results in LQG has been the emergence of nonclassical spacetime structures from simplified analyses relying on effective field theory models for QG. These departures from smooth classical spacetime manifolds can be meaningfully traced back to quantum modifications of so-called hypersurface deformation algebra (HDA). In classical Hamiltonian GR, the HDA is given by the following set of Poisson brackets [12,13]:

$$\begin{aligned} \{D[M^a], D[N^a]\} &= D[\mathcal{L}_{\vec{M}} N^a], \\ \{D[N^a], H[M]\} &= H[\mathcal{L}_{\vec{N}} M], \\ \{H[M], H[N]\} &= D[h^{ab}(M\partial_b N - N\partial_b M)], \end{aligned} \qquad (1)$$

which encodes covariance in the Hamiltonian formulation of classical GR. Here, $D[M^a]$ is the momentum (or spatial) constraint that generates deformations along the three-dimensional hypersurfaces by an amount M^a (with $a = 1, 2, 3$), while $H[N]$ is the Hamiltonian (or time) constraint responsible for translations along the normal direction to these hypersurfaces; finally, h^{ab} are the components of the inverse three metrics. More precisely, given a generic phase-space function $f(h_{ij}, \pi^{ij})$, π^{ij} being the gravitational momentum conjugate to the metric, one has that:

$$\delta_{\vec{M}} f(h_{ij}, \pi^{ij}) = \{f(h_{ij}), D[\vec{M}]\}, \quad \delta_N f(h_{ij}) = \{f(h_{ij}, \pi^{ij}), H[N]\}. \qquad (2)$$

The search for a quantum version of gravitational constraints represents the main objective of the approach known as canonical quantum gravity and, so far, has not been conclusive. However, in spite of the fact that full quantum theory is not available (mainly due to renowned difficulties in the regularization of the Hamiltonian operator), consistency relations implied by the desire to preserve spacetime symmetries can be used to identify an effective formulation of LQG where a consistent set of closed Poisson brackets can be found by introducing restricted and simplified correction functions into the Hamiltonian, which are inspired by the LQG quantization technique (see e.g., Reference [54] and references therein).

Therefore, the first step consists of correcting the classical scalar and diffeomorphism constraints with possible modifications motivated by LQG. There is a certain degree of arbitrariness in the specific choice of these correction functions, greater than what is commonly acknowledged, and this will be discussed in some detail later on (see also Reference [46]). Nonetheless, we can fairly divide them into two broad classes: inverse triad and holonomy corrections. Here, we consider only the latter type of quantum (or, better to say, semiclassical) contributions. They can be motivated by the fact that holonomies of Ashtekar connections are the basic LQG variables and, at the effective level, can be taken into account by replacing the mean connection by a periodic function in the Hamiltonian constraint. The former are the inverse-triad corrections that come from terms in the Hamiltonian constraint that cannot be quantized directly, but only after being re-expressed as a Poisson bracket, a procedure that is usually referred to as the Thiemann trick for the quantization of $H[N]$ [55]. However, as we already said, they are not contemplated here. One then works with (modified) classical phase-space functionals that can be understood as the result of the evaluation of a distribution-valued operator over an orthonormal basis in terms of spin-network states that span the Hilbert space. These quantum corrections may or may not spoil the symmetry of classical theory under diffeomorphisms. Indeed, one has to prove that the quantum-corrected constraints form a closed algebra, thereby eliminating the same number of spurious degrees of freedom as in the classical theory given their role of generators of gauge transformations. This poses the issue of anomaly freedom, which is the focus of the so-called deformed-algebra approach to (effective) LQG [24,26–31,56,57]. The goal consists of introducing these effective quantum corrections into the classical gravitational constraints and then computing the Poisson brackets between them in order to check the compatibility with the symmetry under diffeomorphism. A closure of the HDA despite the presence of holonomy corrections would imply that symmetries are preserved, and it could be regarded as a strong hint that LQG is not anomalous. On the other hand, any kind of modifications to the brackets (1) could signal that diffeomorphism transformations are deformed due to "quantum" effects.

In particular, one starts from polymerizing the angular extrinsic curvature component:

$$K_\phi^2 \to h(K_\phi) = \frac{[\sin(\rho K_\phi)]^2}{\rho^2}, \tag{3}$$

where ρ is related to some scale, usually ℓ_P, as suggested, for instance, by the discrete spectrum of the area operator (ρ is proportional to the square root of the minimum eigenvalue, or the 'area gap' from LQG) or on the size of the loop considered for the definition of holonomies. Clearly, the classical regime is recovered in the limit $\rho \longrightarrow 0$.[1] The above Substitution (3) can be justified as follows. In quantum theory, there is no well-defined operator corresponding to the Ashtekar–Barbero connection A_a^i on the LQG kinematical Hilbert space. Instead, in the loop representation, a well-defined object is the holonomy operator that is defined as parallel transport of the connection:

$$h_\alpha(A) = \mathcal{P} \exp\left(\int_\alpha \dot{e}^a A_a^i \tau_i \right), \tag{4}$$

where $\mathcal{P}$ is the path-ordering operator and $\dot{e}^a$ is the three-vector tangent to the curve α. For our analysis, of particular interest are the holonomies of connections along homogeneous directions, which simplify as [25]:

$$h_j(A) = \exp(\mu A \tau_j) = \cos(\mu A)\mathbb{I} + \sin(\mu A)\sigma_j \tag{5}$$

and do not require a spatial integration, since they transform as scalars. In fact, so far, one knows only how to implement (local) holonomy corrections for connections along homogeneous directions

---

[1] The fact that zero does not belong to the spectrum of the area operator in LQG is precisely the input from the full theory which gives a nontrivial quantum geometrical effect.

(for a negative result concerning implementation of nonlocal (extended) holonomy corrections in spherical symmetry, see Reference [58]). In our case, this is given by $\gamma K_\phi \ (= A_\phi \cos \alpha)$:

$$\begin{aligned} h_\phi(r,\mu) &= \exp(\mu A_\phi \cos \alpha \Lambda_\phi^A) \\ &= \cos(\mu \gamma K_\phi)\mathbb{I} + \sin(\mu \gamma K_\phi)\Lambda \end{aligned} \qquad (6)$$

In order to see how Replacement (3) is implied by Equation (6) one must take into account that the scalar constraint is quantized by utilizing the Thiemann trick $\sqrt{E^r} \propto \{K_\phi, V\}$ (where V is the volume), whose quantum version contains the commutator $h_\phi[h_\phi^{-1}, \hat{V}] = h_\phi h_\phi^{-1}\hat{V} - \hat{V}h_\phi^{-1}\hat{V}h_\phi$. (This is equivalent to regularizing the curvature of the connection by holonomies, with the minimum area being the 'area gap' from LQG.) Using Equation (6) one can easily see that products of holonomies are given by cosine and sine functions of K_ϕ. Finally, it turns out that the resulting quantum or 'effective' (since we are going to ignore operator ordering issues by working in a semiclassical setting, as they are not crucial to our goals) scalar constraint could be obtained simply making the replacement of Equation (3). This justifies the following form of effective Hamiltonian constraint H^Q:

$$H^Q[N] = -\frac{1}{2G}\int_B drN\left[\frac{[\sin(K_\phi\rho)]^2}{\rho^2}E^\phi + 2K_r\frac{\sin(K_\phi\rho)}{\rho}E^r + (1-\Gamma_\phi^2)E^\phi + 2\Gamma_\phi' E^r\right]. \qquad (7)$$

On the other hand, the diffeomorphism constraint remains undeformed, since spatial diffeomorphism invariance translates into vertex-position independence in LQG, which is implemented directly at the kinematical level by unitary operators generating finite transformations[2]. As aforementioned, the crucial point of the deformed algebra approach is to ensure that the resulting algebra of constraints remains consistent so that Poisson brackets between quantum-corrected constraints are proportional to a quantum-corrected constraint. Such a procedure has to be performed "off-shell", i.e., before the quantum-corrected equations have been solved. In the case of gravity, this is the only way to guarantee that quantum theory is fully consistent. With a rather straightforward but lengthy calculation, one can show that gravitational constraints with LQG corrections close the algebra nonpertubatively. Particularly remarkable is the fact that, at least for symmetry-reduced cases, there is a unique solution to the anomaly freedom problem. In fact, the full deformed HDA is given by [24,26–29]:

$$\begin{aligned} \{D[M^a], D[N^a]\} &= D[\mathcal{L}_{\vec{M}}N^a], \\ \{D[N^a], H^Q[M]\} &= H^Q[\mathcal{L}_{\vec{N}}M], \\ \{H^Q[M], H^Q[N]\} &= D[\beta h^{ab}(M\partial_b N - N\partial_b M)]. \end{aligned} \qquad (8)$$

Thus, these modifications amount to a deformation of the brackets closed by the gravitational constraints that generate space and time gauge transformations. Specifically, only the Poisson bracket involving two Hamiltionain constraints is modified by the presence of a deformation function that depends on the phase-space variables, i.e., $\beta = \beta(h_{ij}, \pi^{ij})$ (or, equally, $\beta = \beta(A_i^a, E_b^j)$), whose particular form depends on the specific holonomy corrections considered, as well as on the symmetry reductions implemented and so forth.

The angular component of extrinsic curvature K_ϕ can be consistently quantized and produces the above result. To see that, we have to briefly introduce the spherically symmetric reduction of Hamiltonian gravity in Ashtekar–Barbero variables (see e.g., Reference [60]) in the presence of LQG deformations. In this case, ADM foliation [13] allows a decomposition of the spacetime manifold

---

[2] In fact, there is no well-defined infinitesimal quantum diffeomorphism constraint in LQG for the basis spin network states. Some progress in constructing it has been achieved in Reference [59].

as $\mathcal{M} = \mathbb{R} \times \Sigma = \mathcal{M}_{1+1} \times S^2$, where $\mathcal{M}_{1+1}$ is a two-dimensional manifold spanned by (t,r) and S^2 stands for the two-sphere. Given that, the line element reads:

$$ds^2 = -N^2 dt^2 + h_{rr}(dr + N^r dt)^2 + h_{\theta\theta}[d\theta^2 + (\sin(\theta))^2 d\phi^2], \tag{9}$$

where the shift vector is purely radial, i.e., $N^i = (N^r, 0, 0)$, due to spherical symmetry, and, consequently, we are left only with radial diffeomorphisms generated by $D[N^r] = \int dr N^r \mathcal{H}_r$ (where $\mathcal{H}_r$ is the only nonvanishing component of the momentum density) and time transformations, generated by $H[N] = \int dr N \mathcal{H}$ (where $\mathcal{H}$ is the Hamiltonian density). The components of the spatial metric $(h_{rr}, h_{\theta\theta})$ can be written in terms of rotationally invariant densitized triads that are given by:

$$E = E^a_i \tau^i \frac{\partial}{\partial x^a} = E^r(r)\tau_3 \sin\theta \frac{\partial}{\partial r} + E^\phi(r)\tau_1 \sin\theta \frac{\partial}{\partial \theta} + E^\phi(r)\tau_2 \frac{\partial}{\partial \phi}, \tag{10}$$

where $\tau_j = -\frac{1}{2} i \sigma_j$ represent $SU(2)$ generators. The densitized triads are canonically conjugate to the extrinsic curvature components, which, in the presence of spherical symmetry, are conveniently described as follows:

$$K = K^i_a \tau_i dx^a = K_r(r)\tau_3 dr + K_\phi(r)\tau_1 d\theta + K_\phi(r)\tau_2 \sin\theta d\phi. \tag{11}$$

As a result, the components of the three metrics are:

$$h_{\theta\theta} = E^r(r), \quad h_{rr} = \frac{(E^\phi(r))^2}{E^r(r)}. \tag{12}$$

At this point, one can show that the bracket $\{H^Q[N], H^Q[M]\}$ in Equation (8) reads:

$$\{H^Q[N], H^Q[M]\} = D[\beta(\rho K_\phi) \frac{E^r}{(E^\phi)^2}(N\partial_r M - M\partial_r N)], \tag{13}$$

where β is related to the second derivative of the holonomy-correction function, i.e., $\beta = h''/2$. In particular, for the simplest case including only local holonomy corrections as in Equation (3) (see also References [61,62] for a detailed construction and the related discussion), with $\gamma \in \mathbb{R}$ and $j = 1/2$, deformation β takes the form:

$$h = \frac{[\sin(\rho K_\phi)]^2}{\rho^2} \implies \beta = \cos(2\rho K_\phi). \tag{14}$$

However, more complicated expressions are possible and are discussed in the next section. As shown explicitly in Reference [48], given the modified HDA, one can then obtain Einstein-like equations of motion with LQG corrections from

$$\dot{F} = \{F, H^Q[N] + D[M^r]\}, \tag{15}$$

with $F = (E^r, E^\phi, K_\phi, K_r)$. For instance, the equations of motion for the two independent triads, extrinsic curvature K_ϕ, and the Hamiltonian constraint (that can be used to find K_r) are:

$$\dot{E}^r = N\sqrt{E^r} h'(K_\phi) + M^r \partial_r E^r,$$
$$\dot{E}^\phi = \frac{N}{2}\left(\sqrt{E^r} K_r h''(K_\phi) + \frac{E^\phi}{\sqrt{E^r}} h'(K_\phi)\right) + \partial_r(M^r E^\phi),$$
$$\dot{K}_\phi = -\frac{N}{2\sqrt{E^r}}[1 + f(K_\phi)], \tag{16}$$
$$h'(K_\phi) E^r K_r + (1 + h(K_\phi)) E^\phi = 0.$$

In Reference [48], the above LQG-corrected Einstein equations have been solved explicitly for the interior of a static black hole. The solutions for the triads read:

$$E^r = t^2, \quad E^\phi = \frac{r_S}{2} \frac{h'(K_\phi)}{1 + h(K_\phi)}, \tag{17}$$

and that of the extrinsic curvature K_ϕ is:

$$h(K_\phi) = \frac{r_S}{t} - 1. \tag{18}$$

where r_S is the Schwarzschild radius.[3] Finally, the LQG-modified line element is:

$$ds^2 = -\frac{1}{F(t)} dt^2 + F(t) dr^2 + t^2 d\Omega^2, \tag{19}$$

with

$$F(t) = \left(2 \frac{dh^{-1}}{dx}\bigg|_{x=\frac{r_S}{t}-1}\right)^{-2}. \tag{20}$$

3. Effective Rainbow Metric

In general, one has as the solution a deformed metric that depends on the spacetime co-ordinates and on deformation parameter ρ. However, recently, deformation function $h(K_\phi)$ gained a different role. It was shown that such a function, in fact, deforms the Lorentz algebra of the spacetime found in the flat version of the HDA described above (see Reference [41] and references therein. See also Reference [42] for a different analysis leading to similar outcomes, i.e., deformed Poincaré symmetries in the Minkowski limit of (8)).

For our purposes, it is of pivotal importance to find a way to write β in terms of symmetry generators (see also References [41–43]) and, to this end, it is valuable to notice that observables of the Brown–York momentum [63],

$$P = 2 \int_{\partial \Sigma} d^2z\, v_b (n_a \pi^{ab} - \bar{n}_a \bar{\pi}^{ab}), \tag{21}$$

can be identified by extrinsic curvature components. In Equation (21), we have that $v_a = \partial/\partial x^a$, n_a is the conormal of the boundary of spatial region Σ, and π^{ab} plays the role of gravitational momentum (while the overbarred symbols in the above equation are the same functions but evaluated at the boundary). From this, it is possible to establish that the radial Brown–York momentum P_r is related to extrinsic curvature component K_ϕ in the following way (see, e.g., Reference [43])

$$P_r = -\frac{K_\phi}{\sqrt{|E^r|}}. \tag{22}$$

The flat case was discussed in Reference [41] in the context of DSR symmetries. In that case, since E^r is a constant, it was possible to set parameter $\rho \propto |E^r|^{-1/2}$, which allows to relate deformation function β to the generator of radial translations P_r:

$$\beta = \cos(\lambda P_r), \tag{23}$$

---

[3] We omitted the solution for K_r because it is not used in our analysis.

where λ is a parameter of the order of the Planck length ($\lambda \sim \ell_P \sim 1/M_P$).[4] Such identification allowed the authors to derive deformed relations for the symmetry generators of the flat spacetime.

That is, on one hand, this approach traces a map between DSR and the Minkowski limit of the HDA from the point of view of deformed symmetries. On the other hand, an exact solution was recently found of the field equations derived from a HDA for the curved case of the black-hole interior. In principle, these two approaches are independent, i.e., there is not yet a local DSR description of the symmetries of the deformed metric, or a metric description that emerges from this DSR proposal.

A metric description that is able to encode these aspects of the formalism is still unknown. In this paper, we aim to contribute to this subject by describing the curved effective metric in the light of the discovered relation between HDA and DSR. In fact, a relevant approach to the metric description inspired by the DSR scenario conjectures that the spacetime metric determined by an observer by measurements done with an energetic particle depends on the particle's energy as measured by that observer. The deformed relativistic metric description should be given in terms of a rainbow metric [50]. Therefore, we wanted to study whether the intuition of rainbow gravity finds support in this recently found effective curved metric from HDA, when one uses the prescription that relates HDA and DSR.

3.1. Rainbow Gravity

In this subsection, we review the main aspects of the standard rainbow gravity as proposed in Reference [50]. In this case, consider an MDR of the type:[5]

$$m^2 = E^2 f_1^2(\ell_P E) - p^2 f_2^2(\ell_P E), \tag{24}$$

that can be represented by a simple norm $m^2 = \eta^{\mu\nu} U[p]_\mu U[p]_\nu$, where U is the map in momentum space

$$U[p]_\mu = (U[p]_0, U[p]_i) = (E f_1(\ell_P E), p_i f_2(\ell_P E)), \tag{25}$$

where Greek indices like (μ, ν), run from $0, ..., 4$, and Latin indices like (i, j), run from $1, ..., 3$.

The idea is to write this dispersion relation with an energy-dependent metric $\tilde{\eta}^{\mu\nu}(\ell_P E)$, such that $\eta^{\mu\nu} U[p]_\mu U[p]_\nu = \tilde{\eta}^{\mu\nu}(\ell_P E) p_\mu p_\nu$, which could also be generalized for a curved spacetime. A simple way for achieving this consists in transforming the orthonormal frame as $\tilde{e}_A^\mu = (f_1(\ell_P E) e_0^\mu, f_2(\ell_P E) e_I^\mu)$, such that

$$\eta^{\mu\nu} U[p]_\mu U[p]_\nu = \eta^{AB} \tilde{e}_A^\mu \tilde{e}_B^\nu p_\mu p_\nu, \tag{26}$$

which defines an energy-dependent metric $\tilde{\eta}^{\mu\nu}(\ell_P E) = \eta^{AB} \tilde{e}_A^\mu \tilde{e}_B^\nu$. Here, indices like (A, B) run from $0, ..., 4$ and ones like (I, J) run from $1, ..., 3$.

This construction can be directly generalized to curved vielbeins. In fact, if one uses the same definition above, it is possible to construct a metric

$$\tilde{g}^{\mu\nu}(\ell_P E) = \eta^{AB} \tilde{e}_A^\mu \tilde{e}_B^\nu, \tag{27}$$

whose inverse is given by

$$\tilde{g}_{\mu\nu}(\ell_P E) = \eta_{AB} \tilde{e}^A_\mu \tilde{e}^B_\nu, \tag{28}$$

where $\tilde{e}^A_\mu = \left((f_1(\ell_P E))^{-1} e^0_\mu, (f_2(\ell_P E))^{-1} e^I_\mu\right)$: this is a rainbow metric.[6] This way, one can use this kind of metric as an input into the Einstein equations as an ansatz for the so-called rainbow gravity.

---

[4] Keep in mind that its exact value also depends on quantization ambiguities [46].
[5] We are considering $c = \hbar = 1$, which implies having the Planck length as the inverse of the Planck energy $\ell_P = E_P^{-1}$.
[6] Energy-momentum dependent metrics, like in curved momentum space have been originally considered in [64].

For instance, a known solution [50] of the Einstein equation for the static and spherically symmetric case is the metric

$$ds^2 = -\frac{1-2M/r}{(f_1(\ell_P E))^2}dt^2 + \frac{(1-2M/r)^{-1}}{(f_2(\ell_P E))^2}dr^2 + \frac{r^2}{(f_2(\ell_P E))^2}d\Omega^2. \tag{29}$$

So, this is a deformation of the Schwarzschild line element by functions that depend on the energy of the particles that probe such spacetime. Since this is a static spacetime, the energy of a test particle is, in fact, a conserved quantity and corresponds to the generator of time translations in this manifold, thus implying that the Schwarzschild metric is being essentially deformed by the time-translation generator.

When crossing the horizon, the roles of the radial and the time co-ordinates change. Such modification takes the metric from a static configuration to a purely time-dependent tensor, which also implies that the energy acquires the role of the conserved radial momentum, i.e., the generator of radial translations. In fact, the metric assumes the form

$$ds^2 = -\frac{(2M/t-1)^{-1}}{(f_2(\ell_P P_r))^2}dt^2 + \frac{2M/t-1}{(f_1(\ell_P P_r))^2}dr^2 + \frac{t^2}{(f_2(\ell_P P_r))^2}d\Omega^2. \tag{30}$$

In the next section, we compare this rainbow metric inside the event horizon of a black hole with the one found from the HDA.

3.2. Momentum-Dependent Metric

Using Equation (18), we can write Equation (20) as

$$F(t) = \left[2\frac{dK_\phi(r_s/t-1)}{d(r_s/t-1)}\right]^{-2}. \tag{31}$$

However, if we define $G(r_s/t-1) \doteq dK_\phi(r_s/t-1)/d(r_s/t-1)$, which, using Equation (18), allows us to define K_ϕ-dependent function $\widehat{G}(K_\phi) \doteq G \circ h(K_\phi)$.

Recalling the relation between the extrinsic curvature and the radial momentum (22), we are able to define a metric that presents P_r-dependent corrections. It should be stressed that, in this case, P_r corresponds to the quasilocal radial gravitational momentum, which means that it presents the information of the test particle in this spacetime (as described in References [41,65] for deforming the Poincaré symmetry where there is no gravitational field) and of the gravitational interaction (that was absent in the flat case).

To illustrate this construction, let us consider some examples.

3.2.1. First Case

The most natural choice to begin our analysis is the one exemplified in Reference [48]. In this case

$$h(K_\phi) = \frac{[\sin(\rho K_\phi)]^2}{\rho^2} = \frac{r_s}{t} - 1 \tag{32}$$

with

$$\beta(K_\phi) = h''(K_\phi)/2 = \cos(2\rho K_\phi). \tag{33}$$

From Equation (31), we can derive the co-ordinate dependence of the metric function $F(t)$ that was found in Reference [48]:

$$F(t) = \left(\frac{r_s}{t}-1\right)\left[1-\rho^2\left(\frac{r_s}{t}-1\right)\right]. \tag{34}$$

However, in order to analyze this effective metric in light of rainbow gravity, we propose to take a step back and realize that, in fact, the Schwarzschild metric gets deformed due to parameter ρ, and that such deformation is proportional to $r_s/t - 1$, which, on the other hand, equals to function $h(K_\phi)$ (by Equation (18)), which, in turn, is related to radial momentum P_r by (22). Combining these expressions, we have:

$$F(t, P_r) = \left(\frac{r_s}{t} - 1\right) [\cos(\rho t \, P_r)]^2, \tag{35}$$

which implies in a rainbow-like metric:

$$ds^2 = -\left(\frac{r_s}{t} - 1\right)^{-1} [\cos(\rho t \, P_r)]^{-2} dt^2 + \left(\frac{r_s}{t} - 1\right) [\cos(\rho t \, P_r)]^2 dr^2 + t^2 d\Omega^2. \tag{36}$$

In this case, the second horizon occurs in the phase space for $\rho t \, P_r = (2n+1)\pi/2$, which corresponds to $t_h = \rho^2 r_s/(1+\rho^2)$. In fact, this metric presents the same Penrose diagram, as pointed out in Reference [48]. According to References [47,48], due to the deformation of the Hamiltonian constraint in Equation (8), the time reparametrization of the theory also needs to be modified, leading to a rescaling of lapse function N in Equation (9) as:

$$N \to \beta(K_\phi) N = \cos(2\rho t P_r) N, \tag{37}$$

which leads to a Euclideanization of the metric for $\rho t P_r = (2n+1)\pi/4$. For details, see Reference [47].

This rainbow metric presents some differences with respect to the usual approach presented before (Equation (30)). For instance, there is no rainbow function in the angular sector of the line element; this rainbow metric presents contributions from the single-particle momentum and from the gravity sector; and the momentum P_r is multiplied by ρt, instead of the usual Planck length ℓ_P. Such features are repeated in the next examples.

Indeed, we noted that holonomy corrections can be implemented in different ways. Specifically, the polymerization function (i.e., $K \mapsto f(K)$) depends on some choices we can make such as: the value of the Barbero–Immirzi parameter, the internal gauge group, and finally the spin representation of the group. How these choices affect symmetry deformation in Equation (8) and, perhaps, lead to different phenomenological predictions for the form of the MDR, has been recently discussed [46]. Here, following that line of reasoning, we briefly discuss how, as the reader could easily expect, these formal ambiguities affect the shape of these effective rainbow metrics, too.

3.2.2. Second Case

A second, rather natural, choice is represented by the complex Ashtekar variables that, once we turn to the associated effective quantum corrections, give rise to a similar deformation function through a sort of "Wick rotation" $\rho \mapsto i\rho$ (see Reference [43]) of the standard $SU(2)$ polymerization Function (3), i.e.,

$$h(K_\phi) = \rho^{-2} [\sinh(\rho K_\phi)]^2, \tag{38}$$

producing the deformed rainbow metric

$$ds^2 = -\left(\frac{r_s}{t} - 1\right)^{-1} [\cosh(\rho t \, P_r)]^{-2} dt^2 + \left(\frac{r_s}{t} - 1\right) [\cosh(\rho t \, P_r)]^2 dr^2 + t^2 d\Omega^2. \tag{39}$$

In this case, since the hyperbolic cosine is never null, there is just the usual horizon for this black hole.

3.2.3. Third Case

Although complex connection formulations of LQG are receiving restored attention in the recent literature, it is well known that they also raise major difficulties (for instance, in the analysis of the observables of the theory, which need to be real valued operators) for which nobody has been able to fully and satisfactorily account. Partial progress is given by the proposal of an "analytic continuation" procedure (see e.g., Reference [66]), which has the advantage of preserving the reality of the spectrum of the area operator. We direct interested readers to Reference [66]. From References [43,67], we have

$$h(K_\phi) = -\frac{[\sinh(\rho K_\phi)]^2}{\rho^2} \frac{3}{s(s^2+1)\sinh(\theta_\phi)} \frac{\partial}{\partial \theta_\phi}\left(\frac{\sin(s\theta_\phi)}{\sinh(\theta_\phi)}\right), \tag{40}$$

where

$$\sinh\left(\frac{\theta_\phi}{2}\right) = \left[\sinh\left(\frac{\rho K_\phi}{2}\right)\right]^2. \tag{41}$$

Leading to

$$ds^2 \approx -\left(\frac{r_s}{t}-1\right)^{-1}\left[1+(\rho t)^2 P_r^2 - \frac{(3s^2+4)}{24}(\rho t)^4 P_r^4\right]^{-1} dt^2 \tag{42}$$

$$+\left(\frac{r_s}{t}-1\right)\left[1+(\rho t)^2 P_r^2 - \frac{(3s^2+4)}{24}(\rho t)^4 P_r^4\right] dr^2 + t^2 d\Omega^2.$$

Until the first order, this result coincides with the second case of Section 3.2.2, which is coherent with results reported in Reference [43] in the flat case. We considered just the second-order deformation due to the complexity of this deformation function.

3.2.4. Fourth Case

Another possibility is represented by higher spin representations of the internal $SU(2)$ group. For instance, in this quantization approach to effective LQG, from Reference [46], i.e., $j = 1$ HR (holonomy regularization) scheme for regularization, one has:

$$\beta(K_\phi) = [\cos(\rho K_\phi)]^3 - [\sin(\rho K_\phi)]^4 - \frac{7}{4}\sin(\rho K_\phi)\sin(2\rho K_\phi) + \frac{3}{4}[\sin(2\rho K_\phi)]^2, \tag{43}$$

where $\beta(K_\phi) = h''(K_\phi)/2$. Then

$$ds^2 \approx -\left(\frac{r_s}{t}-1\right)^{-1}\left[1-(\rho t)^2 P_r^2 - \frac{7}{24}(\rho t)^4 P_r^4\right]^{-1} dt^2 \tag{44}$$

$$+\left(\frac{r_s}{t}-1\right)\left[1-(\rho t)^2 P_r^2 - \frac{7}{24}(\rho t)^4 P_r^4\right] dr^2 + t^2 d\Omega^2.$$

We also considered just a second-order approximation, which was sufficient for our discussions.

3.2.5. Fifth Case

Now, we consider the case $j = 1$, but in the connection regularization (CR) scheme. In this case, following Reference [46], we have

$$\beta(K_\phi) = [\cos(\rho K_\phi)]^4 - [\sin(\rho K_\phi)]^4 - \frac{3}{2}[\sin(2\rho K_\phi)]^2. \tag{45}$$

Following the same procedures as the previous cases, we are led to the following line element:

$$ds^2 \approx -\left(\frac{r_s}{t}-1\right)^{-1}\left[1-4(\rho t)^2 P_r^2 + \frac{16}{3}(\rho t)^4 P_r^4\right]^{-1} dt^2 \tag{46}$$

$$+\left(\frac{r_s}{t}-1\right)\left[1-4(\rho t)^2 P_r^2 + \frac{16}{3}(\rho t)^4 P_r^4\right] dr^2 + t^2 d\Omega^2.$$

4. Comparison with Previous Definitions

Now that we have discussed these three cases motivated by three different ways to introduce LQG-inspired corrections into Hamiltonian GR, we recognize that, in general, we obtain metric deformations of the type

$$F(t) \approx \left(\frac{r_s}{t} - 1\right) [1 + \tilde{\zeta}_1 (\rho t)^2 P_r^2 + \tilde{\zeta}_2 (\rho t)^4 P_r^4 + \mathcal{O}(\rho t)^6], \tag{47}$$

where $\tilde{\zeta}_i$ are real numbers. Obviously, the first and second cases could be exactly solved, but they also match this form by performing a Taylor expansion. Therefore, we can characterize each of the solutions derived from the HDA by parameters $\tilde{\zeta}_i$, as can be seen in Table 1 for different j-representations:

Table 1. j-representations and their $\tilde{\zeta}_i$-parameters.

j	$\tilde{\zeta}_1$	$\tilde{\zeta}_2$
$1/2$	-1	$1/3$
$\sim i/2$	1	$1/3$
$\frac{1}{2}(-1 + is)$	1	$-\frac{3s^2 + 4}{24}$
1 (HR)	-1	$-\frac{7}{24}$
1 (CR)	-4	$\frac{16}{3}$

The cases analyzed so far do not present deformations as odd functions; therefore, terms with odd powers of tP_r cannot appear, i.e., the first-order correction appears quadratically, the second-order correction appears in the fourth power, and so on. This could be the consequence of some symmetry principle underlying the LQG construction that, for instance, would preserve the parity of the MDR under transformation $P_r \mapsto -P_r$.

Originally, rainbow gravity was introduced by the so-called rainbow functions of the particle's energy $f_{1,2}(\ell_P E)$, which, in the case of the Schwarzschild metric inside the black hole, reads as Equation (29). There are some fundamental differences with respect to our case:

- Usual rainbow function f_2 deforms the angular sector of the metric, i.e., the line element of unit sphere $\mathbb{S}^2$ is momentum-dependent in usual rainbow gravity;
- our deforming function $F(t, P_r)$ depends on $\rho t P_r$ instead of the usual $\ell_P P_r$;
- momentum P_r consists of the momentum of the single test particle and the momentum of the gravitational field.

The last two points deserve further discussion. Regarding the second point, we are led to speculate whether the rainbow metric inspired by the HDA is deformed by an effective Planck length given by[7]

$$\ell_P^{\text{eff}} = \rho t. \tag{48}$$

This is being generated by the presence of a deformation function on the brackets (Equation (8)). Therefore, a possible direction that we could investigate consists of searching for a representation of scalar-tensor theories in rainbow gravity, where Newton's constant is a scalar field, which would induce a variable Planck length, comparable to what we found in the present paper.

As a matter of fact, if fundamental constants like $\hbar$, G, and c are functions of spacetime co-ordinates, this behavior could be explained, as long as

$$\ell_P(t) = \sqrt{\frac{\hbar(t) G(t)}{c^3(t)}} = \rho t. \tag{49}$$

---

[7] In the present case, co-ordinate time must satisfy $t < r_s$.

This is an important difference with respect to previous approaches that build bridges between energy-momentum-dependent metrics and quantum gravity, since, for the first time, we see a deformation "parameter" that is co-ordinate-dependent in this particular context. These possible phenomenological possibilities deserve further investigation.

Such dependence relies on the relation between radial momentum and the extrinsic curvature given by Equation (22). For the flat case, there is no co-ordinate dependence since the triad E^r is constant, which explains why this feature did not appear in previous analysis of the HDA and DSR, like Reference [41]. In that case, the term ρt is replaced by a dimensionful parameter λ of the order of the Planck length.

In this regard, we notice that the possibility of scale-dependence of the characteristic regime at which we should expect QG effects to be relevant is not something new in the literature. Indeed, some kind of running of Planck-scale physics is at the cornerstones of many approaches to the QG problem. Among them, we can count causal dynamical triangulation [68], asymptotic safety [69], and multifractional geometries [70]. In particular, working within this latter approach, one of us [71,72] found that the multifractional scale (i.e., the ultraviolet scale at which the spacetime dimension changes, as it happens by construction in multifractional geometries), ℓ_*, is related to the scale of the observation at which the measurement is being performed, s, i.e., $\ell_* = \ell_P^2/s$. Within a completely different scenario and framework, here we obtained a similar outcome. Such an interesting suggestion could be worth exploring elsewhere.

The third point, by itself, also deserves a deeper investigation about whether it is possible to uncouple momenta contributions coming from the gravity and test particle sectors, in order to approximate this new effective metric to the usual one from rainbow gravity, probably similarly to what was done in Reference [73] in the context of Palatini $f(R,Q)$ gravity (where R is the usual Ricci scalar and $Q = R^{\mu\nu} R_{\mu\nu}$).

We close this section with a remark concerning how one could coherently make contact with the aforementioned Minkowski limit of the deformed HDA. Since the original Schwarzschild metric already violates Lorentz invariance, in our approach we do not need to consider deformations of the Lorentz symmetry. We would need to be concerned about this issue if we had a Minkowski limit of this metric. However, following the procedures of Reference [48], we cannot simply place $r_s = 0$, because function $h(K_\phi) = r_s/t - 1$ should be a positive definite function; hence, the no-gravity limit needs to be carefully treated, in order to work on the effective spacetime symmetries of this metric description. However, this will be the subject for future investigations.

5. Final Remarks

Based on the recently found black-hole solution inside the event horizon from deformations of GR due to quantum gravitational corrections [48], and on the link between the hypersurface deformation algebra and deformed Poincaré algebra in the flat limit [41], we connected these two perspectives of the same problem using the so-called rainbow metrics. In the present case, we found a metric description for the solutions found in Reference [48] based on the relation between the radial triad, the extrinsic curvature and the radial momentum given by $P_r = -K_\phi/\sqrt{|E^r|}$, which is on the very basis of the linearization of the HDA in terms of DSR symmetries. Such a metric assumes the form of a rainbow metric, in the sense that it depends on spacetime co-ordinates and on momentum P_r.

We analyzed different realizations of this quantization scheme and realized that a pattern emerged for the general form of the rainbow metric. We only have even functions of dimensionless quantity $\rho t \, P_r$, that we expanded in a Taylor series and collected the first two terms in Table 1.

Important differences with respect to the usual rainbow metric ansatz were found, like the absence of a rainbow function in the line element of sphere $\mathbb{S}^2$. The presence of a variable, effective Planck length that governs the deformation $\ell_P^{\text{eff}} = \rho t$, which is a novelty in attempts to find rainbow metrics from quantum-gravity considerations, and the dependence of the metric on the gravitational and single particle momenta.

We should stress that, although effective metrics can already be found from the solution of Reference [48], we here showed a new ansatz of rainbow metrics, given by Equations (19) and (47), inspired by this approach. Alternative formulations of the rainbow-gravity initial proposal have been proposed [74–82] and the issue is still under debate. Since the exact form of the semiclassical spacetime description from quantum gravity is not yet known, we should rely on phenomenological possibilities driven by deformation functions, like the HDA approach or rainbow-gravity models.

Another key issue on rainbow-gravity and quantum-gravity phenomenology, in general, concerns the deformed trajectories of test particles, i.e., the geodesics of a quantum spacetime. Following an approach similar to ours, some efforts have been pushed forward in Reference [83], and MDRs in flat spacetime have been considered in References [41,43,46], which could, on principle, allow us to find trajectories from the Hamilton equations. However, for our purposes, it is of pivotal importance to find exterior or near-horizon metric solutions in order to check deviations of the geodesic equations from GR in the direction of confronting our findings with observations and with the near-horizon phenomenology that has been recently developed (see, for instance, Reference [84]).

As discussed in References [41,65], the deformation of the hypersurface algebra induces a deformation of the Poincaré algebra. On the other hand, we found that the effective metric description found in References [47,48] resembles rainbow metrics, which are historically related to the DSR program. Therefore, we wonder whether our approach could be useful for discovering an effective metric description of the DSR algebraic formalism, such that trajectories found from deformed Hamilton equations are geodesics, and the deformed symmetries are generated by Killing vectors of the metric.

Coherently passing from the "gravity-on" to the "gravity-off" geometric description while preserving the aforementioned structures would be an important step toward a "quantum equivalence principle", in which the relations between the geometrical quantities in such emergent spacetime are preserved even when considering quantum corrections.

In the future, we intend to explore this metric no-gravity limit in order to find a coherent relativistic metric description of DSR, and to better understand the transition from curved to flat metrics in this semiclassical approach.

Author Contributions: All authors contributed equally to this work.

Funding: This article is based upon work from COST Action CA15117, supported by COST (European Cooperation in Science and Technology). This study was financed in part by the Coordenação de Aperfeiçoamento de Pessoal de Nível Superior—Brasil (CAPES)—Finance Code 001.

Acknowledgments: The authors thank Suddhasattwa Brahma for reading a preliminary version of the manuscript and his useful comments.

Conflicts of Interest: The authors declare no conflict of interest.

Abbreviations

The following abbreviations are used in this manuscript:

QG	Quantum Gravity
GR	General Relativity
LQG	Loop Quantum Gravity
ADM	Arnowitt–Deser–Misner
DSR	Doubly (deformed) Special Relativity
MDR	Modified Dispersion Relation
FRW	Friedmann–Robertson–Walker
RG	Rainbow Gravity
HDA	Hypersurface Deformation Algebra
CR	Connection Regularization
HR	Holonomy Regularization

References

1. Oriti, D. (Ed.) *Approaches to Quantum Gravity*; Cambridge University Press: Cambridge, UK, 2009.
2. Smolin, L. What are we missing in our search for quantum gravity? *arXiv* **2017**, arXiv:1705.09208.
3. Amelino-Camelia, G. Quantum-Spacetime Phenomenology. *Living Rev. Relativ.* **2013**, *16*, 5. [CrossRef] [PubMed]
4. Rovelli, C. Loop Quantum Gravity. *Living Rev. Relativ.* **1998**, *1*, 1. [CrossRef] [PubMed]
5. Alvarez, E. *Planck Scale Effects in Astrophysics and Cosmology*; Lecture Notes in Physics; Kowalski-Glikman, J., Amelino-Camelia, G., Eds.; Springer: Berlin/Heidelberg, Germany, 2005; Volume 669, pp. 31–58. ISBN 978-3-540-25263-4.
6. Nicolai, H.; Peeters, K.; Zamaklar, M. Loop quantum gravity: An outside view. *Class. Quantum Gravity* **2005**, *22*, R193. [CrossRef]
7. Nicolai, H. Quantum Gravity: The View From Particle Physics. In *General Relativity, Cosmology and Astrophysics. Fundamental Theories of Physics*; Bičák, J., Ledvinka, T., Eds.; Springer: Cham, Switzerland, 2013; Volume 177, pp. 369–387. ISBN 978-3-319-06348-5.
8. Amelino-Camelia, G.; Smolin, L. Prospects for constraining quantum gravity dispersion with near term observations. *Phys. Rev. D* **2009**, *80*, 084017. [CrossRef]
9. Carballo-Rubio, R.; Di Filippo, F.; Liberati, S. Minimally modified theories of gravity: A playground for testing the uniqueness of general relativity. *J. Cosmol. Astropart. Phys.* **2018**, *1806*, 026. [CrossRef]
10. Carballo-Rubio, R.; Di Filippo, F.; Liberati, S.; Visser, M. Phenomenological aspects of black holes beyond general relativity. *arXiv* **2018**, arXiv:1809.08238.
11. Mattingly, D. Modern Tests of Lorentz Invariance. *Living Rev. Relativ.* **2005**, *8*, 5. [CrossRef]
12. Dirac, P.A.M. An extensible model of the electron. *Proc. R. Soc. Lond. A* **1962**, *268*, 57. [CrossRef]
13. Arnowitt, R.L.; Deser, S.; Misner, C.W. Republication of: The dynamics of general relativity. *Gen. Relativ. Gravit.* **2008**, *40*, 1997. [CrossRef]
14. Thiemann, T. *Modern Canonical Quantum General Relativity*; Cambridge University Press: Cambridge, UK, 2008; ISBN 9780511755682.
15. Bojowald, M. *Canonical Gravity and Applications: Cosmology, Black Holes, and Quantum Gravity*; Cambridge University Press: Cambridge, UK, 2010; ISBN 9780521195751.
16. Corichi, A.; Reyes, J.D. The gravitational Hamiltonian, first order action, Poincaré charges and surface terms. *Class. Quantum Gravity* **2015**, *32*, 195024. [CrossRef]
17. Rovelli, C. Covariant Loop Gravity. In *Quantum Gravity and Quantum Cosmology*; Lecture Notes in Physics; Calcagni, G., Papantonopoulos, L., Siopsis, G., Tsamis, N., Eds.; Springer: Berlin/Heidelberg, Germany, 2013; Volume 863, pp. 57–66. ISBN 978-3-642-33035-3.
18. Perez, A. Spin foam models for quantum gravity. *Class. Quantum Gravity* **2003**, *20*, R43. [CrossRef]
19. Alexandrov, S.; Roche, P. Critical Overview of Loops and Foams. *Phys. Rept.* **2011**, *506*, 41. [CrossRef]
20. Perez, A. Regularization ambiguities in loop quantum gravity. *Phys. Rev. D* **2006**, *73*, 044007. [CrossRef]
21. Corichi, A.; Singh, P. Is loop quantization in cosmology unique? *Phys. Rev. D* **2008**, *78*, 024034. [CrossRef]
22. Cailleteau, T.; Mielczarek, J.; Barrau, A.; Grain, J. Anomaly-free scalar perturbations with holonomy corrections in loop quantum cosmology. *Class. Quantum Gravity* **2012**, *29*, 095010. [CrossRef]
23. Ashtekar, A.; Lewandowski, J.; Marolf, D.; Mourao, J.; Thiemann, T. Quantization of diffeomorphism invariant theories of connections with local degrees of freedom. *J. Math. Phys.* **1995**, *36*, 6456. [CrossRef]
24. Bojowald, M.; Paily, G.M. Deformed general relativity and effective actions from loop quantum gravity. *Phys. Rev. D* **2012**, *86*, 104018. [CrossRef]
25. Bojowald, M.; Hossain, G.M.; Kagan, M.; Shankaranarayanan, S. Anomaly freedom in perturbative loop quantum gravity. *Phys. Rev. D* **2008**, *78*, 063547. [CrossRef]
26. Thiemann, T. The Phoenix Project: Master constraint programme for loop quantum gravity. *Class. Quantum Gravity* **2006**, *23*, 2211. [CrossRef]
27. Cuttell, P.D.; Sakellariadou, M. Fourth order deformed general relativity. *Phys. Rev. D* **2014**, *90*, 104026. [CrossRef]

28. Bojowald, M.; Brahma, S.; Reyes, J.D. Covariance in models of loop quantum gravity: Spherical symmetry. *Phys. Rev. D* **2015**, *92*, 045043. [CrossRef]
29. Bojowald, M.; Brahma, S. Covariance in models of loop quantum gravity: Gowdy systems. *Phys. Rev. D* **2015**, *92*, 065002. [CrossRef]
30. Bojowald, M.; Brahma, S.; Buyukcam, U.; D'Ambrosio, F. Hypersurface-deformation algebroids and effective spacetime models. *Phys. Rev. D* **2016**, *94*, 104032. [CrossRef]
31. Wu, J.P.; Bojowald, M.; Ma, Y. Anomaly freedom in perturbative models of Euclidean loop quantum gravity. *arXiv* **2018**, arXiv:1809.04465.
32. Calcagni, G.; Ronco, M. Deformed symmetries in noncommutative and multifractional spacetimes. *Phys. Rev. D* **2017**, *95*, 045001. [CrossRef]
33. Bojowald, M.; Brahma, S.; Buyukcam, U.; Ronco, M. Extending general covariance: Moyal-type noncommutative manifolds. *Phys. Rev. D* **2018**, *98*, 026031. [CrossRef]
34. Amelino-Camelia, G. Relativity in space-times with short-distance structure governed by an observer-independent (Planckian) length scale. *Int. J. Mod. Phys. D* **2002**, *11*, 35. [CrossRef]
35. Magueijo, J.; Smolin, L. Lorentz invariance with an invariant energy scale. *Phys. Rev. Lett.* **2002**, *88*, 190. [CrossRef]
36. Magueijo, J.; Smolin, L. Generalized Lorentz invariance with an invariant energy scale. *Phys. Rev. D* **2003**, *67*, 044017. [CrossRef]
37. Amelino-Camelia, G. Limits on the Measurability of Space-time Distances in (the Semi-classical Approximation of) Quantum Gravity. *Mod. Phys. Lett. A* **1994**, *9*, 3415. [CrossRef]
38. Amelino-Camelia, G. Testable scenario for relativity with minimum length. *Phys. Lett. B* **2001**, *510*, 255. [CrossRef]
39. Majid, S.; Ruegg, H. Bicrossproduct structure of κ-Poincaré group and non-commutative geometry. *Phys. Lett. B* **1994**, *334*, 348. [CrossRef]
40. Lukierski, J.; Ruegg, H.; Zakrzewski, W.J. Classical and Quantum Mechanics of Free κ-Relativistic Systems. *Ann. Phys.* **1995**, *243*, 90. [CrossRef]
41. Amelino-Camelia, G.; da Silva, M.M.; Ronco, M.; Cesarini, L.; Lecian, O.M. Spacetime-noncommutativity regime of loop quantum gravity. *Phys. Rev. D* **2017**, *95*, 024028. [CrossRef]
42. Mielczarek, J. Loop-deformed Poincaré algebra. *EPL* **2014**, *108*, 40003. [CrossRef]
43. Brahma, S.; Ronco, M.; Amelino-Camelia, G.; Marcianò, A. Linking loop quantum gravity quantization ambiguities with phenomenology. *Phys. Rev. D* **2017**, *95*, 044005. . [CrossRef]
44. Ronco, M. On the UV dimensions of Loop Quantum Gravity. *Adv. High Energy Phys.* **2016**, *2016*, 9897051. . [CrossRef]
45. Mielczarek, J.; Trześniewski, T. Spectral dimension with deformed spacetime signature. *Phys. Rev. D* **2017**, *96*, 024012. [CrossRef]
46. Brahma, S.; Ronco, M. Constraining the loop quantum gravity parameter space from phenomenology. *Phys. Lett. B* **2018**, *778*, 184. [CrossRef]
47. Bojowald, M.; Brahma, S.; Yeom, D.H. Effective line elements and black-hole models in canonical loop quantum gravity. *Phys. Rev. D* **2018**, *98*, 046015. [CrossRef]
48. Ben Achour, J.; Lamy, F.; Liu, H.; Noui, K. Polymer Schwarzschild black hole: An effective metric. *EPL (Europhys. Lett.)* **2018**, *123*, 20006. [CrossRef]
49. Bianchi, E.; Magliaro, E.; Perini, C. Coherent spin-networks. *Phys. Rev. D* **2010**, *82*, 024012. [CrossRef]
50. Magueijo, J.; Smolin, L. Gravity's rainbow. *Class. Quantum Gravity* **2004**, *21*, 1725. [CrossRef]
51. Galan, P.; Mena Marugan, G.A. Quantum time uncertainty in a gravity's rainbow formalism. *Phys. Rev. D* **2004**, *70*, 124003. [CrossRef]
52. Ali, A.F. Black hole remnant from gravity's rainbow. *Phys. Rev. D* **2014**, *89*, 104040. [CrossRef]
53. Heydarzade, Y.; Rudra, P.; Darabi, F.; Ali, A.F.; Faizal, M. Vaidya spacetime in massive gravity's rainbow. *Phys. Lett. B* **2017**, *774*, 46. [CrossRef]
54. Barrau, A.; Bojowald, M.; Calcagni, G.; Grain, J.; Kagan, M. Anomaly-free cosmological perturbations in effective canonical quantum gravity. *JCAP* **2015**, *2015*, 1505. [CrossRef]
55. Giesel, K.; Thiemann, T. Consistency check on volume and triad operator quantization in loop quantum gravity: I. *Class. Quantum Gravity* **2006**, *23*, 5667. [CrossRef]

56. Bojowald, M.; Brahma, S. Signature change in two-dimensional black-hole models of loop quantum gravity. *Phys. Rev. D* **2018**, *98*, 026012. [CrossRef]
57. Brahma, S. Spherically symmetric canonical quantum gravity. *Phys. Rev. D* **2015**, *91*, 124003. [CrossRef]
58. Bojowald, M.; Paily, G.M.; Reyes, J.D. Discreteness corrections and higher spatial derivatives in effective canonical quantum gravity. *Phys. Rev. D* **2014**, *90*, 025025. [CrossRef]
59. Varadarajan, M. The diffeomorphism constraint operator in loop quantum gravity. *J. Phys. Conf. Ser.* **2012**, *360*, 012009. [CrossRef]
60. Olmedo, J. Brief Review on Black Hole Loop Quantization. *Universe* **2016**, *2*, 12. [CrossRef]
61. Wilson-Ewing, E. Holonomy corrections in the effective equations for scalar mode perturbations in loop quantum cosmology. *Class. Quantum Gravity* **2012**, *29*, 085005. [CrossRef]
62. Cailleteau, T.; Barrau, A.; Grain, J.; Vidotto, F. Consistency of holonomy-corrected scalar, vector, and tensor perturbations in loop quantum cosmology. *Phys. Rev. D* **2012**, *86*, 087301. [CrossRef]
63. Brown, J.D.; York, J.W. Quasilocal energy and conserved charges derived from the gravitational action. *Phys. Rev. D* **1993**, *47*, 1407. [CrossRef]
64. Born, M. A Suggestion for Unifying Quantum Theory and Relativity. *Proc. R. Soc. Lond. A* **1938**, *165*, 291. [CrossRef]
65. Bojowald, M.; Paily, G.M. Deformed General Relativity. *Phys. Rev. D* **2013**, *87*, 044044. [CrossRef]
66. Ben Achour, J.; Mouchet, A.; Noui, K. Analytic Continuation of Black Hole Entropy in Loop Quantum Gravity. *JHEP* **2015**, *1506*, 145. [CrossRef]
67. Ben Achour, J.; Grain, J.; Noui, K. Loop Quantum Cosmology with Complex Ashtekar Variables. *Class. Quantum Gravity* **2015**, *32*, 025011. [CrossRef]
68. Ambjørn, J.; Jurkiewicz, J.; Loll, R. The Spectral Dimension of the Universe is Scale Dependent. *Phys. Rev. Lett.* **2005**, *95*, 171301. [CrossRef] [PubMed]
69. Niedermaier, M.; Reuter, M. The Asymptotic Safety Scenario in Quantum Gravity. *Living Rev. Relativ.* **2006**, *9*, 5. [CrossRef] [PubMed]
70. Calcagni, G. Multifractional theories: An unconventional review. *JHEP* **2017**, *1706*, 020. [CrossRef]
71. Amelino-Camelia, G.; Calcagni, G.; Ronco, M. Imprint of quantum gravity in the dimension and fabric of spacetime. *Phys. Lett. B* **2017**, *774*, 630. [CrossRef]
72. Calcagni, G.; Ronco, M. Dimensional flow and fuzziness in quantum gravity: Emergence of stochastic spacetime. *Nucl. Phys. B* **2017**, *923*, 144. [CrossRef]
73. Olmo, G.J. Palatini Actions and Quantum Gravity Phenomenology. *J. Cosmol. Astropart. Phys.* **2011**, *1110*, 018. [CrossRef]
74. Girelli, F.; Liberati, S.; Sindoni, L. Planck-scale modified dispersion relations and Finsler geometry. *Phys. Rev. D* **2007**, *75*, 064015. [CrossRef]
75. Amelino-Camelia, G.; Barcaroli, L.; Gubitosi, G.; Liberati, S.; Loret, N. Realization of doubly special relativistic symmetries in Finsler geometries. *Phys. Rev. D* **2014**, *90*, 125030. [CrossRef]
76. Lobo, I.P.; Loret, N.; Nettel, F. Investigation of Finsler geometry as a generalization to curved spacetime of Planck-scale-deformed relativity in the de Sitter case. *Phys. Rev. D* **2017**, *95*, 046015. [CrossRef]
77. Lobo, I.P.; Loret, N.; Nettel, F. Rainbows without unicorns: Metric structures in theories with Modified Dispersion Relations. *Eur. Phys. J. C* **2017**, *77*, 451. [CrossRef]
78. Barcaroli, L.; Brunkhorst, L.K.; Gubitosi, G.; Loret, N.; Pfeifer, C. Hamilton geometry: Phase space geometry from modified dispersion relations. *Phys. Rev. D* **2015**, *92*, 084053. [CrossRef]
79. Loret, N. Exploring special relative locality with de Sitter momentum-space. *Phys. Rev. D* **2014**, *90*, 124013. [CrossRef]
80. Assanioussi, M.; Dapor, A.; Lewandowski, J. Rainbow metric from quantum gravity. *Phys. Lett. B* **2015**, *751*, 302. [CrossRef]
81. Lewandowski, J.; Nouri-Zonoz, M.; Parvizi, A.; Tavakoli, Y. Quantum theory of electromagnetic fields in a cosmological quantum spacetime. *Phys. Rev. D* **2017**, *96*, 106007. [CrossRef]
82. Weinfurtner, S.; Jain, P.; Visser, M.; Gardiner, C.W. Cosmological particle production in emergent rainbow spacetimes. *Class. Quantum Gravity* **2009**, *26*, 065012. [CrossRef]

83. Vakili, B. Classical polymerization of the Schwarzschild metric. *Adv. High Energy Phys.* **2018**, *2018*, 3610543. [CrossRef]
84. Giddings, S.B.; Psaltis, D. Event Horizon Telescope Observations as Probes for Quantum Structure of Astrophysical Black Holes. *Phys. Rev. D* **2018**, *97*, 084035. [CrossRef]

© 2018 by the authors. Licensee MDPI, Basel, Switzerland. This article is an open access article distributed under the terms and conditions of the Creative Commons Attribution (CC BY) license (http://creativecommons.org/licenses/by/4.0/).

Article

Dynamical Properties of the Mukhanov-Sasaki Hamiltonian in the Context of Adiabatic Vacua and the Lewis-Riesenfeld Invariant

Max Joseph Fahn †, Kristina Giesel *,† and Michael Kobler †

Institute for Quantum Gravity, Department of Physics, FAU Erlangen-Nürnberg, 91058 Erlangen, Germany
* Correspondence: kristina.giesel@gravity.fau.de
† These authors contributed equally to this work.

Received: 1 February 2019; Accepted: 10 July 2019; Published: 13 July 2019

Abstract: We use the method of the Lewis-Riesenfeld invariant to analyze the dynamical properties of the Mukhanov-Sasaki Hamiltonian and, following this approach, investigate whether we can obtain possible candidates for initial states in the context of inflation considering a quasi-de Sitter spacetime. Our main interest lies in the question of to which extent these already well-established methods at the classical and quantum level for finitely many degrees of freedom can be generalized to field theory. As our results show, a straightforward generalization does in general not lead to a unitary operator on Fock space that implements the corresponding time-dependent canonical transformation associated with the Lewis-Riesenfeld invariant. The action of this operator can be rewritten as a time-dependent Bogoliubov transformation, where we also compare our results to already existing ones in the literature. We show that its generalization to Fock space has to be chosen appropriately in order to not violate the Shale-Stinespring condition. Furthermore, our analysis relates the Ermakov differential equation that plays the role of an auxiliary equation, whose solution is necessary to construct the Lewis-Riesenfeld invariant, as well as the corresponding time-dependent canonical transformation, to the defining differential equation for adiabatic vacua. Therefore, a given solution of the Ermakov equation directly yields a full solution of the differential equation for adiabatic vacua involving no truncation at some adiabatic order. As a consequence, we can interpret our result obtained here as a kind of non-squeezed Bunch-Davies mode, where the term non-squeezed refers to a possible residual squeezing that can be involved in the unitary operator for certain choices of the Bogoliubov map.

Keywords: quantum cosmology; cosmological perturbation theory; Lewis-Riesenfeld invariant; Bogoliubov transformation; adiabatic vacua

1. Introduction

In the framework of linear cosmological perturbation theory the Mukhanov-Sasaki equation plays a central role. It encodes the dynamics of the Mukhanov-Sasaki variable, which is a linearized and gauge invariant quantity that is built from a specific combination of matter and gravitational perturbations such that the resulting expression is gauge invariant up to linear order. A way to derive this equation is to consider the Einstein-Hilbert action together with a scalar field minimally coupled to gravity and expand this action up to second order in the perturbations around an FLRW background. One decomposes the perturbations into scalar, vector and tensor perturbations since these decouple at linear order. In the scalar sector, we are left with one physical degree of freedom that can for instance be expressed in terms of the Mukhanov-Sasaki variable denoted by $v(\eta, \mathbf{x})$. Given this, we can express the scalar part of the perturbed action entirely in terms of the Mukhanov-Sasaki variable and the corresponding equation of motion takes the following form [1]:

$$v''(\eta,\mathbf{x}) - \left(\Delta + \frac{z''(\eta)}{z(\eta)}\right)v(\eta,\mathbf{x}) = 0, \qquad z(\eta) = \frac{a}{\mathcal{H}}\frac{d\bar{\phi}}{d\eta}, \qquad \eta := \int^t \frac{d\tau}{a(\tau)},$$

where Δ is the spatial Laplacian, η denotes conformal time, a the scale factor, $\bar{\phi}(\eta)$ the isotropic background scalar field and $\mathcal{H} := \frac{a'}{a}$ the Hubble parameter with respect to conformal time. Contrary to the background quantities, the linear perturbations carry a position dependence breaking the spatial symmetries of the FLRW background spacetime. Throughout this article we will work with the Fourier transform of this differential equation. For each Fourier mode $v_\mathbf{k}(\eta)$, this leads to a differential equation given by:

$$v_\mathbf{k}''(\eta) + \left(\|\mathbf{k}\|^2 - \frac{z''(\eta)}{z(\eta)}\right)v_\mathbf{k}(\eta) = 0, \tag{1}$$

where quantities with **k**-label corresponds to the associated Fourier transforms. The quantity in the brackets of the Fourier transformed equation is called the Mukhanov-Sasaki frequency $\omega_\mathbf{k}(\eta)$ and reflects the backreaction of the matter degrees of freedom with the background spacetime. Further commonly used gauge invariant quantities in the context of linear cosmological perturbation theory are the Bardeen potential Φ_B as well as the comoving curvature perturbation $\mathcal{R}$. The latter is related to the Mukhanov-Sasaki variable v by $v = z\mathcal{R}$. Whether one considers a specific gauge invariant quantity is often influenced by the choice of a particular gauge in which these variables simplify and have an obvious physical interpretation. For the Bardeen potential this is the longitudinal gauge, whereas the Mukhanov-Sasaki variable naturally arises in the spatially flat gauge, where it is directly related to the perturbations of the inflaton scalar field. More details about the construction of these gauge invariant variables as well as the derivation of their dynamics from the perturbed Einstein equations in the Lagrangian framework can for instance be found in Reference [2]. A similar derivation in the canonical approach is for example presented in References [3–6]. Here we will not work in a particular gauge but take the form of the Mukhanov-Sasaki equation in (1) as our starting point. As far as a comparison with experimental data is concerned, the relevant quantity is the power spectrum that is defined as the (dimensionless) Fourier transform of the real space two-point correlation function, that is in the case of the quantized Mukhanov-Sasaki variable $\langle 0|\,\hat{v}(\eta,\mathbf{x}),\hat{v}(\eta,\mathbf{y})\,|0\rangle$.

Obviously the power spectrum can only be determined if some initial state has been chosen with respect to which the correlation functions are defined. The most common choice for the initial state is the Bunch-Davies vacuum that can be uniquely selected by the conditions that it is de Sitter invariant and satisfies the Hadamard condition. The latter requires that the corresponding two-point function has a specific behavior in the ultraviolet, that is for short distances. If we drop the Hadamard condition, we obtain the family of so-called α-vacua that include the Bunch-Davies vacuum. Other choices for the initial conditions than the ones for the Bunch-Davies vacuum have been considered and their possible fingerprints on the power spectrum have been investigated, see for instance References [7–9] and references therein. The Bunch-Davies vacuum is selected by requiring that in the limit of $\eta \to -\infty$ the mode functions take the form of the usual Minkowski mode functions. Another method to choose an initial state is the so-called Hamiltonian diagonalization method, where one minimizes the expectation value $\langle 0_{\eta_0}|\,\hat{H}(\eta_0)\,|0_{\eta_0}\rangle$ of the Mukhanov-Sasaki Hamiltonian at one moment in time, say η_0. Hamiltonian diagonalization refers to the fact that at η_0 the coefficients of the off-diagonal terms involving second powers of annihilation and creation operators, respectivley, vanish for all modes. That is, at η_0 the Mukhanov-Sasaki Hamiltonian is given by the field theoretical generalization of the standard harmonic oscillator. Considering this, a natural question to ask is whether such a Hamiltonian diagonalization can be obtained not only instantaneously but for each moment in time and particularly how this aspect is related to the choice of initial states. The usual form of Hamiltonian diagonalization has been critizised in the literature, see for instance Reference [10]. In the framework considered in our work this corresponds to the question whether there exists a canonical transformation that maps the

Mukhanov-Sasaki Hamiltonian to the time-independent harmonic oscillator for each moment in time. In order to work into that direction we take into account that the Mukhanov-Sasaki equation represents a time-dependent harmonic oscillator in each Fourier mode, whereas the specific form of the time dependence reflects the properties of the expanding background spacetime. What we are aiming at is a transformation that maps the time-dependent harmonic oscillator to the time-independent harmonic oscillator for each mode and all times. Defining such a transformation will only work if we consider time-depedendent canonical transformations, that are adapted specifically to the two systems of the time-dependent and time-independent harmonic oscillator, respectively. This is conveniently done in the extended phase space framework outlined below.

There has been considerable interest in the study of the time-dependent harmonic oscillator, both in a purely classical and quantum mechanical context. A distinct role in all of these considerations is played by the Lewis-Riesenfeld invariant, which is a constant of motion with respect to the evolution governed by a time-dependent harmonic oscillator. At the classical level, this invariant has been considered in the context of a canonical transformation in the extended phase space [11,12] that involves time and its momentum as canonical phase space variables among the usual position and momentum variables. Such an extended phase space provides a convenient platform to implement time-dependent canonical transformations. The obtained canonical transformation allows to map the system of the time-dependent harmonic oscillator onto the system of a harmonic oscillator with constant frequency and thus completely removes the time dependence of the Hamiltonian, which drastically simplifies the task of finding solutions of the equations of motion after applying the transformation. As shown in Reference [13], the invariant can also be defined in the context of quantum mechanics. In this case the eigenstates of the invariant can be used to construct solutions of the Schrödinger equation involving the original time-dependent Hamiltonian. Further application are to construct coherent states of the time-dependent harmonic oscillator by means of the eigenstates of the Lewis-Riesenfeld invariant as for instance discussed in References [13,14].

If we aim at relating the framework of the Lewis-Riesenfeld invariant to the notion of initial states associated with the Mukhanov-Sasaki equation, we need to generalize this approach to the field theory context. There exists already some work in this direction, see for example in References [14,15] and references therein, although with a slightly different focus than we want to consider here, because both of them do not apply this techniques directly to the Mukhanov-Sasaki equation in the framework of the extended phase space, meaning that they consider different time-dependent frequencies in general and particularly the generalization to field theory was not analyzed in very much detail in Reference [15]. The strategy we want to follow in our work is that first we consider the Lewis-Riesenfeld invariant and the corresponding canonical transformation at the classical level for finitely many degrees of freedom in the extended phase space, building on former work of References [11,12], who however did not consider the quantization of the canonical transformation. In order to be able to implement the corresponding unitary map at the quantum level, we also construct the corresponding generator of the canonical transformation. For the reason that in the extended phase space the physical system of the time-dependent harmonic oscillator is described as a constrained system, we construct Dirac observables and use the technique of reduced phase space quantization to implement this unitary map on the physical Hilbert space in a quantum mechanical setting, where it can also be formulated in terms of a time-dependent Bogoliubov transformation. Given this setup, we could take the vacuum of the time-independent harmonic oscillator, apply the constructed unitary map to it and obtain a in this sense natural candidate for a vacuum state for the time-dependent harmonic oscillator, that has then been determined directly by means of the unitary map.

The question we want to address in this article is whether we can carry this idea over from finitely many degrees of freedom to field theory and use the Lewis-Riesenfeld invariant approach to obtain possible candidates for initial states. In particular, we are interested in the physical properties of such initial states and their relation to the Bunch-Davies vacuum and other adiabatic vacua. As we will show, the most straightforward generalization to field theory is not possible because the so constructed map involves an infrared divergence, hence the Shale-Stinespring condition is violated. As we will discuss, a suitable modification of the map in the infrared range can be obtained to cure the infrared divergenes. Furthermore, as we will show, if this map is not chosen carefully for all but the infrared modes it can also involve ultraviolet divergences strictly permitting a unitary implementation on Fock space. Interestingly, in the context of the Mukhanov-Sasaki Hamiltonian, different choices at this level can be related to different choices for the initial conditions of the associated mode functions. Moreover it becomes clear that we can recover the defining differential equation for adiabatic vacua from the Ermakov equation, where the latter is an auxiliary differential equation whose solution is needed to explicitly construct the Lewis-Riesenfeld invariant and the corresponding canonical transformation. This allows us to interpret the initial conditions and the result for the Fourier modes we obtain using the method of the Lewis-Riesenfeld invariant in the context of adiabatic vacua.

This article is structured as follows: In Section 2 we introduce the framework of the extended phase space and rederive the canonical transformation that maps the system of the time-dependent harmonic oscillator to the time-independent one generalizing the approach in Reference [12]. The time-rescaling that is involved in this canonical transformation naturally occurs in the extended phase space and the physical interpretation of the Lewis-Riesenfeld invariant can be easily understood. In order to deal with the constrained system in the extended phase space later on, we want to choose reduced phase space quantization and thus derive the reduced phase space in terms of Dirac observables. Their dynamics is generated by the Dirac observable associated with the time-dependent Hamiltonian. As the next step in Section 3, we consider the quantization of the system and show that the canonical transformation can be implemented as a unitary map on the one-particle physical Hilbert space, where our results agree with already existing results in the literature for finitely many degrees of freedom. In order to simplify the actual application of the unitary operator we perform a generalized Baker-Campbell-Hausdorff decomposition by means of which we then rewrite the unitary transformation as a time-dependent Bogoliubov map.

Afterwards we consider the generalization of our results obtained so far to field theory, discussing the two most common cases in the literature, where one maps from a time-dependent harmonic oscillator to a harmonic oscillator with either frequency $\omega_k = k$ or $\omega_k = 1$. As far as the implementation on Fock space is considered, the first choice can be implemented unitarily, whereas the second cannot due to an ultraviolet divergence. This ultraviolet divergence is caused by a residual squeezing operation by which the two maps differ. To avoid issues that occur for the infrared modes, we discuss a possible modification of the map using the Arnold transformation discussed in Reference [16]. Section 6 presents practical applications of this formalism by considering the case of a quasi-de Sitter spacetime and the corresponding Mukhanov-Sasaki equation in a slow-roll approximation. We construct the Lewis-Riesenfeld invariant and, in the context of a quantum mechanical toy model, compute the lowest and next to lowest eigenvalue eigenstates associated to it and analyze their properties. Finally we summarize and conclude in Section 7.

2. Extended Phase Space Formulation and Time-Dependent Canonical Transformations

A convenient framework for implementing time-dependent canonical transformations is the extended phase space in which also time and its conjugate momentum are treated as phase space variables and thus transformations of them can be naturally formulated. Motivated by the Mukhanov-Sasaki equation, we first investigate a single mode of the equation in a classical context. This corresponds to a harmonic oscillator with time-dependent frequency. We will consider the single mode Mukhanov-Sasaki Hamiltonian as a mechanical toy model and later generalize the results obtained in this case to the field theory context. Our goal is to remove this explicit time dependence by a time-dependent canonical transformation. This transformation will be defined on the extended phase space as a symplectic map that also includes the time variable and its associated conjugate momentum as phase space degrees of freedom.

2.1. Time-Dependent Hamiltonians on Extended Phase Space

As a first step we reformulate the dynamics encoded in the single mode Mukhanov-Sasaki Hamiltonian on the extended phase space, where it becomes a constrained system. Let us consider a system with finitely many degrees of freedom where we denote all configuration variables as $\mathbf{q} = (q^1, \cdots, q^n)$ and the configuration space by Σ. A time-dependent Lagrangian is then defined as a function $L: T\Sigma \times \mathbb{R} \to \mathbb{R}$. Because we want to include time among the elementary configuration variables, closely following the work in References [11,12], we extend the configuration manifold Σ to $M := \Sigma \times \mathbb{R}$ and rewrite the action as

$$S[L] = \int_{\mathbb{R}} ds\, L\left(\tilde{\mathbf{q}}(s), t(s), \left(\frac{dt}{ds}\right)^{-1}\frac{d\tilde{\mathbf{q}}}{ds}\right)\frac{dt(s)}{ds} \qquad (2)$$

$$:= \int_{\mathbb{R}} ds\, \mathcal{L}\left(\tilde{\mathbf{q}}(s), t(s), \left(\frac{dt}{ds}\right)^{-1}\frac{d\tilde{\mathbf{q}}}{ds}, \frac{dt(s)}{ds}\right) =: S[\mathcal{L}],$$

where we will refer to $\mathcal{L}$ as the *extended* Lagrange function now understood as a function on the extended tangent bundle TM that is even-dimensional and associated to the extended configuration manifold, including the former system evolution parameter commonly referred to as time. A non-degenerate symplectic structure on the corresponding cotangent bundle T^*M, whose elementary variables are $(\tilde{\mathbf{q}}, t, \tilde{\mathbf{p}}, p_t)$ can be defined as usual. This allows to establish a one-to-one correspondence between smooth phase space functions and Hamiltonian vector fields. In complete analogy to the conventional case, one can formulate the Euler-Lagrange equations in terms of the extended variables by means of the variational principle, which results in equivalent equations of motion as derived from the original action $S[L]$. The equations of motion for the time variable are just given by $\frac{dt}{ds} = \lambda(s)$ where $\lambda(s)$ is an arbitrary real parameter reflecting the rescaling symmetry of the action, this reflects the arbitrary parametrization of time and has no physical significance. If we perform a Legendre transform, we realize that $p_t = -H(\tilde{\mathbf{q}}, \tilde{\mathbf{p}}, t)$ becomes a primary constraint since it cannot be solved for the velocities $\frac{dt}{ds}$ with[1]

$$H(\tilde{\mathbf{q}}, \tilde{\mathbf{p}}, t) = \frac{\tilde{\mathbf{p}}^2}{2} + \frac{1}{2}\omega^2(t)\tilde{\mathbf{q}}^2,$$

where the Hamiltonian is a function $H: T^*M \to \mathbb{R}$ that is independent of p_t. We denote this constraint by $C := p_t + H(\tilde{\mathbf{q}}, \tilde{\mathbf{p}}, t)$. Therefore, we apply the Legendre transform for singular systems and obtain the following Hamiltonian on the extended phase space T^*M:

$$\mathfrak{H} = \tilde{p}_a\frac{d\tilde{q}^a}{ds} + p_t\frac{dt}{ds} - \mathcal{L}\Big|_{\dot{q}^a(\tilde{\mathbf{q}},\tilde{\mathbf{p}},t,\lambda),\frac{dt}{ds}=\lambda} = \left(H(\tilde{\mathbf{q}}, \tilde{\mathbf{p}}, t) + p_t\right)\lambda(s) = \lambda(s)C \approx 0,$$

where we used $\approx$ to denote weak equivalence and used the definition of $\lambda(s)$ from above. Due to reparametrization invariance of the extended action, there is no true Hamiltonian but a Hamiltonian constraint C. From now on we will neglect the tilde on the top of the variables $\mathbf{q}, \mathbf{p}$ to keep our notation more compact. For the time-dependent harmonic oscillator the so-called Lewis-Riesenfeld invariant I_{LR} has played a pivotal role, particularly in the construction of solutions for the corresponding equations of motion. I_{LR} is a phase space function being quadratic in the elementary variables $(\mathbf{q}, \mathbf{p})$ and its time dependence is encoded in a function $\xi: I \subseteq \mathbb{R} \to \mathbb{R}$. Explicitly, it is given by:

$$I_{LR}(\mathbf{q}, \mathbf{p}, t) := \frac{1}{2}\left((\xi(t)\mathbf{p} - \dot{\xi}(t)\mathbf{q})^2 + \frac{\omega_0^2 \mathbf{q}^2}{\xi^2(t)}\right). \qquad (3)$$

---

[1] In general we could also take into account a time dependent mass in the Hamiltonian, however in the case of the Mukhanov-Sasaki equation it is sufficient to set the mass parameter m equal to $m = 1$.

Since I_{LR} is an invariant, it has to commute with the constraint C on the extended phase space[2]:

$$\{I_{LR}, C\}_{ext} = \{I_{LR}, H(t)\} + \frac{\partial I_{LR}}{\partial t} = 0. \quad (4)$$

This carries over to a condition on the function ξ that has to satisfy the following non-linear, ordinary, second-order differential equation

$$\left(\frac{d^2}{dt^2} + \omega(t)^2\right)\xi(t) - \omega_0^2\,\xi(t)^{-3} = 0, \quad (5)$$

known as the Ermakov equation. It has been shown that I_{LR} is an invariant both at the classical level [11] and at the quantum level [13,17]. In the following, the explicit form of I_{LR} will be our guiding line for finding an extended canonical transformation that removes the time dependence from the Hamiltonian of the harmonic oscillator with a time-dependent frequency.

2.2. Extended Canonical Transformations and Hamiltonian Flows

In the framework of the extended phase space formalism we can now regard time as a configuration degree of freedom and consequently apply a canonical transformation to implement a time-rescaling. It is worth noting that the Mukhanov-Sasaki equation in *conformal* time (commonly denoted η) takes the form of a time-dependent harmonic oscillator, however we shall refer to the time variable as t in the context of the classical and one-particle quantum theory, respectively. We aim at finding a symplectic map Φ such that the explicitly time-dependent Mukhanov-Sasaki Hamiltonian is mapped into an autonomous one, that is, one with time-independent frequency that we denote by ω_0. During this procedure, the Hamiltonian constraint together with the Poisson structure on the extended phase space remain invariant by construction, that is:

$$\Phi: T^*M \to T^*M, \quad \mathfrak{H} \mapsto \Phi^*\mathfrak{H} = \mathfrak{H}. \quad (6)$$

Correspondingly, the symplectic form Ω on T^*M is invariant under the Hamiltonian flow of the associated Hamiltonian vector field of Φ that infinitesimally generates this transformation. In order to apply this procedure to the case of the single-mode Mukhanov-Sasaki Hamiltonian, we need to impose conditions on the explicit form of Φ. We would like to preserve the functional dependence of the Hamiltonian constraint on the one hand and keep the quadratic order in both momentum and configuration variables on the other hand. For this purpose, we make the following ansatz, closely related to the work presented in Reference [12]:

$$\Phi: \begin{pmatrix} q^a \\ p_a \\ t \\ p_t \end{pmatrix} \mapsto \begin{pmatrix} Q^a(\mathbf{q}, t) \\ F(\mathbf{q}, t)p_a + G_a(\mathbf{q}, t) \\ T(\mathbf{q}, t) \\ P_T(\mathbf{q}, t, \mathbf{p}, p_t) \end{pmatrix} \quad \text{s.t.} \quad \Phi^*\Omega = \Omega \quad (7)$$

Note the ansatz $\mathbf{p} \propto \mathbf{P} + \mathbf{G}(\mathbf{q}, t)$ ensures that the transformed Hamiltonian is again quadratic in the new momentum, whereas the prefactor allows for a time-rescaling of the momentum variable. Additionally, the only variable that carries a dependence on p_t is the new momentum conjugate to T denoted by P_T, which is a choice that preserves the form of the Hamiltonian constraint being linear in the conjugate momentum of the time variable. We employ the ansatz in (7) for the symplectic map Φ and from subsequent comparison of coefficients of the two-form basis elements we obtain a set

---

[2] The symplectic form associated with the Poisson bracket $\{.,.\}_{ext}$ on the extended phase space has the form $\Omega = d\tilde{q}^a \wedge d\tilde{p}_a + dt \wedge dp_t$.

of five coupled differential equations that determine the form of Φ to be a canonical transformation. This system of differential equations corresponds to a generalization to $n+1$ configuration degrees of freedom of the set of equations presented in Reference [12], where only the case for $n = 1$ was presented. It explicitly reads:

$$\frac{\partial Q^a}{\partial t}\left(p_a\frac{\partial F}{\partial q^b} + \frac{\partial G_a}{\partial q^b}\right) + \frac{\partial P_T}{\partial q^b}\frac{\partial T}{\partial t} = \frac{\partial Q^a}{\partial q^b}\left(p_a\frac{\partial F}{\partial t} + \frac{\partial G_a}{\partial t}\right) + \frac{\partial P_T}{\partial t}\frac{\partial T}{\partial q^b},$$

$$F\frac{\partial Q^a}{\partial t} + \frac{\partial P_T}{\partial p_a}\frac{\partial T}{\partial t} = 0, \qquad \frac{\partial P_T}{\partial p_a}\frac{\partial T}{\partial q^b} + F\frac{\partial Q^a}{\partial q^b} = \delta^a{}_b, \qquad (8)$$

$$\frac{\partial P_T}{\partial p_t}\frac{\partial T}{\partial q^a} = 0, \qquad \frac{\partial P_T}{\partial p_t}\frac{\partial T}{\partial t} = 1.$$

We can get a first hint how a solution could look like when we consider the Lewis-Riesenfeld invariant I_{LR} from Equation (3) in Section 2 above. Hence, there is a natural starting point for finding the favored canonical transformation we are aiming at, by fixing the transformations of $\mathbf{q}$ and $\mathbf{p}$ according to a factorization of I_{LR}. This leads to

$$Q^a(\mathbf{q}, t) := \frac{q^a}{\xi(t)} \quad \Longleftrightarrow \quad q^a(\mathbf{Q}, T) = \xi(t(T))Q^a, \qquad (9)$$

$$P_a(\mathbf{q}, \mathbf{p}, t) := \xi(t)p_a - \dot{\xi}(t)q_a \quad \Longleftrightarrow \quad p_a(\mathbf{Q}, \mathbf{P}, T) = \frac{P_a}{\xi(t(T))} + \dot{\xi}(t(T))Q_a, \qquad (10)$$

where $\dot{\xi} = \partial_t \xi$ is the derivative with respect to the dynamical time variable. In order to proceed, we need to find a suitable transformation for the time variable $T(t)$ that is consistent with the system of Equation (8) previously found. A convenient possibility is to use Euler's time scaling transformation for the three-body problem, recently introduced by Struckmeier [11] in the context of the time-dependent harmonic oscillator. However, the approach in Reference [11] differs from the one outlined in this work in the sense that we derive the explicit form of the transformation instead of making use of the corresponding generating function. The relevant transformation of t is given by:

$$T(t) := \int_{t_0}^{t} \frac{d\tau}{\xi^2(\tau)} \quad \Longleftrightarrow \quad \frac{\partial T}{\partial t} = \frac{1}{\xi^2(t)}, \qquad (11)$$

where $\xi(t) \in C^2(\mathbb{R})$ is an up-to-now arbitrary function with the only restriction that the above integral needs to be well-defined. Given the explicit form of (11), we can require mutual consistency of the transformations in (8) in order to fix the form of the transformed canonical momentum P_T. Solving the first equation in (8) for $\partial_b P_T$ and subsequently integrating the obtained expression yields the following result:

$$P_T(\mathbf{q}, \mathbf{p}, t, p_t) = \xi^2(t)p_t + \xi(t)\dot{\xi}(t)\mathbf{q} \cdot \mathbf{p} - \frac{1}{2}\left(\xi(t)\ddot{\xi}(t) + \dot{\xi}^2(t)\right)\mathbf{q}^2, \qquad (12)$$

with the term $\xi^2(t)p_t$ arising from an arbitrary additive constant with respect to $\mathbf{q}$ and the requirement of inverse scaling behavior between t and p_t according to (8). Now that we have fixed the transformation to the new canonical coordinates, we can use the invariance of the Hamiltonian constraint $\mathfrak{H}$ under the change of canonical coordinates to derive an autonomous Hamiltonian from the original, time-dependent one:

$$\Phi^*\mathfrak{H} = \left(\Phi^*H + P_T\right)\frac{dT}{ds} = \left(\Phi^*H + P_T\right)\frac{\partial T}{\partial t}\frac{dt}{ds} = \left(H(\mathbf{q}, \mathbf{p}, t) + p_t\right)\frac{dt}{ds} = \mathfrak{H}. \qquad (13)$$

In fact, using the one before the last equality sign in (13) we find an expression for $\Phi^* H$:

$$H_0 := \Phi^* H = \xi^2(t)\left(H(\mathbf{q},\mathbf{p},t) + p_t\right)\Big|_{(\Phi)(\mathbf{q},\mathbf{p},t)} - P_T, \tag{14}$$

with $\mathbf{q}, \mathbf{p}$ and t considered as functions of the new variables $\mathbf{Q}, \mathbf{P}$ and T via the extended canonical transformation $\Phi(\mathbf{q},\mathbf{p},t)$ defined in Equation (7). Analogous to the treatment displayed in Reference [11], we would also like to point out the crucial property that not the bare constraint C but the product with the Lagrange multiplier $\mathfrak{H} = \lambda C(\mathbf{q},\mathbf{p},t,p_t)$ is invariant under this transformation by construction. As a consequence, the canonical momentum P_T drops out in H_0. If we evaluate all expressions using the inverse of Φ to express $\mathbf{q},\mathbf{p}$ in terms of $\mathbf{Q},\mathbf{P}$, we finally obtain:

$$H_0(\mathbf{Q},\mathbf{P},T) = \left[\frac{\xi^2(t)}{2}\left(\mathbf{p}^2 + \omega(t)^2 \mathbf{q}^2\right) - \xi(t)\dot{\xi}(t)\mathbf{q}\cdot\mathbf{p} + \frac{1}{2}\left(\xi(t)\ddot{\xi}(t) + \dot{\xi}^2(t)\right)\mathbf{q}^2\right]\Big|_{(\Phi^{-1})(\mathbf{Q},\mathbf{P},T)}$$

$$= \frac{\xi^2}{2}\left(\frac{\mathbf{P}^2}{\xi^2} + 2\frac{\dot{\xi}}{\xi}\mathbf{Q}\cdot\mathbf{P} + \dot{\xi}^2\mathbf{Q}^2 + \omega(t(T))^2\xi^2\mathbf{Q}^2\right) - \xi\dot{\xi}\mathbf{Q}\cdot\mathbf{P} - \frac{1}{2}\left(\xi^2\dot{\xi}^2 - \xi^3\ddot{\xi}\right)\mathbf{Q}^2 \tag{15}$$

$$= \frac{1}{2}\left(\mathbf{P}^2 + \xi^3\left(\ddot{\xi} + \omega(t(T))^2\xi\right)\mathbf{Q}^2\right) = \frac{1}{2}\left(\mathbf{P}^2 + \omega_0^2\mathbf{Q}^2\right),$$

where we designed the symplectic map Φ in such a way that the requirement that the term in the brackets multiplying $\mathbf{Q}^2$ in Equation (15) equals $\omega_0^2 \in \mathbb{R}$ is respected. This leads to the condition that $\xi(t)$ needs to satisfy the Ermakov differential equation, which we already encountered during the discussion of the Lewis-Riesenfeld invariant I_{LR} in Section 2 in (5). The so constructed map Φ describes a time-dependent canonical transformation that maps a harmonic oscillator with time-dependent frequency $\omega(t)$ onto a time-independent harmonic oscillator with constant frequency ω_0. The explicit form of the map of course depends on the time dependence of $\omega(t)$ but can be determined from the Ermakov equation once $\omega(t)$ is given. While in principle we could fix the frequency ω_0 to one, as it has been done for the form of the Ermakov equation for instance in References [13,14], we would like our transformation Φ to correspond to the identity for an already time-independent harmonic oscillator Hamiltonian. This can only be achieved if not all time-dependent frequencies are mapped to unity, as even a constant ω_0 would then be transformed non trivially, resulting in a residual transformation analogous to a squeezing operation in quantum theory.

2.3. The Reduced Phase Space Associated with T^*M and the Infinitesimal Generator of Φ

In this section we want to derive the infinitesimal generator corresponding to the finite canonical transformation Φ on the extended phase space T^*M that we presented in the last section. This will be relevant later on when we discuss the implementation of Φ in the quantum theory. As we have discussed, the system under consideration can be understood as a constrained system in the context of the extended phase space. Consequently, we have two options for handling the constraint, either we solve it in the quantum theory via Dirac quantization or we reduce with respect to this constraint already classically and quantize the reduced phase space only. In the first place, both approaches are equally justified from the physical perspective, so this is a choice one makes for each given model. In our case this goes along with the selection whether we want to implement the canonical transformation Φ on the extended or reduced phase space, respectively. Firstly, as the transformation from t to $T(t)$ in (11) involves a time-rescaling in form of an integral, if we are not able to obtain the antiderivative of the integrand in closed form, it will be problematic to formulate this kind of canonical transformation in the quantum theory based on the extended phase space where t becomes an operator. Secondly, following Dirac quantization, we need to construct a physical inner product for physical states and this is non-trivial if the constraint is of the form $C = p_t + H(\mathbf{q},\mathbf{p},t)$ with H being explicitly time-dependent, a similar situation that occurs in loop quantum cosmology if we consider the inflaton as reference matter. The final physical sector of the theory should be related in both approaches and in the best case yield the same physical predictions. This might not be the case in general but yields some restrictions on possible choices in the quantization procedure to match the models based on Dirac and reduced quantization respectively. In the following we choose the reduced phase space approach for which the initial phase space $T^*\Sigma$ can be naturally identified with the reduced phase space of our system. In order to show this we construct Dirac observables for our constrained system by means of the formalism presented in References [18,19] and references therein, that is based on the relational formalism originally introduced in References [20,21]. In the extended phase space, we consider the configuration variable t as the reference field (clock) for time and introduce the following gauge fixing condition $G_\tau := t - \tau \approx 0$. G_τ together with the first class constraint C build a second class pair since $\{G_\tau, C\} = 1$. The Dirac observables for all degrees of freedom except the clock degrees of freedom (t, p_t) are given by

$$\mathcal{O}^C_{q^a,t}(\tau) = \sum_{n=0}^{\infty} \frac{G_\tau^n}{n!} \{C(\mathbf{q},\mathbf{p},t), q^a\}_{(n)}, \quad \mathcal{O}^C_{p_a,t}(\tau) = \sum_{n=0}^{\infty} \frac{G_\tau^n}{n!} \{C(\mathbf{q},\mathbf{p},t), p_a\}_{(n)}, \tag{16}$$

where $\{A, B\}_{(n)}$ denotes the iterated Poisson bracket defined via $\{A, B\}_{(n)} := \{A, \{A, B\}_{(n-1)}\}$ and $\{A, B\}_{(0)} := B$ and we have used that q^a and p_a both commute with the conjugate momentum p_t. The observable map can also be applied to the clock degrees of freedom, leading to

$$\mathcal{O}^C_{t,t}(\tau) = \sum_{n=0}^{\infty} \frac{G_\tau^n}{n!} \{C(\mathbf{q},\mathbf{p},t,p_t), t\}_{(n)} = \tau, \quad \mathcal{O}^C_{p_t,t}(\tau) = \sum_{n=0}^{\infty} \frac{G_\tau^n}{n!} \{C(\mathbf{q},\mathbf{p},t,p_t), p_t\}_{(n)}. \tag{17}$$

We realize that the clock t is mapped to the parameter τ as expected, whereas contrary to the deparametrized models presented in References [22–29], the physical Hamiltonian retains its time dependence, hence p_t is not yet a Dirac observable by itself. Using the properties of the observable map we have that $p_t = -\mathcal{O}^C_{H(\mathbf{q},\mathbf{p},t),t} = -H(\mathcal{O}^C_{\mathbf{q},t}, \mathcal{O}^C_{\mathbf{p},t}, \tau)$ and hence p_t can be expressed as a function of $\mathcal{O}^C_{\mathbf{q},t}, \mathcal{O}^C_{\mathbf{p},t}$ only, where we introduced the abbreviation $\mathcal{O}^C_{\mathbf{q},t} := (\mathcal{O}^C_{q^1,t}, \cdots, \mathcal{O}^C_{q^n,t})$ and likewise for the momenta. This shows that $(\mathcal{O}^C_{\mathbf{q},t}, \mathcal{O}^C_{\mathbf{p},t})$ are the elementary variables of the reduced phase space and the degrees of freedom encoded in (t, p_t) have been reduced, which leaves us with $2n$ true degrees of freedom in the physical sector of the phase space. As a consequence, we can identify the reduced phase space with $T^*\Sigma$ and the Hamiltonian can be understood as a function from $T^*\Sigma \times \mathbb{R}$ to the real numbers. In order to analyze the Poisson algebra of the observables we have to construct the corresponding Dirac bracket, denoted by $\{.,.\}^*$, associated to the second class system (G_τ, C). However, for the reason that all variables $(\mathbf{q}, \mathbf{p})$ commute with the gauge fixing condition, their Dirac bracket reduces to the usual Poisson bracket. Given this and considering the result in Reference [19], the algebra of our Dirac observables reads:

$$\{\mathcal{O}^C_{q^a,t}(\tau), \mathcal{O}^C_{p_b,t}(\tau)\} = \mathcal{O}^C_{\{q^a,p_b\}^*,t}(\tau) = \delta^a_b.$$

Thus, the kinematical Poisson algebra of $(\mathbf{q}, \mathbf{p})$ and the algebra of their corresponding Dirac observables are isomorphic, which is a big advantage for finding representations of the observable algebra in the context of the quantum theory in Section 3. The observable map applied to a generic phase space function f returns the values of f at those values where the clock takes the value τ. Therefore, the natural evolution parameter for these Dirac observables is τ. If the constraint is linear in the clock momenta as in our case where $C = p_t + H(\mathbf{q}, \mathbf{p}, t)$, then as shown in References [19,26] the so-called physical Hamiltonian generating the τ-evolution is given by the Dirac observable corresponding to $H(\mathbf{q},\mathbf{p},t)$. Thus, in our case the evolution on the reduced phase space is given by the following Hamilton's equations:

$$\frac{d}{d\tau}\mathcal{O}^C_{q^a,t} = \{\mathcal{O}^C_{q^a,t}, H(\mathcal{O}^C_{\mathbf{q},t}, \mathcal{O}^C_{\mathbf{p},t}, \tau)\}, \quad \frac{d}{d\tau}\mathcal{O}^C_{p_a,t} = \{\mathcal{O}^C_{p_a,t}, H(\mathcal{O}^C_{\mathbf{q},t}, \mathcal{O}^C_{\mathbf{p},t}, \tau)\}. \tag{18}$$

Lastly, by an abuse of notation we replace τ by t as well as $\mathcal{O}^C_{q^a,t}$ by q^a and $\mathcal{O}^C_{p_a,t}$ by p_a in order to be closer to the notation used in previous works in the literature and emphasize that the generator of Φ acts as a one-parameter family of transformations on configuration and momentum degrees of freedom in $T^*\Sigma$. When we have a look at the form of Φ, we immediately recognize that the generator $\mathcal{G} \in C^\infty(T^*\Sigma \times \mathbb{R})$ needs to be a polynomial of second order in the original configuration and momentum variables, where $T^*\Sigma \times \mathbb{R}$ corresponds to the presymplectic space for explicitly time-dependent systems as for instance used in Reference [11]. This ensures that the action of the associated Hamiltonian vector field $X_\mathcal{G}$ with $X_\mathcal{G}(f) := \{\mathcal{G}, f\}$ onto the elementary phase space variables $\mathbf{q}$ and $\mathbf{p}$ results in a linear combination of those quantities. The explicit form of Φ suggests an ansatz in order to find $\mathcal{G}$, which naturally depends on $\xi, \dot\xi$, incorporating the parametric dependence on t:

$$\mathcal{G}(\xi, \dot\xi, \mathbf{q}, \mathbf{p}) := f(\xi, \dot\xi)\mathbf{q} \cdot \mathbf{p} + \tfrac{1}{2}g(\xi, \dot\xi)\mathbf{q}^2, \tag{19}$$

where the factor in front of $g(\xi, \dot\xi)$ was introduced for later convenience. Application of the exponentiated Hamiltonian vector field $X_\mathcal{G}$ onto $\mathbf{q}$ and $\mathbf{p}$ leads to the following results:

$$\exp\{X_\mathcal{G}\}q^a := \sum_{n=0}^\infty \frac{1}{n!}\{\mathcal{G}, q^a\}_{(n)} = \sum_{n=0}^\infty \frac{(-1)^n}{n!}(f(\xi,\dot\xi))^n q^a = e^{-f(\xi,\dot\xi)} q^a, \tag{20}$$

$$\exp\{X_\mathcal{G}\}p_a := \sum_{n=0}^\infty \frac{1}{n!}\{\mathcal{G}, p_a\}_{(n)} = e^{f(\xi,\dot\xi)} p_a + \frac{1}{2}\left(e^{f(\xi,\dot\xi)} - e^{-f(\xi,\dot\xi)}\right) \frac{g(\xi,\dot\xi)}{f(\xi,\dot\xi)} q_a, \tag{21}$$

with the iterated Poisson bracket defined as above. A direct comparison of the results in (20) and (21) to the solutions of the system of equations in (8) yields the dependencies of $f(\xi,\dot\xi)$ and $g(\xi,\dot\xi)$ on ξ and $\dot\xi$, respectively:

$$f(\xi,\dot\xi) = \ln(\xi), \quad g(\xi,\dot\xi) = \frac{2\ln(\xi)\xi\dot\xi}{1-\xi^2}. \tag{22}$$

Finally, we are able to explicitly write down the generator of the extended canonical transformation Φ restricted to the constraint hypersurface $T^*\Sigma \times \mathbb{R}$, that is the physical sector. We call this restriction of Φ, which is a time-dependent canonical transformation on the reduced phase space, Γ_ξ from now on. In a convenient notation, it has the following form:

$$\mathcal{G}(\xi,\dot\xi,\mathbf{q},\mathbf{p}) = \frac{1}{2}\ln(\xi)\left(\mathbf{q}\cdot\mathbf{p} + \mathbf{p}\cdot\mathbf{q} + h(\xi,\dot\xi)\mathbf{q}^2\right), \quad h(\xi,\dot\xi) := \frac{2\xi\dot\xi}{1-\xi^2}. \tag{23}$$

In fact, this classical generator precisely corresponds to the exponential operator found in Reference [17] for a quantized version of the time-dependent harmonic oscillator. It is worth noting that, regardless of the choice of coordinates, $\mathcal{G}$ takes the same form in either $\mathbf{q},\mathbf{p}$ or $\mathbf{Q},\mathbf{P}$, that is it holds that $\mathcal{G}(\mathbf{q}(\mathbf{Q},\mathbf{P}),\mathbf{p}(\mathbf{Q},\mathbf{P})) = \mathcal{G}(\mathbf{Q},\mathbf{P})$. Not surprisingly, we can switch between the autonomous Hamiltonian and the Lewis-Riesenfeld invariant in this framework, using the action of Γ_ξ on (analytical) phase space functions, leading to:

$$H_0(\Gamma_\xi(\mathbf{q}),\Gamma_\xi(\mathbf{p})) = H_0(e^{X_\mathcal{G}}\mathbf{q}, e^{X_\mathcal{G}}\mathbf{p}) = \frac{1}{2}\left((\xi(t)\mathbf{p} - \dot\xi(t)\mathbf{q})^2 + \frac{\omega_0^2 \mathbf{q}^2}{\xi(t)^2}\right) =: I_{LR}, \tag{24}$$

Of course this was how Γ_ξ or rather Φ was constructed in the first place. However, relation (24) will be of importance in the quantum theory, where it is part of the time evolution operator (i.e., the Dyson series) associated to the time-dependent Hamiltonian. Furthermore, this will allow us to make contact to previous work and strictly derive the phase factor that was introduced by hand in Reference [13] in order to construct eigenfunctions of the time-dependent Schrödinger equation. Referring to the relational formalism outlined in for example, References [18,19], we reconsider the fact that the Lewis-Riesenfeld invariant strongly commutes with the constraint C as shown in (3). Hence, in this language I_{LR} is a strong Dirac observable with respect to the constraint $C(\mathbf{q}, \mathbf{p}, t, p_t)$ if and only if $\xi(t)$ satisfies the Ermakov Equation (5), connecting to the results presented in Reference [13] in the context of quantization. As a concluding remark, let us introduce $e_+ := \frac{1}{2}\mathbf{p}^2$, $e_- := -\frac{1}{2}\mathbf{q}^2$ and $h := q^a p_a$, which amount to the generators of the classical canonical transformation Γ_ξ we derived in the preceding section. Then these three generators form a basis of the $\mathfrak{sl}(2, \mathbb{R})$ algebra, which is evident due to the structure constants of their Poisson brackets. Hence, the exponential of these generators (or a subset thereof) constitutes a group element of $SL(2, \mathbb{R})$ and consequently the classical canonical transformation Γ_ξ is a real representation of $SL(2, \mathbb{R})$ on the space of phase space polynomials or everywhere-analytic phase space functions, respectively.

Let us briefly summarize what we have established in the previous section. Starting from an explicitly time-dependent Hamiltonian and its associated Lewis-Riesenfeld invariant, we systematically constructed a time-dependent canonical transformation on an extended phase space, which removes the time dependence of the original Hamiltonian. Let us stress at this point that $H(t)$, $I_{LR}(t)$ and H_0 are in fact the *same object* in *different coordinates* on the extended phase space. Consequently, we were able to construct the associated infinitesimal generator of this symplectic map and established the notion of a reduced phase space with the prospect of a corresponding unitary transformation in the one-particle quantum theory. The construction of the latter will be the content of the next section.

3. Quantization: One-Particle Hilbert Space

In this section we will present the quantization of the time-dependent canonical transformation derived in the last section on the one-particle Hilbert space. This allows to transform each mode of the single-mode Mukhanov-Sasaki Hamiltonian into a harmonic oscillator with constant frequency. In Section 4 we will discuss in which sense the results obtained in this section can be generalized to field theories. The unitary implementation of the symplectic transformation we considered can be used for constructing an analytic solution to the time-dependent Schrödinger equation in the form of a unitary time evolution operator.

3.1. Canonical Quantization of the Time-Dependent Canonical Transformation

From the classical theory, the relevant algebra is $\mathcal{P} = (C^\infty(T^*\mathbb{R}^d), \{.,.\}, \cdot)$ equipped with the Poisson bracket and pointwise multiplication, which is the algebra of elementary variables of a classical point-particle in d-dimensional Euclidean space. This algebra can be further extended by an involution operation leading to the Poisson *-algebra that will be our starting point for the canonical quantization. In the following we can restrict our discussion to the case d=1 which is sufficient for the quantization of the single mode Mukhanov-Sasaki system. As a first step we define a quantization map $\mathcal{Q}$ that maps elements of $\mathcal{P}$ into an abstract operator algebra $\mathcal{Q}(\mathcal{P})$. Given any two smooth phase space functions $f, g \in \mathcal{P}$ we have

$$\mathcal{Q} : \mathcal{P} \to \mathcal{Q}(\mathcal{P}), \qquad \{f, g\} \mapsto \mathcal{Q}(\{f, g\}) = -i\big[\mathcal{Q}(f), \mathcal{Q}(g)\big] \in \mathcal{Q}(\mathcal{P}), \qquad (25)$$

where we have set $\hbar = 1$. Requiring that $\mathcal{Q}$ is function-preserving, that is $\mathcal{Q}(F(q, p)) = F(\mathcal{Q}(q), \mathcal{Q}(p))$ for any real function F as usually required for any quantization map, we can now directly write down the quantum version of the generator for the one-parameter family (i.e., time-dependent) of canonical transformations Γ_ξ on $T^*\Sigma$ and its exponential:

$$\mathcal{Q}(\mathcal{G}) := \hat{\mathcal{G}} = \frac{1}{2}\ln(\xi)\left(\hat{q}\hat{p} + \hat{p}\hat{q} + h(\xi,\dot{\xi})\hat{q}^2\right), \tag{26}$$

where $\hat{q}, \hat{p}$ denote elements of the abstract operator algebra $\mathcal{Q}(\mathcal{P})$. For later convenience we quantize the inverse of Γ_ξ and hence the inverse map, that is due to the minus sign in the quantization prescription and to be closer to existing results in the literature, since the mapping to the autonomous Hamiltonian is classically achieved by the inverse of Φ:

$$\mathcal{Q}(\Gamma_\xi^{-1}) = \exp\{i[\mathcal{Q}(\mathcal{G}),.]\} = \exp\{i[\hat{\mathcal{G}},.]\} = \exp\{i\,\mathrm{ad}_{\hat{\mathcal{G}}}\} =: \mathrm{Ad}_{\hat{\Gamma}_\xi},$$

with 'ad' and 'Ad' denoting the adjoint representation of the Lie algebra and the corresponding Lie group, respectively. Now since we want to define the action of $\hat{\Gamma}_\xi$ on some Hilbert space we need a representation that maps the abstract operators into the set of linear operators on a Hilbert space respecting the commutator relations of the abstract algebra, that is $\pi : \mathcal{Q}(\mathcal{P}) \to \mathcal{L}(\mathcal{H})$ such that $\pi(\mathcal{Q}(\mathbb{1}_\mathcal{P})) = \mathbb{1}_\mathcal{H}$, $\pi(\mathcal{Q}(\{q,p\})) = -i[\pi(\mathcal{Q}(q)), \pi(\mathcal{Q}(p))]$ as well as $\pi(\mathcal{Q}(\{q,q\})) = -i[\pi(\mathcal{Q}(q)), \pi(\mathcal{Q}(q))]$ and $\pi(\mathcal{Q}(\{p,p\})) = -i[\pi(\mathcal{Q}(p)), \pi(\mathcal{Q}(p))]$. If not otherwise stated we will work with the standard Schrödinger position representation given by $(\pi, \mathcal{H} = L_2(\mathbb{R},dx))$ with

$$\pi_q(\mathcal{Q}(q)) = \pi_q(\hat{q}) : \mathcal{S}(\mathbb{R}) \to \mathcal{S}(\mathbb{R}), \quad (\pi_q(\hat{q})\Psi)(q) = q\Psi(q),$$
$$\pi_q(\mathcal{Q}(p)) = \pi_q(\hat{p}) : \mathcal{S}(\mathbb{R}) \to \mathcal{S}(\mathbb{R}), \quad (\pi_q(\hat{p})\Psi)(q) = -i\frac{d\Psi}{dq}(q).$$

Here $\mathcal{S}(\mathbb{R})$ denotes the space of Schwartz functions on $\mathbb{R}$. Given the representation we can define the action of $\hat{\Gamma}_\xi$ on both operators and elements Ψ in $\mathcal{S}(\mathbb{R})$ lying dense in $L_2(\mathbb{R},dq)$ according to the prescription:

$$\pi_q(\hat{O}) \mapsto \mathrm{Ad}_{\hat{\Gamma}_\xi}(\pi_q(\hat{O})) := \hat{\Gamma}_\xi \pi_q(\hat{O}) \hat{\Gamma}_\xi^\dagger, \quad \Psi \mapsto \hat{\Gamma}_\xi \Psi := \sum_{n=0}^\infty \frac{(i\pi_q(\hat{\mathcal{G}}))^n}{n!}\Psi, \tag{27}$$

where we used the abbreviation $\hat{\Gamma}_\xi := \pi_q(\hat{\Gamma}_\xi)$ to keep our notation compact. Let us briefly check that the the adjoint action of $\hat{\Gamma}_\xi$ on $\pi_q(\hat{q})$ and $\pi_q(\hat{p})$ is consistent. We have:

$$\mathrm{Ad}_{\hat{\Gamma}_\xi}(\pi_q(\hat{q})) = \sum_{n=0}^\infty \frac{(-i^2)^n}{n!}\left(\ln(\xi)\right)^n \pi_q(\hat{q}) = \xi\pi_q(\hat{q}), \tag{28}$$

where the iterated commutator $[\pi_q(\hat{A}), \pi_q(\hat{B})]_{(n)}$ is defined similarly to the iterated Poisson bracket with an identity at the zeroth order. For $\pi_q(\hat{p})$ we get as expected:

$$\mathrm{Ad}_{\hat{\Gamma}_\xi}(\pi_q(\hat{p})) = \sum_{n=0}^\infty \frac{i^{2n}}{n!}\left(\ln(\xi)\right)^n \pi_q(\hat{p}) + \sum_{n=0}^\infty \frac{(i^2)^{2n+1}}{(2n+1)!}\left(\ln(\xi)\right)^{2n+1} h(\xi,\dot{\xi})\pi_q(\hat{q}) = \frac{\pi_q(\hat{p})}{\xi} + \dot{\xi}\pi_q(\hat{q}), \tag{29}$$

which precisely corresponds to the inverse of the transformation of q and p generated by the classical Hamiltonian vector field $X_\mathcal{G}$. As discussed in Section 2.3, the dynamics of the classical theory is generated by the physical Hamiltonian $H(\mathbf{q},\mathbf{p},t)$. Thus, we can directly consider the corresponding Schrödinger equation in the one dimensional case that is given by

$$i\frac{\partial}{\partial t}\Psi(q,t) = \frac{1}{2}\left(\pi_q(\hat{p})^2 + \omega^2(t)\pi_q(\hat{q})^2\right)\Psi(q,t)$$

and unitarily equivalent to the corresponding Heisenberg equations for $\pi_q(q)$ and $\pi_q(p)$. If we apply the transformation induced by $\hat{\Gamma}_\xi$ on the Hamiltonian and Ψ, which is the natural choice since classically, the replacement of $\mathbf{q},\mathbf{p}$ in terms of $\mathbf{Q},\mathbf{P}$ (the inverse extended map Φ) achieved our aim of mapping $H(t)$ to H_0, we end up with:

$$\hat{\Gamma}_\xi \left(\frac{1}{2}\left(\pi_q(\hat{p})^2 + \omega^2(t)\pi_q(\hat{q})^2\right) - i\frac{\partial}{\partial t}\right)\hat{\Gamma}_\xi^\dagger \hat{\Gamma}_\xi \Psi(q,t) = 0$$

$$\Longleftrightarrow \left[\frac{1}{2}\hat{\Gamma}_\xi\left(\pi_q(\hat{p})^2 + \omega^2(t)\pi_q(\hat{q})^2\right)\hat{\Gamma}_\xi^\dagger - i\hat{\Gamma}_\xi\frac{\partial \hat{\Gamma}_\xi^\dagger}{\partial t} - i\frac{\partial}{\partial t}\right]\hat{\Gamma}_\xi \Psi(q,t) = 0$$

$$\Longleftrightarrow \left[\frac{1}{2}\left(\frac{\pi_q(\hat{p})^2}{\xi^2} + \xi\left(\omega^2(t)\xi + \ddot{\xi}\right)\pi_q(\hat{q})^2\right) - i\frac{\partial}{\partial t}\right]\hat{\Gamma}_\xi \Psi(q,t) = 0$$

$$\Longleftrightarrow \left[\frac{1}{2\xi^2}\left(\pi_q(\hat{p})^2 + \omega_0^2\pi_q(\hat{q})^2\right) - i\frac{\partial}{\partial t}\right]\hat{\Gamma}_\xi \Psi(q,t) = 0 \tag{30}$$

$$\Longleftrightarrow \left[\frac{1}{\xi^2}\hat{H}_0 - i\frac{\partial}{\partial t}\right]\hat{\Gamma}_\xi \Psi(q,t) = 0, \tag{31}$$

with $\hat{H}_0 := \frac{1}{2}(\pi_q(\hat{p})^2 + \omega_0^2\pi_q(\hat{q})^2)$. In the second step, we used the t-derivative of the one-parameter family of transformations $\hat{\Gamma}_\xi^\dagger$, which has already been derived in References [14,17]. We can rediscover their result by using the explicit form of the generator $\pi_q(\hat{\mathcal{G}})$ using (26) and a Baker-Campbell-Hausdorff decomposition of $\hat{\Gamma}_\xi$ in the position representation. Later a similar but slightly generalized procedure for the occupation number representation will be discussed in Section 3.2. We realize that $\hat{H}_0$ in Equation (31) does *not* carry any explicit time dependence, hence we can construct a solution of the Schrödinger equation in (31) by integration. Further note that the inverse square of the time scaling function $\xi(t)$ precisely corresponds to the Lagrange multiplier $\lambda(s) = dt/ds$ that is involved in the extended classical Hamiltonian (14). Given this result we can now give an explicit solution of the Schrödinger equation as was already shown in Reference [17]:

$$\Psi(q,t) = \hat{\Gamma}_\xi^\dagger \exp\left\{-i\pi_q(\hat{H}_0)\int_{t_0}^t \frac{d\tau}{\xi^2(\tau)}\right\}\hat{\Gamma}_{\xi,0}\Psi(q,t_0), \quad \Psi(q,t_0) \in \mathcal{S}(\mathbb{R})_{t_0}, \tag{32}$$

with $\mathcal{S}(\mathbb{R})_{t_0}$ denoting a one-parameter family of Schwarz spaces, each corresponding to a different initial time t_0. In a cosmological context, this behavior is a very natural one, as the *instantaneous vacuum* on cosmological backgrounds shows an analogous behavior. Using that $\hat{I}_{LR}(t) = \hat{\Gamma}_\xi^\dagger \hat{H}_0 \hat{\Gamma}_\xi$, the time evolution in Equation (32) can also be rewritten as:

$$\hat{U}(t_0,t) = \hat{\Gamma}_\xi^\dagger \exp\left\{-i\pi_q(\hat{H}_0)\int_{t_0}^t \frac{d\tau}{\xi^2(\tau)}\right\}\hat{\Gamma}_{\xi,0} = \exp\left\{-i\pi_q(\hat{I}_{LR})\int_{t_0}^t \frac{d\tau}{\xi^2(\tau)}\right\}\hat{\Gamma}_\xi^\dagger\hat{\Gamma}_{\xi,0} \tag{33}$$

At this point let us further discuss the result in the quantum theory: Firstly, the integrand in the exponential corresponds exactly to our time-rescaling transformation in (11) that we naturally obtained in the extended phase space approach of the classical theory. Secondly, if we compare the result here to that in Reference [13], they use the eigenstates of the Lewis-Riesenfeld invariant multiplied by a phase factor to construct the solutions of the time-dependent Schrödinger equation. Now, if $\hat{\Gamma}_\xi^\dagger\hat{\Gamma}_{\xi,0}\Psi(q,t_0)$ corresponds to an eigenstate of the Lewis-Riesenfeld invariant, then this reproduces precisely the phase factor that was introduced in Reference [13] in a rather ad hoc manner. In fact, it can be easily shown that $\hat{\Gamma}_\xi^\dagger\Psi_0(q,t_0)$ for the time-independent vacuum Ψ_0 corresponds to the time-dependent vacuum state of the Lewis-Riesenfeld invariant as we will see later. The expression in Equation (33) corresponds to the unique time evolution operator, that is the Dyson series associated to the time-dependent Hamiltonian $\hat{H}(t)$, since it satisfies identical initial conditions. Moreover, $\hat{U}(t_0,t)$ is closely related to the unitary operator found in Reference [14] (see the equation above (3.15) in that reference). In our framework, it is very natural to find the time-independent Hamiltonian in the central exponential operator on the left-hand-side of (33). The reason for this is twofold: Firstly, the extended canonical transformation Φ maps the Hamiltonian $\hat{H}(t)$ into the time-independent one $\hat{H}_0$ by transforming the Schrödinger equation via $\hat{\Gamma}_\xi$. Secondly, the time-rescaling that is used in the extended phase space appears as a Lagrange multiplier in the extended Hamiltonian constraint and consequently as the integrand in the time-evolution operator. Lastly, let us mention that compared to Reference [14] we use a slightly different Ermakov equation here because the prefactor of $\xi^{-3}(t)$ in the Ermakov equation in (5) corresponds to the squared frequency ω_0^2 of the time-independent oscillator. In the prospects of a field theoretical treatment of this transformation, it is rather unnatural to map every time-dependent mode $\omega_\mathbf{k}(t)$ onto the Minkowski case $\omega_\mathbf{k}^{(0)} = 1$ for all $\mathbf{k}$ as done in Reference [14]. As we will discuss later on, our choice of mapping $\omega_\mathbf{k}(t)$ onto $\omega_\mathbf{k}^{(0)} = k$ is of advantage when we analyze the implementation of the unitary map on the bosonic Fock space in Section 4.

3.2. Baker-Campbell-Hausdorff Decomposition

Explicit calculations involving the evolution operator derived in the last section turn out to be rather tedious, even for simple initial conditions. This is due to the structure of the exponential in $\hat{\Gamma}_\xi^\dagger$ and the associated generator, respectively. As we will show in this section, we can perform a generalized Baker-Campbell-Hausdorff (BCH) decomposition of the operator $\hat{\Gamma}_\xi$ that brings it into a form that is more suitable for actual practical computations. For this purpose it is of advantage to change the representation and henceforth work in the occupation number basis, that is with the usual ladder operators defined as (where we omit the explicit mentioning of the representation from now on):

$$\pi_q(\hat{q}) = \frac{1}{\sqrt{2\omega_0}}(\hat{A}^\dagger + \hat{A}), \quad \pi_q(\hat{p}) = i\sqrt{\frac{\omega_0}{2}}(\hat{A}^\dagger - \hat{A}), \quad [\hat{A}, \hat{A}^\dagger] = \mathbb{1}_\mathcal{H}, \tag{34}$$

where we set as before $\hbar = 1$ and $m = 1$. Inserting these identities into the generator $\pi_q(\hat{\mathcal{G}})$ from (26) and $\hat{\Gamma}_\xi$, we obtain (again without the explicit representation):

$$\begin{aligned}
\hat{\Gamma}_\xi &= \exp\left\{\frac{i}{2}\ln(\xi)\left(i(\hat{A}^\dagger\hat{A}^\dagger - \hat{A}\hat{A}) + \frac{h(\xi)}{2\omega_0}(\hat{A}^\dagger\hat{A}^\dagger + \hat{A}^\dagger\hat{A} + \hat{A}\hat{A}^\dagger + \hat{A}\hat{A})\right)\right\} \\
&= \exp\left\{\frac{1}{2}\ln(\xi)\left(\left(1 + \frac{ih(\xi)}{2\omega_0}\right)\hat{A}\hat{A} - \left(1 - \frac{ih(\xi)}{2\omega_0}\right)\hat{A}^\dagger\hat{A}^\dagger + \frac{ih(\xi)}{\omega_0}\left(\hat{A}^\dagger\hat{A} + \frac{1}{2}\right)\right)\right\} \\
&= \exp\left\{\bar{\alpha}(\xi)\frac{\hat{A}\hat{A}}{2} - \alpha(\xi)\frac{\hat{A}^\dagger\hat{A}^\dagger}{2} + i\lambda(\xi)\left(\hat{A}^\dagger\hat{A} + \frac{1}{2}\right)\right\} \\
&=: \exp\left\{\bar{\alpha}(\xi)\hat{o}_- - \alpha(\xi)\hat{o}_+ + i\lambda(\xi)\hat{o}_3\right\},
\end{aligned} \tag{35}$$

where we made the following redefinitions for later notational convenience:

$$\hat{o}_+ := \frac{1}{2}\hat{A}^\dagger\hat{A}^\dagger, \quad \hat{o}_- := \frac{1}{2}\hat{A}\hat{A}, \quad \hat{o}_3 := \hat{A}^\dagger\hat{A} + \frac{1}{2}, \quad [\hat{o}_3, \hat{o}_\pm] = \pm 2\hat{o}_\pm, \quad [\hat{o}_-, \hat{o}_+] = \hat{o}_3. \tag{36}$$

The coefficients are in fact explicitly time-dependent functions $\alpha(\xi)$, $\lambda(\xi)$, where the time dependency is carried by the solution ξ of the Ermakov Equation (5) as we have seen in the discussion of the classical setup. They are defined as:

$$\alpha = \ln(\xi)\left(1 - \frac{ih(\xi)}{2\omega_0}\right), \quad \lambda = \frac{h(\xi)}{2\omega_0}\ln(\xi), \quad |\alpha|^2 > \lambda^2 \,\forall\, \xi : \mathbb{R} \supseteq I \to \mathbb{R}. \tag{37}$$

After this replacement the resulting expression for $\hat{\Gamma}_\xi$ takes the form of a generalized, time-dependent squeezing operation. The commutation relations in Equation (36) are those of $\mathfrak{sl}(2,\mathbb{R})$, which was already evident in the classical sector of the theory. It is straightforward to see that the standard Baker-Campbell-Hausdorff decomposition does not work, since the iterated commutator structure leads to infinitely many non-vanishing contributions in the well-known formula. However, a BCH decomposition of $SL(2,\mathbb{R})$ elements has been performed using analytic techniques as shown in Reference [30]. This was done by introducing a parametric rescaling of $\hat{\Gamma}_\xi$ and allowing a corresponding dependence of the coefficient functions in the decomposition on this parameter. In our case, a rescaling of the original $\hat{\Gamma}_\xi$ leads to:

$$\hat{\Gamma}_\xi(\mu) := \exp\left\{\mu\hat{\mathcal{G}}\right\} = \exp\left\{\mu\left(\bar{\alpha}(\xi)\hat{\sigma}_- - \alpha(\xi)\hat{\sigma}_+ + i\lambda(\xi)\hat{\sigma}_3\right)\right\}, \tag{38}$$

with an arbitrary rescaling by some parameter $\mu \in \mathbb{R}$. Let us denote the decomposed version of $\hat{\Gamma}_\xi(\mu)$ by $\tilde{\Gamma}_\xi(\mu)$, with a semicolon representing a parametric dependence:

$$\tilde{\Gamma}_\xi(\mu) =: \exp\left\{\beta_+(\xi;\mu)\hat{\sigma}_+\right\}\exp\left\{\gamma(\xi;\mu)\hat{\sigma}_3\right\}\exp\left\{\beta_-(\xi;\mu)\hat{\sigma}_-\right\} \tag{39}$$

Then we aim at determining the coefficient functions $\beta_+(\xi;\mu)$, $\gamma(\xi;\mu)$ and $\beta_-(\xi;\mu)$ such that we have $\hat{\Gamma}_\xi(\mu) = \tilde{\Gamma}_\xi(\mu)$. This rescaling allows us to differentiate $\hat{\Gamma}_\xi(\mu)$ and $\tilde{\Gamma}_\xi(\mu)$ with respect to μ. Considering this we start with a consistency requirement for $\hat{\Gamma}_\xi(\mu)$ and $\tilde{\Gamma}_\xi(\mu)$ given by:

$$\left(\frac{\partial}{\partial\mu}\hat{\Gamma}_\xi(\mu)\right)\left(\hat{\Gamma}_\xi(\mu)\right)^\dagger = \left(\frac{\partial}{\partial\mu}\tilde{\Gamma}_\xi(\mu)\right)\left(\tilde{\Gamma}_\xi(\mu)\right)^\dagger. \tag{40}$$

In the next step we will omit the arguments of the coefficient functions for the sake of a more compact notation. Explicitly evaluating the differentials and using the unitarity of $\hat{\Gamma}_\xi$, we end up with three contributions. A closer look reveals that these contributions contain the adjoint action of $\hat{\Gamma}_\xi$ onto the three generators of the algebra in (36), which can be easily computed due to the simple structure of their commutators. The linear independence of the generators then leads to a coupled system of differential equations for the coefficient functions:

$$\bar{\alpha} = \exp\{-2\gamma\}\frac{\partial\beta_-}{\partial\mu} \tag{41}$$

$$i\lambda = \frac{\partial\gamma}{\partial\mu} - \beta_+\exp\{-2\gamma\}\frac{\partial\beta_-}{\partial\mu} \tag{42}$$

$$\alpha = 2\beta_+\frac{\partial\gamma}{\partial\mu} - \frac{\partial\beta_+}{\partial\mu} - \beta_+^2\exp\{-2\gamma\}\frac{\partial\beta_-}{\partial\mu}. \tag{43}$$

Performing a number of substitutions, this system of differential equations can be cast into the form of a complex Riccati-type ordinary differential equation, for more details we refer the reader to the explicit computations done in Reference [30]. An appropriate ansatz for this equation yields a solution, subsequent resubstitution then leads to the desired BCH coefficient functions of the normal-ordered decomposition of $\hat{\Gamma}_\xi$. A similar procedure can be performed for the normal and anti-normal ordering of both $\hat{\Gamma}_\xi$ and $\hat{\Gamma}_\xi^\dagger$, respectively, while we have chosen that $\hat{\sigma}_3$ remains in the middle for computational convenience. Although depending on the given initial state $\Psi(q,t_0)$ at our disposal, the most useful forms of the coefficients (or operator orderings, respectively) regarding computational convenience are given by:

$$\delta_+(\mu) = +\frac{\alpha\,\text{sh}(\Delta\mu)}{\Delta\,\text{ch}(\Delta\mu) + i\lambda\,\text{sh}(\Delta\mu)} \qquad \tau_+(\mu) = +\frac{\alpha\,\text{sh}(\Delta\mu)}{\Delta\,\text{ch}(\Delta\mu) - i\lambda\,\text{sh}(\Delta\mu)}$$

$$\delta_-(\mu) = -\frac{\bar{\alpha}\,\text{sh}(\Delta\mu)}{\Delta\,\text{ch}(\Delta\mu) + i\lambda\,\text{sh}(\Delta\mu)} \qquad \tau_-(\mu) = -\frac{\bar{\alpha}\,\text{sh}(\Delta\mu)}{\Delta\,\text{ch}(\Delta\mu) - i\lambda\,\text{sh}(\Delta\mu)}$$

$$\underbrace{\nu(\mu) = -\ln\left(\text{ch}(\Delta\mu) + \frac{i\lambda}{\Delta}\text{sh}(\Delta\mu)\right),}_{\text{normal ordering of }\hat{f}^\dagger_\zeta} \qquad \underbrace{\rho(\mu) = \ln\left(\text{ch}(\Delta\mu) - \frac{i\lambda}{\Delta}\text{sh}(\Delta\mu)\right).}_{\text{anti-normal ordering of }\hat{f}^\dagger_\zeta}$$

$$\beta_+(\mu) = -\frac{\alpha\,\text{sh}(\Delta\mu)}{\Delta\,\text{ch}(\Delta\mu) - i\lambda\,\text{sh}(\Delta\mu)} \qquad \varepsilon_+(\mu) = -\frac{\alpha\,\text{sh}(\Delta\mu)}{\Delta\,\text{ch}(\Delta\mu) + i\lambda\,\text{sh}(\Delta\mu)}$$

$$\beta_-(\mu) = +\frac{\bar{\alpha}\,\text{sh}(\Delta\mu)}{\Delta\,\text{ch}(\Delta\mu) - i\lambda\,\text{sh}(\Delta\mu)} \qquad \varepsilon_-(\mu) = +\frac{\bar{\alpha}\,\text{sh}(\Delta\mu)}{\Delta\,\text{ch}(\Delta\mu) + i\lambda\,\text{sh}(\Delta\mu)}$$

$$\underbrace{\gamma(\mu) = -\ln\left(\text{ch}(\Delta\mu) - \frac{i\lambda}{\Delta}\text{sh}(\Delta\mu)\right),}_{\text{normal ordering of }\hat{f}_\zeta} \qquad \underbrace{\iota(\mu) = \ln\left(\text{ch}(\Delta\mu) + \frac{i\lambda}{\Delta}\text{sh}(\Delta\mu)\right).}_{\text{anti-normal ordering of }\hat{f}_\zeta}$$

with $\Delta^2 := |\alpha|^2 - \lambda^2$ and $\Delta^2 > 0$ for all real solutions $\xi(t)$ of the Ermakov Equation (5). By fixing the parameter $\mu = 1$, we recover the unitary transformation we initially started with. Let us note that for an initially time-independent Hamiltonian, the decomposed transformation reproduces the identity operator, as expected. This is due to the fact that in this case $\xi(t) = 1$, which in turn leads to a vanishing generator. Furthermore one can explicitly check that the adjoints of the decompositions of $\hat{f}_\zeta$ and $\hat{f}^\dagger_\zeta$ are the decompositions of the adjoints, which illustrates mutual consistency and conservation of unitarity among the obtained results. To briefly summarize this chapter, we have used analytical techniques to perform a decomposition of the exponentiated generator in (35) into three individual contributions. Due to the fact that we are working with unitary representations of the algebra of non-compact Lie group with mutually non-commuting elements, this result is nontrivial and enables the realization of computations in a compact form. Examples of applications of the Baker-Campbell-Hausdorff decomposition of $\hat{f}_\zeta$ can be found in Sections 3.3 and 6, respectively.

3.3. Time-Dependent Bogoliubov Maps

In this section we will show that the transformation induced by the operator $\hat{f}_\zeta$ can be understood as a time-dependent Bogoliubov transformation when applied to the ladder operators. Given the action of $\hat{f}_\zeta$ on the elementary position and momentum operators, it can naturally be extended to the ladder operators as well. The same applies also to the adjoint action, which is however tedious to evaluate in the original form of the generator. Due to the possibility of decomposing the operator $\hat{f}_\zeta$ and its adjoint, we can take advantage of the result in the last section and compute the action on $\hat{A}$ and $\hat{A}^\dagger$ with a normal and anti-normal ordered decomposition, respectively.

Using the commutator structure of the generators in (36), we obtain:

$$\mathrm{Ad}_{\hat{f}_{\xi}}(\hat{A}) = e^{-\gamma(\xi)}\left(\hat{A} - \beta_{+}(\xi)\hat{A}^{\dagger}\right), \qquad (44)$$

$$\mathrm{Ad}_{\hat{f}_{\xi}}(\hat{A}^{\dagger}) = e^{\iota(\xi)}\left(\hat{A}^{\dagger} + \varepsilon_{-}(\xi)\hat{A}\right), \qquad (45)$$

$$\mathrm{Ad}_{\hat{f}_{\xi}^{\dagger}}(\hat{A}) = e^{-\nu(\xi)}\left(\hat{A} - \delta_{+}(\xi)\hat{A}^{\dagger}\right), \qquad (46)$$

$$\mathrm{Ad}_{\hat{f}_{\xi}^{\dagger}}(\hat{A}^{\dagger}) = e^{\rho(\xi)}\left(\hat{A}^{\dagger} + \tau_{-}(\xi)\hat{A}\right), \qquad (47)$$

with the corresponding coefficient functions derived in the preceding section. These functions carry an explicit time dependence via $\xi(t)$, which is a solution of the Ermakov Equation (5) with the time-dependent frequency of the initial Hamiltonian. In fact, the transformations of the ladder operators in (47) look already close to that of a time-dependent Bogoliubov transformation. Whether this is indeed the case depends on the coefficient functions involved and will be analyzed in the following. For this purpose, let us rewrite the action of $\hat{\Gamma}_{\xi}$ on these operators as a 2×2 matrix representation, considering the Equations (44)–(47):

$$\begin{pmatrix} g(\xi,\dot{\xi}) & \overline{h}(\xi,\dot{\xi}) \\ h(\xi,\dot{\xi}) & \overline{g}(\xi,\dot{\xi}) \end{pmatrix} \begin{pmatrix} \hat{A} \\ \hat{A}^{\dagger} \end{pmatrix} = \begin{pmatrix} g(\xi,\dot{\xi})\hat{A} + \overline{h}(\xi,\dot{\xi})\hat{A}^{\dagger} \\ h(\xi,\dot{\xi})\hat{A} + \overline{g}(\xi,\dot{\xi})\hat{A}^{\dagger} \end{pmatrix} := \begin{pmatrix} \hat{B} \\ \hat{B}^{\dagger} \end{pmatrix}, \qquad (48)$$

with the additional requirement that if $[\hat{A}, \hat{A}^{\dagger}] = \mathbb{1}_{\mathcal{H}}$, then similarly $[\hat{B}, \hat{B}^{\dagger}] = \mathbb{1}_{\mathcal{H}}$ needs to hold, usually required for a Bogoliubov transformation. In order to achieve this, we need to impose the condition that the determinant of the matrix on the left-hand side of (48) involving $g(\xi,\dot{\xi})$ and $h(\xi,\dot{\xi})$ is equal to one, which amounts to:

$$[\hat{B}, \hat{B}^{\dagger}] = \mathbb{1}_{\mathcal{H}} \quad \Longleftrightarrow \quad |g(\xi,\dot{\xi})|^{2} - |h(\xi,\dot{\xi})|^{2} = 1.$$

Applying this to the transformation in (44), we get:

$$\begin{pmatrix} g(\xi,\dot{\xi}) & \overline{h}(\xi,\dot{\xi}) \\ h(\xi,\dot{\xi}) & \overline{g}(\xi,\dot{\xi}) \end{pmatrix} = \begin{pmatrix} e^{-\gamma(\xi)} & -e^{-\gamma(\xi)}\beta_{+}(\xi) \\ -e^{-\overline{\gamma}(\xi)}\overline{\beta}_{+}(\xi) & e^{-\overline{\gamma}(\xi)} \end{pmatrix} \quad \Longrightarrow \quad e^{-(\gamma+\overline{\gamma})}\left(1 - |\beta_{+}|^{2}\right) \stackrel{!}{=} 1.$$

Given the explicit functional form of the BCH coefficients, it can be easily shown that $\hat{\Gamma}_{\xi}$ indeed describes a time-dependent Bogoliubov transformation and the expression above equals one. For all remaining cases this can be also shown using the same method. We are now in a situation where we can formulate the time evolution of $\hat{A}, \hat{A}^{\dagger}$ in the Heisenberg picture, using the time-evolution operator $\hat{U}(t_0, t)$. It consists of the aforementioned Bogoliubov map together with an exponential operator involving the Lewis-Riesenfeld invariant or the autonomous Hamiltonian, respectively. The additional exponent also carries the information of the time-rescaling encoded in the function $\xi(t)$ and hence is sensitive to the underlying spacetime geometry. In the following, we introduce the following notation for the coefficient functions $\beta_{+}(\xi(t)) = \beta_{+}(\xi)$ and $\beta_{+}(\xi(t_0)) := \beta_{+}(\xi_0)$ involved in the decomposition of $\hat{\Gamma}_{\xi}$ and $\hat{\Gamma}_{\xi,0}$, respectively. Carefully applying $\hat{U}(t_0, t)$ and collecting everything together, we obtain:

$$\hat{A}_H(t_0,t) = \exp\left\{-\left(\gamma(\xi) + i\omega_0 \int_{t_0}^t \frac{d\tau}{\xi^2(\tau)} + \nu(\xi_0)\right)\right\}\left(\hat{A} - \delta_+(\xi_0)\hat{A}^\dagger\right)$$
$$- \exp\left\{-\left(\gamma(\xi) - i\omega_0 \int_{t_0}^t \frac{d\tau}{\xi^2(\tau)} + \bar{\nu}(\xi_0)\right)\right\}\beta_+(\xi)\left(\hat{A}^\dagger - \bar{\delta}_+(\xi_0)\hat{A}\right), \quad (49)$$

$$\hat{A}_H^\dagger(t_0,t) = \exp\left\{-\left(\bar{\gamma}(\xi) - i\omega_0 \int_{t_0}^t \frac{d\tau}{\xi^2(\tau)} + \bar{\nu}(\xi_0)\right)\right\}\left(\hat{A}^\dagger - \bar{\delta}_+(\xi_0)\hat{A}\right)$$
$$- \exp\left\{-\left(\bar{\gamma}(\xi) + i\omega_0 \int_{t_0}^t \frac{d\tau}{\xi^2(\tau)} + \nu(\xi_0)\right)\right\}\bar{\beta}_+(\xi)\left(\hat{A} - \delta_+(\xi_0)\hat{A}^\dagger\right). \quad (50)$$

Although the expressions (49) and (50) look rather complicated at first, as expected they reduce to $\hat{A}, \hat{A}^\dagger$ in the limit $t \to t_0$. This can be seen by replacing t by t_0 in the expression above and using the definitions of the Baker-Campbell-Hausdorff coefficients from Section 3.2. One finally observes that all contributions apart from $\hat{A}$ or $\hat{A}^\dagger$, respectively, cancel upon inserting the definition of Δ and using the fact that $\cosh^2(\Delta) - \sinh^2(\Delta) = 1$. In principle, these results allow to compute expectation values for various initial conditions and investigate the behavior of these operators for the single-mode Mukhanov-Sasaki equation. We will discuss some application of this framework in Section 6, where we consider the derived unitary map for the single-mode Mukhanov-Sasaki equation in the context of quasi-de Sitter spacetimes. Prior to that, in the next section we will discuss whether the results obtained so far can be carried over to field theory, that is whether the obtained unitary map can be extended to the bosonic Fock space.

4. Implementing the Time-Dependent Canonical Transformation as a Unitary Map on the Bosonic Fock Space

For the reason that we were able to construct a unitary map for the toy model of the single-model Mukhanov-Sasaki Hamiltonian, the next obvious step is to aim at a unitary implementation of the time evolution operator $\hat{U}(t_0,t)$ on the full Fock space $\mathcal{F}$. Since every mode of the Mukhanov-Sasaki equation is a time-dependent harmonic oscillator, we need to treat every mode separately and with a different frequency, depending on the absolute value of $\mathbf{k}$. Hence it is natural to equip the solution of the Ermakov equation, which also differs from mode to mode for precisely this reason, with a corresponding mode label, which in turn carries over to the time-dependent Bogoliubov transformation $\hat{\Gamma}_\xi$. In the conventional formalism, the Mukhanov-Sasaki Hamiltonian and the mode expansion of the Mukhanov-Sasaki variable and its conjugate momentum are of the form:

$$\hat{H}(\eta) = \frac{1}{2}\int d^3x \left(\hat{\pi}_v^2(\eta,\mathbf{x}) + (\partial_a \hat{v}(\eta,\mathbf{x}))(\partial^a \hat{v}(\eta,\mathbf{x})) - \frac{z''(\eta)}{z(\eta)}\hat{v}^2(\eta,\mathbf{x})\right), \quad (51)$$

$$\hat{v}(\eta,\mathbf{x}) = \int \frac{d^3k}{(2\pi)^3}\left(v_\mathbf{k}(\eta)\hat{a}_\mathbf{k}\exp\{i\mathbf{k}\cdot\mathbf{x}\} + \bar{v}_\mathbf{k}(\eta)\hat{a}_\mathbf{k}^\dagger\exp\{-i\mathbf{k}\cdot\mathbf{x}\}\right), \quad (52)$$

$$\hat{\pi}_v(\eta,\mathbf{x}) = \int \frac{d^3k}{(2\pi)^3}\left(\partial_0 v_\mathbf{k}(\eta)\hat{a}_\mathbf{k}\exp\{i\mathbf{k}\cdot\mathbf{x}\} + \partial_0\bar{v}_\mathbf{k}(\eta)\hat{a}_\mathbf{k}^\dagger\exp\{-i\mathbf{k}\cdot\mathbf{x}\}\right), \quad (53)$$

with ∂_0 denoting a derivative with respect to conformal time η, $a(\eta)$ is the scale factor, $\mathcal{H}$ is the conformal Hubble function and $\bar{\phi}$ stands for the homogeneous and isotropic part of the inflaton scalar field. Given the canonical commutator $[\hat{v}(\eta,\mathbf{x}), \hat{\pi}_v(\eta,\mathbf{y})] = i\delta^{(3)}(\mathbf{x},\mathbf{y})\mathbb{1}_\mathcal{H}$ together with the mode expansion of $\hat{v}(\eta,\mathbf{x})$ and $\hat{\pi}_v(\eta,\mathbf{x})$ as well as the following choice for the Wronskian

$$W(v_\mathbf{k},\bar{v}_\mathbf{k}) := v_\mathbf{k}\bar{v}_\mathbf{k}' - v_\mathbf{k}'\bar{v}_\mathbf{k} = i, \quad (54)$$

the corresponding annihilation and creation operators satisfy the commutator algebra

$$[\hat{a}_\mathbf{k}, \hat{a}_\mathbf{m}^\dagger] = (2\pi)^3 \delta^{(3)}(\mathbf{k}, \mathbf{m}) \mathbb{1}_\mathcal{H},$$

where all remaining commutators vanish. Compared to the one-particle case we obtain an additional factor of $(2\pi)^3$ here, which in principle needs to be considered when deriving the corresponding Bogoliubov coefficients in (44)–(47) for the field theory case. In order to avoid to include appropriate powers of 2π in the derivation of the Bogoliubov coefficients, as an intermediate step we rescale the creation and annihilation operators such that they satisfy a commutator algebra that involves just the δ-function. This yields:

$$\hat{A}_\mathbf{k} := (2\pi)^{-\frac{3}{2}} \hat{a}_\mathbf{k}, \quad \hat{A}_\mathbf{k}^\dagger := (2\pi)^{-\frac{3}{2}} \hat{a}_\mathbf{k}^\dagger, \quad [\hat{A}_\mathbf{k}, \hat{A}_\mathbf{m}^\dagger] = \delta^{(3)}(\mathbf{k}, \mathbf{m}) \mathbb{1}_\mathcal{H}. \tag{55}$$

Note that we consider a quantization of the inflaton perturbation in the context of quantum field theory on a curved background, where the background quantities are considered as external quantities and we thus neglect any backreaction effects.

Now we can let the Bogoliubov transformation act on the rescaled operators $\hat{A}_\mathbf{k}, \hat{A}_\mathbf{k}^\dagger$ and all results obtained in the previous Section 3.3 can be easily carried over to the field theoretic case, where the Bogoliubov transformation maps $\hat{A}_\mathbf{k}, \hat{A}_\mathbf{k}^\dagger$ to a new set of creation and annihilation operators $\hat{B}_\mathbf{k}, \hat{B}_\mathbf{k}^\dagger$ that fulfill the same rescaled commutation relation. Due to the linearity of the Bogoliubov transformation, the rescaling affects both sides of the equation and thus can be easily removed and the standard algebra we started with is restored. In the field theory the generator in the exponential of $\hat{f}_\xi^\mathbf{k}$ is smeared with the Baker-Campbell-Hausdorff coefficient functions that act as the smearing functions. The action on the operators $\hat{A}_\mathbf{k}, \hat{A}_\mathbf{k}^\dagger$ is then diagonal, because at each order of the iterated commutator, the to be found Dirac distributions can be absorbed into the integral involved due to the smearing. Hence, the generalization of Bogoliubov coefficients we obtained in the one-particle case in (48) to the field theory case just consists of equipping them with a mode label. The questions that still needs to answered is whether the so defined extension of $\hat{f}_\xi^\mathbf{k}$ to Fock space describes a unitary map on the latter. Fortunately, there exists a criterion whether a given Bogoliubov transformation can be unitarily implemented on Fock space, called the *Shale-Stinespring* condition. A review on the Shale-Stinespring condition with a sketched proof can be for example found in Reference [31]. The theorem essentially states that the anti-linear part of the Bogoliubov transformation under consideration needs to be a Hilbert-Schmidt operator. In our case, this condition carries over to the product of the off-diagonal coefficients in Equation (48) being bounded when integrated over all of $\mathbb{R}^3$:

$$\int_{\mathbb{R}^3} d^3k\, h_\mathbf{k}(\xi, \xi') \overline{h}_\mathbf{k}(\xi, \xi') < \infty, \quad \overline{h}_\mathbf{k}(\xi, \xi') = -\exp\{-\gamma(\xi_\mathbf{k})\} \beta_+(\xi_\mathbf{k}), \tag{56}$$

where $\gamma(\xi_\mathbf{k})$ and $\beta_+(\xi_\mathbf{k})$ are the Baker-Campbell-Hausdorff coefficients from the decomposition in Section 3.2, $\xi_\mathbf{k}(t)$ is the mode-dependent solution of the Ermakov equation and $\xi'_\mathbf{k}$ is the derivative with respect to conformal time. At this point the advantage of rescaling the operator algebra becomes evident, since we can copy our results from previous computations of the Bogoliubov coefficients. A discussion on the initial conditions regarding the solutions $\xi_\mathbf{k}(t)$ can be found in Section 6 below, which will provide the basis for the investigations of the finiteness of the integral over the anti-linear part of $\hat{f}_\xi^\mathbf{k}$. Explicitly inserting the coefficients while still keeping $\xi_\mathbf{k}(\eta)$ in the arguments and considering the rescaled operators such that they satisfy the same commutator algebra as in Section 3.3 leads to:

$$|h_\mathbf{k}(\xi, \xi')|^2 = \left(\text{ch}^2(\Delta_\mathbf{k}) + \frac{\lambda_\mathbf{k}^2}{\Delta_\mathbf{k}^2} \text{sh}^2(\Delta_\mathbf{k})\right) \frac{|\alpha_\mathbf{k}|^2 \text{sh}^2(\Delta_\mathbf{k})}{\Delta_\mathbf{k}^2 \text{ch}^2(\Delta_\mathbf{k}) + \lambda_\mathbf{k}^2 \text{sh}^2(\Delta_\mathbf{k})} = \frac{|\alpha_\mathbf{k}|^2 \text{sh}^2(\Delta_\mathbf{k})}{|\alpha_\mathbf{k}|^2 - \lambda_\mathbf{k}^2}.$$

Given the former definition of α, λ and Δ we consider the extension of these quantities to the multi-mode case given by $\Delta_{\mathbf{k}} = \sqrt{|\alpha_{\mathbf{k}}|^2 - \lambda_{\mathbf{k}}^2}$. Inserting the explicit form of α and λ from Equation (37) we end up with:

$$\Delta_{\mathbf{k}}^2 = |\alpha_{\mathbf{k}}|^2 - \lambda_{\mathbf{k}}^2 = \left| \ln(\tilde{\zeta}_{\mathbf{k}}) \left(1 - \frac{ih(\tilde{\zeta}_{\mathbf{k}})}{2(\omega_{\mathbf{k}}^{(0)})^2} \right) \right|^2 - \left(\frac{h(\tilde{\zeta}_{\mathbf{k}})}{2(\omega_{\mathbf{k}}^{(0)})^2} \ln(\tilde{\zeta}_{\mathbf{k}}) \right)^2 = \ln^2(\tilde{\zeta}_{\mathbf{k}}).$$

Note that $\ln^2(\tilde{\zeta}_{\mathbf{k}}) > 0$ for all modes $\mathbf{k} \in \mathbb{R}^3$ with $\|\mathbf{k}\| \neq 0$ and all conformal times $\eta \in \mathbb{R}_- \setminus \{0\}$, from which we can conclude that $\Delta_{\mathbf{k}} = \ln(\tilde{\zeta}_{\mathbf{k}})$ since we already know that $\Delta_{\mathbf{k}} > 0$ holds. Explicitly substituting the definitions of α_k, λ_k and Δ_k into $v_k(\xi, \dot{\xi})$, we arrive at the following integral for the de Sitter case with $\tilde{\zeta}(\eta)$ as derived in the succeeding Section 6:

$$\int_{\mathbb{R}^3} d^3k \, |h_{\mathbf{k}}(\xi, \xi')|^2 = \int_{\mathbb{R}^3} d^3k \left(1 + \frac{1}{(\omega_{\mathbf{k}}^{(0)})^2} \left[\frac{\tilde{\zeta}_{\mathbf{k}}(\eta) \tilde{\zeta}'_{\mathbf{k}}(\eta)}{1 - \tilde{\zeta}_{\mathbf{k}}^2(\eta)} \right]^2 \right) \left(\frac{1}{2} \frac{\tilde{\zeta}_{\mathbf{k}}^2(\eta) - 1}{\tilde{\zeta}_{\mathbf{k}}(\eta)} \right)^2$$

$$= \frac{1}{4} \int_{\mathbb{R}^3} d^3k \left[\left(\frac{\tilde{\zeta}_{\mathbf{k}}^2(\eta) - 1}{\tilde{\zeta}_{\mathbf{k}}(\eta)} \right)^2 + \left(\frac{\tilde{\zeta}'_{\mathbf{k}}(\eta)}{\omega_{\mathbf{k}}^{(0)}} \right)^2 \right]$$

$$= \frac{1}{4} \int_{\mathbb{R}^3} d^3k \left[\left(\frac{(k\eta)^2}{1 + (k\eta)^2} \frac{1}{(k\eta)^4} \right) + \frac{1}{k^2} \left(\frac{(k\eta)^2}{1 + (k\eta)^2} \frac{1}{k^4 \eta^6} \right) \right].$$

This expression allows us to consider a simple power-counting procedure of the individual contributions. For large k, the first term behaves as k^{-4} whereas the second term decays as k^{-6}, so there is no divergence in the ultraviolet. For small k we observe the first term to be proportional to k^{-2} and the second contribution to k^{-4}, which leads to an infrared divergence of the latter, which in turn shows that the integral above is not finite. Consequently, the Shale-Stinespring condition is not satisfied in our case and the Bogoliubov transformation $\hat{\Gamma}_{\xi}$ cannot be unitarily implemented on Fock space by simply extending the toy model of the single-mode case to the multi-mode case due to 'infinite particle production' between mutually different vacuum states. Interestingly, there is no issue with the ultraviolet here but just in the infrared sector, showing that next to the large k behavior one also needs to check whether there are occurring singularities in the infrared, as they can equally add a diverging contributions to the number operator expectation value with respect to different vacua. It is not obvious to us that this aspect has been considered in the recent work of Reference [15], where a similar Bogoliubov transformation is used on Fock space. As can be seen form our analysis, the behavior of the BCH coefficients is different for small k than it is for large k, hence it is not obvious that even if the Bogoliubov coefficients are finite for large k this is a sufficient check in order to conclude that the Bogoliubov transformation under consideration can be unitarily implemented on Fock space.

As discussed before at the end of Section 3.1, the most common transformation in the literature in the context of the Lewis-Riesenfeld invariant is the one where the time-dependent Hamiltonian is mapped to the Hamiltonian of a harmonic oscillator with frequency $\omega_{\mathbf{k}}^{(0)} = 1$. For this reason we also analyze what happens to the Shale-Stinespring condition if we do not require the time-independent frequency to be just $\omega_{\mathbf{k}}^{(0)} = k$ but unity instead. This changes the solution $\tilde{\zeta}_{\mathbf{k}}(\eta)$ by an additional factor of $k^{-\frac{1}{2}}$ if we impose similar initial conditions, that is $\tilde{\zeta}_{\mathbf{k}}^{(sq)}(\eta) = k^{-\frac{1}{2}} \tilde{\zeta}_{\mathbf{k}}(\eta)$ and leads to the residual squeezing transformation in $\hat{\Gamma}_{\xi}^{\mathbf{k}}$ in the limit of past conformal infinity already mentioned in Section 3.1. Considering this modification in $\tilde{\zeta}_{\mathbf{k}}^{sq}$ compared to $\tilde{\zeta}_{\mathbf{k}}$, we can also analyze whether the Shale-Stinespring conditions is satisfied here. We have:

$$\int_{\mathbb{R}^3} d^3k\, |h_{\mathbf{k}}(\zeta_{\mathbf{k}}^{(sq)}, \zeta_{\mathbf{k}}^{(sq)\prime})|^2 = \frac{1}{4} \int_{\mathbb{R}^3} d^3k \left[\left(\frac{(\zeta_{\mathbf{k}}^{(sq)}(\eta))^2 - 1}{\zeta_{\mathbf{k}}^{(sq)}(\eta)} \right)^2 + \left(\zeta_{\mathbf{k}}^{(sq)\prime}(\eta) \right)^2 \right]$$

$$= \frac{1}{4} \int_{\mathbb{R}^3} d^3k \left[(\zeta_{\mathbf{k}}^{(sq)})^{-2} \left(\frac{1}{k^3\eta^2} - \frac{k-1}{k} \right)^2 + \frac{1}{k} \left(\frac{(k\eta)^2}{1+(k\eta)^2} \frac{1}{k^4\eta^6} \right) \right]$$

$$= \frac{1}{4} \int_{\mathbb{R}^3} d^3k \left[k \left(\frac{(k\eta)^2}{1+(k\eta)^2} \right) \left(\frac{1}{k^3\eta^2} - \frac{k-1}{k} \right)^2 + \frac{1}{k} \left(\frac{(k\eta)^2}{1+(k\eta)^2} \frac{1}{k^4\eta^6} \right) \right].$$

We apply a similar power counting to the two terms involved in the last line separately. For small k the second summand in that line decays as k^{-3}, whereas it is proportional to k^{-5} in the limit of large k, which yields a finite contribution in the ultraviolet and an infrared divergence. Similarly, in the small k region at lowest order, the first summand behaves as k^{-3} and thus is divergent in the infrared. Furthermore, it increases linearly in k in the large k limit causing a divergence in the ultraviolet. As a consequence, also the transformation associated with $\zeta_{\mathbf{k}}^{(sq)}$ is not unitarily implementable on Fock space, just as it was the case with $\zeta_{\mathbf{k}}$. However, there is a subtle distinction between the two cases. For $\zeta_{\mathbf{k}}$ we found that the infrared modes lead to a divergence, whereas this time both small and large values of $\|\mathbf{k}\|$ are problematic. Let us understand a bit more in detail why it is expected that the infrared modes can be problematic in the case of the map corresponding to $\zeta_{\mathbf{k}}$. This map transforms the time-dependent harmonic oscillator Hamiltonian with frequency $\omega_{\mathbf{k}}^2(\eta) = k^2 - \frac{z''(\eta)}{z(\eta)}$ into the Hamiltonian of the harmonic oscillator with constant frequency $\omega_{\mathbf{k}}^{(0)} = k$. Hence, for $\mathbf{k} = 0$ the latter corresponds to the Hamiltonian of a free particle because here the frequency just vanishes. This aspect has not been carefully taken into account in the map constructed so far. Therefore, in the next section we will discuss how the map constructed up to now could be modified for the low $\|\mathbf{k}\|$ modes such that the infrared singularity can be avoided. For the map corresponding to $\zeta_{\mathbf{k}}$, this attempt is possible because the problematic behavior of these modes constitutes only a compact domain in the space of modes, in contrast to the additional ultraviolet divergence involved in the map associated with $\zeta_{\mathbf{k}}^{(sq)}$. For this purpose, we will introduce the so-called Arnold transformation that has been used already in Reference [16] at the quantum level, which is designed to perform a mapping to the free particle Hamiltonian.

Proposal of a Modified Map for the Infrared Modes: The Arnold Transformation

In the previous section we have observed that the map $\hat{\Gamma}_{\zeta}$ that maps the time-dependent harmonic oscillator system onto the system of a harmonic oscillator with constant frequency $\omega_{\mathbf{k}}^{(0)} = k$ with $k := \|\mathbf{k}\|$ is not a unitary operator on Fock space due to an infrared divergence that occurs in the off-diagonal trace of the Bogoliubov coefficients for the infrared modes. Given the fact that no ultraviolet singularities arise, the strategy we will follow in this section is to consider a modification of the map induced by $\hat{\Gamma}_{\zeta}$ for a finite spherical neighbourhood $1 \gg \|\mathbf{k}_\epsilon\| > 0$ of the infrared modes in such a way that no infrared singularities occur. As mentioned above, the natural target Hamiltonian we should map to in the case of the zero mode is the Hamiltonian of a free particle. At the classical level this so-called Arnold transformation [32] was introduced in order to transform a generic second-order differential equation, that physically describes a driven harmonic oscillator with time-dependent friction coefficient and time-dependent frequency, into the differential equation corresponding to the motion of a free particle. Its implementation as a unitary map at the quantum level has been investigated for instance in Reference [16]. From our approach we can make an immediate connection to this formalism by going back to Equation (15) that has played an important role in deriving our classical transformation. Now if we aimed at mapping the original time-dependent Hamiltonian $H(\eta)$ onto the free particle Hamiltonian, $\zeta(\eta)$ would need to satisfy the harmonic equation of motion with the frequency $\omega_{\mathbf{k}}(\eta)$ for each mode instead of the Ermakov equation, as it was presented in our case before. As already discussed in Reference [16], one can recover the Ermakov equation when considering three physical systems, a time-dependent and a time-independent harmonic oscillator together with the free particle. Then one constructs the two Arnold transformations that relate the time-dependent and the time-independent harmonic oscillator to the free particle. From combining one of these Arnold transformation with the inverse of the second one, one obtains a map that relates the systems of the time-dependent harmonic oscillator with the time-independent one via a time-rescaling. For more details regarding this aspect we refer the reader to the presentation in Reference [16]. In order to be able to discuss the approach from Reference [16] and ours in parallel, we will denote the time-rescaling function associated with the Arnold transformation by $\Theta_{\mathbf{k}_\epsilon}(\eta)$. The non-zero $\|\mathbf{k}_\epsilon\|$ modes lead to the well-known solutions of the Mukhanov-Sasaki equation for finite $\|\mathbf{k}\|$, whereas for the $\mathbf{k} = 0$ mode we need to find appropriate solutions. By construction, $\Theta_{\mathbf{k}_\epsilon}(\eta)$ satisfies the Ermakov equation with vanishing $\omega_{\mathbf{k}}^{(0)}$, that is the time-dependent harmonic oscillator equation of the associated mode, given by:

$$\Theta_{\mathbf{k}_\epsilon}''(\eta) + \omega_{\mathbf{k}_\epsilon}^2(\eta)\Theta_{\mathbf{k}_\epsilon}(\eta) = 0,$$

where we are interested in the case where $\omega_{\mathbf{k}_\epsilon}(\eta)$ is determined by the Mukhanov-Sasaki equation. If we compare the time rescaling in Equation (11) with the one given in Reference [16], we obtain an exact agreement if we take into account that the Wronskians of two solutions of the time-dependent and time-independent harmonic oscillator, respectively, are constant and can be chosen to be identical. Since in our work the physical system under consideration is described by a time-dependent harmonic oscillator (that is, the Mukhanov-Sasaki equation), let us first consider this equation for arbitrary modes $\mathbf{k}$ and prior to any gauge-fixing:

$$v_{\mathbf{k}}''(\eta) + \left(\|\mathbf{k}\|^2 - \frac{z''(\eta)}{z(\eta)}\right)v_{\mathbf{k}}(\eta) = 0, \tag{57}$$

For the particular case of a quasi-de Sitter spacetime, the explicit form of this equation can be given in terms of the so-called slow-roll parameters. The Friedmann equations together with the Klein-Gordon equation describing the dynamics of the background scalar field ϕ on a slow-rolling quasi-de Sitter background can be used to rewrite the Mukhanov-Sasaki equation in a convenient way. For this purpose we define a set of three slow-roll parameters ε, τ and κ, that describe the fractional change of $\dot{H}$ per Hubble time, the fractional chance of ε per Hubble time as well as the fractional change of τ per Hubble time, respectively:

$$\varepsilon = -\frac{\dot{H}}{H^2}, \quad \tau = \frac{\dot{\varepsilon}}{\varepsilon H}, \quad \kappa = \frac{\dot{\tau}}{\tau H},$$

where a dot denotes a derivative with respect to cosmological time in these expressions. Inserting the Klein-Gordon equation and using the Friedmann equations along with an assumed subdominance of the second-order derivatives of ϕ, we can rewrite $z(\eta)$ and its derivatives in terms of ε, τ and κ. By truncating the resulting expressions after the first order in the slow-roll parameters, the time-dependent part of the frequency $\omega_k(\eta)$ becomes:

$$z = \frac{a^2\varepsilon}{4\pi}, \quad \frac{z'}{z} = \mathcal{H}\left(1 + \frac{\tau}{2}\right), \quad \frac{z''}{z} = \frac{1}{\eta^2}(1+\varepsilon)^2\left(2 - \varepsilon + \frac{3\tau}{2}\right) \approx \frac{4\nu^2 - 1}{4\eta^2}, \quad \nu = \frac{3}{2} + \varepsilon + \frac{\tau}{2}.$$

Thus, the Mukhanov-Sasaki equation up to first order in the slow-roll parameters for a quasi-de Sitter background reads:

$$v_{\mathbf{k}}''(\eta) + \left(\|\mathbf{k}\|^2 - \frac{4v^2 - 1}{4\eta^2} \right) v_{\mathbf{k}}(\eta) = 0. \tag{58}$$

Given the equation above we can read off the time-dependent frequency that we considered for the time-dependent harmonic oscillator in our single-mode toy model approach. This is also precisely the equation that $\Theta(\eta)$ needs to satisfy for a given but finite $\|\mathbf{k}_\epsilon\|$. Let us emphasize that the solutions to Equation (58) need to be computed separately for vanishing and non-vanishing $\|\mathbf{k}\|$, respectively. The real-valued solutions for $\|\mathbf{k}\| > 0$ are given by the Bessel functions of first and second kind, for details the reader is referred to Section 6. If we consider the limiting case of $\mathbf{k} = 0$ in the context of the quantum Arnold transformation, we obtain the following linear differential equation with time-dependent coefficients for the rescaling function $\Theta_{\mathbf{k}_\epsilon}(\eta)$, omitting the label for the zero mode:

$$\Theta''(\eta) - \frac{4v^2 - 1}{4\eta^2} \Theta(\eta) = 0 \quad \Longleftrightarrow \quad \eta^2 \Theta''(\eta) - \left(v^2 - \frac{1}{4} \right) \Theta(\eta) = 0 \quad \text{for} \quad \|\mathbf{k}\| = 0.$$

This differential equation with time-dependent coefficients can be transformed into an equation with constant coefficients, which then again can be solved by means of the substitution $y = \ln(|\eta|)$ and an exponential ansatz of this new variable incorporating the dependence on the effective slow-roll parameter v. The general solution of this differential equation is given by:

$$\Theta(\eta) = c_1 |\eta|^r + c_2 |\eta|^s, \quad \text{with} \quad r, s = \frac{1}{2}\left(1 \pm \sqrt{4v^2}\right) \quad \text{for} \quad v^2 > 0 \tag{59}$$

From this solution we readily obtain two linearly independent solutions Θ_1, Θ_2 that can be used to construct the Arnold transformation for the $\mathbf{k} = 0$ mode. Note that the differential equation above can be also solved for $v = 0$ or $v^2 < 0$, respectively. However, according to the parameter space of the slow-roll parameters in Reference [33], this range is not physically reasonable and hence we only use the result for strictly positive, real-valued slow-roll parameters. Due to the range of conformal time ($\eta \in \mathbb{R}_- \setminus \{0\}$), the relevant solution here is the growing branch proportional to $|\eta|^s$ with $s < 0$, since the decreasing branch diverges in the limit of past conformal infinity. This then coincides with the choice of the final time-rescaling transformation suggested in Reference [16]. We reconsider the form of the transformation $\hat{\Gamma}_{\tilde{\zeta}}$ and insert the corresponding solution for $\Theta_{\mathbf{k}_\epsilon}(\eta)$ to obtain an analogous transformation $\hat{\Gamma}_{\Theta_{\mathbf{k}_\epsilon}}$ by means of which we can transform the Schrödinger equation similar to Equation (31) and according to:

$$i\frac{\partial}{\partial t} \Psi(q,t) = \frac{1}{2}\left(\pi_q(\hat{p})^2 + \omega_{\mathbf{k}_\epsilon}^2(t) \pi_q(\hat{q})^2 \right) \Psi(q,t) \quad \Longleftrightarrow \quad \left(\frac{1}{2\Theta_{\mathbf{k}_\epsilon}^2} \pi_q(\hat{p})^2 - i\frac{\partial}{\partial t} \right) \hat{\Gamma}_{\Theta_{\mathbf{k}_\epsilon}} \Psi(q,t) = 0.$$

This means that $\hat{\Gamma}_{\Theta_{\mathbf{k}_\epsilon}}$ maps the time-dependent Hamiltonian for the $\mathbf{k}_\epsilon$ mode into the time-independent Hamiltonian of the free particle modulo a rescaling of the momentum operator. It is important to emphasize that this unitary map can only be performed at the level of the full Schrödinger equation, as otherwise the spectrum of the two related operators would have to be equivalent, which is clearly not the case for the time-dependent harmonic oscillator and the free particle. It is the time-derivative in the Schrödinger equation that is crucial for removing the term proportional to $\pi_q(\hat{q})^2$ altogether. Furthermore we would like to stress that the time rescaling function $\Theta_{\mathbf{k}_\epsilon}$ is different for $\|\mathbf{k}_\epsilon\| > 0$ and $\|\mathbf{k}_\epsilon\| = 0$, respectively. In the first case, depending on the imposed initial conditions, it is given by the Bessel functions of first and second kind $J_v(-\mathbf{k}_\epsilon \eta)$ and $Y_v(-\mathbf{k}_\epsilon \eta)$. In the latter case, $\Theta_{\mathbf{k}_\epsilon}$ corresponds to the above power-law solution. Unfortunately, there are some drawbacks of the implementation of the quantum Arnold transformation with the help of $\hat{\Gamma}_{\Theta_{\mathbf{k}_\epsilon}}$. Firstly, the limit of past conformal infinity is not well-defined in terms of the generator $\hat{\mathcal{G}}$ as depicted in (26), for neither of the two cases. This especially means that we do not get an asymptotic identity map for an already free particle (i.e., the $\|\mathbf{k}_\epsilon\| = 0$ case in the limit of past conformal infinity) as we do with the initially time-independent harmonic oscillator in the case of the original transformation $\hat{T}_\xi$. Secondly, the attempt to relate the free particle with a Lewis-Riesenfeld type invariant does not work as smoothly as in the case of the previous map. If we construct a similar invariant in this case here for the initially time-dependent oscillator Hamiltonian, it can be trivially factorized and has the following form:

$$I_{LR} = \frac{1}{2}\hat{T}^\dagger_{\Theta_{\mathbf{k}_\epsilon}} \pi_q(\hat{p})^2 \hat{T}_{\Theta_{\mathbf{k}_\epsilon}} = \frac{1}{2}(\Theta_{\mathbf{k}_\epsilon} \pi_q(\hat{p}) - \Theta'_{\mathbf{k}_\epsilon} \pi_q(\hat{q}))^2 = \hat{a}^\dagger_{\mathbf{k}_\epsilon} \hat{a}_{\mathbf{k}_\epsilon}, \quad \hat{a}_{\mathbf{k}_\epsilon} = \frac{i}{\sqrt{2}}(\Theta_{\mathbf{k}_\epsilon} \pi_q(\hat{p}) - \Theta'_{\mathbf{k}_\epsilon} \pi_q(\hat{q})). \quad (60)$$

This quantity has for example been already obtained in Reference [17] as a quantum invariant based on orthogonal functions in a similar context. It is immediate that the above factorization is in this sense pathological, as one can immediately see that the occurring operators $\hat{a}_{\mathbf{k}_\epsilon}, \hat{a}^\dagger_{\mathbf{k}_\epsilon}$ can not be interpreted as ladder operators due to $[\hat{a}_{\mathbf{k}_\epsilon}, \hat{a}^\dagger_{\mathbf{k}_\epsilon}] = 0$, since they only differ by a global sign. This can be also seen by looking at the original invariant (24) which has an additional term proportional to $\hat{q}^2 \xi^{-2}$ that is absent in the case of the Arnold transformation for $\mathbf{k}_\epsilon$ by construction, simply because of the lack of a term proportional to $\pi_q(\hat{q})$ in the transformed free particle Hamiltonian. Nevertheless, the transformation $\hat{T}_{\Theta_{\mathbf{k}_\epsilon}}$ is unitary for all finite times η and all considered modes. However, due to the non-preservation of the commutator structure between $\hat{a}_{\mathbf{k}_\epsilon}$ and $\hat{a}^\dagger_{\mathbf{k}_\epsilon}$, it is not a Bogoliubov transformation, hence it does not qualify as an infrared continuation of the map $\hat{T}_\xi$ used throughout this work.

In summary, it was not possible to find a transformation similar to $\hat{T}_\xi$ for the infrared modes. Regarding predictions in inflationary comsology, we are naturally interested in the large k modes, which are properly implementable in the context of our symplectic transformation. Hence, as an alternative to the proposed maps for the infrared, we suggest the identity map as a proper choice, that is:

$$\hat{T}_\xi := \begin{cases} \exp\left(-i\left[\int_{V_\epsilon} d^3k\, \hat{\mathcal{G}}(\xi_\mathbf{k}, \dot{\xi}_\mathbf{k}), \cdot\right]\right) & \text{for} \quad \|\mathbf{k}\| > \|\mathbf{k}_\epsilon\|, \\ \mathbb{1}_\mathcal{H} & \text{for} \quad \|\mathbf{k}\| \leq \|\mathbf{k}_\epsilon\|, \end{cases} \quad (61)$$

where $V_\epsilon := \{\mathbf{k} \in \mathbb{R}^3 : \|\mathbf{k}\| > \|\mathbf{k}_\epsilon\|\}$ is the smearing domain and $\hat{\mathcal{G}}_\mathbf{k}$ denotes the mode-dependent generator of the Bogoliubov transformation depicted in Equation (26) and especially in (35) in terms of annihilation and creation operators, respectively. This is possible and well-defined since the occurring coefficients in the generator are smooth for $\|\mathbf{k}\| > \|\mathbf{k}_\epsilon\|$ and moreover lie in $L^1(V_\epsilon, d^3k)$ as can be checked by explicit integration. The reasons for choosing the identity map are twofold. Firstly, this trivially constitutes a Bogoliubov transformation with the off-diagonal coefficients in (48) vanishing, rendering the Shale-Stinespring integral finite, thus allowing for unitary implementability of $\hat{T}^\mathbf{k}_\xi$ on Fock space. Secondly, the functions multiplying the off-diagonal elements in the Mukhanov-Sasaki Hamiltonian remain unchanged compared to the standard case, which means that they can be neglected for sufficiently early times. This is due to the fact that the effective friction term in the equation of motion for these functions is subdominant in this regime. For details the reader is referred to the discussion in the succeeding section.

5. Relation of the Lewis-Riesenfeld Invariant Approach to the Bunch-Davies Vacuum and Adiabatic Vacua

In the context of the results of the previous sections, it is a natural question whether there exists a relation of the mode functions obtained in the framework of the formalism in this work and the ones obtained in the standard approach in cosmology. As we will show by taking time-rescaling transformation into account, we can relate the solutions $\xi_\mathbf{k}$ of the Ermakov equation to the mode functions associated with the Bunch-Davies vacuum and other adiabatic vacua. For this purpose we consider the following form of the Mukhanov-Sasaki mode function

$$v_k(\eta) = N_k \xi_k(\eta) \exp\left\{-i\omega_k^{(0)} \int^\eta \frac{d\tau}{\xi_k^2(\tau)}\right\}, \qquad (62)$$

corresponding to a polar representation of the complex mode v_k into a real function ξ_k and a complex phase that was in a similar form already mentioned in Reference [15]. N_k is time-independent for each mode, $\xi_k(\eta)$ remains arbitrary at this point and $\omega_k^{(0)}$ can take the values k or 1 depending on the choice of map that is considered. We want to show that the Mukhanov-Sasaki equation

$$v_k''(\eta) + \omega_k^2(\eta) v_k(\eta) = 0 \qquad (63)$$

expressed in terms of the polar representation exactly coincides with the Ermakov equation. Starting from this polar representation of the mode functions we compute the second derivative and reinsert it into the Mukhanov-Sasaki equation to obtain:

$$v_k''(\eta) = N_k \exp\left\{-i\omega_k^{(0)} \int^\eta \frac{d\tau}{\xi_k^2(\tau)}\right\}\left(\xi_k'' - i\omega_k^{(0)} \frac{\xi_k'}{\xi_k^2} + i\omega_k^{(0)} \frac{\xi_k'}{\xi_k^2} - \frac{(\omega_k^{(0)})^2}{\xi_k^3}\right). \qquad (64)$$

We realize that summands involving ξ_k' cancel each other and the Mukhanov-Sasaki equation can be rewritten as:

$$\xi_k'' + \omega_k^2(\eta)\xi_k - \frac{(\omega_k^{(0)})^2}{\xi_k^3} = 0 \quad \Longleftrightarrow \quad v_k''(\eta) + \omega_k^2(\eta) v_k(\eta) = 0. \qquad (65)$$

That is, we recover the Ermakov equation for the radial part of the polar representation in Equation (62). The polar representation of the mode functions can also be obtained if we consider how the Fourier modes transform under the time-dependent canonical transformation that relates the time-dependent and time-independent harmonic oscillator. The mode functions in the system of the harmonic oscillator written as a function of conformal time are given by $u_k(\eta) = N_k \exp\left(-i\omega_k^{(0)} \int^\eta \frac{d\tau}{\xi_k^2(\tau)}\right)$. Here $u_k(T)$ satisfies the standard harmonic oscillator differential equation with respect to the time variable T. Using that for each mode we have $T_k' = \xi_k^{-2}$, one can easily derive the corresponding differential equation that $u_k(\eta)$ fulfills with respect to conformal time η. Now the time-dependent canonical transformation rescales the spatial coordinate by ξ_k^{-1}. Considering this as well as the fact that the mode $u_k(\eta)$ depends on k only, the corresponding Fourier mode after the transformation is given by $v_k(\eta) = \xi_k u_k(\eta)$, yielding again the polar representation of the Fourier mode shown in (62), where we used how the Fourier transform changes under a scaling of the coordinates.

A second way to obtain this result is via the explicit form of the Bogoliubov transformation associated with the time-dependent canonical transformation. We denote the time-dependent annihilation and creation operators of the harmonic oscillator system by $\hat{b}_k(T) = u_k(T)\hat{b}_k$ and $\hat{b}_k^\dagger(T) = \bar{u}_k(T)\hat{b}_k^\dagger$ respectively, where the time-dependent annihilation and creation operators satisfy the Heisenberg equation associated with the Hamiltonian of the harmonic oscillator. Once more considering the relation between T and η for each mode, we can also understand $\hat{b}_k(\eta)$ and $\hat{b}_k^\dagger(\eta)$ as operator-valued functions of conformal time η. The mode expansion in the system of the time-dependent harmonic oscillator can be written in terms of time-dependent annihilation and creation operators $\hat{a}_k(\eta) = v_k(\eta)\hat{a}_k$ and $\hat{a}_k^\dagger(\eta) = \bar{v}_k(\eta)\hat{a}_k^\dagger$ which both satisfy the Heisenberg equation associated to the Mukhanov-Sasaki Hamiltonian. As shown in Section 4, the time-dependent canonical transformation corresponds to a time-dependent Bogoliubov map at the quantum level. In the notation of the last section, this relates the two sets of annihilation and creation operators as follows[3]:

$$\hat{a}_{\mathbf{k}}(\eta) = g_{\mathbf{k}}(\xi,\xi')\hat{b}_{\mathbf{k}}(\eta) + \overline{h}_{\mathbf{k}}(\xi,\xi')\hat{b}^{\dagger}_{\mathbf{k}}(\eta), \quad \hat{a}^{\dagger}_{\mathbf{k}}(\eta) = \overline{g}_{\mathbf{k}}(\xi,\xi')\hat{b}^{\dagger}_{\mathbf{k}}(\eta) + h_{\mathbf{k}}(\xi,\xi')\hat{b}_{\mathbf{k}}(\eta).$$

The explicit form of these coefficients is given by:

$$g_{\mathbf{k}}(\xi,\xi') = \frac{1}{2}\left(\xi_{\mathbf{k}} + \frac{1}{\xi_{\mathbf{k}}}\right) + \frac{i}{2\omega_{\mathbf{k}}^{(0)}}\xi', \quad h_{\mathbf{k}}(\xi,\xi') = \frac{1}{2}\left(\xi_{\mathbf{k}} - \frac{1}{\xi_{\mathbf{k}}}\right) - \frac{i}{2\omega_{\mathbf{k}}^{(0)}}\xi'. \tag{66}$$

Given this time-dependent Bogoliubov map, the Fourier modes in the two systems are related via

$$v_{\mathbf{k}}(\eta) = \left(g_{\mathbf{k}}(\xi,\xi') + h_{\mathbf{k}}(\xi,\xi')\right)u_{\mathbf{k}} = \xi_{\mathbf{k}} u_{\mathbf{k}}(\eta) = N_{\mathbf{k}}\xi_{\mathbf{k}} \exp\left\{-i\omega_{\mathbf{k}}^{(0)}\int^{\eta}\frac{d\tau}{\xi_{\mathbf{k}}^{2}(\tau)}\right\}. \tag{67}$$

Hence, we again recover the polar representation of the Fourier mode. At this point we did not yet clarify the purpose of the $N_{\mathbf{k}}$, which is intricately connected with the commutator algebra of annihilation and creation operators as we will see. Recall the well-known (off-diagonal) form of the Mukhanov-Sasaki Hamiltonian if we insert the mode expansions into the Hamiltonian density:

$$H = \int \frac{d^3k}{(2\pi)^3}\left[F_{\mathbf{k}}(\eta)\hat{a}_{\mathbf{k}}\hat{a}_{-\mathbf{k}} + \overline{F}_{\mathbf{k}}(\eta)\hat{a}^{\dagger}_{\mathbf{k}}\hat{a}^{\dagger}_{-\mathbf{k}} + E_{\mathbf{k}}(\eta)\left(2\hat{a}^{\dagger}_{\mathbf{k}}\hat{a}_{\mathbf{k}} + (2\pi)^3\delta^{(3)}(0)\right)\right], \tag{68}$$

where we used the isotropy of the mode functions due to the high degree of symmetry of the spacetime, the invariance of the measure under reflection and the following definitions:

$$F_{\mathbf{k}}(\eta) := (v'_{\mathbf{k}})^2 + \omega_{\mathbf{k}}^2(\eta)v_{\mathbf{k}}^2, \quad E_{\mathbf{k}}(\eta) := v'_{\mathbf{k}}\overline{v}'_{\mathbf{k}} + \omega_{\mathbf{k}}^2(\eta)v_{\mathbf{k}}\overline{v}_{\mathbf{k}}. \tag{69}$$

Regarding the normalization of the mode functions $v_{\mathbf{k}}$, we can transfer this condition to the polar representation given in Equation (62) by just inserting the definition into the Wronskian. This removes the dependence on $\xi_{\mathbf{k}}$ completely and we can explicitly give a relation between $N_{\mathbf{k}}$ and the Wronskian of the original mode functions:

$$W(v_{\mathbf{k}},\overline{v}_{\mathbf{k}}) = 2i\omega_{\mathbf{k}}^{(0)} N_{\mathbf{k}}^2. \tag{70}$$

This is not surprising upon closer inspection. Recall that the $\omega_{\mathbf{k}}^{(0)}$ in the Ermakov equation corresponds to the time-independent frequency in the transformed Schrödinger equation. We conveniently chose to map the Mukhanov-Sasaki frequency into just the k-dependent part, completely removing the time dependence. This has the effect that $\hat{\Gamma}_{\xi}$ becomes the identity transformation for the case of an initially time-independent oscillator, whereas we would obtain a residual squeezing if we mapped every mode to unity. This freedom of choice is reflected in the explicit form of the normalization constant $N_{\mathbf{k}}$, which depends on the choice of the oscillator frequency in the target system in order to preserve the normalization of the mode functions and hence the standard commutator algebra of annihilation and creation operators. Given the Mukhanov-Sasaki Hamiltonian in the form of annihilation and creation operators in (68), we can discuss the assumptions for the initial condition regarding the Fourier modes associated with the Bunch-Davies vacuum and

---

[3] Note that the roles of $\hat{a}_{\mathbf{k}}, \hat{a}^{\dagger}_{\mathbf{k}}$ and $\hat{b}_{\mathbf{k}}, \hat{b}^{\dagger}_{\mathbf{k}}$ are interchanged in comparison to the one-particle case considered in (48) for notational convenience, whereas the coefficients are named analogously. Here the first set of operators belongs to the Mukhanov-Sasaki Hamiltonian, whereas the second set is associated to the time-independent harmonic oscillator. In contrast, in (48) the operators $\hat{B}, \hat{B}^{\dagger}$ belong to the time-dependent system, whereas $\hat{A}, \hat{A}^{\dagger}$ are associated with the time-independent harmonic oscillator.

the ones obtained in our work and compare them, consequently. First, we rewrite the Fourier mode associated to the Bunch-Davies vacuum given by

$$v_k^{BD}(\eta) = \frac{1}{\sqrt{2k}}\left(1 - \frac{i}{k\eta}\right)e^{-ik\eta} \tag{71}$$

in the polar representation as shown in (62) for our general solution. This yields:

$$v_k^{BD}(\eta) = i|v_k^{BD}|\exp\left\{-i\omega_k^{(0)}\int^\eta \frac{d\tau}{\xi_k^2(\tau)}\right\} = \frac{i}{\sqrt{2k}}\sqrt{1 + \frac{1}{(k\eta)^2}}\exp\left\{-i\omega_k^{(0)}\int^\eta \frac{d\tau}{\xi_k^2(\tau)}\right\}, \tag{72}$$

which corresponds exactly to the $\xi_k^{(sq)}$ that we obtained from the Ermakov equation by requiring appropriate initial conditions for ξ_k which carry over to initial conditions on the Fourier mode and the additional factor i comes from the phase of (71) compared to the one arising from the integral.

As far as the Hamiltonian diagonalization (HD) of the Mukhanov-Sasaki Hamiltonian is concerned, one diagonalizes this Hamiltonian instantaneously at some time η_0 which requires the coefficients F_k and $\bar{F}_k$ to vanish at η_0. In addition it can be shown that the state satisfying this requirement also minimizes the energy at that time η_0, so that requiring both does not yield to further conditions on the state. If the requirement $F_k = 0$ and the normalization of the Wronskian $W(v_k, \bar{v}_k)$ to $W(v_k, \bar{v}_k) = i$ is satisfied, we choose the following initial conditions for the model:

$$\text{HD (I)} \quad |v_k|(\eta_0) = \frac{1}{\sqrt{2\omega_k(\eta_0)}} \quad \text{and} \quad \text{HD (II)} \quad v_k'(\eta_0) = -i\omega_k(\eta_0)v_k(\eta_0). \tag{73}$$

If we consider the specefic choice $\eta_0 \to -\infty$ in this context we exactly end up with the initial conditions usually chosen to obtain the Bunch-Davies vacuum:

$$\text{BD (I)} \quad |v_k|(-\infty) = \frac{1}{\sqrt{2k}} \quad \text{and} \quad \text{BD (II)} \quad v_k'(-\infty) = -ikv_k(-\infty), \tag{74}$$

where we used that $\omega_k = k$ at $\eta_0 \to -\infty$, meaning that the modes become the standard Minkowski modes in this limit. Looking closer into the condition $F_k = 0$ we can rewrite this non-linear differential equations as:

$$F_k = 0 \iff v_k'(\eta)\left(v_k'' - \left(\frac{\omega_k'(\eta)}{\omega_k(\eta)}\right)v_k'(\eta) + \omega_k^2(\eta)v_k(\eta)\right) = 0. \tag{75}$$

We realize that $F_k = 0$ at all times η requires that v_k satisfies a differential equations that looks like the Mukhanov-Sasaki equation but with an additional friction term included. For a constant frequency ω_k the friction term vanishes, which for the case of de Sitter where $\omega_k^2(\eta) = k^2 - \frac{2}{\eta^2}$ is given in the limit of large k. For de Sitter the friction coefficient reads $\frac{\omega_k'}{\omega_k} = \frac{2}{\eta}\frac{1}{(k\eta)^2-2}$ and thus, depending on the values of k and η it will not always be negligible, which is the reason why in the case of Bunch-Davies one can only achieve an instantaneous Hamiltonian diagonalization. This is due to the fact that v_k satisfies the Mukhanov-Sasaki equation and at the same time needs to fulfill $F_k = 0$, generally being in conflict already for the simple case of a de Sitter universe. Note that in our work the Hamiltonian diagonalization of the Mukhanov-Sasaki Hamiltonian can be obtained for each instant in time and is not obtained by setting $F_k(\eta)$ equal to zero but by a time-dependent unitary transformation that involves also a time-rescaling. Now since we fixed our initial condition in the limit $\eta \to -\infty$ and, as we will show below, the solution we obtained satisfies the differential equation for adiabatic vacua without any approximation, it is very natural that our initial conditions at $\eta_0 = -\infty$ are given by the following:

$$\text{LR (I)} \quad |v_\mathbf{k}(\eta_0)| = |N_\mathbf{k}\tilde{\zeta}_\mathbf{k}(\eta_0)| = \frac{1}{\sqrt{2k}} \quad \text{and} \quad \text{LR (II)} \quad v'_\mathbf{k}(\eta_0) = -i\omega_\mathbf{k}^{(0)} v_\mathbf{k}(\eta_0), \qquad (76)$$

where we used again the same normalization of the Wronskian for the condition LR (II) and that $\lim_{\eta_0 \to -\infty} \tilde{\zeta}_\mathbf{k}(\eta_0) = 1$. Thus the initial conditions obtained here coincide with the initial conditions one chooses for adiabatic vacua to any order as well as the ones chosen for the Bunch-Davies vacuum where we fix them in the large k limit and for $\eta_0 \to -\infty$. However, in our work the latter was necessary in order that the unitary operator that implements the Bogoliubov transformation (see Equation (35)) becomes the identity operator for an already time-independent harmonic oscillator and is hence considerably natural. Now let us discuss how the results obtained in our work are related to the notion of adiabatic vacua. In the framework of adiabatic vacua one uses the following ansatz for the mode functions:

$$v_\mathbf{k}(\eta) = \frac{1}{\sqrt{W_\mathbf{k}}} \exp\left\{-i \int^\eta d\tilde{\eta} W_\mathbf{k}(\tilde{\eta})\right\}, \qquad (77)$$

where $W_\mathbf{k}(\eta)$ is defined through the following differential equation

$$W_\mathbf{k}^2(\eta) = \omega_\mathbf{k}^2(\eta) - \frac{1}{2}\left(\frac{W''_\mathbf{k}(\eta)}{W_\mathbf{k}(\eta)} - \frac{3}{2}\left(\frac{W'_\mathbf{k}(\eta)}{W_\mathbf{k}(\eta)}\right)^2\right), \qquad (78)$$

where $\omega_\mathbf{k}(\eta)$ is the time-dependent frequency, so in our case the one involved in the Mukhanov-Sasaki Hamiltonian. If we compare the ansatz in (77) with the form of the solution for the Mukhanov-Sasaki equation in (62), we realize that we can map the two expression for $v_\mathbf{k}$ into each other by the substitution $\tilde{\zeta}_\mathbf{k} := (\omega_\mathbf{k}^{(0)})^{\frac{1}{2}} W_\mathbf{k}^{-\frac{1}{2}}$, where we choose $\omega_\mathbf{k}^{(0)} = k$ and $\omega_\mathbf{k}^{(0)} = 1$, respectively to consider the case where the Mukhanov-Sasaki Hamiltonian is mapped to the harmonic oscillator with frequency k and 1, respectively. As shown above, rewritten in terms of $\tilde{\zeta}_\mathbf{k}$ the Mukhanov-Sasaki equation merges into the Ermakov equation. Hence, if we express the Ermakov equation in terms of $W_\mathbf{k}$ we can rewrite the Mukhanov-Sasaki equation in terms of $W_\mathbf{k}$. For this we consider the second derivative $\tilde{\zeta}''_\mathbf{k}$ expressed in terms of $W_\mathbf{k}$. We obtain:

$$\tilde{\zeta}''_\mathbf{k} = -\frac{\sqrt{\omega_\mathbf{k}^{(0)}}}{2}\left(\frac{W''_\mathbf{k}(\eta)}{W_\mathbf{k}^{\frac{3}{2}}(\eta)} - \frac{3}{2}\frac{(W'_\mathbf{k}(\eta))^2}{W_\mathbf{k}^{\frac{5}{2}}(\eta)}\right).$$

Reinserting this back into the Ermakov equation yields:

$$-\frac{\sqrt{\omega_\mathbf{k}^{(0)}}}{2}\left(\frac{W''_\mathbf{k}(\eta)}{W_\mathbf{k}^{\frac{3}{2}}(\eta)} - \frac{3}{2}\frac{(W'_\mathbf{k}(\eta))^2}{W_\mathbf{k}^{\frac{5}{2}}(\eta)}\right) + \omega_\mathbf{k}^2(\eta)\frac{\sqrt{\omega_\mathbf{k}^{(0)}}}{W_\mathbf{k}^{\frac{1}{2}}(\eta)} - \frac{(\omega_\mathbf{k}^{(0)})^2}{(\omega_\mathbf{k}^{(0)})^{\frac{3}{2}}} W_\mathbf{k}^{\frac{3}{2}}(\eta) = 0. \qquad (79)$$

Multiplying the entire equation by $(\omega_\mathbf{k}^{(0)})^{-\frac{1}{2}} W_\mathbf{k}^{\frac{1}{2}}$ we end up with:

$$W_\mathbf{k}^2 = \omega_\mathbf{k}^2 - \frac{1}{2}\left(\frac{W''_\mathbf{k}(\eta)}{W_\mathbf{k}(\eta)} - \frac{3}{2}\left(\frac{W'_\mathbf{k}(\eta)}{W_\mathbf{k}(\eta)}\right)^2\right), \qquad (80)$$

and this agrees precisely with the defining differential equation for $W_\mathbf{k}$ in (78). The adiabatic condition required for the modes in this context carries over to a condition on the large k behavior of the function $\tilde{\zeta}_\mathbf{k}$, being a solution of the Ermakov equation. As usual for adiabatic vacua, they do depend on the chosen extension to the infraed sector. In the formalism presented in this work this arbitrariness is encoded in the choice of how the unitary transformation is modified for the modes $\mathbf{k}$ with $||\mathbf{k}|| \leq$

$\|\mathbf{k}_\epsilon\|$. From this we can conclude that the ansatz for adiabatic vacua and the framework of the Lewis-Riesenfeld invariant leads to equivalent solutions for possible vacuum states if one reformulates the adiabatic condition in terms of the the solution ξ of the Ermakov equation. Furthermore, we can understand our solution obtained for quasi-de Sitter and de Sitter in this context now. For the modes associated with the Mukhanov-Sasaki equation on a de Sitter background, the adiabatic condition needs to be satisfied for $k^2 \gg \eta^{-2}$, that is $k\eta \gg 1$. Using the explicit solution for ξ in the case of de Sitter given by $\xi_\mathbf{k} = \left(1 + \frac{1}{(k\eta)^2}\right)^{\frac{1}{2}}$ we obtain $\lim_{k\eta \to \infty} \xi_\mathbf{k} = 1$. This corresponds to $\lim_{k\eta \to \infty} W_\mathbf{k} = \omega_\mathbf{k}^{(0)} = k$, where we only considered the map with $\omega_\mathbf{k}^{(0)} = k$ here because the second one with $\omega_\mathbf{k}^{(0)} = 1$ was not unitarily implementable on Fock space. In the case of de Sitter, the integral can be easily computed and the solution is given by

$$v_\mathbf{k}(\eta) = \frac{1}{\sqrt{2k}} \sqrt{1 + \frac{1}{(k\eta)^2}} e^{-ik\eta} e^{i \arctan(k\eta)}. \tag{81}$$

In case the solution for $W_\mathbf{k}$ cannot be determined in a simple manner, one uses a WKB approximation for the integral involved in the adiabatic ansatz in (77), yielding adiabatic vacua of a certain order at which the expansion is truncated, see for instance [34,35] for applications. However, since we have determined an analytical solution for the Ermakov equation for $\omega_\mathbf{k}^{(0)} = k$ we did not get an approximate solution for $W_\mathbf{k}$ up to some adiabatic order and obtained the full solution for $W_\mathbf{k}$. This way of relating the two formalisms also provides the possibility to have a very clear interpretation of the Fourier mode associated with the Bunch-Davies vacuum in the Lewis-Riesenfeld invariant formalism. Now comparing the phase factors of Fourier modes associated with the Bunch-Davies vacuum with the ones obtained from the ansatz for the adiabatic vacua in (77), we realize the following: The Fourier modes we obtain from the Lewis-Riesenfeld invariant formalism, that agree with the conventional one, can be understood as an adiabatic vacuum of non-linear adiabatic order, that is without any truncation, using the relation between the Ermakov equation and the defining differential equation for adiabatic vacua. Considering the solution in (81) in the limit $k\eta \gg 1$, we realize that these modes merge into the standard Minkowski modes up to an irrelevant phase and thus satisfy the adiabatic condition. Note that we have chosen the normalization of the Wronskian in such a way that the final mode functions $v_\mathbf{k}$ agree, regardless of whether we chose the map that relates the MS system with a harmonic oscillator to have frequency $\omega_\mathbf{k}^{(0)} = k$ or $\omega_\mathbf{k}^{(0)} = 1$, respectively. However, our analysis shows that on Fock space, the map that intertwines between the harmonic oscillator with $\omega_\mathbf{k}^{(0)} = 1$ and the Mukhanov-Sasaki equation cannot be implemented unitarily due to ultraviolet divergences and thus the latter choice cannot be obtained in a natural way in the Lewis-Riesenfeld formalism. For the reason that the solution in (81) was obtained from a unitary transformation that maps the Mukhanov-Sasaki Hamiltonian into the harmonic oscillator Hamiltonian for all modes $\mathbf{k}$ with $\|\mathbf{k}\| > \|\mathbf{k}_\epsilon\|$, we can interpret this adiabatic vacuum as the natural one associated to this unitary transformation.

We summarize these results of the last two sections in Figure 1 below. We have seen that we can obtain a solution of the Mukhanov-Sasaki equation at the level of the mode functions (and find the associated vacuum) by means of the solution of the Ermakov equation $\xi_\mathbf{k}(\eta)$ combined together with a time-dependent phase that corresponds to the time rescaling from the classical theory in Equation (11). In our formalism we have the freedom of choosing the target frequency $\omega_\mathbf{k}^{(0)}$ as we map our Hamiltonian, where we considered two different choices in this work here. One natural choice is to just remove the time dependence and keep the time-independent $\mathbf{k}^2$ term in the frequency, which gives a transformation that is implementable for all but the infrared modes, where one can choose to modify the map appropriately as has been discussed above. It is in this sense natural to do so, since in the limit at past conformal infinity, this transformation is the identity as one would expect. Contrary to that, mapping all frequencies to unity results in a residual squeezing at very early times and most importantly in an ultraviolet divergence in the integral of the Shale-Stinespring condition. Using our results it can be shown that the non-squeezed adiabatic vacua are unitarily inequivalent to the generalized Bunch-Davies vacuum because the time-independent squeezing map that relates a harmonic oscillator with frequency $\omega_{\mathbf{k}}^{(0)} = k$ to the one with frequency $\omega_{\mathbf{k}}^{(0)} = 1$ cannot be implemented as a unitary operator on Fock space.

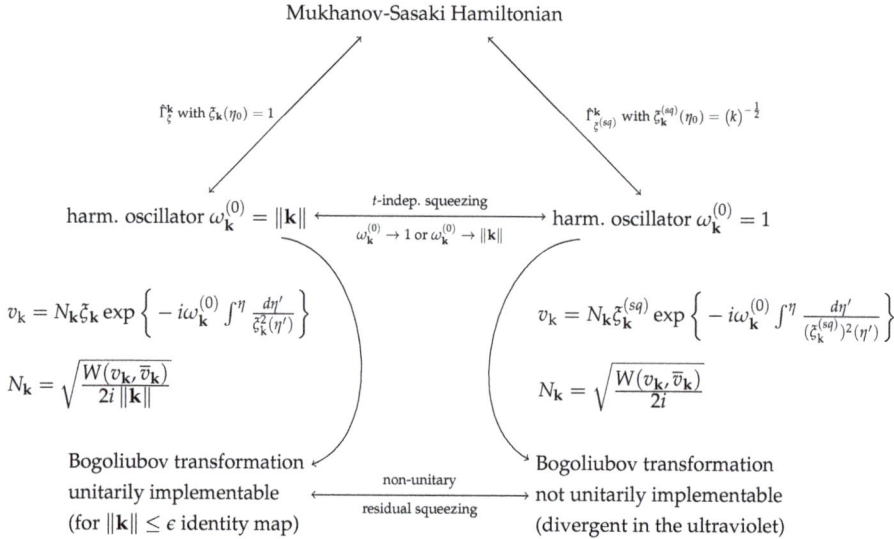

Figure 1. Graphical summary of the two different maps analyzed on Fock space.

6. Applications

6.1. Solution of the Ermakov Equation on Quasi-de Sitter Spacetime

In the following we will derive and investigate a specific solution to the Ermakov equation on a quasi-de Sitter background. This leaves us with the opportunity to simplify this solution to the case of de Sitter, where the solution is known and perform a quantitative comparison of the behavior of $\xi(\eta)$ for these two spacetimes. Our starting point is again the Mukhanov-Sasaki equation shown in (58), which will be restated for the reader's convenience:

$$v_{\mathbf{k}}''(\eta) + \left(\|\mathbf{k}\|^2 - \frac{4\nu^2 - 1}{4\eta^2} \right) v_{\mathbf{k}}(\eta) = 0.$$

Naturally, in the limit of vanishing slow-roll parameters ε, τ and κ the equation merges into the Mukhanov-Sasaki equation on de Sitter, that is, we find that $\nu^2 - \frac{1}{4} \to 2$ as expected. Next, we will bring the equation into a slightly different but also frequently used form. This is done by multiplying the entire equation by η^2, which is possible since η ranges from $-\infty$ to 0, both obviously excluded. Further, we introduce new functions $w(-k\eta)$ with $k = \|\mathbf{k}\|$ that are related to the original mode functions by $w(-k\eta) = \frac{v_{\mathbf{k}}}{\sqrt{-\eta}}$. This leads to the following differential equation for $w(-k\eta)$:

$$\chi^2 \frac{d^2 w(-\chi)}{d\chi^2} + \chi \frac{dw(-\chi)}{d\chi} + (\chi^2 - \nu^2) w(-\chi) = 0, \quad \chi = k\eta, \tag{82}$$

This is an advantage because (82) precisely corresponds to the generalized Bessel differential equation with a well-studied framework of solution techniques. Primarily, we are interested in a set of two linearly independent solutions of this equation in order to construct a solution for the Ermakov equation following the path taken in Reference [36]. The most general solution to the Bessel equation is given in terms of Bessel functions $J_\nu(-k\eta), Y_\nu(-k\eta)$ of the first and second kind, respectively. These can be rewritten in terms of Hankel functions $H_\nu^{(1)}(-k\eta), H_\nu^{(2)}(-k\eta)$ of the first and second kind, which are given by:

$$H_\nu^{(1)}(-k\eta) := J_\nu(-k\eta) + iY_\nu(-k\eta), \quad H_\nu^{(2)}(-k\eta) := J_\nu(-k\eta) - iY_\nu(-k\eta). \tag{83}$$

These functions form a linearly independent set of solutions to Equation (82). Introducing constants $\alpha, \beta \in \mathbb{C}$ we can give a general solution to the Mukhanov-Sasaki equation on quasi-de Sitter by means of resubstituting $v_\mathbf{k} = \sqrt{-\eta}\, w(-k\eta)$ and inserting the previously found basis of solutions for the Bessel equation:

$$v_\mathbf{k}(\eta) = \sqrt{-\eta}\left(\alpha H_\nu^{(1)}(-k\eta) + \beta H_\nu^{(2)}(-k\eta)\right)$$

Now we follow Reference [36] and can start to construct a (unique) solution of the Ermakov equation with either the use of $J_\nu(-k\eta), Y_\nu(-k\eta)$ or $H_\nu^{(1)}(-k\eta), H_\nu^{(2)}(-k\eta)$, respectively. This can be achieved by the following procedure, which can be straightforwardly verified by direct computation. We have

$$\xi_\mathbf{k}(\eta) = \sqrt{A_\mathbf{k} u_\mathbf{k}^2 + 2B_\mathbf{k} u_\mathbf{k} v_\mathbf{k} + C_\mathbf{k} v_\mathbf{k}^2}, \quad A_\mathbf{k} C_\mathbf{k} - B_\mathbf{k}^2 = \|\mathbf{k}\|^2 W(u_\mathbf{k}, v_\mathbf{k})^{-2}, \tag{84}$$

where $u_\mathbf{k}, v_\mathbf{k}$ are two linearly independent solutions of the Ermakov equation and $W(u_\mathbf{k}, v_\mathbf{k})$ denotes the Wronskian determinant. As an additional 'initial' condition next to the Wronskian, we impose the well-definedness of the solution in the limit of past conformal infinity where for each mode the Mukhanov-Sasaki equation reduces to an harmonic oscillator with constant frequency. The function $\xi_\mathbf{k}$ should also solve the Ermakov equation in this limiting case of constant frequency. Consequently, we can insert the linear independent solutions (83) into formula in (84) and investigate the behavior for $\eta \to -\infty$. Analyzing the asymptotic behavior of the Bessel functions (and correspondingly Hankel functions) according to Reference [37], we get:

$$H_\nu^{(1)}(-k\eta) \sim \sqrt{-\frac{2}{\pi k\eta}} \exp\left\{-i\left(k\eta + \frac{\pi}{4}(2\nu+1)\right)\right\} \quad \text{for} \quad |\eta| \gg 1,$$

$$H_\nu^{(2)}(-k\eta) \sim \sqrt{-\frac{2}{\pi k\eta}} \exp\left\{+i\left(k\eta + \frac{\pi}{4}(2\nu+1)\right)\right\} \quad \text{for} \quad |\eta| \gg 1.$$

We realize that in this limit the summand under the square root in (84) is only well-defined for vanishing coefficients $A_\mathbf{k}, C_\mathbf{k}$ such that only the mixed term remains. In order to determine the coefficient $B_\mathbf{k}$ we need to find an expression for the Wronskian determinant of Hankel functions, which is non-trivial to obtain in a straightforward manner. However, we know that the Wronskian of solutions of the harmonic oscillator equation is constant in time and we have an relation for the asymptotic behavior of the Hankel functions. Given this we have

$$W(\sqrt{-\eta} H_\nu^{(1)}(-k\eta), \sqrt{-\eta} H_\nu^{(2)}(-k\eta)) = -\eta\, W(H_\nu^{(1)}(-k\eta), H_\nu^{(2)}(-k\eta)). \tag{85}$$

As the next step, let us rewrite the derivative with respect to conformal time of $W(H_\nu^{(1)}(-k\eta), H_\nu^{(2)}(-k\eta))$ in terms of a differential equation by the use of its anti-symmetry and the Bessel differential Equation (82) obeyed by $H_\nu^{(1)}, H_\nu^{(2)}$:

$$W'(H_\nu^{(1)}, H_\nu^{(2)}) = -\frac{1}{\eta} W(H_\nu^{(1)}, H_\nu^{(2)}) \implies W(H_\nu^{(1)}(-k\eta), H_\nu^{(2)}(-k\eta)) \propto \frac{D_k}{\eta}, \quad (86)$$

where we allowed for that the constant D_k can vary for each mode. Note that the proportionality of the Wronskian of the Hankel functions in (86) is in accordance with the fact that it is conserved on solutions of the Mukhanov-Sasaki equation, as seen in Equation (85). Finally, after insertion of the asymptotic behavior of the Hankel functions, we find:

$$W(H_\nu^{(1)}(-k\eta), H_\nu^{(2)}(-k\eta)) \sim \frac{4i}{\pi\eta} \quad \text{for } |\eta| \gg 1 \implies D_k = \frac{4i}{\pi}.$$

Note that the Wronskian is purely imaginary, which is expected due to the negative sign of the B_k^2 term in the condition presented in the second equation in (84) for the coefficients. If we had chosen a different route and had taken Bessel instead of Hankel functions, we would have to choose $B_k = 0$ for consistency and with a corresponding purely real Wronskian determinant. As a final result, we can determine B_k:

$$W(\sqrt{-\eta} H_\nu^{(1)}(-k\eta), \sqrt{-\eta} H_\nu^{(2)}(-k\eta)) = -\eta \frac{4i}{\pi\eta} = -\frac{4i}{\pi} = \text{const} \implies B_k = -\frac{k\pi}{4} \quad (87)$$

Due to the requirement that the transformation induced by $\Gamma_{\bar{\zeta}}$ should be unitary, we need B_k to be chosen such that the final solution $\zeta_k(\eta)$ is real, which is always possible in this case due to the involved squares:

$$\zeta_k(\eta) = \sqrt{-\frac{k\pi\eta}{2} H_\nu^{(1)}(-k\eta) H_\nu^{(2)}(-k\eta)} = \sqrt{-\frac{k\pi\eta}{2} \left((J_\nu(-k\eta))^2 + (Y_\nu(-k\eta))^2 \right)} \quad (88)$$

Another important aspect is the correct limit at past conformal infinity, which we can immediately deduce from the asymptotic forms of the Hankel functions above. This suggests that for each Fourier mode $\zeta_k(\eta)$ solves the Ermakov equation in the case where the Mukhanov-Sasaki frequency becomes a constant $\omega_k^{(0)} := \lim_{\eta \to -\infty} \omega_k(\eta) = \|k\|$, that is:

$$\lim_{\eta \to -\infty} \zeta_k(\eta) = \lim_{\eta \to -\infty} \sqrt{-\frac{k\pi\eta}{2} \frac{2}{\pi k|\eta|}} = \lim_{\eta \to -\infty} \sqrt{-\text{sgn}(\eta)} = 1.$$

At this point we still need to investigate whether given the solution $\zeta_k(\eta)$ on quasi-de Sitter we can rediscover the solution for de Sitter in the case of vanishing slow-roll parameters. For this purpose, we consider the half-integer expressions for the Bessel functions:

$$J_{n+\frac{1}{2}}(x) = (-1)^n \sqrt{\frac{2}{\pi}} x^{n+\frac{1}{2}} \left(\frac{d}{xdx} \right)^n \frac{\sin(x)}{x} \quad \forall n \in \mathbb{N},$$

$$Y_{n+\frac{1}{2}}(x) = (-1)^{n+1} \sqrt{\frac{2}{\pi}} x^{n+\frac{1}{2}} \left(\frac{d}{xdx} \right)^n \frac{\cos(x)}{x} \quad \forall n \in \mathbb{N}.$$

Form this we obtain an expression for $\zeta_k(\eta)$ on de Sitter where $\nu = 3/2$:

$$\zeta_k^{(dS)} = \sqrt{-\frac{k\pi\eta}{2} \left(\left(\sqrt{\frac{2}{\pi}} \frac{k\eta \cos(k\eta) - \sin(k\eta)}{(-k\eta)^{\frac{3}{2}}} \right)^2 + \left(\sqrt{\frac{2}{\pi}} \frac{k\eta \sin(k\eta) + \cos(k\eta)}{(-k\eta)^{\frac{3}{2}}} \right)^2 \right)} = \sqrt{1 + \frac{1}{(k\eta)^2}},$$

which of course retains the same limit at past conformal infinity as the more complicated solution for non-vanishing slow-roll parameters. The solution $\zeta_k^{(dS)}(\eta)$ can be obtained in full analogy to the procedure outlined above using the well-known solution for the Mukhanov-Sasaki frequency on de Sitter. Depending on the particular choice of basis for the space of solutions, one needs to eliminate either of the coefficients in (84) due to the required well-definedness of the limiting case $|\eta| \to \infty$. The outcome precisely corresponds to $\zeta_k^{(dS)}(\eta)$ found in the limit above.

6.2. Eigenstates of the Lewis-Riesenfeld Invariant

As a test scenario for the formalism outlined in this work we construct and analyze the explicitly time-dependent eigenstates of the Lewis-Riesenfeld invariant. This will happen at the level of a quantum mechanical toy-model and serve the purpose of exhibiting the mathematical convenience of the formalism as well as the (squeezing) properties of the unitary transformation obtained in the context of the Lewis-Riesenfeld invariant. These eigenstates can be easily found by applying the previously obtained (inverse) Bogoliubov transformation $\hat{\Gamma}_\xi^\dagger$ to the defining property of the vacuum, that is $\hat{A}|0\rangle = 0$. We obtain:

$$\hat{\Gamma}_\xi^\dagger \hat{A} \hat{\Gamma}_\xi \hat{\Gamma}_\xi^\dagger |0\rangle = \mathrm{Ad}_{\hat{\Gamma}_\xi^\dagger}(\hat{A}) \hat{\Gamma}_\xi^\dagger |0\rangle = e^{-\nu(\xi)}(\hat{A} - \delta_+(\xi)\hat{A}^\dagger)\hat{\Gamma}_\xi^\dagger|0\rangle =: \hat{B}\hat{\Gamma}_\xi^\dagger|0\rangle = 0,$$

with the BCH coefficients $\nu(\xi)$ and $\delta_+(\xi)$ determined in Section 3.2. That is, the vacuum state of the Bogoliubov transformed annihilation operator $\hat{B}$ corresponds to the unitarily transformed initial vacuum state. Recall that $\hat{\Gamma}_\xi$ was capable of relating the time-independent Hamiltonian $\hat{H}_0$ and the Lewis-Riesenfeld invariant $\hat{I}_{LR}$ via the adjoint action, in other words, the Lewis-Riesenfeld invariant factorizes in terms of $\hat{B}, \hat{B}^\dagger$. Reexpressing the operators above in position representation we end up with a first-order differential equation for the transformed vacuum state. Understandably, this equation contains explicitly time-dependent coefficients due to the explicit time dependence of the Bogoliubov transformation. We obtain the following solution for the ground state $\Psi_0(q, \eta)$:

$$\Psi_0(q,\eta) = \left(\frac{\omega_0}{\pi \xi^2(\eta)}\right)^{\frac{1}{4}} \exp\left\{\left(\frac{i}{2}\frac{\xi'(\eta)}{\xi(\eta)} - \frac{\omega_0}{2\xi^2(\eta)}\right)q^2\right\}, \quad (89)$$

where $\xi'(\eta)$ denotes the derivative with respect to conformal time, we again used that the mass $m = 1$ here and conveniently have set $\hbar = 1$ as before. The first excited state can be obtained from $\mathrm{Ad}_{\hat{\Gamma}_\xi^\dagger}(\hat{A}^\dagger)\hat{\Gamma}_\xi^\dagger|0\rangle = \hat{B}^\dagger \hat{\Gamma}_\xi^\dagger|0\rangle$ and is found to be:

$$\Psi_1(q,\eta) = \left(\frac{\omega_0}{\pi \xi^2(\eta)}\right)^{\frac{1}{4}} \sqrt{\frac{2\omega_0}{\xi^2(\eta)}}\, q \exp\left\{\left(\frac{i}{2}\frac{\xi'(\eta)}{\xi(\eta)} - \frac{\omega_0}{2\xi^2(\eta)}\right)q^2\right\}. \quad (90)$$

Note that for a time-independent frequency $\omega(\eta) = \omega_0$ the solution merges into the standard quantum harmonic oscillator since $\hat{I}_{LR}$ and $\hat{H}_0$ coincide in this limit by construction due to $\xi(\eta) = \xi_0 = 1$. The details of the underlying spacetime, that is, what determines the values of the various slow-roll parameters enters through the solution $\xi(\eta)$ of the Ermakov equation, which is sensitive to the background via the Mukhanov-Sasaki frequency $\omega(\eta)$ and consequently through the real index ν of the Hankel and Bessel functions in the final solution in Equation (88). The plots in Figure 2 display the absolute squares of the solutions in Equations (89) and (90), respectively, at two different conformal times.

Considering the explicit form of the generator of the Bogoliubov transformation $\hat{\Gamma}_\xi$, we realize that it represents a generalized squeezing operator with explicitly time-dependent coefficient functions. These coefficient functions on the other hand are sensitive to the background spacetime via the Ermakov equation and consequently the Mukhanov-Sasaki frequency $\omega_k(\eta)$ involved in Equation (58). In this way it is expected that the eigenstates of the Lewis-Riesenfeld invariant, which are, up to a phase, eigenstates of the single-mode time-dependent Mukhanov-Sasaki Hamiltonian, show a time-dependent spread which approaches the time-independent case for very large absolute values of conformal time $|\eta| \gg 1$, that is, close to the Big Bang.

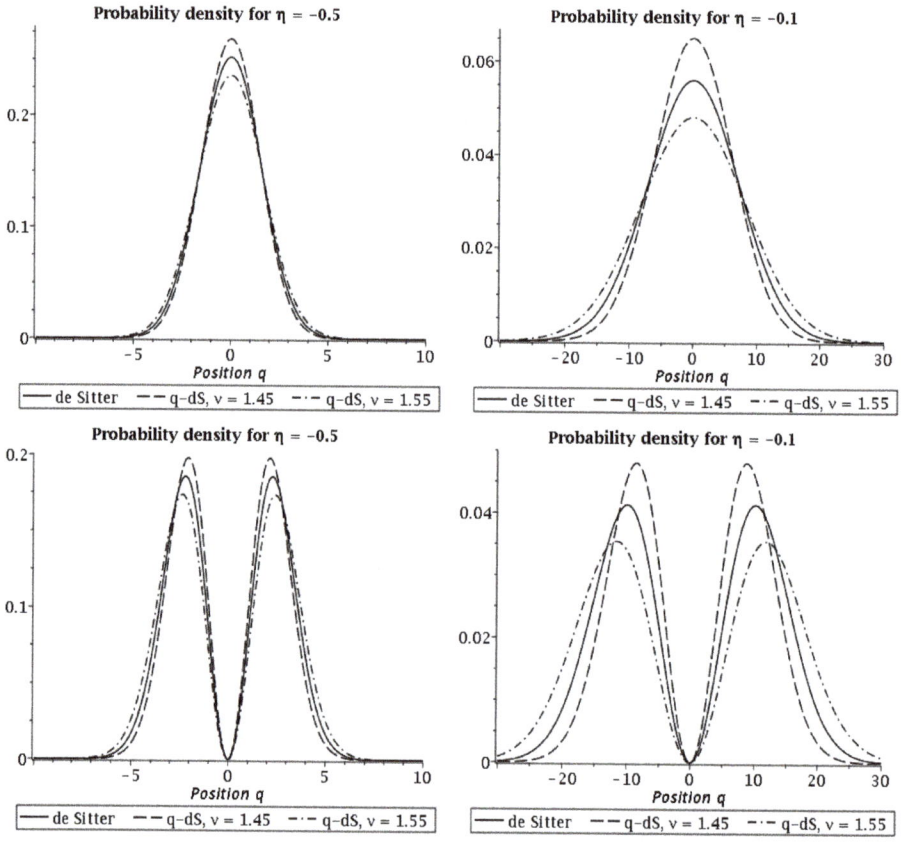

Figure 2. Single-mode probability densities $|\Psi_0(q,\eta)|^2$ (upper line) and $|\Psi_1(q,\eta)|^2$ (lower line) according to the solutions in (89) and (90) on quasi-de Sitter for three different values of the effective slow-roll parameter ν from Equation (58), including de Sitter with $\nu = 3/2$ at two different conformal times and with $\omega_0 = k = 1$. The used slow-roll parameters are to be understood as an example, consider Reference [33] for the allowed parameter space and constraints on them according to the Planck mission.

A comparison with the work in Reference [38] bears a strong resemblance to the eigenstate of the Lewis-Riesenfeld invariant found there, however let us compare the results from Reference [38] and ours in more detail. Firstly, the derivation in Reference [38] is performed in cosmological time, whereas we have made the transition to conformal time beforehand, so the explicit occurrences of the scale factor are absent in our work. Secondly, when having a closer look at the Ermakov equation in Reference [38] it becomes evident that in the context of the canonical transformation $\Gamma_{\tilde{\xi}}$, the time-independent frequency of the Hamiltonian H_0 is unity. As a consequence, this means for the case of field theory that every mode with $\omega_k(\eta)$ would be mapped to exactly the *same* frequency $\omega_k^{(0)} = 1$, which modifies the solution of the Ermakov equation by an additional $k^{-\frac{1}{2}}$, leading to an ultraviolet divergence in the Shale-Stinespring condition (56) and diminishing the ability to implement it by a separate treatment of the infrared modes with $\|\mathbf{k}\| \le \|\mathbf{k}_e\|$. Thirdly, the authors in Reference [38] claim that the creation- and annihilation operators that decompose the time *dependent* Hamiltonian are related to the ones associated with the Lewis-Riesenfeld invariant via a Bogoliubov transformation. According to our analysis, while this is true, their relation is more subtle: As it is true that $H(t)$ can be mapped into I_{LR} in the classical theory by means of an extended symplectic map (which in a sense corresponds to a Bogoliubov transformation quantum mechanically), this might not be straightforwardly implementable in the quantum theory even on the one-particle Hilbert space. It can be implemented if and only if the time-rescaling function is chosen such that $\xi(t)^{-2}$ has an analytic anti-derivative, which is for example the case on a de Sitter background. The exponential sandwiched between $\hat{\Gamma}_\xi^\dagger$ and $\hat{\Gamma}_{\xi,0}$ in (33) can then be rewritten as the exponential of an analytic function in the time *operator*, conjugate to the momentum operator $\hat{p}_t$. This is the reason why we perform a reduced phase space quantization of Dirac observables, where this problem is absent, rather than Dirac quantization. Consequently, the transformation $\hat{\Gamma}_\xi$ acts as a one-parameter (i.e., time-dependent) family of unitary transformations on the reduced (physical) phase space. This transformation is suitable for transforming the Hamiltonian within the Schrödinger equation into the independent one $\hat{H}_0$, which again can be related to the invariant $\hat{I}_{LR}$ by means of a time-dependent Bogoliubov transformation.

7. Conclusions

In this work we used the method of the Lewis-Riesenfeld invariant in order to analyze in which sense the dynamical properties of the Mukhanov-Sasaki equation select possible candidates for initial states in the context of inflation. We started in the classical theory and rederived a time-dependent canonical transformation that relates the system of a time-dependent harmonic oscillator to the one of a time-independent harmonic oscillator, where in our case the explicit time-dependence is determined by the frequency involved in the Mukhanov-Sasaki equation. Using this map, the entire time dependence of the oscillator can be removed leading to a simplification as far as finding solutions for the time-dependent system is concerned. As a first step, this was done for systems with finitely many degrees of freedom using an extended phase space in which time and its corresponding momentum are part of the phase space, following the work in Reference [11]. This has the advantage that the time rescaling involved in this transformation can be naturally embedded in the framework of the extended phase space, whereas in early work such as Reference [13], the corresponding phase factor needs to be introduced with a less clear physical motivation. The ansatz we used employed techniques introduced in Reference [11] and was based on an idea from Reference [12], which was generalized to arbitrary even and finite phase space dimensions. This transformation revealed the relationship between the Lewis-Riesenfeld invariant, the time-dependent and time-independent harmonic oscillator Hamiltonian already implicitly used in Reference [13].

Since at the level of the extended phase space the system of the time-dependent harmonic oscillator is a constrained system, we could choose between Dirac and reduced quantization in order to later extract the physical sector of the quantum theory. Because the time rescaling included in the canonical transformation involves an integral over a time interval, a Dirac quantization might be problematic if an analytic expression for the anti-derivative does not exist. In order to circumvent this problem, we chose reduced phase space quantization for which it was necessary to construct Dirac observables and consider the physical phase space according to the methods introduced in References [18,19]. Fortunately, the Dirac observables satisfy the standard canonical algebra so that representations thereof can be easily found. Their dynamics is generated by the Dirac observable associated with the time-dependent Hamiltonian that consequently becomes the physical Hamiltonian of the system. As a preparation for the quantum theory, we constructed the associated generator of the canonical transformation on the physical phase space, giving rise to a corresponding flow that represents the canonical transformation on the physical degrees of freedom. A crucial ingredient for the construction of the Lewis-Riesenfeld invariant as well as the corresponding canonical transformation that removes the time dependence from the Hamiltonian is a time-dependent auxiliary function $\xi(t)$ that has to satisfy the Ermakov differential equation. This requirement ensures that the Lewis-Riesenfeld invariant is a quadratic polynomial in the elementary configuration and momentum variables, that can be interpreted as a Dirac observable in the extended phase space because it commutes with the first class constraint. Given a solution of the Ermakov equation, we can use it to construct the canonical transformation on the finite dimensional physical phase space, whose generator is shown in (23).

For the quantum theory we first restricted our analysis to the case of finitely many degrees of freedom and as a special case considered the one-particle Hilbert space. All results obtained in this context can be easily generalized to more but finite degrees of freedom. On the one-particle Hilbert space, the time-dependent canonical transformation can be implemented as a unitary operator $\hat{\Gamma}_\xi$ whose explicit form is given in (27). With the help of $\hat{\Gamma}_\xi$ we can remove the explicit time dependence of the Schrödinger equation, analogous to the treatment in Reference [17] and map it to a Schrödinger equation involving a time-independent harmonic oscillator Hamiltonian. In this way we further obtain a time evolution operator $\hat{U}(t_0, t)$ which can be shown to correspond to the Dyson series of the time-dependent theory, consisting of three individual unitary operators. When having a closer look at the occurring exponentials, there is an obvious explanation of the required additional phase factor in the solutions of the Schrödinger equation in Reference [13], amounting to some function multiplying either the exponentiated time-independent Hamiltonian or the Lewis-Riesenfeld invariant, depending on the relative ordering we chose in $\hat{U}(t_0, t)$. The immediate effect of the time-rescaling $\xi(t)$ becomes evident in this exponential, as the relative phase is sensitive to the background spacetime. Each of the operators in $\hat{U}(t_0, t)$ corresponds to an exponentiated unitary representation of the $\mathfrak{sl}(2, \mathbb{R})$ Lie algebra, as can be shown by explicitly evaluating the Lie brackets of the generators. For practical computations and later applications, we used a generalized Baker-Campbell-Hausdorff decomposition of unitary representations of non-compact groups shown in Reference [30] to decompose $\hat{\Gamma}_\xi$ into normal- or anti-normal ordered contributions, respectively. This gave us the possibility to rewrite the unitary transformation $\hat{\Gamma}_\xi$ on the one-particle Hilbert space as an explicitly time-dependent Bogoliubov transformation, where the time dependence enters through the solution $\xi(t)$ of the Ermakov equation.

This was an important preparation for the generalization to the field theory context we were mainly interested in in this work. The crucial criterion for the existence of a unitary implementation of a Bogoliubov transformation on the Fock space is the Shale-Stinespring condition [31], which essentially denotes that the product of off-diagonal entries of the Bogoliubov map needs to be Hilbert-Schmidt. As our results show a straightforward generalization to Fock space where the time-dependent oscillator is described by the Mukhanov-Sasaki equation and the target system is for each mode a harmonic oscillator with constant frequency does not work because either infrared or infrared and ultraviolet divergences occur, leading to a violation of the Shale-Stinespring condition. Here we considered two common choices used in the existing literature, where the constant frequency is either $\omega_\mathbf{k}^{(0)} = k$ or $\omega_\mathbf{k}^{(0)} = 1$ respectively, which corresponds to two slightly different Ermakov equations in our case. Both choices yield an infrared divergence caused by the infrared modes, whereas for the second choice in addition an ultraviolet divergence occurs. If we compare our results obtained with the existing results in the literature, the work in Reference [15] takes as the starting point a charged massive scalar field in a de Sitter space time and hence the Ermakov equation in this case includes an additional friction term and cannot directly be compared to our result. The author of Reference [15] also uses the map to a harmonic oscillator with frequency $\omega_\mathbf{k}^{(0)} = k$ and also obtains no ultraviolet divergences for his slightly different map. However, as far as we can see, a careful analysis of the Shale-Stinespring condition is not presented in Reference [15] and thus we expect that, similar to our case, infrared singularities are present. In Reference [14] the map with constant frequency $\omega_\mathbf{k}^{(0)} = 1$ was considered and in agreement with our results, they also obtain an ultraviolet divergence for the operator $\hat{\Gamma}_\xi$. In their work the theory is defined on a torus allowing them to isolate the zero mode and exclude it from their analysis, hence no infrared divergences occur. Thus, we can conclude that the second choice, where the target frequency is chosen to be $\omega_\mathbf{k}^{(0)} = 1$, cannot be implemented unitarily on Fock space, whereas for the first choice there might be a chance to find a unitary implementation for the first case with target frequency $\omega_{\mathbf{k}}^{(0)} = k$ if we are able to consistently modify the map for the infrared modes such that the infrared divergence are no longer present. One possibility can be to also formulate a model where the spatial slices have the topology of a torus chosen in such a way that experimentally one cannot distinguish between a model whose spatial slices have the topology of a torus and one with non compact spatial slices. In this case we could also exlucde the zero mode and modify the map for this specific mode in a way that the Shale-Stinespring condition is satisfied. The corresponding Bolgoliubov transformation can then be defined for all but the zero mode. Note that this case further allows to identify the background with the zero mode, as it is for example usually done in hybrid loop quantum cosmology, see for instance Reference [39]. However, if we stick to non-compact spatial slices we have to consider a slightly different strategy.

As possible solutions in this direction we discussed the quantum Arnold transformation introduced in Reference [16]. The goal was to apply it to the modes below a certain infrared cutoff, that is for the sphere with $\|\mathbf{k}\| \leq \|\mathbf{k}_\epsilon\|$ in Fourier space. We found that the Arnold transformation cannot be understood as a Bogoliubov map, since it renders the creation and annihilation operators in the transformed picture equal up to a global sign. A closer look at the involved terms shows that the reason for this pathological behavior is the absence of a q^2 contribution in the Lewis-Riesenfeld invariant. It is precisely this aspect that disrupts the commutator algebra and hence does not qualify as an infrared extension of $\hat{\Gamma}_\xi$. Our proposal for an infrared extension is to use the identity map within the cutoff region, the reasons are twofold. Firstly, the identity map can be trivially regarded as a Bogoliubov transformation that exists on this sector of the Fock space. Secondly, by not altering the form of the Hamiltonian in the infrared regime, the off-diagonal terms remain as in the standard case, which means that they are subdominant for very early times where $k\eta \ll 1$. Hence, by adopting this strategy we are able to define a unitarily implementable Bogoliubov transformation on the entire Fock space as depicted in Equation (61), that performs a Hamiltonian diagonalization on all modes with norm greater than the infrared cutoff $\|\mathbf{k}_\epsilon\|$.

In Section 5 we showed how the solution of the Ermakov equation can be used to construct a solution of the Mukhanov-Sasaki equation and as expected, the time-rescaling plays a pivotal role here. If we rewrite the solution to the Mukhanov-Sasaki equation in a polar representation as shown in (62), the Mukhanov-Sasaki equation requires that the real part in the polar representation needs to be a solution of the Ermakov equation, opening a clear connection between the two formalisms. Following this route further, we can also recover the defining differential equation for adiabatic vacua from the Ermakov equation, meaning that if we have a solution of the Ermakov equation given, from this we can easily construct a non-linear solution of the adiabatic vacua differential equation, where non-linear refers to the fact that it is a full solution without truncating the solution at any adiabatic order. The adiabatic condition usually required in this context carries over to a condition on the solution of the Ermakov equation for each mode. This in turn can be directly related to specific properties of the unitary map corresponding to the time-dependent canonical transformation between the time-dependent and time-indepdendent harmonic oscillator. Hence, there is an interesting interplay between the characteristic properties of the unitary map and the choice of adiabatic vacua. Considering this and the fact that we set our initial condition at the limit of conformal past infinity, the Lewis-Riesenfeld method leads to mode functions that can be interpreted as a non-squeezed adiabatic vacuum of non-linear order, that is without performing any truncation. The property of being non-squeezed reflects our freedom of choice of mapping to a target frequency $\omega_{\mathbf{k}}^{(0)} = k$ that causes no residual squeezing if we apply the unitary operator to an already time-independent harmonic oscillator. Furthermore, the time rescaling involved in the mode function becomes $e^{-ik\eta}$ in the limit of large (negative) conformal times, showing that the mode function obtained here are compatible with the condition used in the Bunch-Davies case.

Finally, in Section 6 we have illustrated how the formalism we used throughout this work can be used in terms of computing eigenstates of the Lewis-Riesenfeld invariant associated to a particular system. In our case, this was a time-dependent harmonic oscillator corresponding to the

Mukhanov-Sasaki equation on a quasi-de Sitter spacetime for a single mode. Together with the construction of these eigenstates via $\hat{\Gamma}_\xi$, we outlined how to find a solution $\xi(\eta)$ of the Ermakov equation with this particular time-dependent frequency $\omega(\eta)$ corresponding to the background geometry of quasi-de Sitter. This was done by using the asymptotic behavior of the Hankel functions (which solve the Mukhanov-Sasaki equation on quasi-de Sitter analytically) and the fact that the Wronskian of two linearly independent solutions of this equation is a constant. Finally, we provided a visualization of the time-dependent squeezing operation $\hat{\Gamma}_\xi$ in terms of the probability densities of two time-dependent eigenstates of $\hat{I}_{LR}$ for different values of the slow-roll parameters, which precisely correspond to the probability densities of solutions of the Schrödinger equation of the associated time-dependent Hamiltonian $H(t)$. As far as the computation of the power spectrum is concerned, we do not expect new insights from our obtained results because what enters into the computation is the final Fourier mode that we constructed in both cases in such a way that the results agree with the standard result for the Mukhanov-Sasaki mode. Our results however, give new insights on whether there exists a time-independent harmonic oscillator Hamiltonian associated with the Mukhanov-Sasaki Hamiltonian that is unitarily equivalent in the field theory context. In future work we want to analyze applications of this formalism to other than quasi de Sitter spacetimes. This requires in particular to find solutions of the Ermakov equation in this more general case and analyze whether the corresponding transformation can be implemented unitarily. Furthermore, we plan to investigate in future research how the transformation in the classical theory on the extended phase space can be lifted to the field theory context. This might be realizable in the framework of the Gaussian dust model presented in Reference [26] where the dust fields can be used as reference fields for physical temporal and spatial coordinates.

Author Contributions: All authors contributed equally to this work.

Funding: This research was partly funded by the Heinrich-Böll Foundation.

Acknowledgments: M.K. and K.G. would like to thank Hanno Sahlmann for illuminating and productive discussions during the project as well as Beatriz Elizaga Navascués and Thomas Thiemann for fruitful discussions towards the end of this work.

Conflicts of Interest: The authors declare no conflict of interest.

References

1. Mukhanov, V.F.; Feldman, H.A.; Brandenberger, R.H. Theory of cosmological perturbations. Part 1. Classical perturbations. Part 2. Quantum theory of perturbations. Part 3. Extensions. *Phys. Rept.* **1992**, *215*, 203–333. [CrossRef]
2. Mukhanov, V. *Physical Foundations of Cosmology*; Cambridge University Press: Cambridge, UK, 2005.
3. Langlois, D. Hamiltonian formalism and gauge invariance for linear perturbations in inflation. *Class. Quantum Gravity* **1994**, *11*, 389–407. [CrossRef]
4. Giesel, K.; Herzog, A. Gauge invariant canonical cosmological perturbation theory with geometrical clocks in extended phase-space—A review and applications. *Int. J. Mod. Phys. D* **2018**, *27*, 1830005. [CrossRef]
5. Giesel, K.; Herzog, A.; Singh, P. Gauge invariant variables for cosmological perturbation theory using geometrical clocks. *Class. Quantum Gravity* **2018**, *35*, 155012. [CrossRef]
6. Giesel, K.; Singh, P.; Winnekens, D. Dynamics of Dirac observables in canonical cosmological perturbation theory. *Class. Quantum Gravity* **2019**, *36*, 085009. [CrossRef]
7. Danielsson, U.H. On the consistency of de Sitter vacua. *J. High Energy Phys.* **2002**, *2002*, 025. [CrossRef]
8. Armendariz-Picon, C.; Lim, E.A. Vacuum choices and the predictions of inflation. *J. Cosmol. Astropart. Phys.* **2003**, *2003*, 006. [CrossRef]
9. Handley, W.; Lasenby, A.; Hobson, M. Novel quantum initial conditions for inflation. *Phys. Rev. D* **2016**, *94*, 024041. [CrossRef]
10. Fulling, S.A. Remarks on positive frequency and hamiltonians in expanding universes. *Gen. Relativ. Gravit.* **1979**, *10*, 807–824. [CrossRef]

11. Struckmeier, J. Hamiltonian dynamics on the symplectic extended phase space for autonomous and non-autonomous systems. *J. Phys. A Math. Gen.* **2005**, *38*, 1257. [CrossRef]
12. Garcia-Chung, A.; Ruiz, D.G.; Vergara, J.D. Dirac's method for time-dependent Hamiltonian systems in the extended phase space. *arXiv preprint* **2017**, arXiv:1701.07120.
13. Hartley, J.G.; Ray, J.R. Coherent states for the time-dependent harmonic oscillator. *Phys. Rev. D* **1982**, *25*, 382. [CrossRef]
14. Gómez-Vergel, D.; Villaseñor, E. The time-dependent quantum harmonic oscillator revisited: Applications to Quantum Field Theory. *Ann. Phys.* **2009**, *324*, 1360–1385. [CrossRef]
15. Robles-Perez, S. Invariant vacuum. *Phys. Lett. B* **2017**, *774*, 608–615. [CrossRef]
16. Guerrero, J.; López-Ruiz, F.F. The quantum Arnold transformation and the Ermakov–Pinney equation. *Phys. Scr.* **2013**, *87*, 038105. [CrossRef]
17. Guasti, M.F.; Moya-Cessa, H. Solution of the Schrödinger equation for time-dependent 1D harmonic oscillators using the orthogonal functions invariant. *J. Phys. A Math. Gen.* **2003**, *36*, 2069. [CrossRef]
18. Dittrich, B. Partial and complete observables for Hamiltonian constrained systems. *Gen. Relativ. Gravit.* **2007**, *39*, 1891–1927. [CrossRef]
19. Thiemann, T. Reduced phase space quantization and Dirac observables. *Class. Quantum Gravity* **2006**, *23*, 1163–1180. [CrossRef]
20. Rovelli, C. Partial Observables. *Phys. Rev. D* **2002**, *65*, 124013. [CrossRef]
21. Rovelli, C. What is observable in classical and quantum gravity? *Class. Quantum Gravity* **1991**, *8*, 297. [CrossRef]
22. Giesel, K.; Hofmann, S.; Thiemann, T.; Winkler, O. Manifestly Gauge-Invariant General Relativistic Perturbation Theory. I. Foundations. *Class. Quantum Gravity* **2010**, *27*, 055005. [CrossRef]
23. Giesel, K.; Hofmann, S.; Thiemann, T.; Winkler, O. Manifestly Gauge-invariant general relativistic perturbation theory. II. FRW background and first order. *Class. Quantum Gravity* **2010**, *27*, 055006. [CrossRef]
24. Giesel, K.; Thiemann, T. Algebraic quantum gravity (AQG). IV. Reduced phase space quantisation of loop quantum gravity. *Class. Quantum Gravity* **2010**, *27*, 175009. [CrossRef]
25. Domagala, M.; Giesel, K.; Kaminski, W.; Lewandowski, J. Gravity quantized: Loop Quantum Gravity with a Scalar Field. *Phys. Rev. D* **2010**, *82*, 104038. [CrossRef]
26. Giesel, K.; Thiemann, T. Scalar Material Reference Systems and Loop Quantum Gravity. *Class. Quantum Gravity* **2015**, *32*, 135015. [CrossRef]
27. Husain, V.; Pawlowski, T. Time and a physical Hamiltonian for quantum gravity. *Phys. Rev. Lett.* **2012**, *108*, 141301. [CrossRef]
28. Han, Y.; Giesel, K.; Ma, Y. Manifestly gauge invariant perturbations of scalar–tensor theories of gravity. *Class. Quantum Gravity* **2015**, *32*, 135006. [CrossRef]
29. Giesel, K.; Vetter, A. Reduced Loop Quantization with four Klein-Gordon Scalar Fields as Reference Matter. *Class. Quantum Gravity* **2019**, *36*, 145002. [CrossRef]
30. Truax, D.R. Baker-Campbell-Hausdorff relations and unitarity of SU(2) and SU(1,1) squeeze operators. *Phys. Rev. D* **1985**, *31*, 1988–1991. [CrossRef]
31. Nam, P.T.; Napiórkowski, M.; Solovej, J.P. Diagonalization of bosonic quadratic Hamiltonians by Bogoliubov transformations. *J. Funct. Anal.* **2016**, *270*, 4340–4368. [CrossRef]
32. Arnold, V.I. Supplementary chapters to the theory of ordinary differential equations. In *Geometrical Methods in the Theory of Ordinary Differential Equations*; Nauka: Moscow, Russia, 1978; English translated by Springer: New York, NY, USA; Berlin, Germany, 1983.
33. Akrami, Y.; Arroja, F.; Ashdown, M.; Aumont, J.; Baccigalupi, C.; Ballardini, M.; Banday, A.J.; Barreiro, R.B.; Bartolo, N.; Basak, S.; et al. Planck 2018 results. X. Constraints on inflation. *arXiv preprint* **2018**, arXiv:1807.06211.
34. Winitzki, S. Cosmological particle production and the precision of the WKB approximation. *Phys. Rev. D* **2005**, *72*, 104011. [CrossRef]
35. Casadio, R.; Finelli, F.; Luzzi, M.; Venturi, G. Improved WKB analysis of cosmological perturbations. *Phys. Rev. D* **2005**, *71*, 043517. [CrossRef]
36. Leach, P.; Andriopoulos, K. The Ermakov equation: A commentary. *Appl. Anal. Discret. Math.* **2008**, *2*, 145–157. [CrossRef]
37. Watson, G.N. *A Treatise on the Theory of Bessel Functions*; Cambridge University Press: Cambridge, UK, 1995.

38. Bertoni, C.; Finelli, F.; Venturi, G. Adiabatic invariants and scalar fields in a de Sitter space-time. *Phys. Lett. A* **1998**, *237*, 331–336. [CrossRef]
39. Elizaga Navascués, B.; Martín-Benito, M.; Mena Marugán, G.A. Hybrid models in loop quantum cosmology. *Int. J. Mod. Phys. D* **2016**, *25*, 1642007. [CrossRef]

© 2019 by the authors. Licensee MDPI, Basel, Switzerland. This article is an open access article distributed under the terms and conditions of the Creative Commons Attribution (CC BY) license (http://creativecommons.org/licenses/by/4.0/).

MDPI
St. Alban-Anlage 66
4052 Basel
Switzerland
Tel. +41 61 683 77 34
Fax +41 61 302 89 18
www.mdpi.com

Universe Editorial Office
E-mail: universe@mdpi.com
www.mdpi.com/journal/universe